Digital Mobile Communications and the TETRA System

Digital Mobile Communications and the TETRA System

John Dunlop
Demessie Girma
James Irvine
University of Strathclyde, Glasgow, Scotland

JOHN WILEY & SONS, LTD
Chichester · New York · Weinheim · Brisbane · Singapore · Toronto

Baffins Lane, Chichester,
West Sussex PO19 1UD, England
National 01243 779777
International (+44) 1243 779777
e-mail (for orders and customer service enquiries): cs-books@wiley.co.uk
Visit our Home Page on http://www.wiley.co.uk or http://www.wiley.com

Reprinted, with corrections, February 2000, May 2000

Other Wiley Editorial Offices

John Wiley & Sons, Inc., 605 Third Avenue,
New York, NY 10158-0012, USA

WILEY-VCH Verlag GmbH, Pappelallee 3,
D-69469 Weinheim, Germany

Jacaranda Wiley Ltd, 33 Park Road, Milton,
Queensland 4064, Australia

Singapore • Toronto

Library of Congress Cataloging-in-Publication Data
Dunlop, John
Digital mobile communications and the TETRA system / John Dunlop, Demessie Girma, James Irvine.
p. cm.
Includes bibliographical references.
ISBN 0-471-98792-1 (alk. paper)
1. TETRA (standard) 2. Mobile communication systems—Standards.
3. Digital communication systems—Standards. I. Girma, Demessie.
II. Irvine, James. III. Title.
TK5103.488.D86 1999
621.3845'0218—dc21 99-32468
CIP

British Library Cataloguing in Publication Data
A catalogue record for this book is available from the British Library

ISBN 0-471-98792-1

Printed and bound by Antony Rowe Ltd, Eastbourne

Contents

Preface

The digital mobile telephone is now a ubiquitous object but it is still predominantly voice based. This will change radically in the near future, with a huge emphasis being placed on wireless access to the Internet and the provision of wireless multimedia services by all mobile telephone networks. This movement is being accommodated in the private mobile radio environment by the introduction of the Terrestrial Trunked Radio Standard (TETRA) which has been expressly designed to accommodate voice and data services.

This book has been written to specifically explain the capabilities of the TETRA specifications and to give some insight into the wide range of services which digital technology can support in private mobile radio networks. The book is based on courses which the authors have presented to engineers engaged in the design and procurement of TETRA based networks and emphasises the relationship between the TETRA specifications and the layered structure of the open systems interconnection (OSI) model. The book assumes a basic understanding of the concepts of analogue modulation systems but begins by introducing some of the fundamental concepts of digital signal transmission. Therefore the non-TETRA specialist will find this text a useful introduction to digital mobile communications systems.

Chapter 1 begins with the principles of digital mobile radio and introduces concepts which are essential to the understanding of digital systems. It considers, in detail, the properties of amplitude, phase and frequency modulation of a carrier with digital signals with particular emphasis on bandwidth requirements. This is followed by an introduction to the properties of radio channels which is essential to the understanding of the compromises which are part of the design of all radio systems. This chapter also presents an overview of cellular topics with examples of the design constraints which cellular systems must address.

Chapter 2 considers the public digital radio environment and describes examples of time division cellular radio (GSM), and cordless telephone systems (DECT). The emphasis in this chapter is on the demands of public systems and the way in which these demands influence design choices. This chapter also considers the basics of spread spectrum mobile communications systems with respect to cdmaOne™, which is essentially a second generation system. Direct sequence CDMA systems will be the basis of 3rd generation mobile systems with which TETRA will have to co-exist and therefore it is important that specialists in TETRA have some understanding of the differences between CDMA and TDMA systems.

Chapter 3 deals with the private mobile environment and highlights the different requirements of private network mobile users and operators compared with the public equivalents. The different types of private mobile network are described, and a review of alternatives to TETRA is presented. This includes a brief description of the other digital PMR systems which are available, as well as a discussion of other mobile and cordless radio solutions.

Chapter 4 is an overview of the TETRA system and concentrates on the explanation of logical channels and how these relate to the OSI model. Particular attention is given to trunked and direct mode operation which emphasises the way in which TETRA meets the demands of the private mobile radio environment as outlined in Chapter 3. This chapter also introduces the concepts of the layered OSI model which is explored in more detail in subsequent chapters.

Chapter 5 describes TETRA system architecture and components in more detail. TETRA reference configurations are introduced with the identification of important interfaces, which are considered in subsequent chapters, in the context of interfaces and protocols. The description of system components and network services in this chapter is intended to provide a high-level overview of TETRA as a background to the more technical chapters that follow. First time readers will therefore find browsing this chapter useful before the TETRA specific technical descriptions in later chapters.

Chapter 6 describes the TETRA physical layer, and the choices made for modulation and transmission. The nature of the PMR requirements makes these choices significantly different from those made for public cellular systems, and the chapter describes the justification the design parameters used in the TETRA physical layer. Other functions and features of the physical layer are described, such as the delays introduced to the transmitted bit stream, synchronisation, power control and coverage techniques.

Chapter 7 is concerned with Layer 2 of the TETRA system, the data link layer. Layer 2, and in particular the MAC sub-layer, contains key functionality for the efficient and successful operation of the system. The chapter covers the operation of the MAC, including coding, random access, logical channel routing and stealing, as well as a number of support functions the MAC undertakes for upper layers in respect of encryption and channel quality measurement. This chapter also details the logical link control sub-layer, which is responsible for ensuring that the information transmitted on behalf of the upper layers is received without error, and arranging for its retransmission if necessary.

Chapter 8 deals with network layer functions of TETRA as a follow on from the physical and data link layers discussed in Chapter 6 and Chapter 7. It introduces general network layer concepts at the outset and develops into more detailed treatment of TETRA specific network layer protocols. Radio link management and mobility management constitute the major part of the TETRA network layer functions, and this chapter attempts to present these two important topics (and other network layer protocols) in a clear and an orderly manner. Where appropriate, *protocol data units* are used to illustrate network control procedures and at the same time highlight important parameters associated with the network protocols. This chapter also presents an overview of packet mode data services that are supported by TETRA.

Chapter 9 collects together additional topics that are essential to the operation of TETRA over and above the network functions addressed in Chapters 6 to 8. Three important topics are identified for this chapter, namely, network security management, inter-system interface signalling, and network management, each described under a separate section. Network security management is an important feature of TETRA and particular attention is given to this topic for clear descriptions of the protocol mechanisms which are specified for authentication and encryption key management.

There is very significant potential for TETRA based systems to introduce widespread use of information technology into what has been a voice communications based environment. Whilst this text has not dealt in depth with these new IT-base services it does provide sufficient coverage, in a single volume, to stimulate engineers and service providers to expand the potential of TETRA into a range of exciting new application areas.

List of Abbreviations and Symbols

ABBREVIATIONS

A

AC	authentication code, authentication centre
ACCH	Associated Control CHannel
ACELP	Algebraic CELP
ADC	analogue to digital conversion
ADPCM	adaptive differential pulse code modulation
AGCH	Access Grant CHannel
AI	air interface
AM	amplitude modulation
AMPS	Advanced Mobile Phone System
ANF	additional network feature
APCO	Association of Public-safety Communications Officers
ARQ	Automatic Repeat reQuest
ASCI	Advanced Speech Call Items (GSM)
ASK	amplitude shift keying
ASN	abstract syntax notation
AT	ATtention (command set for modems)
ATDM	asynchronous time division multiplexing
ATM	asynchronous transfer mode
ATSI	alias TETRA subscriber identity
AuC	authentication centre (also AC)

B

BBK	Broadcast BlocK
BCCH	Broadcast Control CHannel
BCCH	Broadcast Control CHannel
BER	bit error rate
BLCH	Base station Linearisation CHannel
BNCH	Broadcast Network Control CHannel
BPSK	binary phase shift keying
BS	base station
BSC	base station controller
BSCH	Broadcast Synchronisation CHannel

BSS	base station subsystem
BST	base station transceiver
BT	time bandwidth product of pulse shaping filter
BTS	base transceiver station

C

CB	Control Burst
CC	call control
CCCH	Common Control CHannel
CCH	Control CHannel
CCK	common cipher key
CDMA	code division multiple access
CDPD	Cellular Digital Packet Data
CEPT	Conférence Européenne Postes des et Télécommunication
CFP	Cordless Fixed Part
CIR	carrier to interference ratio
CK	cipher key
CLCH	Common Linearisation CHannel
CMCE	Circuit Mode Control Entity
CMIP	common management information protocol
CNLS	connectionless (packet data service)
CNM	central network management
CONP	connection-oriented network protocol
CPP	Cordless Portable Part
CRC	cyclic redundancy check
CSMA/CD	carrier sense multiple access with collision detection

D

DAWS	Digital Advanced Wireless System
dB	decibel
dBm	decibel with reference to a milli-watt power
DCCH	Dedicated Control CHannel
DCE	data communication equipment
DCK	derived cipher key
DDB	distributed database
DECT	Digital Enhanced Cordless Telecommunication
DES	data encryption standard
DM	Direct Mode
DMCC	direct mode call control
DM-MS	Direct Mode Mobile Station
DMO	Direct Mode Operation
DM-REP	Direct Mode Repeater
DPSK	differential phase shift keying
DQPSK	differential quaternary phase shift keying
DSB-AM	double sideband amplitude modulation
DSRR	Digital Short Range Radio system
DTE	data terminal equipment

DTMF	dual tone multi frequency
DW-MS	Dual Watch-Mobile Station

E

ECCH	Extended Common Control CHannel
ECMA	European Computer Manufacturers Association
EDACS	Enhanced Digital Access Communication System
EIRP	effective isotropic radiated power
EKSG	End-to-end Key Stream Generator
EMC	ElectroMagnetic Compatibility
ERP	effective radiated power
ESN	electronic serial number
ETSI	European Telecommunications Standards Institute

F

FAC	final assembly code
FACCH	Fast Associated Control CHannel
FCCH	Frequency Correction CHannel
FCS	frame check sequence
FDM	frequency division multiplex
FDMA	frequency division multiple access
FEC	Forward Error Correction
FHMA	frequency hopping multiple access
FM	frequency modulation
FSK	frequency shift keying

G

GCK	group cipher key
GFP	generic functional protocol
GMSK	Gaussian minimum shift keying
GoS	grade of service
GPRS	General Packet Radio Service
GSM	Global System for Mobile communications
GSM-R	Global System for Mobile communications for Railways
GTSI	group TETRA subscriber identity

H

HAC	home authentication centre
HDB	home database
HLR	home location register
HSCSD	High Speed Circuit Switched Data
HT200	Hilly Terrain environment at 200kph

I

IDEA	International Data Encryption Algorithm
iDEN	Integrated Digital Enhanced Technology
IDFT	inverse discrete Fourier transform

IF intermediate frequency
IMEI International Mobile Equipment Identity
IMSI International Mobile Subscriber Identity
IN Intelligent Network
IP Internet Protocol
ISDN integrated services digital network
ISI inter-system interface
ISO International Standards Organisation
ISSI individual short subscriber identity
ITSI individual TETRA subscriber identity
IV initial value (authentication algorithm)
IVN InterVening Network

K

kHz kilo Hertz (10^3 cycles/s)
KSG Key Stream Generator

L

LA location area
LAC location area code
LACC location area country code
LAN local area network
LANC location area network code
LAP link access protocol (or procedure)
LAP-B link access protocol for B channel (or - balanced)
LAP-D local access protocol for D Channel
LB Linearisation Burst
LCH Linearisation Channel
LCH Linearisation CHannel
LLC logical link control
LMN land mobile network
LMR land mobile radio
LNM local network management
LS line (-connected) station
LTR logic trunked radio
LTU line termination unit

M

MAC medium access control
MAF mutual authentication flag
MAN metropolitan area network
MCC Mobile Country Code
MCCH Main Control CHannel
MER Message Error Rate
MFA management functional area
MHz Mega hertz (10^6 cycles/s)
MIB management information base

MLE	Mobile/base Link Entity
MM	mobility management
MMI	man machine interface
MNC	Mobile Network Code
MNI	Mobile Network Identity
MoU	Memorandum of Understanding
MPT	Ministry of Post and Telecommunications (standards, e.g., MPT1327)
MS	mobile station
MSC	mobile switching centre
MSK	minimum shift keying
MT	mobile termination
MTU	mobile termination unit

N

NDB	Normal Downlink Burst
NMU	network management unit
NSAP	Network-layer Service Access Point
NT	network termination
NUB	Normal Uplink Burst

O

OMC	operations & management centre
OQPSK	offset quaternary phase shift keying
OSI	open systems interconnection
OTAR	Over The Air Re-keying

P

PA	power amplifier
PAD	packet assembler/disassembler
PAMR	public access mobile radio
PBR	private business radio
PC	personal computer; protocol control
PCH	Paging CHannel
PCM	pulse code modulation
PD	packet data
PDN	public data network
PDO	packet data optimised
PDU	protocol data unit
PEI	peripheral equipment interface
PICS	protocol implementation conformance statement
PIN	personal identity number
PINX	Private Integrated Network eXchange
PISN	private integrated services network
PLMN	public land mobile network
PLP	packet level protocol
PM	phase modulation
PMR	private mobile radio

PN	Pseudo raNdom
POTS	Plain Old Telephone System
PPP	point-to-point protocol
PSK	phase shift keying
PSTN	public switched telephone network
PTT	press to talk; post, telegraphy & telecommunications
PVC	permanent virtual circuit

Q

QoS	quality of service
QPSK	quaternary phase shift keying
QSIG	Q-reference point SIGnalling

R

RACH	Random Access CHannel
RAND	RANDom challenge (on authentication)
RES	Radio Equipment and System; RESponse value (authentication)
RF	radio frequency
ROSE	remote operations service entity
RS	random seed
RSSI	received signal strength indicator

S

SACCH	Slow Associated Control CHannel
SAGE	Security Algorithms Group of Experts
SAP	service access point
SB	Synchronisation Burst
SCCH	Secondary Common Control CHannel
SCH	Signalling CHannel (TETRA), Synchronisation CHannel (GSM)
SCH/F	Signalling CHannel/ Full slot
SCH/H	Signalling CHannel/ Half slot
SCH/S	Synchronisation CHannel (TETRA Direct Mode)
SCK	static cipher key
SCLNP	Specific ConnectionLess Network Protocol
SDCCH	Stand alone Dedicated Control CHannel
SDS	short data service
SIM	subscriber identity module
SIR	signal to interference ratio
SMI	Short Management Identity
SMR	specialist mobile radio
SMS	short message service
SNAF	sub-network access function
SNMP	simple network management protocol
SNR	signal to noise ratio
SRBR	short range business radio
SS	supplementary service
SS-C	supplementary services control

SSI	short subscriber identity
STCH	STealing CHannel
STE	signalling terminal exchange
SVC	switched virtual call
SwMI	Switching and Management Infrastructure

T

TA	terminal adapter (or adapting)
TAC	type approval code
TCH	Traffic CHannel
TCH/S	Traffic Channel/ Speech
TCP	transport control protocol
TDD	time division duplex
TDM	time division multiplexing
TDMA	time division multiple access
TE	terminal equipment
TEI	TETRA Equipment Identity
TETRA	TErrestrial Trunked RAdio
TMI	TETRA Management Identity
TMN	telecommunications management network
TNMM	TETRA network mobility management (protocol access point)
TP	Traffic Physical channel
TSI	TETRA Subscriber Identity
TU50	Typical Urban channel at 50kph

U

UAK	user authentication key
UDP	user datagram protocol
UIC	Union Internationale des Chemins de Fer (International Rail Union)

V

V+D	voice plus data
VAC	visitor authentication centre
VAD	voice activity detector
VDB	visitor database

W

WT	Waiting Time

SYMBOLS

α	roll off factor for pulse shaping
β	modulation index
Δf	elemental bandwidth
Δf_c	carrier deviation
Δf_n	noise bandwidth
Δt_d	delay spread
γ	attenuation coefficient
η	bandwidth efficiency
λ	free space wavelength
ρ	channel occupancy or utilisation factor (teletraffic load)
σ_m	standard deviation of shadow fading
τ	dummy time variable
$\tau(R)$	average fade duration (vertical monopole antenna)
A	mean offered teletraffic load, in Erlang
A_e	effective aperture of an antenna
B	bandwidth of a signal or channel
$B(m)$	m^{th} binary digit
B_c	coherence bandwidth
B_{ch}	bandwidth of a mobile communications channel
B_{tot}	allocated spectrum
c	velocity of light in free space
C	trunk size in number of channels
$C1, C2$	TETRA pathloss parameter
C_n	nth harmonic in a Fourier series
D	frequency re-use distance
$D_k(P)$	decryption function
$D_\phi(k)$	phase transition in $\pi/4$–DQPSK
$E_k(P)$	encryption function
f_0	fundamental frequency of a periodic wave
f_c	cut-off frequency, carrier frequency
f_m	frequency of modulating waveform
$F_v(v)$	cumulative distribution function
G	normalised offered traffic
$G(f)$	frequency response between transmitter and receiver
$G(f)$	power spectral density
$G(x)$	Gaussian shaping function
$G_r(f)$	frequency response of receiver filter
$G_t(f)$	frequency response of transmitter filter
H	magnetic field
h_b	height of base station antenna
h_b, h_m	height of an antenna

h_m	height of mobile station antenna
$i(k)$	in-phase component
I_k	interference power
K	cluster size, encryption key
L	pathloss
m	depth of modulation
m_L	sector mean
$N(R)$	level crossing rate
n_{ch}	number of channels per cell
P	power in a signal
$p(t)$	impulse response of a network
P_a	power per unit area
P_B	blocking probability
P_Q	queuing probability
P_r	received power
P_t	transmitted power
$p_V(v)$	probability density function
q	co-channel interference reduction factor
Q	number of calls in the busy hour
$q(k)$	quadrature phase component
S	normalised throughput
S_c	normalised carrier power
T	duration of bit or symbol; call holding time in minutes
t_d	mean packet delivery time
$v(t)$	general function of time
$v_c(t)$	carrier voltage as a function of time
$v_m(t)$	voltage of modulating waveform as a function of time
$X(k)$, $Y(k)$	k^{th} dibit combination

1

Principles of Digital Mobile Radio

1.1 INTRODUCTION

TETRA is a digital mobile radio system based on a modulation technique known as π/4–DQPSK. In order to interpret this statement fully it is necessary to examine the underlying concepts on which digital transmission systems, and TETRA in particular, are based. The starting point is to review the main properties of modulation schemes which form the basis of all radio systems.

1.2 MODULATION METHODS

The characteristics of the channel over which the signal is to be transmitted may be specified in terms of a frequency and phase response. For efficient transmission to occur, the parameters of the signal must match the characteristics of the channel.

Frequency multiplexing is commonly used in long-distance telephone transmission, in which many narrowband voice channels are accommodated in a wideband coaxial cable. The bandwidth of such a cable is typically 4 MHz, and the bandwidth of each voice channel is about 3 kHz. The 4 MHz bandwidth is divided up into intervals of 4 kHz, and one voice channel is transmitted in each interval. Hence each voice channel must be processed (modulated) in order to shift its amplitude spectrum into the appropriate frequency slot. This form of processing is termed frequency division multiplexing.

For efficient radiation of electromagnetic energy to occur from an antenna, the wavelength of the radiated waveform must be comparable with the physical dimensions of the antenna. The wavelength of a carrier of frequency 300 MHz is approximately 1 metre. For audio-frequency signals (3 kHz), antennas of several hundred kilometres length would be required, which is a practical impossibility. For convenient antenna dimensions the radiated waveform must be of a very high frequency and varied (modulated) in proportion to the information which it is required to transmit.

The general expression for a sinusoidal carrier is

$$v_c(t) = A\cos(2\pi f_c t + \phi) \tag{1.1}$$

The three parameters A_c, f_c and ϕ may be varied for the purpose of transmitting information giving respectively *amplitude*, *frequency* and *phase* modulation. Frequency is essentially the rate of change of phase of a carrier, therefore frequency and phase modulation are very closely linked.

Amplitude modulation occurs when A_c is made proportional to the amplitude of the low frequency information signal $v_m(t)$. A typical example in which $A_c = K + v_m(t)$ and $v_m(t) = a\cos 2\pi f_m t$ is shown in Figure 1.1. The modulated carrier is then

$$v_c(t) = K[1 + mv_m(t)]\cos(2\pi f_c t + \phi) \tag{1.2}$$

where $m = a/K$ is the depth of modulation. The amplitude spectrum of this waveform is shown in Figure 1.2, the important point to note is that the bandwidth of the modulated carrier is twice the bandwidth of the waveform $v_m(t)$.

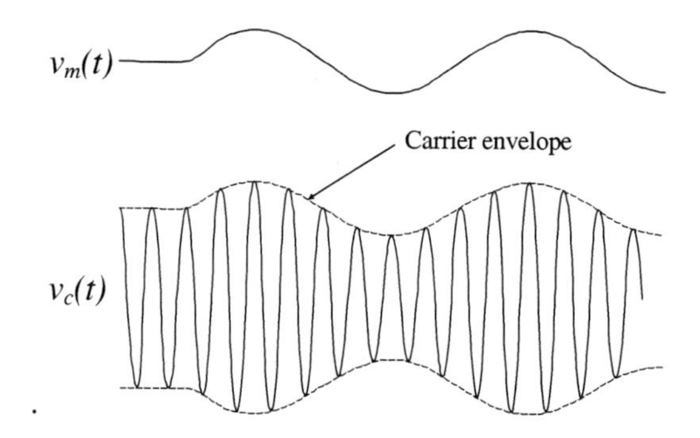

Figure 1.1 Double sideband amplitude modulation

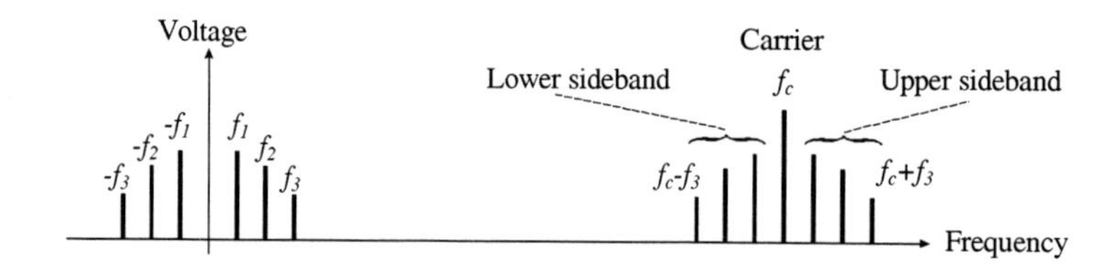

Figure 1.2 Amplitude spectrum of double sideband amplitude modulation

Frequency modulation occurs when the instantaneous frequency of the carrier is made proportional to the amplitude of $v_m(t)$. A typical example is shown in Figure 1.3.

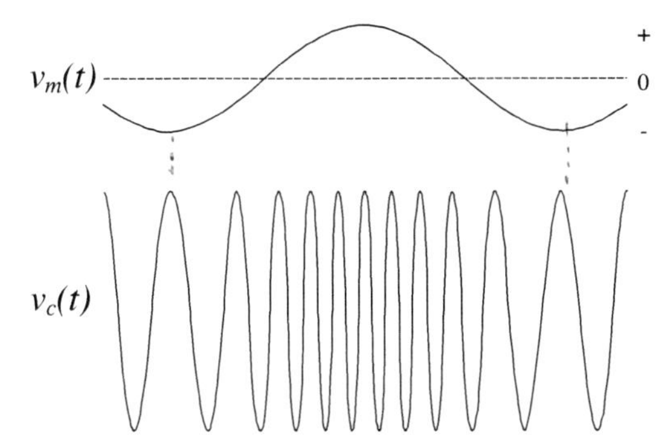

Figure 1.3 Frequency modulated carrier

If the carrier frequency is f_c this will be increased to $f_c + \Delta f_c$ when the amplitude of $v_m(t)$ has its most positive value and will be decreased to $f_c - \Delta f_c$ when the amplitude of $v_m(t)$ has its most negative value. The parameter Δf_c is known as the *carrier deviation* and a further parameter β, known as the *modulation index*, is defined as $\beta = \Delta f_c/f_m$. The expression for the frequency modulated carrier when $v_m(t) = a \cos 2\pi f_m t$ is

$$v_c(t) = A\cos(2\pi f_c t + \beta \sin 2\pi f_m t) \tag{1.3}$$

It should be noted that β depends both on the carrier deviation Δf_c, which is linearly proportional to the amplitude of $v_m(t)$, and also on the frequency of $v_m(t)$. The bandwidth of a frequency modulated carrier is not constant but varies with the value of β. For values of $\beta < 1/\sqrt{3}$ the spectrum of the frequency modulated carrier is very similar to that of Figure 1.2 and the bandwidth is twice the bandwidth of the waveform $v_m(t)$. This is known as narrowband FM. When $\beta >> 1$ the bandwidth is approximately $2 \times \Delta f_c$ and the resulting modulation is known as wideband FM [1].

For commercial FM broadcasts $\Delta f_c = 75$ kHz and $f_m = 15$ kHz which results in a bandwidth of 240 kHz. The bandwidth of modulated carrier is thus much wider than the original modulating signal bandwidth (16 × in the example given). In such cases it can be shown that some gain in signal to noise ratio can usually be achieved when the signal is demodulated. This gain does not occur with amplitude modulation but the effect is very marked with frequency modulation and is known as the capture effect. The capture effect is particularly useful in cellular mobile communications and is illustrated in Figure 1.4 which shows the locus of the resultant phase of the frequency modulated carrier for both high signal to interference ratio (SIR) and low SIR.

In this diagram $x(t)$ represents the in-phase interference component and $y(t)$ represents the quadrature phase interference component. When the SIR is high, the locus of the resultant is confined to an area close to the tip of the carrier and the angle $\phi(t) \cong y(t)/A_c$. As the SIR decreases the locus of the resultant occasionally traverses the origin which gives a rapid change in $\phi(t)$ of 2π radians. This produces a spike in the output of the demodulator, with an area of approximately $\pm 2\pi$ radians. As the SIR decreases further the spikes become more frequent and dominate the output of the demodulator. Thus frequency modulation exhibits a marked threshold, or capture, effect. The presence of modulation has a secondary effect on the occurrence of these noise spikes.

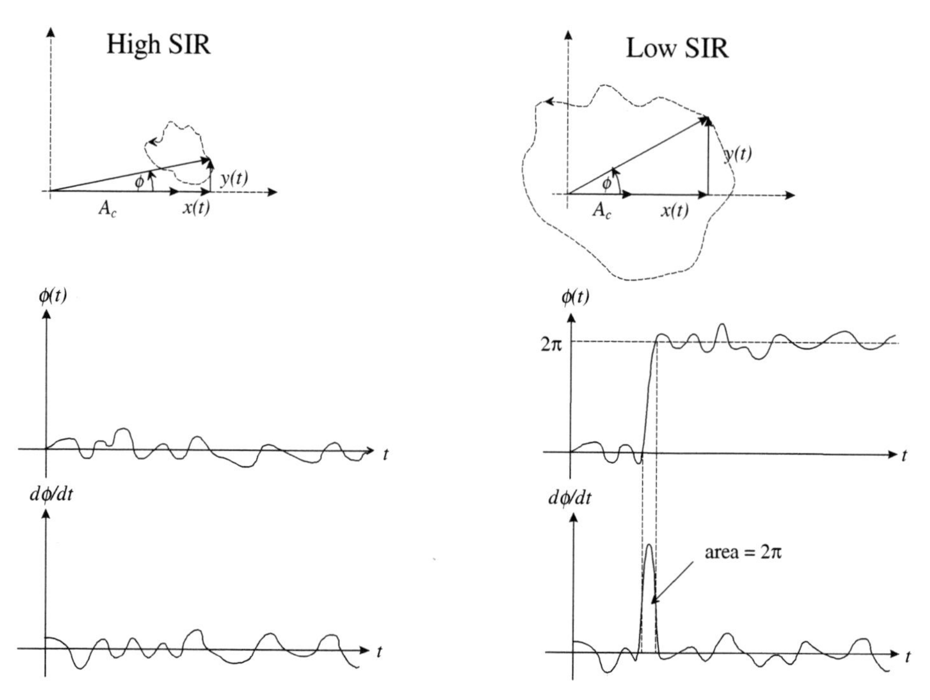

Figure 1.4 Capture effect of FM

Phase modulation occurs when the instantaneous phase of the carrier is made proportional to the amplitude of $v_m(t)$. The expression for the phase modulated carrier when $v_m(t) = a \cos 2\pi f_m t$ produces a maximum phase shift of $\Delta\phi$ is

$$v_c(t) = A\cos(2\pi f_c t + \Delta\phi \cos 2\pi f_m t) \tag{1.4}$$

Analogue phase modulation has similar properties to frequency modulation, but uses the allocated bandwidth less efficiently [1]. The distinction between frequency and phase modulation is less obvious when the modulating waveforms are digital in nature.

1.3 FREQUENCY AND TIME DIVISION MULTIPLEXING

Frequency division multiplexing is shown schematically in Figure 1.5. Each signal occupies a different frequency band and this is achieved by modulating a different carrier frequency with the individual signals. The bandwidths occupied by the modulated carriers are then stacked one above the other throughout the allocated spectrum. These bandwidths are separated by a guard frequency Δf, which is necessary to allow separation of the individual modulated carriers at the receiver.

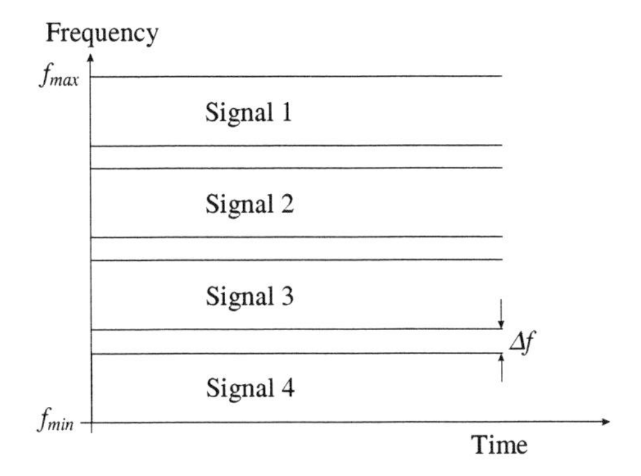

Figure 1.5 Frequency division multiplexing (FDM)

In FDMA the transmitted signals occupy part of the transmission bandwidth for the whole of the transmission time. In cellular systems it is important that the spectrum allocations occupied by different carriers used in the same location do not overlap. This is an issue for frequency planning and is considered in more detail in Section 1.20.

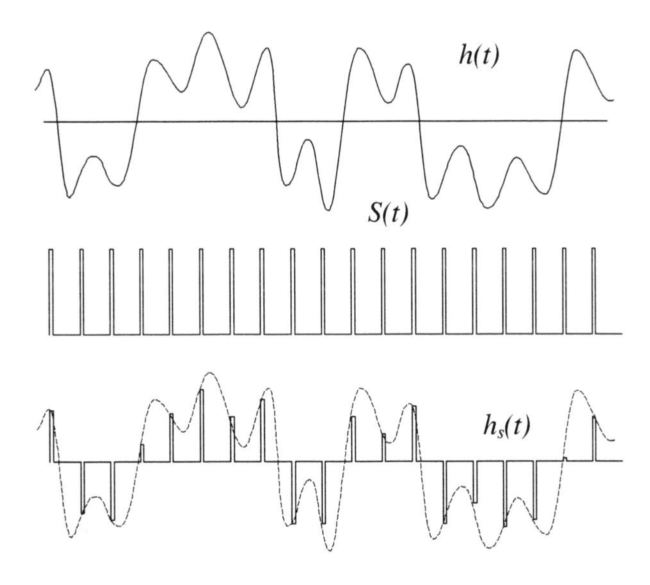

Figure 1.6 The sampling process

The principle of time division multiplexing is based on the sampling theorem. This states that any waveform with a maximum frequency of W Hz is completely defined by samples taken at intervals of $1/2W$ seconds. This is illustrated in Figure 1.6 in which the continuous waveform $h(t)$ is multiplied by the waveform $S(t)$ to produce the sampled waveform $h_s(t)$. This figure reveals that the duration of the pulses in $S(t)$ is much less than the pulse period. (In fact the sampling theorem may be extended to bandlimited waveforms, in which case a waveform with a bandwidth B is completely defined by samples taken at a rate of $2B$ per second.)

When a waveform is sampled by narrow pulses there are large intervals between the samples in which no signal exists. It is possible during these intervals to transmit the samples of other signals. This process is shown in Fig 1.7 and is called time division multiplex (TDM) transmission. Since each sampled signal gives rise to a continuous signal after filtering, TDM allows simultaneous transmission of several signals over a single wideband link. It is therefore an alternative to FDM transmission.

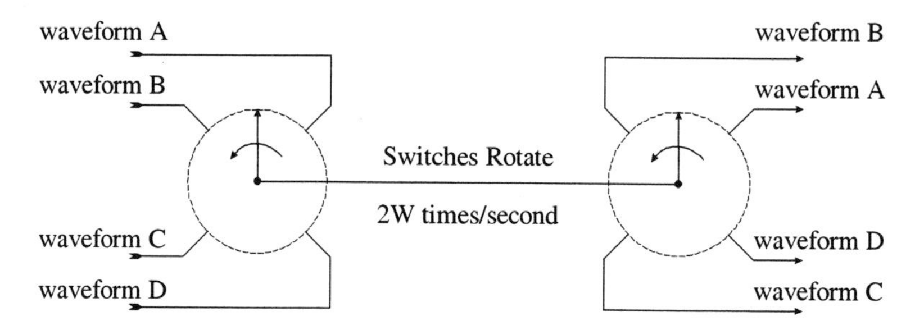

Figure 1.7 TDM transmission

The switches at transmitter and receiver in a TDM system (which would be solid state devices) are synchronised and perform the sampling and interlacing. The samples themselves are very narrow and consequently have a large bandwidth. When transmitted over a link with a fixed bandwidth the samples are spread and can overlap adjacent samples. To minimise this, guard bands Δt are allowed between adjacent signals. In the case of TDM transmission each signal therefore occupies the whole of the transmission bandwidth for part of the transmission time as shown in Figure 1.8

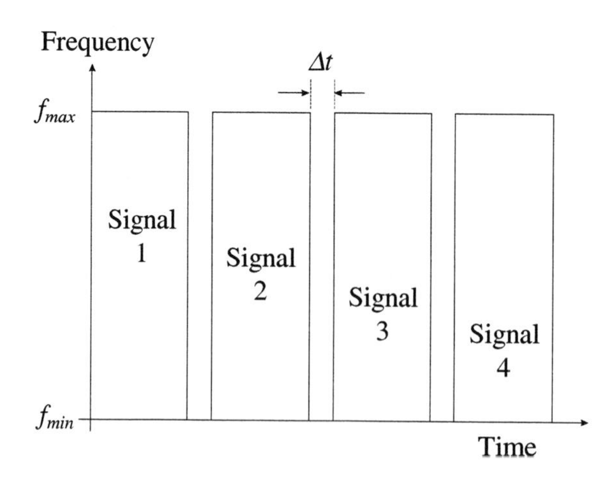

Figure 1.8 Frequency/time diagram for TDM

1.4 ANALOGUE TO DIGITAL CONVERSION

Second generation mobile radio systems (GSM, TETRA, DECT) are essentially digital transmission systems. Therefore analogue signals such as voice must first be converted to digital format. Each mobile system has its own particular method of coding voice, however the starting point for each one is pulse code modulation.

This process is essentially one of converting each of the samples of Figure 1.6 into binary format which effectively means that each has been quantised into one of a fixed number of levels. The greater the number of quantisation levels the greater is the accuracy of the quantised representation, but also the greater is the number of binary digits that are required to represent the sample. Since more digits require a higher transmission bandwidth a balance must be struck between accuracy and bandwidth. The quantisation process is illustrated in Figure 1.9.

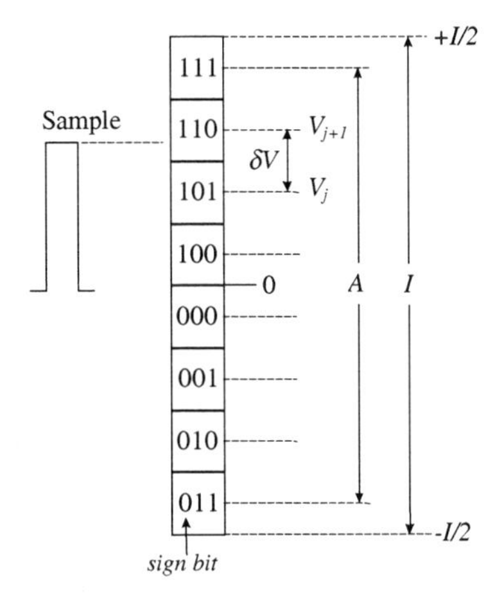

Figure 1.9 Linear quantisation (8 level)

It is clear that once quantised the precise amplitude of the original sample cannot be restored. This gives rise to an error in the recovered analogue signal, known as the quantisation error which is random in nature and is called quantisation noise. The signal to quantisation noise (SQNR) for a sinusoidal waveform is given by

$$\text{mean SQNR} = (1.8 + 6m)\ \text{dB} \tag{1.5}$$

In this equation m is the number of digits representing the sample amplitude. The larger the value of m the smaller the quantisation error, but the larger the bandwidth required to transmit the digital waveform. A typical quantisation error waveform is shown Figure 1.10

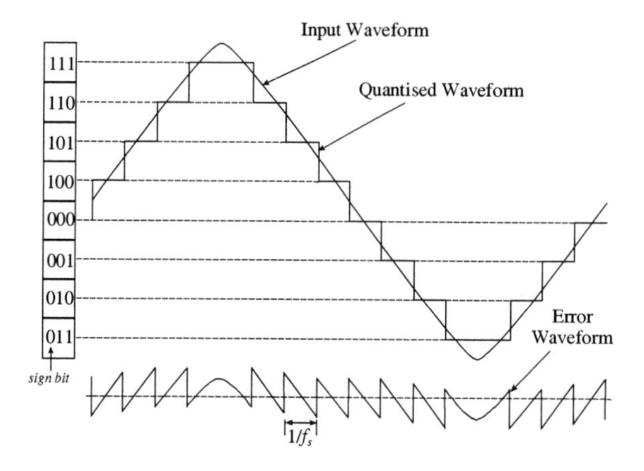

Figure 1.10 Quantisation error waveform

Once the signal has been converted into digital form there are significant advantages as the receiver then simply has to determine the level of the transmitted signal. This process is illustrated in Figure 1.11

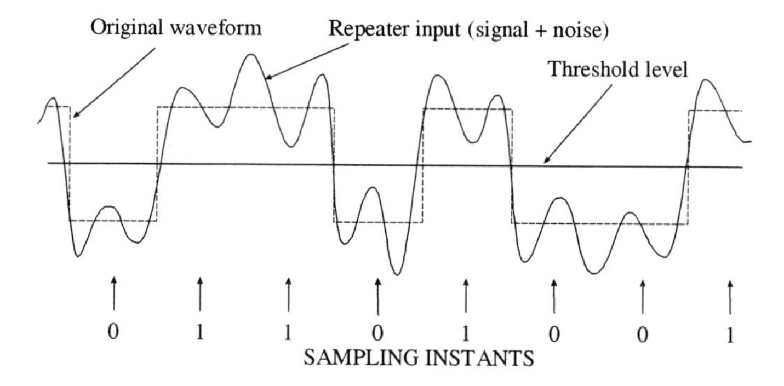

Figure 1.11 Binary waveform corrupted by noise

1.5 SPECTRAL PROPERTIES OF DIGITAL SIGNALS

In order to specify the most appropriate form of data transmission over communication channels it is necessary consider the spectral properties of data signals. A typical data signal will consist of a random sequence of pulses of binary 1s and 0s. The power spectral density of such a signal is given by the Fourier transform of its autocorrelation function. A random binary signal with pulse amplitudes of 0 or A volts and pulse duration T seconds has an amplitude spectrum given by

$$H(f) = AT\frac{\sin \pi fT}{\pi fT}$$

It can be seen from Figure 1.12 that most of the energy in the spectral envelope is confined to frequencies below $f = 1/T$ Hz. The bandwidth of the data signal is therefore usually approximated by the reciprocal of the pulse width. However, it will be appreciated from this figure, that the spectrum of the pulse actually extends beyond this value and this fact must be taken into account when considering digital modulation systems.

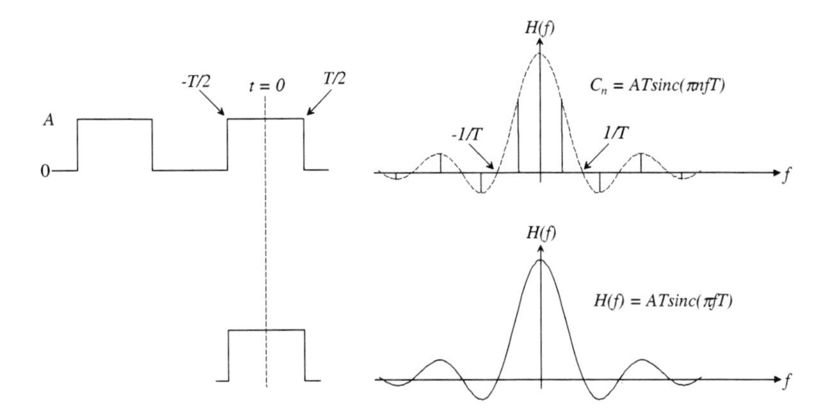

Figure 1.12 Amplitude spectrum of data signal waveforms

In Figure 1.12 the transmitted pulse (symbol) has 2 levels and the signal which results is a binary waveform. Each pulse in a binary waveform is called a *b*inary dig*it*, which is often shortened to *bit*. This is actually a misnomer as the *bit* is specifically a unit of information. A binary pulse contains 1 bit of information only if it has 2 levels which are equi-probable. The amount of information contained in a pulse is increased when the number of levels is increased.

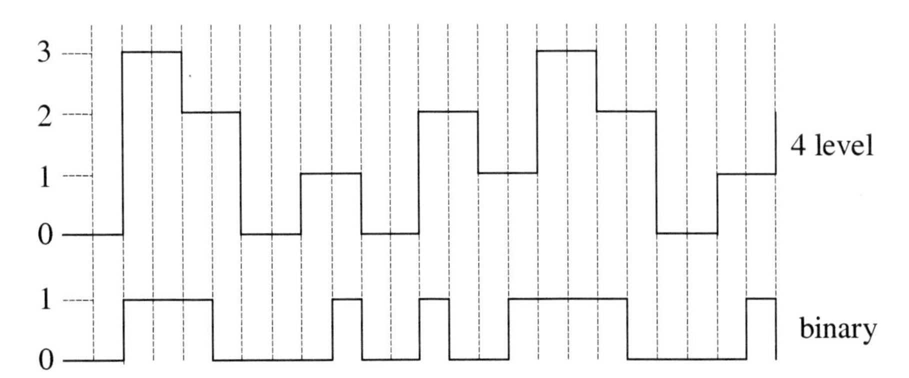

Figure 1.13 Multi-level digital pulses

Two binary digits (*dibits*) have 4 possible combinations 00, 01, 10, and 11. Since each combination is effectively 2 bits of information, it becomes clear that a pulse with 4 possible levels also contains 2 bits as each level is equivalent to one of the possible *dibit* combinations. A 4 level pulse (symbol) is shown in Figure 1.13. The signalling speed of a communication link, measured in Baud, is equal to the number of symbols/second. The information rate (measured in bits/second) would be twice the symbol rate if each symbol has 4 equiprobable levels. Hence one way of increasing the information rate, without increasing the bandwidth of the waveform, is to increase the number of levels per symbol.

1.6 PULSE SHAPING

It is theoretically possible to increase the number of levels beyond 4, e.g. 8, 16, 32 etc. and hence achieve progressively higher bit rates without increasing the pulse rate. There is a limit to this process which is caused by the problem of inter-symbol interference and also noise. Figure 1.12 indicates that the bandwidth of a pulse waveform is essentially infinite. It is therefore necessary to limit the bandwidth of such a signal before transmission and this is done by a process known as *pulse shaping*. This is equivalent to passing the waveform through a low pass filter. The amount of pulse shaping depends on the ratio of the filter bandwidth (B) and modulating bit period (T). Some typical examples are shown in Figure 1.14, which illustrates the effect of passing the pulse through a Gaussian filter. In this figure a value of $BT = \infty$ corresponds to no filtering and the pulse has a duration T. After shaping the pulse has a duration greater than the bit period and therefore interferes with adjacent pulses, producing inter-symbol interference (ISI). ISI causes errors to occur in the received symbols. As the value of BT is reduced this results in a narrower bandwidth but a higher induced ISI. A value of $BT = 0.3$ is regarded as an acceptable compromise in GSM between transmission bandwidth and bit error rate.

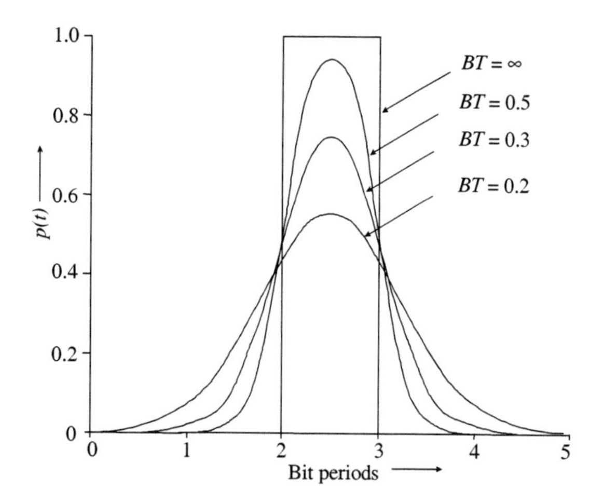

Figure 1.14 Pulse shaping with a Gaussian filter

Inter-symbol interference is a much more significant problem when multi-level pulses are used rather than binary pulses. For example, if a symbol of level 3 is immediately followed by a symbol of level 0 it is more likely to produce a decision error after pulse shaping than would be the case with a binary transmission with the same pulse amplitudes. Hence there is a limit on the number of pulse levels which can usefully be employed and a value of 16 levels is usually regarded as the upper limit. A second issue which must be taken into account with multi-level symbols is that each level must be separated from its neighbour by an amount greater than the noise or interference.

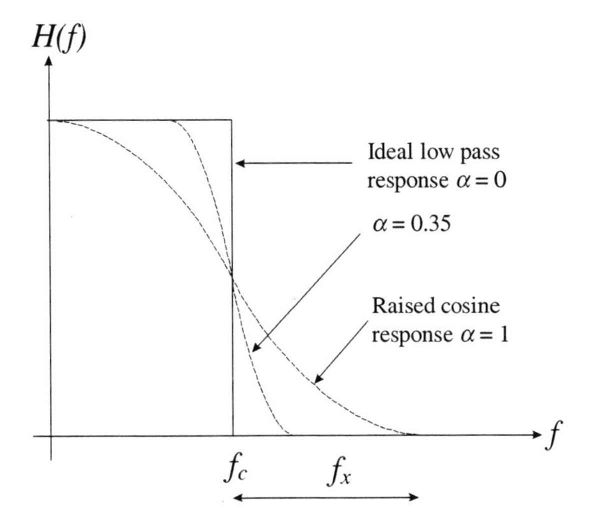

Figure 1.15 Pulse shaping with a raised cosine filter

Alternative pulse shaping filters are also used in digital radio systems, one example of which is a filter with a raised cosine impulse response, which is the filter used in the TETRA system. This impulse response is the Fourier transform of a raised cosine frequency response. This frequency response is defined as

$$\begin{aligned} &G(f) = 1 \text{ for } |f| \leq (1-\alpha)/2T \\ &G(f) = 0.5[1 - \sin(\pi(2|f|T - 1)/2\alpha)] \text{ for } (1-\alpha)/2T \leq |f| \leq (1+\alpha)/2T \\ &G(f) = 0 \text{ for } |f| \geq (1+\alpha)/2T \end{aligned} \quad (1.6)$$

where α is the roll-off factor and is chosen to produce a compromise between pulse bandwidth and inter-symbol interference.

The value of α may be expressed as $\alpha = f_c/f_x$ where f_x is the frequency by which the bandwidth of the raised cosine filter exceeds f_c, the cut-off frequency of the equivalent ideal low pass filter. This is made clear by reference to Figure 1.15. The smaller the value of α the narrower the bandwidth of the transmitted waveform but the greater the inter-symbol interference. The value of roll off factor used in TETRA is $\alpha = 0.35$, which provides a compromise similar to that observed in the case of GSM, with a choice of $BT = 0.3$.

1.7 DIGITAL MODULATION

Carrier modulation is a fundamental component of all digital radio systems. As with analogue systems the three possible forms of modulation are amplitude modulation, frequency modulation and phase modulation. These types of modulation are not always distinct when data signals are considered, as will become apparent from what follows.

1.7.1 Amplitude Shift Keying (ASK)

This is the name given to amplitude modulation when used to transmit data signals. It is not normally used in mobile communications due to the significant variation in received signal strength which can occur and the associated difficulty of fixing a decision threshold. However, ASK will be considered in some detail because it is convenient in the analysis of frequency shift keying and phase shift keying

The ASK signal is generated by multiplying the data signal by a carrier and the expression for the modulated carrier is:

$$v_c(t) = h(t)\cos(2\pi f_c t), \text{ where } h(t) = A \text{ or } 0. \tag{1.7}$$

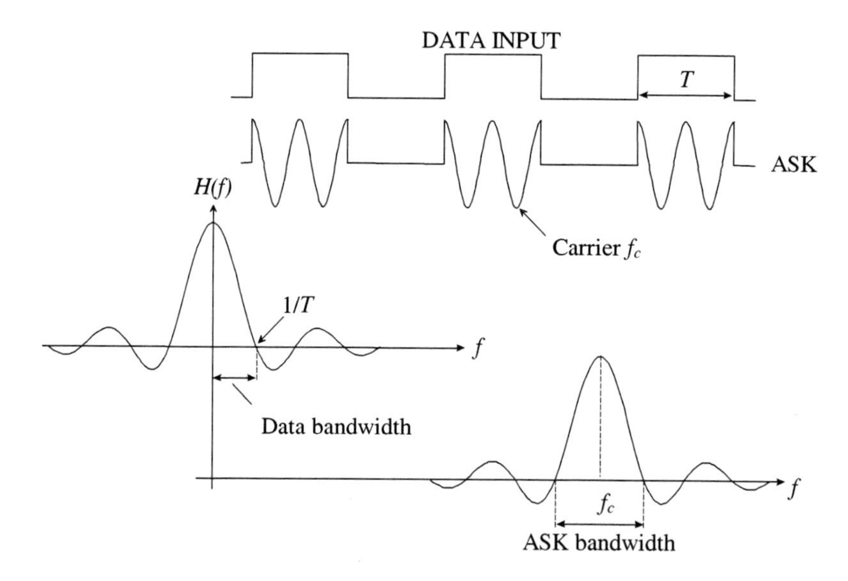

Figure 1.16 Amplitude spectrum of ASK

This multiplication effectively shifts the data spectrum to a centre frequency equal to that of the carrier. The process is shown in Figure 1.16. The bandwidth of the modulated signal is twice the bandwidth of the original data signal. If the symbol duration is T the approximate bandwidth of the ASK waveform is $2/T$ Hz.

ASK ≡ DSBAM ≡ (carrier + upper and lower sidebands)

ASK is decoded using an envelope detector or a coherent detector followed by a decision circuit with a threshold set at a level of $A/2$.

1.7.2 Binary Phase Shift Keying (BPSK)

This is the binary equivalent of analogue phase modulation, the binary symbols being transmitted as a phase shift of 0 or π radians. The expression for the BPSK waveform is

$$v_c(t) = h(t)\cos(2\pi f_c t), \text{ where } h(t) = +A \text{ or } -A \tag{1.8}$$

It is common practice to refer to BPSK simply as phase shift keying or PSK. The PSK waveform and amplitude spectrum are shown in Figure 1.17. The bandwidth of PSK is the same as that of ASK, however because the mean value of $h(t) = 0$ there is no component in the spectrum at the carrier frequency f_c. PSK is demodulated by a coherent detector followed by a decision circuit, with the threshold set at 0 volts. Considerable difficulties can occur in generating the required carrier phase reference for the coherent detector, especially in multipath environments. The problem is avoided by use of differential phase shift keying (DPSK).

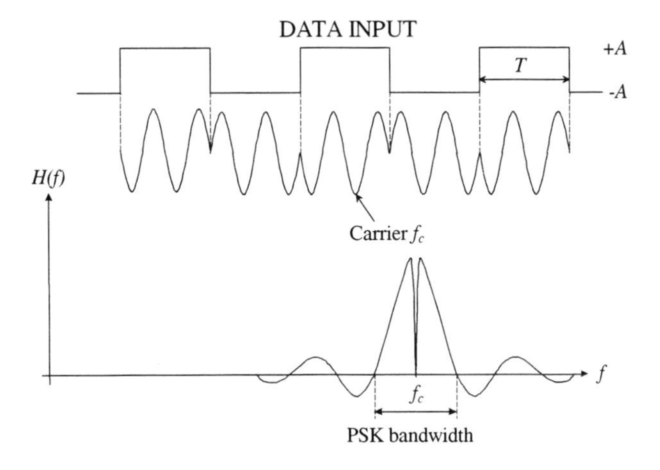

Figure 1.17 Amplitude spectrum of PSK

1.7.3 Differential Phase Shift Keying (DPSK)

DPSK uses the phase of the carrier in the previous digit interval as the reference for the present digit interval. In order to make this possible, a binary 0 is transmitted as the same phase as the previous digit and a binary 1 is transmitted as a change of phase. The relationship between PSK and DPSK is shown in Figure 1.18. The receiver compares (i.e. correlates) the phase of the current digit with the phase of the previous digit. If they are the same the current digit is interpreted as a 0, otherwise it is interpreted as a 1.

DPSK can be produced by pre-coding the data signal which then modulates the carrier as in standard PSK. If A_n is the present input to the differential encoder (A_n is binary) and C_{n-1} is the previous output of the differential encoder truth table for the encoder is

A_n	C_{n-1}	C_n
0	0	0
0	1	1
1	0	1
1	1	0

which will be recognised as the exclusive-OR operation $C_n = A_n \oplus C_{n-1}$

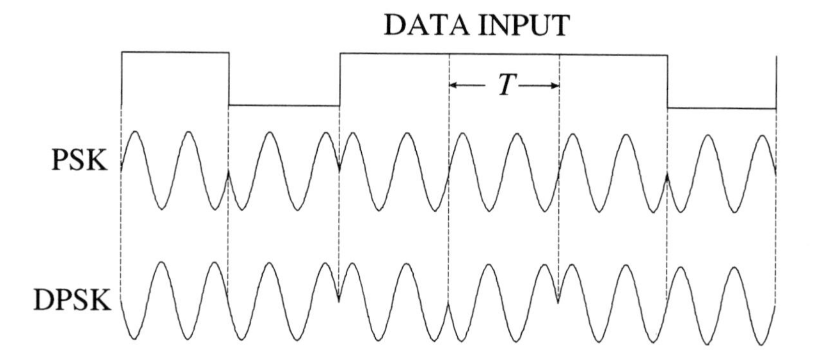

Figure 1.18 Relationship between PSK and DPSK

The PSK and DPSK encoders are shown in Figure 1.19.

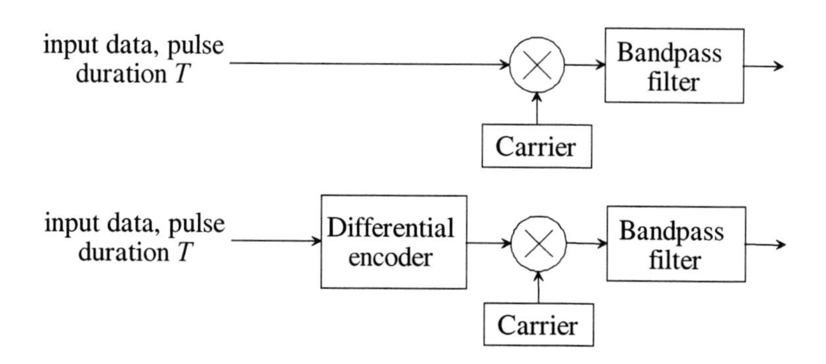

Figure 1.19 Binary PSK and differential binary PSK modulators

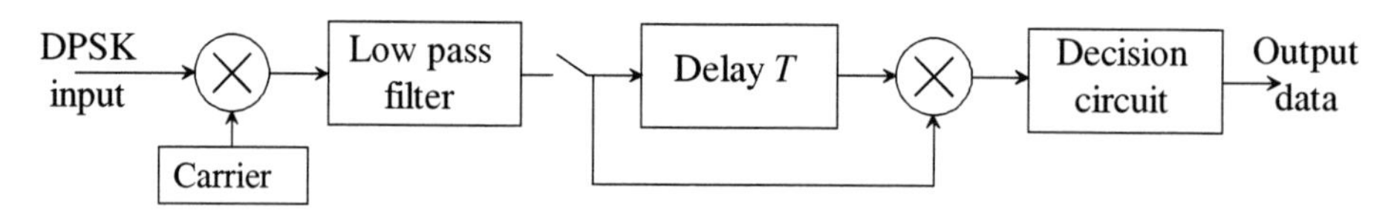

Figure 1.20 DPSK demodulator

The DPSK decoder is shown in Figure 1.20. The output of the coherent detector is filtered and passed to a correlator, realised as a delay circuit and multiplier. If the current phase is the same as the previous phase the correlator output will be positive (which is interpreted as binary 0), if the current phase is the inverse of the previous phase the output of the correlator will be negative (which is interpreted as binary 1). A small phase error in the carrier injected into the coherent detector will not affect the correlator output.

1.7.4 Quaternary Phase Shift Keying (QPSK)

A phase shift keyed waveform is represented by the expression

$$v_c(t) = A_c \cos(2\pi f_c t + \phi_k) \tag{1.9}$$

For binary PSK (BPSK) ϕ_k takes on two possible values separated by 180°, for quaternary PSK (QPSK) ϕ_k takes on four possible values separated by 90°. Thus QPSK is equivalent to phase modulation by a symbol with 4 possible levels of the type shown in Figure 1.13. Equation 1.9 may be expanded to give

$$v_c(t) = i\cos 2\pi f_c t + q\sin 2\pi f_c t \tag{1.10}$$

where $i = A_c\cos\phi_k$ represents the amplitude of the in-phase component and $q = -A_c\sin\phi_k$ represents the amplitude of the quadrature component. The phase modulated carrier is effectively the sum of modulated in-phase and quadrature components. As $v_c(t)$ is a phase modulated waveform the carrier amplitude is constant and this form of modulation is referred to as *constant envelope* modulation.

Table 1.1 Phase shifts in QPSK

Dibit	Phase shift	In phase amplitude ***i***	Quadrature amplitude ***q***
00	$\pi/4$	$+1/\sqrt{2}$	$+1/\sqrt{2}$
01	$3\pi/4$	$-1/\sqrt{2}$	$+1/\sqrt{2}$
11	$-3\pi/4$	$-1/\sqrt{2}$	$-1/\sqrt{2}$
10	$-\pi/4$	$+1/\sqrt{2}$	$-1/\sqrt{2}$

In QPSK the input data stream is grouped into *dibits* each combination of which is used to produce a unique phase shift of the carrier, separated by intervals of $\pi/2$. Table 1.1 lists the possible *dibits* and the values of i and q, in equation 1.10, necessary to produce the required phase shifts. In effect i and q represent pulses of amplitude $\pm 1/\sqrt{2}$, occurring at half the original bit rate, which are used to multiply the in-phase and quadrature carriers. The resulting quaternary PSK can be represented on an I/Q diagram of the type shown in Figure 1.21 which also shows the possible phase transitions which can occur (the maximum *phase transition* is π). The QPSK modulator, based on the dibit principle, is shown in Figure 1.22.

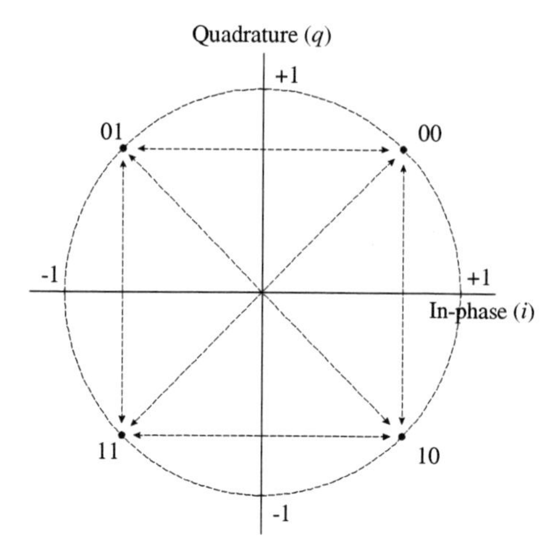

Figure 1.21 I/Q diagram for QPSK

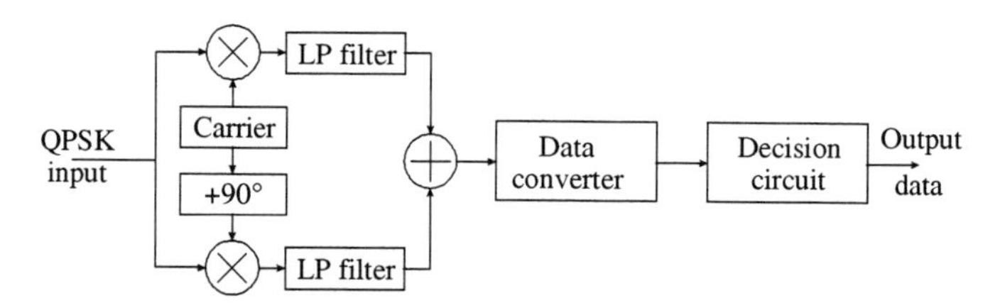

Figure 1.22 QPSK modulator and demodulator

The data signal is recovered from the QPSK waveform by using two coherent detectors supplied with locally generated carriers in phase quadrature. This process produces the in-phase data waveform and the quadrature data waveform each of which has a pulse duration of $2T$. The original input data waveform with a pulse duration of T can then be reproduced. Clearly if a single error is made in the detection process this will affect two bits in the reproduced input data waveform.

1.7.5 Offset QPSK (OQPSK)

In mobile communications systems the carrier frequencies used are in the microwave region. In such systems the microwave amplifiers employed tend to have a non linear characteristics and are most suited to constant envelope modulation systems. (Linear amplifiers are available but tend to be much more costly). Thus QPSK might be regarded as suitable modulation scheme for such applications. However, when pulse shaping is applied to limit the bandwidth of the transmitted waveform this produces amplitude variations in the QPSK carrier envelope. Because of the non-linearity of the microwave amplifiers these amplitude variations are largely suppressed, returning the QPSK signal to an almost constant envelope. This negates the advantage of the pulse shaping operation in reducing the bandwidth of the transmitted waveform.

A special form of QPSK known as Offset QPSK (OQPSK) has been developed to address this problem. In effect the data for the quadrature channel is delayed by the input pulse duration T. This results in the maximum phase transition of the carrier being restricted to 50% of the phase transition of standard QPSK and results in a much smaller envelope variation after bandpass filtering. The OQPSK modulator is shown in Figure 1.23 and the effect of the delay T on allowable phase shifts is shown in Figure 1.24. In effect the phase transitions which pass through the origin are removed by the delay of T thus restricting the maximum phase shift to $\pi/2$.

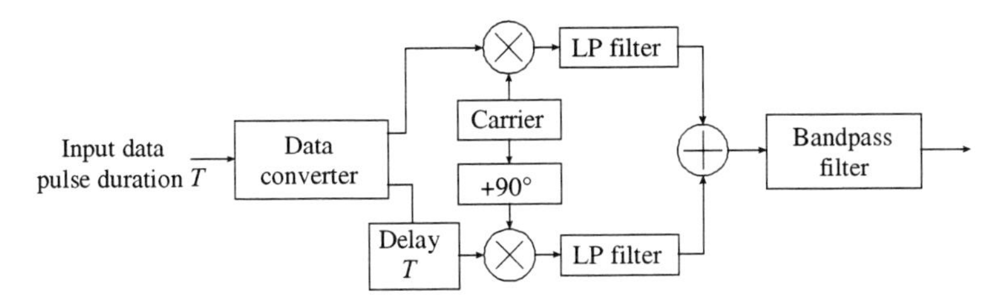

Figure 1.23 Offset QPSK

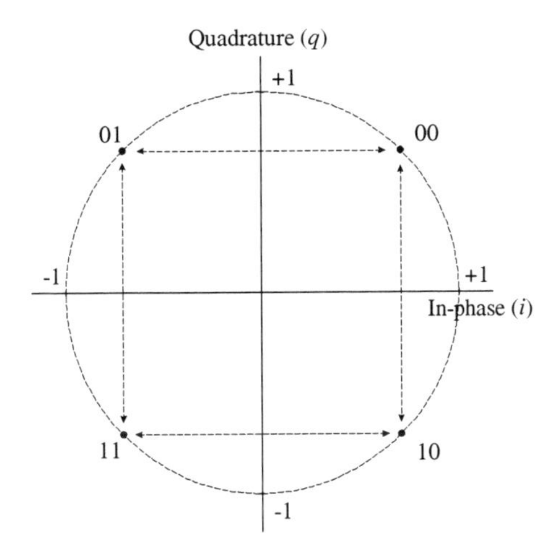

Figure 1.24 Phase transitions in OQPSK

1.7.6 Differential QPSK (DQPSK)

Data recovery from the QPSK waveform requires two quadrature coherent detectors supplied with the appropriate reference carriers. As with BPSK, this is difficult to achieve in a multipath environment and the problem is overcome by the use of differential QPSK. A typical DQPSK modulator and demodulator is shown in Figure 1.25. The DPSK demodulation process reproduces the in-phase and quadrature data waveforms each of which has a pulse duration of $2T$. The original input data with a pulse duration of T can then be derived.

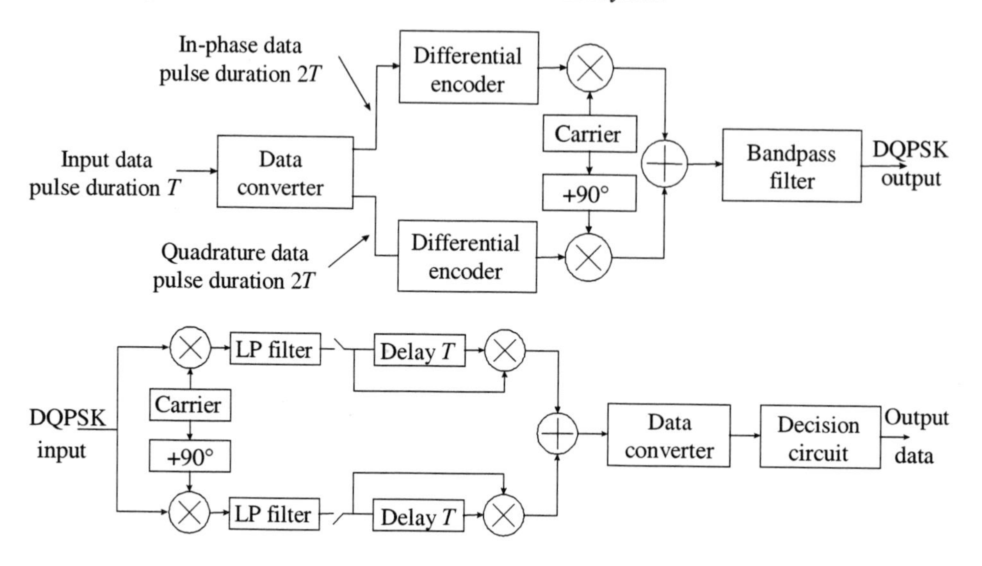

Figure 1.25 DQPSK modulator and demodulator

1.7.7 π/4–DQPSK

This is the form of modulation used in the Terrestrial Trunked Radio Standard TETRA. In the case of π/4–DQPSK each *dibit* combination produces a specified PHASE TRANSITION $D_\phi(k)$ (which is a multiple of π/4). The phase transition is independent of the current carrier phase and has a maximum value of 3π/4, which is less than the maximum phase shift produced by OQPSK. The carrier phase transitions produced in π/4–DQPSK are shown in Table 1.2 and are illustrated in Figure 1.26. Although there are eight separate carrier phases in this diagram it is important to observe that there are only four possible *phase transitions*, which are independent of the current carrier phase.

Table 1.2 Phase transitions in π/4–DQPSK

Dibit [*X(k)*, *Y(k)*]	$D_\phi(k)$
00	+π/4
01	+3π/4
11	−3π/4
10	−π/4

The encoder is shown in Figure 1.27 in which *B(m)* represents the input data sequence which is mapped into *dibits* (symbols) [*X(k)*, *Y(k)*] = [*B(2k-1)*, *B(2k)*]. The in-phase and quadrature components *i(k)*, *q(k)* which are applied to the phase modulator are:

$$i(k) = i(k-1)\cos\left[D_\phi(k)\{X(k),Y(k)\}\right] - q(k-1)\sin\left[D_\phi(k)\{X(k),Y(k)\}\right]$$
$$q(k) = i(k-1)\cos\left[D_\phi(k)\{X(k),Y(k)\}\right] + q(k-1)\sin\left[D_\phi(k)\{X(k),Y(k)\}\right]$$

the values of $D_\phi(k)$ are those specified in Table 1.2 for the corresponding values of $X(k)$ and $Y(k)$.

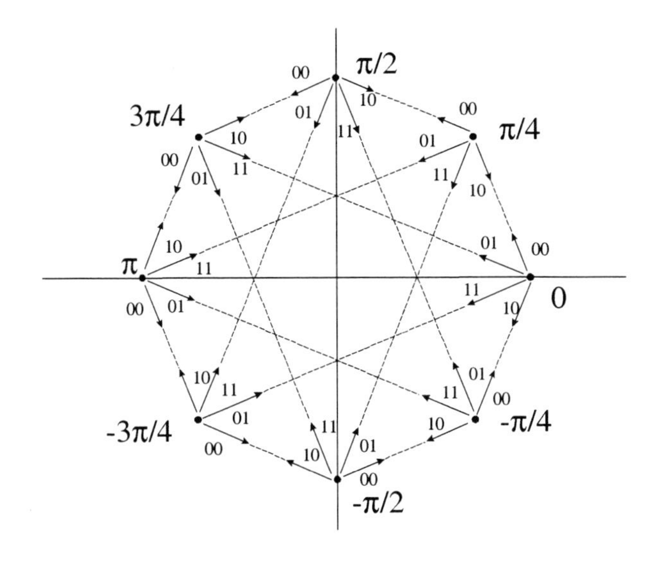

Figure 1.26 Phase transitions in π/4-DQPSK

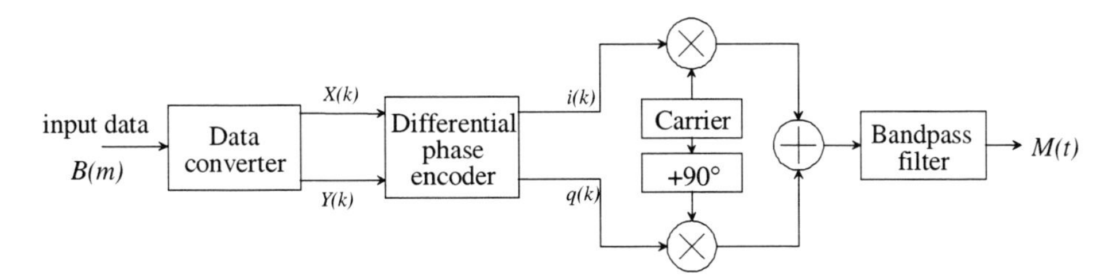

Figure 1.27 Modulator for π/4–DQPSK

At the receiver a particular *dibit* is reproduced by comparing the current carrier phase with the previous carrier phase and measuring the phase difference. This relatively simple operation requires the presence of both in-phase and quadrature phase components in order to differentiate between phase transitions π/4 and –π/4, for instance and results in a relatively complex decoder, which has the form shown in Figure 1.28.

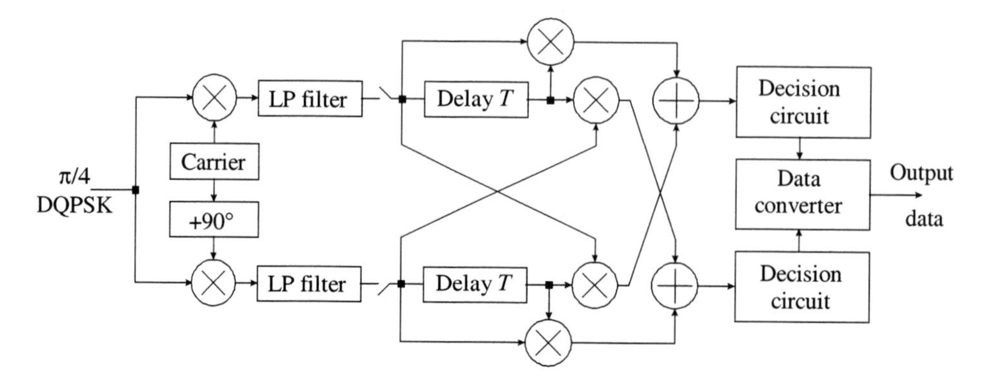

Figure 1.28 π/4–DQPSK demodulator

The carrier phases which result for an input waveform of 010101010101 etc. are shown in Figure 1.29 in which the initial carrier (reference) phase is assumed to be zero. Each *dibit* combination 01 in the input data stream will then produce a phase transition of 3π/4.

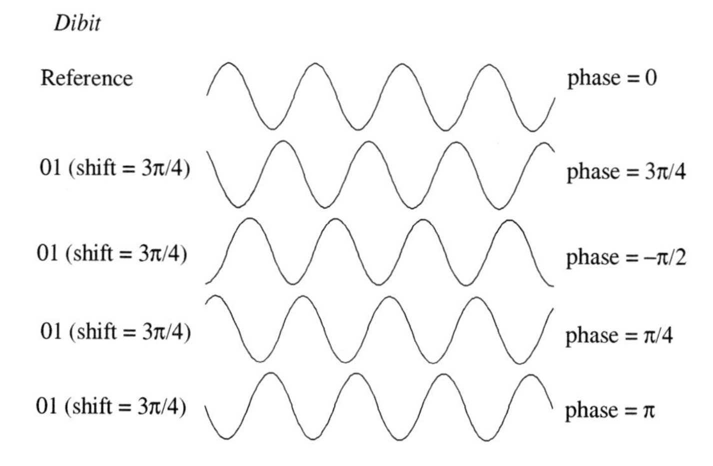

Figure 1.29 Carrier phases produced by input data 01010101

1.7.8 Linear Amplifiers

Although the maximum phase shift in π/4–DQPSK is restricted to 3π/4 the bandwidth is further reduced by pulse shaping. The effect of the pulse shaping on the constant envelope π/4–DQPSK waveform is shown in Figure 1.30. It should be noted from this figure that the sharp discontinuities at the end of each symbol (*dibit*) are smoothed out by the pulse shaping process. The fact that no phase transition passes through the origin (Figure 1.26) assists this process considerably and is one of the major advantages of π/4–DQPSK.

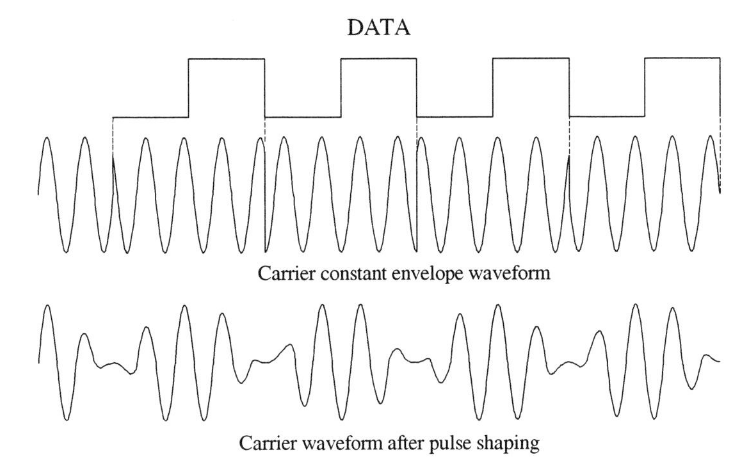

Figure 1.30 Effect of pulse shaping on the transmitted waveform

To avoid losing the advantage of pulse shaping it is necessary to ensure that the power amplifiers used in transmission have an approximately linear characteristic. This is achieved in the TETRA system by linearising the amplifiers at regular intervals and a logical channel is available within the TETRA specification for this purpose, and is discussed in more detail in Chapter 4.

It should be noted that $\pi/4$–DQPSK does have some disadvantages, one being that a higher signal to interference ratio is required than with modulation schemes with 2 levels (1 bit) per symbol .

1.7.9 Frequency Shift Keying (FSK)

This is the binary equivalent of analogue frequency modulation. In this case a binary 0 is transmitted as a frequency f_0 and a binary 1 is transmitted as a frequency f_1. Hence the binary signal effectively modulates the frequency of a "carrier".

FSK may actually be considered as the sum of two ASK waveforms with different carrier frequencies. The spectrum of the FSK wave is thus the sum of the spectra of the two ASK waveforms and is shown in Figure 1.31. Using the frequency modulation analogy, it is possible to define a "carrier frequency" $f_c = f_0 + (f_1 - f_0)/2$ and a "carrier deviation" $\Delta f_c = (f_1 - f_0)/2$. The modulation index β is defined as $\beta = \Delta f_c /B$, where $B = 1/T$ is the bandwidth of the data signal.

Using these definitions the bandwidth of the FSK signal is

$$B_{FSK} = 2B(1 + \beta)$$

This is similar to Carson's rule for continuous frequency modulation. Unlike analogue frequency modulation there is no advantage in increasing Δf_c beyond the value $\Delta f_c = B$ since the receiver is required only to differentiate between the two frequencies f_0 and f_1.

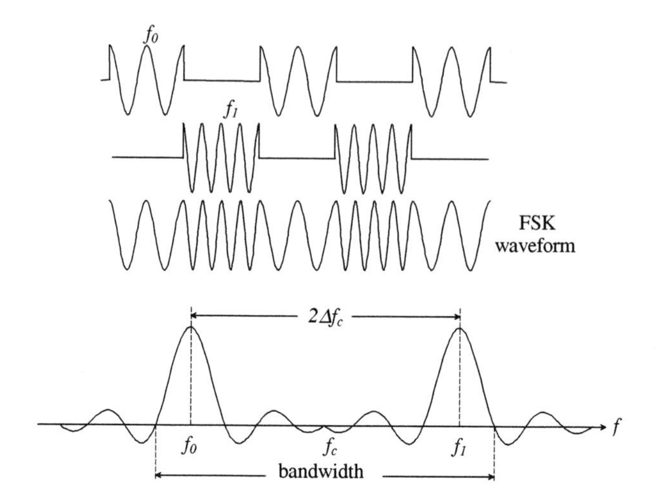

Figure 1.31 Amplitude spectrum of FSK

This is especially important in mobile communications where there is an overriding need to minimise the transmitted bandwidth. Minimum shift keying is a development of FSK which addresses this issue.

1.7.10 Minimum Shift Keying (MSK)

In mobile communication systems there are basically two problems which need to be addressed, these are:

- minimum transmission bandwidth
- minimum error probability

Standard frequency shift keying uses two separate carriers f_0 and f_1 to transmit binary 0 and binary 1. In order to produce the smallest error probability the carriers f_0 and f_1 must be orthogonal, i.e. they must have a correlation coefficient which is zero. In order to minimise the bandwidth of the transmitted signal it is necessary to determine the minimum difference between f_0 and f_1 which will produce orthogonal signals and this is called *minimum shift keying*. If the number of cycles of f_0 in the bit period T is n_0, then f_1 will be orthogonal to f_0 if the number of cycles in the interval T is $n_1 = n_0 + 0.5$, alternatively $(f_1 - f_0) = 1/2T$. Hence MSK is effectively FSK with the minimum frequency difference between f_1 and f_0.

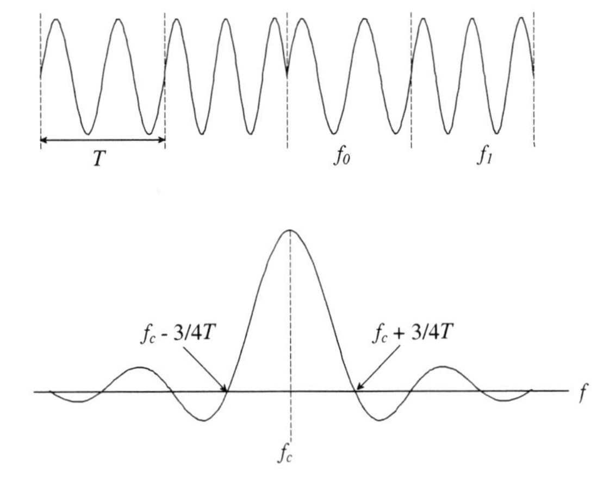

Figure 1.32 MSK waveform and spectrum

If MSK is considered in terms of the modulation of a single carrier frequency f_c the instantaneous frequency is given by

$f_i = f_c + a.\Delta f_c$, where $a = \pm 1$ and Δf_c is the carrier deviation.

For MSK the carrier deviation $\Delta f_c = 1/4T$ which for GSM (Chapter 2), with a data rate of 270.833 kb/s, is $(270.833 \times 10^3)/4 = 67.7$ kHz. Thus in GSM f_1 and f_0 appear as carriers 67.7 kHz above and below the assigned channel frequency. When the keyed frequencies are separated by an amount less than the data bandwidth it is not appropriate to represent the spectrum as the sum of two independent sinc functions as these sinc functions will overlap to a considerable extent and will produce a "composite" spectrum centred at the assigned channel frequency f_c. This spectrum has the form shown in Figure 1.32 and it should be noted that this spectrum has a wider bandwidth than the corresponding ASK (or PSK) spectrum but that the sidelobes decrease at a faster rate.

1.7.11 MSK Considered in Terms of Phase Modulation

The instantaneous frequency of a MSK carrier is $f_i = f_c + a.\Delta f_c$, where $a = \pm 1$ and $\Delta f_c = 1/4T$ is the carrier deviation.

It may be noted that the rate of change of carrier phase is $\dot{\phi}(t) = 2\pi f_i = 2\pi f_c + \dfrac{2\pi a}{4T}$

Hence the carrier phase $\phi(t) = 2\pi f_c t + \dfrac{2\pi a}{4T}\int_0^t dt$

The MSK carrier may thus be written as $v_c(t) = A\cos\left(2\pi f_c t + \dfrac{2\pi a t}{4T}\right)$

This may be expanded to $v_c(t) = A\left[\cos 2\pi f_c t \cos\left(\dfrac{2\pi a t}{4T}\right) - \sin 2\pi f_c t \sin\left(\dfrac{2\pi a t}{4T}\right)\right]$

When $a = 1$, $\sin\left(\frac{2\pi a t}{4T}\right) = \sin\left(\frac{\pi t}{2T}\right)$ and $\cos\left(\frac{2\pi a t}{4T}\right) = \cos\left(\frac{\pi t}{2T}\right)$

When $a = -1$, $\sin\left(\frac{2\pi a t}{4T}\right) = -\sin\left(\frac{\pi t}{2T}\right)$ and $\cos\left(\frac{2\pi a t}{4T}\right) = \cos\left(\frac{\pi t}{2T}\right)$

Hence $v_c(t) = i_n \cos\left(\frac{\pi t}{2T}\right)\cos 2\pi f_c t + q_n \sin\left(\frac{\pi t}{4T}\right)\sin 2\pi f_c t$

When the input data = 0, i_n = -1 and q_n = -1, when the input data = 1, i_n = 1 and q_n = 1

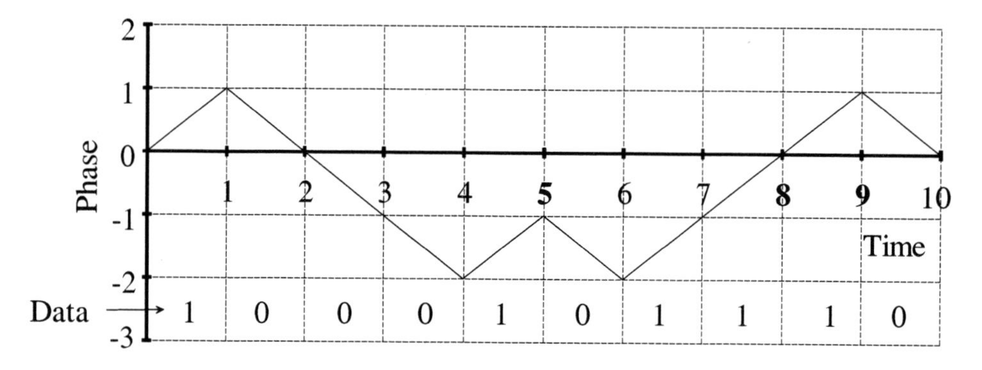

Figure 1.33 Phase shift of a MSK carrier as a function of modulating waveform

If the input is differentially encoded, MSK may be regarded as a form of offset QPSK in which the *I* and *Q* signals are sinusoidally shaped before modulation of the carrier. When considered from this perspective MSK is thus a form of continuous phase modulation.

The expression for the modulated carrier may be written as

$$v_c(t) = A\cos[2\pi f_c t + \phi(t)], \text{ where } \phi(t) = a\int_0^t \Delta f_c dt \qquad (1.11)$$

The phase of the MSK carrier is thus a series of ramps, of duration *T*, as shown in Figure 1.33 which demonstrates that, unlike the phase modulation equivalents described earlier, there is no phase transition at the edge of the binary interval. This means that the bandwidth of MSK is relatively narrow. There is however a change in the slope of the phase which contributes to the bandwidth of MSK and this can be avoided by shaping the data waveform to remove the rapid amplitude transitions between binary 1 and binary 0. This further reduces the bandwidth of the transmitted signal. In GSM the baseband waveform is shaped by a filter with Gaussian impulse response as shown in Figure 1.14. This produces Gaussian minimum shift keying or GMSK. In the case of GSM the Gaussian filter was selected so that the product of filter bandwidth and modulating bit period = 0.3 (BT = 0.3) thus the filter cut off frequency is B = 81.3 kHz (0.3 × 270.833 kHz).

MSK is effectively phase modulation by the integral of the data signal and the need for coherent detection is avoided if differential encoding is employed. GMSK is actually produced from the differentially encoded input with the phase angle $\phi(t)$ being derived according to the relationship of equation 1.12

$$\phi(t) = \phi_0 + \sum_i k_i \Phi(t - iT) \tag{1.12}$$

The value of k_i is related to the input data $B(i)$ so that $k_i = 1$, if $B(i) = B(i-1)$ and $k_i = -1$ if $B(i) \neq B(i-1)$. $\Phi(t)$ is the convolution of a ramp of width 1 bit period (T) and amplitude $\pi/2$, with a Gaussian function $G(x)$, i.e.

$$\Phi(xT) = \frac{\pi}{2}\left[G(x+0.5) - G(x-0.5)\right] \tag{1.13}$$

where $G(x) = x\int_{-\infty}^{x} \frac{1}{\sqrt{2\pi}\sigma} e^{-\frac{t^2}{2\sigma^2}} dt + \frac{\sigma}{\sqrt{2\pi}} e^{-\frac{t^2}{2\sigma^2}}$

In GSM the values of the parameters are $T = 48/13$ μs and $\sigma = \frac{\sqrt{\ln(2)}}{2\pi(0.3)} = 0.441684$

The function $\Phi(t)$ is shown in Figure 1.34 and smooths the phase ramp of the MSK carrier.

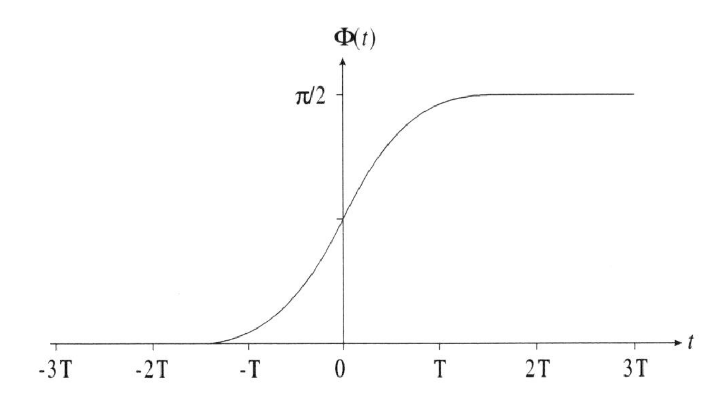

Figure 1.34 Phase characteristic of a GMSK carrier

It should be noted that if the phase ramp was a step function this would be equivalent to OQPSK, which would consequently have a wider spectrum. Thus MSK has a narrower spectrum than OQPSK and the pulse shaping, which further reduces the spectrum, has a minimal effect on the envelope of the modulated carrier. Consequently GMSK is effectively a constant envelope modulation scheme and does not require linear amplifiers. This contrasts with π/4–DQPSK, which does require linear amplifiers.

The differential nature of the modulation is evident when two specific examples of input data are considered.

When $B(i)$ is constant (all equal to 0 or all equal to 1) then all values of k_i = 1 and the phase of the GMSK waveform varies linearly with time,

$$\phi(t) = \phi_0 + \sum_i \Phi(t - iT) = \phi_0 + \frac{\pi t}{2T} \tag{1.14}$$

This produces a sine wave of frequency $f_1 = f_c + 1/4T$ (i.e. $f_c + \Delta f_c$)

When $B(i)$ is an alternating sequence of 0 and 1 then all values of $k_i = -1$ and the phase of the GMSK waveform is

$$\phi(t) = \phi_0 - \sum_i \Phi(t - iT) = \phi_0 - \frac{\pi t}{2T}$$

This produces a sine wave of frequency $f_2 = f_c - 1/4T$ (i.e. $f_c - \Delta f_c$)

This clearly demonstrates the frequency modulation nature of GMSK and also indicates that a pure sine wave may be generated when all data bits = 0. The transmitted frequency for a particular data sequence is essentially derived by exclusive ORing successive digits in the input data stream, the resulting frequency produced in the case of GSM is shown in Table 1.3. This principle is used in GSM to synchronise the oscillators in mobile terminals and is known as a frequency correction burst .

Table 1.3 Differential encoding used in GSM

B_i	B_{i-1}	$B_i \oplus B_{i-1}$	k_i	f (kHz)
0	0	0	+1	$f_c + 67.7$
0	1	1	–1	$f_c - 67.7$
1	0	1	–1	$f_c - 67.7$
1	1	0	+1	$f_c + 67.7$

1.7.12 Bandwidth of GMSK

Figure 1.32 reveals that the bandwidth of a GSM carrier is *approximately* $6/4T$ which translates to 1.5 × *carrier bit rate* = 406.32 kHz. This is considerably larger than the GSM carrier separation of 200 kHz (see Chapter 2) and results in a non negligible spectrum overlap in adjacent frequency bands. This source of interference is minimised by careful frequency planning, which avoids the use of adjacent frequencies in the same geographical area. The modulation spectrum of GMSK, as used in GSM is shown in Figure 1.35.

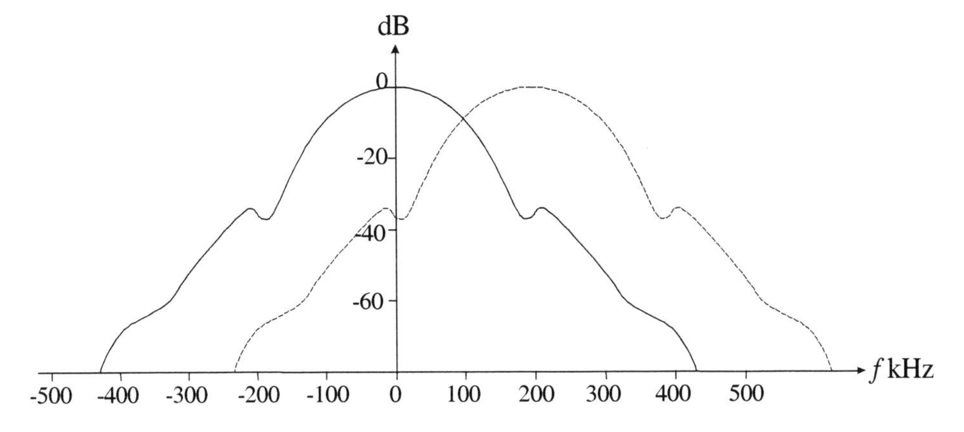

Figure 1.35 Spectrum of the GSM modulated carrier (GMSK)

1.7.13 *m*-ary Modulation

The QPSK signal was characterised by the fact that the coefficients *i* and *q* in equation 1.9 are chosen to give the carrier a constant amplitude. It is also possible to have other values and the resulting signal becomes quadrature amplitude modulation. The I/Q diagram for QAM with 16-level pulses is shown in Figure 1.36. Each individual level is represented by a unique combination of *i* and *q* in equation 1.10. The major difference between QAM and QPSK is that the former does not have a constant envelope and thus linear amplifiers are essential.

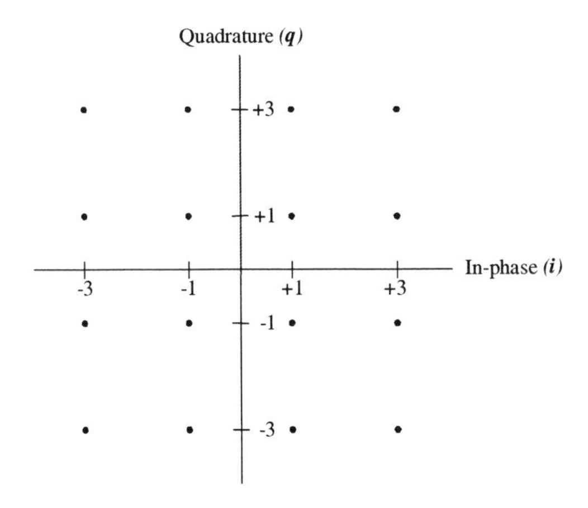

Figure 1.36 I/Q diagram for 16-QAM

1.8 PULSE SHAPING IN CARRIER MODULATED SYSTEMS

The basic principles of pulse shaping have already been covered in Section 1.6 which illustrated the compromise necessary between signal bandwidth and inter-symbol interference. In a digital radio system the function of the receiver is to minimise the probability of error in the received digits, or symbols. Error probability is a function of signal to interference ratio and this is minimised when the signal to interference ratio is maximised. This is a well known-problem in telecommunications and leads to the specification of a filter which maximises signal to interference ratio at a precise instant of time, this filter is known as the matched filter. The derivation of the matched filter for any received wave shape is given in [1].

Essentially the matched filter has an impulse response which is the time reverse of the transmitted wave shape. Hence, if the transmitted wave shape is known the frequency response of the matched filter may be derived by taking the Fourier transform of the impulse response. The situation appropriate to a radio transmission system is shown in Figure 1.37 and it will be noted that this specifies a transmission filter $G_t(f)$ and a receiver filter $G_r(f)$. In this particular case $G_r(f)$ would be the filter designed to maximise the received signal to interference ratio.

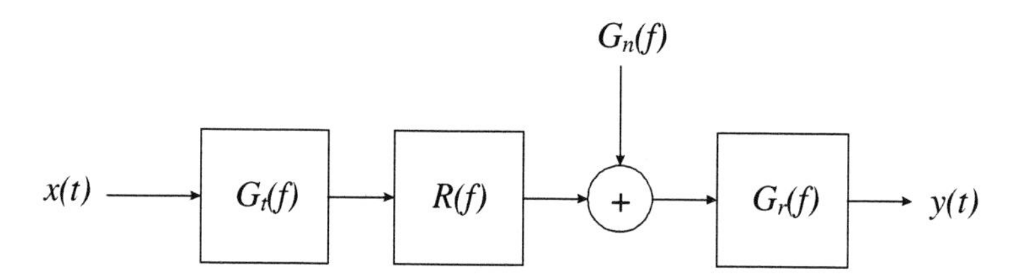

Figure 1.37 Transmitter and receiver filters in digital radio systems

If the radio channel is assumed to have a frequency response $R(f) = 1$ over the bandwidth of the transmitted signal then the frequency response between transmitter and receiver is $G(f) = G_t(f).\ G_r(f)$. If it is further assumed that the interference $G_n(f)$ has a uniform power spectral density over the bandwidth of the transmitted signal, then it can be shown that the signal to interference ratio at the receiver is maximised when $G_r(f) = G_t(f)$. Alternatively $G_r(f)$ is the matched filter for the waveform produced by $G_t(f)$.

In the TETRA system $G(f)$ is the raised cosine response given by equation 1.6. Thus both the TETRA transmitter and receiver are provided with filters with a frequency response $\sqrt{G(f)}$, which is termed a square root raised cosine response.

1.9 BANDWIDTH EFFICIENCY

The bandwidth efficiency of a digital modulation scheme is defined as

$$\eta = \text{bits/s/unit bandwidth} \tag{1.15}$$

This definition requires some care in interpretation as it is clear that the bandwidth of a digital waveform is essentially infinite. When pulse shaping is applied this will affect the spectrum of the modulated waveform. Under such circumstances the transmission bandwidth is defined as the bandwidth which contains a fixed percentage of the power spectrum (e.g. 95%). This definition of bandwidth is considerably tighter than the simple assumption of data signal bandwidth = $1/T$ Hz, where T = symbol duration. Based on the simple definition of it is possible to compare various modulation schemes in an *approximate* way.

The bandwidth required to transmit one symbol by BPSK is approximately $2/T$ Hz. Assuming 1 bit is transmitted by each symbol then number of bits/s = $1/T$.

Hence $\eta = \frac{1}{T} \times \frac{T}{2} = 0.5$ bits/s/Hz

For MSK the bandwidth required to transmit one symbol is approximately $3/2T$ Hz.

Hence $\eta = \frac{1}{T} \times \frac{2T}{3} = 0.66$ bits/s/Hz

For QPSK the bandwidth required to transmit one symbol is approximately $2/T$ Hz, however 2 bits are transmitted by each symbol.

Hence $\eta = \frac{2}{T} \times \frac{T}{2} = 1.0$ bit/s/Hz

Similar results may be defined for other modulation schemes, but these results will be significantly affected by pulse shaping as illustrated in Chapter 3.

1.10 THE RADIO ENVIRONMENT

The radio channel in a cellular system has a major influence on the overall system design. In order to indicate how this relationship comes about it is necessary to consider some basic principles of radiation from an antenna. If it is assumed that an antenna, in free space, radiates energy equally in all directions (isotropic antenna) it is possible to calculate the power density at a distance r from the antenna.

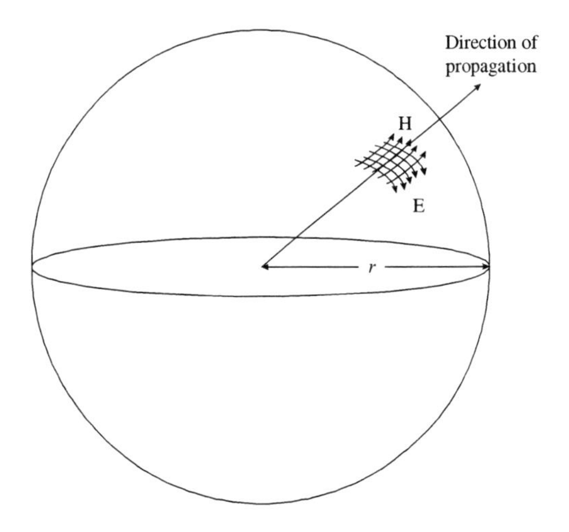

Figure 1.38 Isotropic antenna at the centre of a sphere

The power at any distance r from the antenna is the power passing through the surface of a sphere of radius r, as shown in Figure 1.38. The surface area of the sphere is $4\pi r^2$ and, if the antenna radiates a total power P_t, the power received per unit area is

$$P_a = \frac{P_t}{4\pi r^2} \text{ watts/m}^2 \quad (1.16)$$

At sufficiently large value of r the wave becomes a plane wave. The power received by an antenna placed in this field is

$$P_r = P_a . A_e \quad (1.17)$$

A_e is known as the "effective aperture" of the antenna and is the equivalent power absorbing area of the antenna. The effective aperture of an isotropic antenna when used as a receiver can be shown to be $A_e = \lambda^2/4\pi$, where λ is the wavelength of the received frequency.

(It may be shown that the gain of a half wave dipole, relative to an isotropic radiator is 120/73 = 1.64, thus the effective aperture of a half wave dipole is $A_e = 30\lambda^2/73\pi$. The gain of a unipole antenna, relative to an isotropic radiator is 3, thus the effective aperture of a unipole antenna is $A_e = 3\lambda^2/4\pi$.)

The power received by an isotropic antenna is given by $P_r = P_a \times \frac{\lambda^2}{4\pi}$, but $P_a = \frac{P_t}{4\pi r^2}$

$$P_r = \frac{P_t \lambda^2}{(4\pi r)^2} \quad (1.18)$$

The isotropic antenna has unity gain in both the transmit and receive modes. A non isotropic transmit antenna will have a gain of G_t and the product $P_t G_t$ is known as the effective radiated power (ERP). In mobile radio ERP is used as the standard method of quoting transmitted power. In effect, if the ERP is quoted as 100W (50 dBm) and the antenna gain is 10 dB, the actual transmitted power would be 10W (40 dBm). A non isotropic receive antenna will have a gain of G_r and, in such cases, the received power would be given by

$$P_r = \frac{G_t G_r P_t \lambda^2}{(4\pi r)^2} \quad (1.19)$$

This expression indicates that the attenuation is proportional to $(distance)^2$ and this is the free space propagation equation. In the case of mobile radio it is necessary to consider the height of both transmit and receive antennas above the earth's surface and the effect of the earth as a reflector.

1.11 PLANE EARTH PROPAGATION MODEL

Mobile radio systems are categorised by the fact that the height of the antennas at both transmitter and receiver are usually low compared to the distance of separation. If the height of the base station antenna is h_b and the height of the mobile antenna is h_m the system may be represented as shown in Figure 1.39, where the separation between transmitter and receiver is d. It is assumed that d is small enough to neglect the curvature of the earth. Figure 1.39 shows that there will be both a direct and ground reflected wave.

The direct path length is d_d and the reflected path length is d_r. It may be seen from the geometry of the system that $d_d = \sqrt{d^2 + (h_b - h_m)^2}$.

Using the binomial expansion and noting that $d >> h_b$ or h_m the length of the direct path approximates to $d_d \cong d.\left\{1 + 0.5\left(\frac{h_b - h_m}{d}\right)^2\right\}$

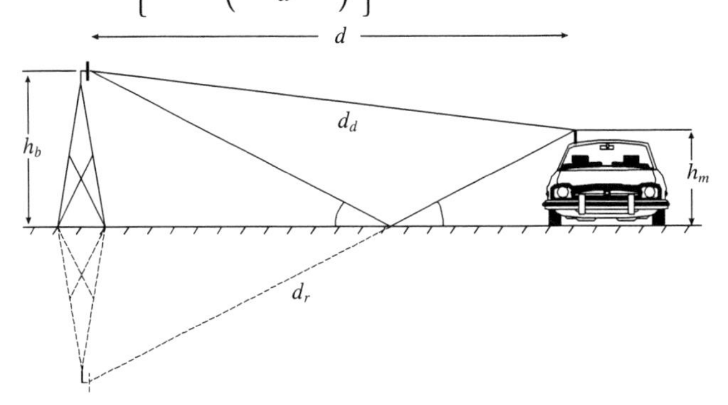

Figure 1.39 Plane earth propagation model

similarly the length of the reflected path is

$$d_r \cong d.\left\{1 + 0.5\left(\frac{h_b + h_m}{d}\right)^2\right\}$$

The path difference is thus $\Delta d = d_r - d_d$, which equates to

$$\Delta d = \frac{2h_b h_m}{d} \tag{1.20}$$

The corresponding phase difference between direct and reflected path is

$$\Delta\phi = \frac{2\pi}{\lambda} \times \frac{2h_b h_m}{d} = \frac{4\pi h_b h_m}{\lambda d} \tag{1.21}$$

The total received power is thus

$$P_r = P_t \cdot \left(\frac{\lambda}{4\pi d}\right)^2 \times \left|1 + \rho e^{j\Delta\phi}\right|^2 \tag{1.22}$$

ρ is the reflection coefficient and, for low angles of incidence, the earth approximates to an ideal reflector with $\rho = -1$.

i.e. $P_r = P_t \cdot \left(\frac{\lambda}{4\pi d}\right)^2 \times \left|1 - e^{j\Delta\phi}\right|^2$

but $1 - e^{j\Delta\phi} = 1 - \cos\Delta\phi - j\sin\Delta\phi$

hence $\left|1 - e^{j\Delta\phi}\right|^2 = (1 - \cos\Delta\phi)^2 + \sin^2\Delta\phi = 2(1 - \cos\Delta\phi) = 4\sin^2(\Delta\phi/2)$, thus

$$P_r = P_t . 4\left(\frac{\lambda}{4\pi d}\right)^2 . \sin^2\left(\frac{2\pi h_b h_m}{\lambda d}\right) \tag{1.23}$$

It should be noted that the received signal power has alternate maxima and minima when the mobile is close to the base station. The last local minimum occurs when $\left(\frac{2\pi h_b h_m}{\lambda d}\right) = \frac{\pi}{2}$.

Assuming that $d >> h_b$ and h_m then $\left(\frac{2\pi h_b h_m}{\lambda d}\right) << 1$.

Noting that $\sin x = x$ for small x, equation 1.23 reduces to

$$P_r = P_t \cdot \left(\frac{h_b h_m}{d^2}\right)^2 \tag{1.24}$$

This is a 4th power law and is known as the plane earth propagation equation. This equation is used as the basis for calculating received signal strength in mobile communication systems. The propagation loss is given by

$$L = 40 \log_{10} d - 20 \log_{10} h_b - 20 \log_{10} h_m \text{ dB} \tag{1.25}$$

This means that the loss increases by 40 dB each time the distance is multiplied by 10 (one decade). It should be noted that this equation is not dependent on the carrier frequency f_c, which is a consequence of assuming that h_b and h_m are much smaller than d and that the earth is flat and perfectly reflecting. If the surface is undulating a correction factor, which is frequency dependent, must be included. Typical values for a mobile communications system are $h_b = 30$ m, $h_m = 3$m. This gives the path loss as $L = 40 \log_{10} d - 39.1$ dB or 40 dB/decade.

It should be noted that although the 4th power law is commonly used in mobile radio calculations significant variations may be observed in practice.

Equation 1.25 gives the *mean loss* as a function of distance. However it is important to realise that the radio channel is subject to *fading*. Fading occurs as a result of multi-path propagation the principle of which is illustrated in Figure 1.40 in which it is assumed that there is no direct path between base station and mobile antennas. This figure shows two separate propagation paths. In practice there will be a large number of such paths and the comments relating to two paths in the following paragraphs should be extended to cover the multiple path situation.

If the phase difference between diffracted and reflected waves in this figure is a whole number of wavelengths the two waveforms will re-reinforce and the amplitude at the receiving antenna will (approximately) double. As the mobile moves the phase difference between the two paths changes.

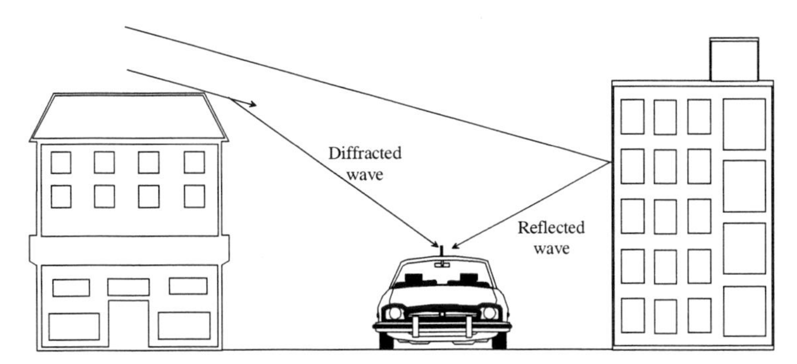

Figure 1.40 Multipath propagation

If the phase difference becomes an odd number of half wavelengths the two waves cancel producing a null. Thus as the mobile moves there are substantial amplitude fluctuations in the received signal known as *fast fading*. (There will also be a Doppler shift associated with the movement.) A typical variation of signal strength with distance is shown in Figure 1.41.

Fast fading (due to local multipath) is also accompanied by a slower variation in mean signal strength know as *slow fading* or *shadow fading*. Fast fading is observed over distances of about half a wavelength and can produce signal strength variations in excess of 30 dB. Slow fading is produced by movement over much longer distances, sufficient to produce gross variations in the overall path between base station and mobile. It should be noted that at the frequencies used in cellular radio a mobile moving at 50 km/h will experience several fast fades/second which will clearly effect the system performance. It should also be noted that fading is a spatially varying phenomenon which becomes a time varying phenomenon only when the mobile moves.

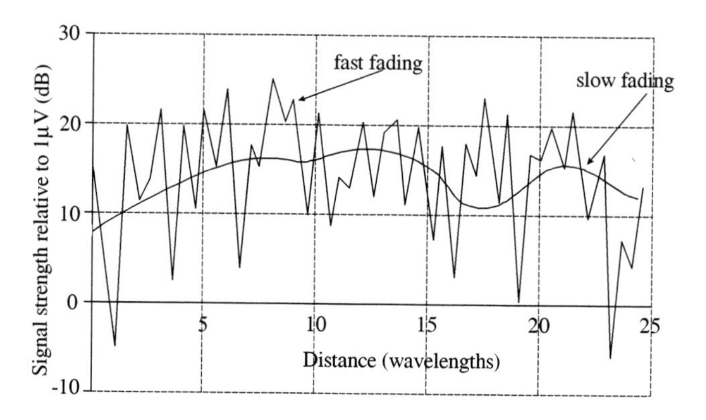

Figure 1.41 Fading due to multipath propagation

It is clear that an exact representation of fading characteristics is not possible because of the effectively infinite number of situations which would have to be considered. Reliance therefore has to be placed on statistical methods which produce general guidelines for system design.

1.12 FAST FADING

When a mobile unit is stationary the received signal strength will be formed by the vector sum of the scattered components of transmitted signal which reach the antenna by different paths, each having a different amplitude and phase. This will result in a received signal of constant amplitude. When the mobile moves the paths of the scattered signals change which is equivalent to introducing random amplitude and phase variations on the constituent signals.

If the kth scattered beam is given by $a_k e^{j(2\pi ft+\theta_j)}$ the received signal at the mobile antenna will be

$$s_k(t) = \sum_k re^{j(2\pi ft+\theta_k)} = re^{j(2\pi ft+\theta)}, \text{ where } re^{j\theta} = \sum_k a_k e^{j\theta_k}$$

The resultant complex phasor may be written as $re^{j\theta} = x + jy$

$$\text{where } x = \sum_k a_k \cos\theta_k \text{ and } y = \sum_k a_k \sin\theta_k$$

If it is assumed that the individual values of θ_k are independent and uniformly distributed the central limit theorem may be applied to show that x and y are Gaussian random variables with zero mean. The received signal may thus be regarded as the sum of two amplitude modulated carriers in phase quadrature with randomly varying envelopes. This leads to the conclusion that the *envelope* of the resultant received carrier has an amplitude which has a Rayleigh distribution given by

$$p_r(r) = \frac{re^{-r^2/2\sigma^2}}{\sigma^2} \tag{1.26}$$

In this expression σ^2 is the variance of the Gaussian variables x and y and is the mean square value of the carrier envelope. The value r is the instantaneous amplitude of the envelope. The distribution function is shown in Figure 1.42. It should be noted that the probability density function has a peak value of $e^{-1/2}/\sigma$ at $r = \sigma$ and that the envelope can have only positive values. Hence σ is the most probable value (mode), the mean (expected) value is $E(r) = \sqrt{\frac{\pi}{2}}\sigma$, and the mean square value is $E(r^2) = 2\sigma^2$.

Further information, relevant to the mobile radio environment, may be derived from the Rayleigh distribution function. In particular, it is possible to calculate the probability that the received signal strength will fall below some pre-determined threshold level R. In Figure 1.42 this is the area under the Rayleigh distribution curve between $r = 0$ and $r = R$.

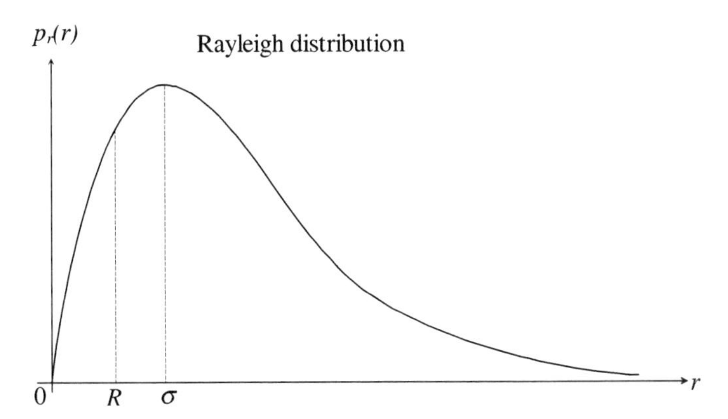

Figure 1.42 Resultant carrier envelope distribution function

The probability that the signal will fade below an arbitrary value R is given by

$$\int_0^R \frac{re^{-r^2/2\sigma^2}}{\sigma^2}dr = 1 - e^{-R^2/2\sigma^2} \tag{1.27}$$

When $R = \sigma$ the probability that the received signal will fade below the most probable value = 0.39. The median of the distribution is determined by noting that, on average, the received signal will be above (and below) this value for 50% of the time. Hence

$$\int_r^{\infty} p_r(r)dr = e^{-r^2/2\sigma^2} = 0.5$$

from which $r = 1.185\sigma$. The fast fading described in this section is often referred to as Rayleigh fading.

When the mobile is moving it is possible to determine the average number of times per second that the received signal envelope fades below a particular level, this is known as the *level crossing rate*. The level crossing rate is related to the velocity of the mobile v and the wavelength of the received carrier λ. For a vertical monopole antenna it can be shown that the level crossing rate is:

$$N(R) = \frac{\sqrt{2\pi}\, v\rho e^{-\rho^2}}{\lambda} \tag{1.28}$$

Where $\rho = \frac{R}{R_{rms}}$ and R is the specified level $(R_{rms} = \sqrt{2}\sigma)$.

The situation is shown in Figure 1.43. At a carrier frequency of 900 MHz and a mobile speed of 48 km/h the level crossing rate at ρ = -3dB is $N(R)$ = 39 per second. (In effect the number of fades per second is 39). A further parameter of importance is the average fade duration. The duration of a fade is the interval of time that the envelope remains below the level R and this is also shown in Figure 1.43.

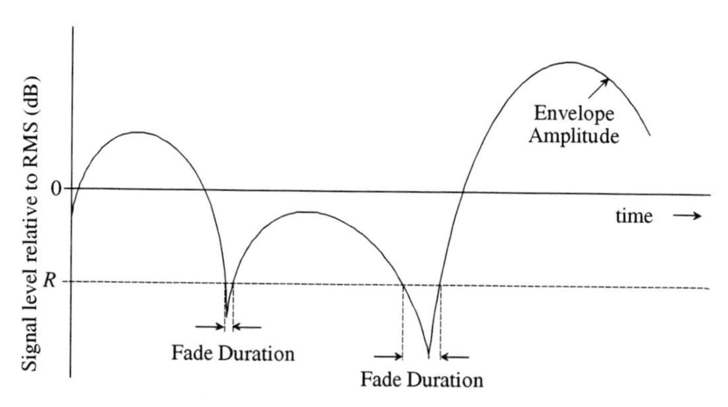

Figure 1.43 Fading experienced by a mobile

The average duration of fades below the level R is $\tau(R) = \frac{\text{prob}[r < R]}{N(R)}$

However, $\text{prob}[r < R] = 1 - e^{-R^2/2\sigma^2} = 1 - e^{-\rho^2}$, hence the average fade duration for a vertical monopole is

$$\tau(R) = \frac{\lambda.\{e^{\rho^2} - 1\}}{v\rho\sqrt{2\pi}} \tag{1.29}$$

Table 1.4 Delay spreads

Environment	Delay spread Δt_d (μs)
Rural area	< 0.2
Suburban area	0.5
Urban area	3.0

It is clear that fading is a frequency selective phenomenon. This effect is also apparent in the time domain and is measured in terms of delay spread. As the signal arriving at the antenna of a mobile (or base station) is the sum of a number of waves of different path lengths, the time of arrival of each of the waves is different. If an impulse is transmitted from the base station then by the time it is received at the mobile it will no longer be an impulse but rather a pulse of width given by the delay spread (dispersion) Δt_d. The delay spread is different for different environments and typical values are given in Table 1.4.

The delay spread is an important parameter for digital mobile systems as it limits the maximum data rate which can be sent. In general the time delay dispersion should be much less than the bit rate in a digital cellular system (without equalisation). Coherence bandwidth is an additional parameter closely related to delay spread. In a wideband signal two closely spaced frequency components will suffer similar multipath effects. However, as the frequency separation increases the differential phase shifts over the various paths become de-correlated and the spectral components in the received signal will not have the same relative amplitudes and phases as in the transmitted signal. This is essentially frequency selective fading and the bandwidth over which the spectral components are affected in a similar way (i.e. are correlated) is known as the coherence bandwidth. It is common practice to define the coherence bandwidth for a correlation coefficient of 0.5, in which case the approximate relationship between coherence bandwidth and delay spread is given by equation 1.30.

$$B_c = \frac{1}{2\pi\Delta t_d} \tag{1.30}$$

1.13 SLOW FADING

Figure 1.44 indicates that if the received signal is averaged over about 20 wavelengths the sector mean m_L is itself subject to variation. This is known as *slow fading* and is due, for example, to the variation in shadowing as the mobile moves between structures of differing heights. Slow fading, also known as *shadow fading*, may be viewed as the error which results when the simple distance based model is applied to field strength calculations. This model is still a useful indicator in predicting interference between cells.

As a mobile moves along a street, the variation in sector means m_L, is treated as a random variable. This produces a normal distribution when received signal strength measured in dB (commonly known as a log-normal distribution). This may be explained by the fact that slow fading is due to a number of multiplicative random processes acting on the signal in sequence. When expressed in dB these processes add and the distribution of the sum of random variables tends to be Gaussian.

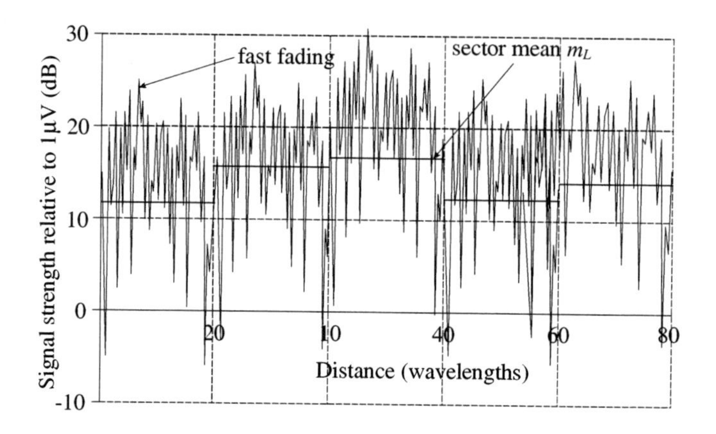

Figure 1.44 Slow fading (variation in sector mean)

The probability density function of m_L, measured in dB, is

$$p(m_{L(dB)}) = \frac{1}{\sqrt{2\pi}\sigma_m} e^{-\left\{\frac{(m_{L(dB)}-\mu_m)^2}{2\sigma_m^2}\right\}} \tag{1.31}$$

where μ_m is the mean value of $m_{L(dB)}$ and is determined by the path loss between the base station and mobile. The standard deviation of this distribution is σ_m and a typical value for macrocells is σ_m = 6 dB at a frequency of 900 MHz. It has been observed that σ_m is independent of the radio path length *d*. Shadow fading has a significant effect on the performance of mobile systems and as a mobile moves the value of will $m_{L(dB)}$ clearly vary. If the correlation between successive values of $m_{L(dB)}$ is measured it is found that this correlation decreases with the distance between the measurements. The decorrelation distance, which is an important parameter in the modelling of mobile communication systems, is the distance between measurements at which the correlation coefficient is 0.5, a typical value for urban macrocells being 30 m.

1.14 PATHLOSS MODELS

Equation 1.24 gives the underlying 4th power relationship between path loss and distance. In the cellular environment the base station antenna is typically above the surrounding buildings whilst the mobile is at street level, thus is unlikely to have a line of sight path (assumed in the 4th power case). Account must be taken of the ground and planar building elements such as walls, floors roofs etc. (trees, foliage etc. also have a significant effect). It is found that γ can be less than 2 when a mobile is close to a building and between 3.5 and 4 when further away [3]. For obstructed paths signal must travel down streets, around corners of high rise buildings and over roofs of low rise buildings. Fast fading is similar to that observed for open areas. However, shadow fading tends to be related to specific situations and the aggregating of measurements undertaken for open areas may not be applicable. Waves incident on the edge of buildings cause the edge to act as a secondary source for radiating cylindrical waves. This is known as diffraction and is the main mechanism for propagation down to street level when the mobile is close to a building. Diffracted power decreases with frequency.

Mathematical models may be derived and are dependent on the relative heights of buildings and base station antennas. Hata has derived parametric fits to large scale measurements reported by Okumura, for various situations. These models have been widely used for system design and are known to be accurate to within 1 dB for values of d between 1 and 20 km. For distances less than 1 km a significant difference in propagation loss occurs. The propagation models derived by Hata are as follows:

$L_{UA} = A + B \log_{10}(d)$ (urban area)
$L_{SA} = A + B \log_{10}(d) - C$ (suburban area)
$L_{OA} = A + B \log_{10}(d) - D$ (open area)

where

$A = 69.55 + 26.16 \log_{10}(f_c) - 13.82 \log_{10}(h_b) - (1.1 \log_{10}(f_c) - 0.7)h_m + (1.56 \log_{10}(f_c) - 0.8)$
$B = 44.9 - 6.55 \log_{10}(h_b)$
$C = 2 [\log_{10}(f_c/28)]^2 + 5.4$
$D = 4.78 (\log_{10}(f_c))^2 - 18.33 \log_{10}(f_c) + 40.94$

In each environment model, when f_c and h_m are fixed the path loss is a function only of the base station antenna height. For example:

When $h_b = 30$ m, $L = 35.22 \log_{10}(d)$ – constant term

i.e. path loss $L = 35.22$ dB/decade

It should be noted that when $h_b = 6$ m, path loss $L = 40$ dB/decade, which is the same as the 4th power law.

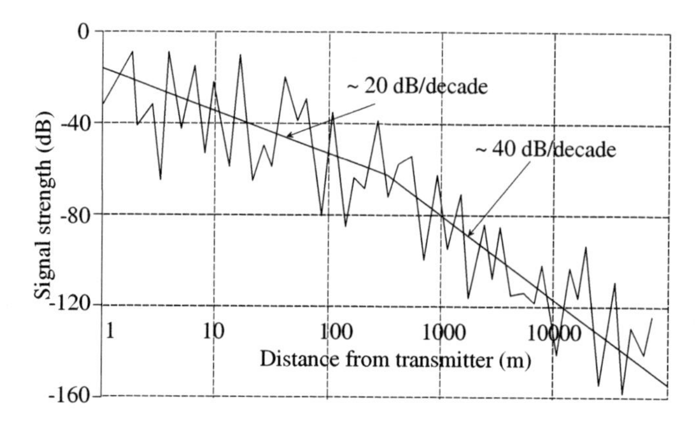

Figure 1.45 Two slope propagation model for microcells

Simple (distance)$^{\gamma}$ based models are valid for distances between 1 km and 35 km and additional correction factors are included to the Okumura-Hata model for large cities with high buildings.

These models must be modified for distances < 1 km, as the plane earth model requires that $\left(\frac{2\pi h_b h_m}{\lambda d}\right) \ll 1$. This condition is not met in microcells when $d < 1$ km (when h_b = 30 m and h_m = 3 m). In such circumstances, then $\gamma < 4$. A two slope attenuation model with a change in attenuation from 20 dB/decade to 35 dB/decade at a distance of about 0.5 km, is more appropriate. This is illustrated in Figure 1.45.

1.15 LINK BUDGETS

This topic is concerned with the allowances which must be made to ensure that a mobile will receive a power above a specified minimum for a high percentage (e.g. 98%) of the time and introduces the concept of slow fading margin and fast fading margin. The concept is illustrated in Figure 1.45 which illustrates that an "effective attenuation", known as the shadow fade margin, must be added to the mean pathloss to accommodate shadow fading and a further effective attenuation, known as the fast fade margin, must be added to accommodate fast fading. The shadow fade margin is derived from the shadow fading (Log Normal) distribution and the fast fade margin is derived from the shadow fading (Rayleigh). The effective loss (or link budget) is then

Loss (dB) = Mean pathloss (dB) + Shadow fade margin (dB) + Fast fade margin (dB)

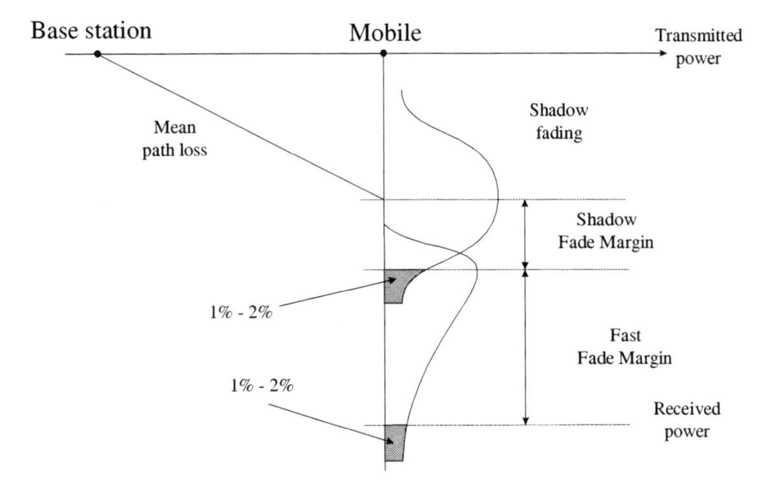

Figure 1.46 Fast and slow fading margins

1.16 FADING IN DIGITAL MOBILE COMMUNICATIONS

Fast fading is a frequency selective phenomenon described in terms of coherence bandwidth. The effect of frequency selective fading on the received signal spectrum is shown in Figure 1.47.

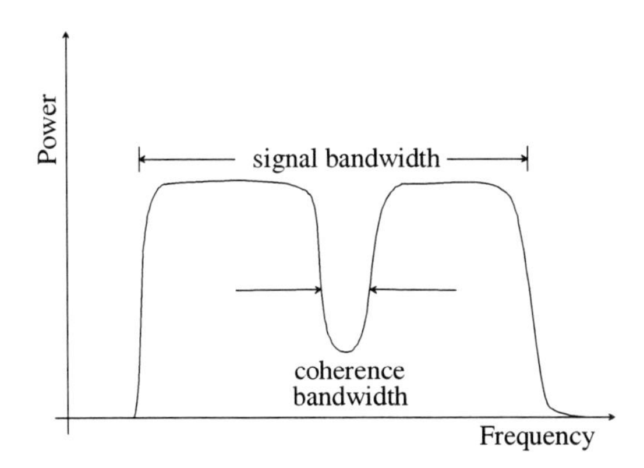

Figure 1.47 Effect of frequency selective fading on received signal spectrum

When the signal bandwidth is much wider than the coherence bandwidth frequency selective fading causes a notch to occur in the signal spectrum. This produces inter-symbol interference. As the multipath characteristics change (due to the movement of the mobile) the position and width of the notch will also change. If the frequency response of the multipath channel can be estimated it is possible to produce the inverse response and to equalise the channel frequency response. To make this possible a known sequence called the *training sequence* is transmitted in each burst. This training sequence is used to update an adaptive equaliser and hence reduce the inter-symbol interference. It should be noted that if the signal bandwidth is less than the coherence bandwidth a flat fade will result and a burst of errors will occur in the received signal. Digital systems employ a variety of error correction mechanisms to overcome this problem and these will be discussed in later chapters.

1.17 THE CELLULAR CONCEPT

The essential feature of all cellular networks is that the final link between the subscriber and fixed network is by radio. This raises a number of issues which must be addressed:

- the amount of spectrum available for mobile communications is strictly limited;
- the radio environment is subject to multipath propagation, fading and interference and is not therefore an ideal transmission medium;
- the subscriber is able to move and this movement must be accommodated by the communications system.

The basic elements of a cellular system designed for public use are shown in Figure 1.48.

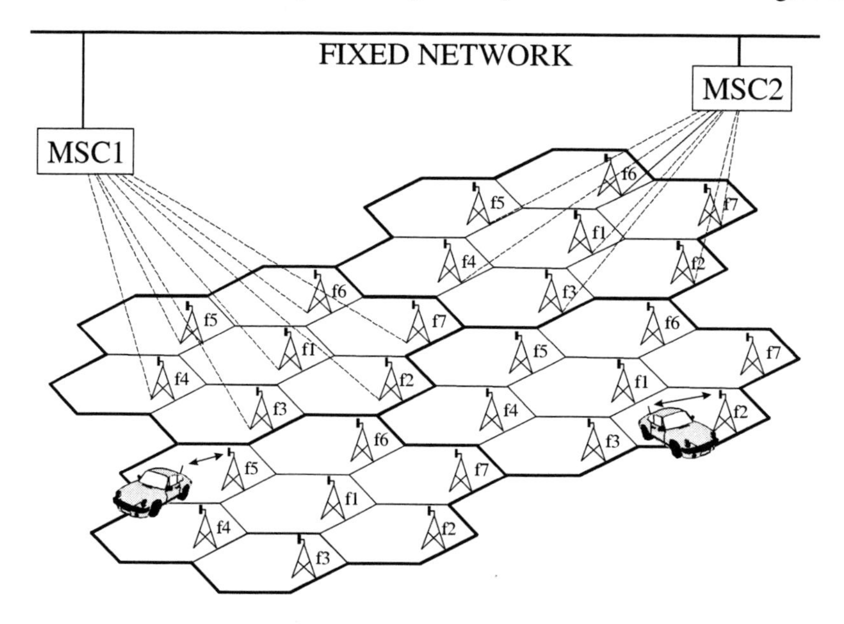

Figure 1.48 Basic elements of a cellular system

The mobile units may be in a vehicle or carried as a portable and are assigned a duplex channel and communicate with an assigned base station. The base stations communicate simultaneously with all mobiles within their area of coverage (or cell) and are connected to mobile switching centres (MSCs). A mobile switching centre controls a number of cells, arranges base stations and channels for the mobiles and handles connections with the fixed public switched telephone network (PSTN).

Figure 1.48 indicates that each base station (in a cluster) is allocated a different carrier frequency and each cell has a usable bandwidth associated with this carrier. Because only a finite part of the radio spectrum is allocated to cellular radio the number of carrier frequencies available is limited. This means that it is necessary to re-use the available frequencies many times in order to provide sufficient channels for the required demand. This introduces the concept of frequency re-use and with it the possibility of interference between cells using the same carrier frequencies.

Clearly with a fixed number of carrier frequencies available the capacity of the system can be increased only by re-using the carrier frequencies more often. This means making the cell sizes smaller. This has two basic consequences:

- It increases the likelihood of interference (known as co-channel interference) between cells using the same frequency.
- If a mobile is moving it will cross cell boundaries more frequently when the cells are small. Whenever a mobile crosses a cell boundary it must change from the carrier of the cell which it is leaving to the carrier of the cell which it is entering, and this process is known as handover.

Handover cannot be performed instantaneously and hence there will be a loss of communication whilst the handover is being processed (in certain circumstances, described in Chapter 2, seamless (soft) handover is possible. If the cell sizes become very small (microcells) handovers may occur at a very rapid rate.

It becomes clear that frequency planning is a major issue in the design of a cellular system which must achieve an acceptable compromise between the efficient utilisation of the available radio spectrum and the problems associated with frequency re-use.

1.18 TYPICAL CELL OPERATION

Each cell has allocated to it a number of channels which can be used for traffic or signalling. When a mobile is active it "registers" with an appropriate base station. The information regarding the validity of the mobile (for charging etc.) and its cellular location is stored in the responsible MSC. When a call is set up either from or to the mobile the control and signalling system assigns a channel (from those available to the base station with which the mobile is registered) and instructs the mobile to use the corresponding channel. This channel may be provided on a frequency division basis (typical of analogue systems), on a time division basis (typical of digital systems) or on a code division basis (also typical of digital systems).

A connection is thus established via the base station to the fixed network. The quality of the channel (i.e. radio link) will be monitored by the base station for the duration of the call and reported to the MSC. The MSC will make decisions concerning the quality and will instruct the mobile and base station accordingly. As the mobile moves around the signal to interference ratio on its allocated channel will vary. This is monitored and if it falls below some threshold, the mobile can be instructed to handover to the strongest base station. The handover algorithm is actually significantly more complicated than this simple treatment suggests. For example it must be able to cope with

- whether the current loss in channel quality is due to short term fading,
- whether a simple increase in power would be sufficient to restore the channel quality (this could however produce an unacceptable co-channel interference in other cells using the same frequency),
- whether the measurements from adjacent cells are valid (averaging is necessary to remove spurious fluctuations),
- whether the cell chosen for handover has spare channels available.

In digital systems both mobiles and base stations take measurements of signal levels and quality and report these to the fixed network for handover decisions (this is known as mobile assisted handover). When handover does occur, communication in the traffic channel is interrupted. This interruption is usually short and largely unnoticed during voice communications. However, it can present serious problems if the mobile is transmitting or receiving data, especially when handover is frequent. When the call is completed the mobile releases the traffic channel which can then be re-allocated to other users.

1.19 CAPACITY OF A CELLULAR SYSTEM

The capacity of a system may be described in terms of the number of available channels, or alternatively in terms of the number of subscribers that the system will support. Currently public mobile systems are dominated by voice traffic and the number of subscribers that a system will support depends on the fact that each call has a mean duration and that not all of the subscribers will be trying to make a call at the same time.

The system capacity depends on:

- the total number of radio channels;
- the size of each cell;
- the frequency re-use factor (or frequency re-use distance);

The total number of voice channels that can be made available to any system depends on the radio spectrum allocated and the bandwidth of each channel. Once this number is defined a frequency re-use pattern must be developed which will allow optimum use of the channels. This, in turn, is closely linked to cell size.

The minimum distance which allows the same frequencies to be re-used will depend on many factors, for example:

- the number of co-channel cells in the vicinity of the centre cell,
- the geography of the terrain,
- the antenna height,
- the transmitted power within each cell.

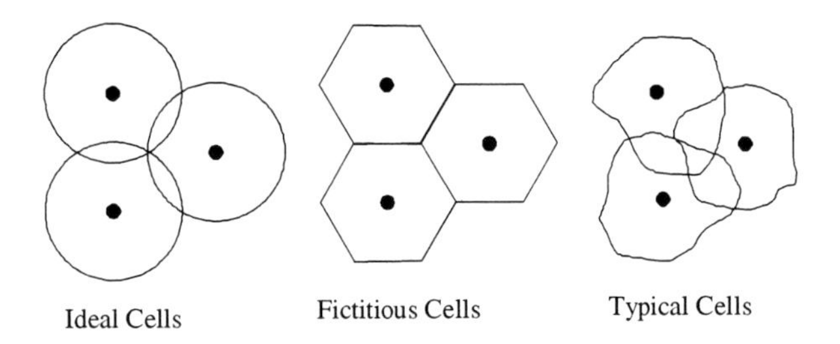

Figure 1.49 Schematic representation of cells

Assuming an omni-directional base station antenna it is appropriate to consider the cells as circles with the base station at the centre. (Such a model is actually only appropriate for a flat terrain with no obstacles.) This means that in order to provide complete coverage of an area the circular cells would overlap and this would make diagrams somewhat confusing. It is therefore common practice to represent the cells as non-overlapping hexagons which would fit into the corresponding circles. This fictitious model with the ideal and real models are shown in Figure 1.49.

1.20 FREQUENCY RE-USE DISTANCE

When calculating the frequency re-use distance this is based on the cluster size K. The cluster size is specified in terms of the offset of the centre of a cluster from the centre of the adjacent cluster. This is made clearer by reference to Figure 1.50. In this figure the cell cluster size is 7 and the centre cell is the cell marked 1. The next cell 1 is offset by i = 2 cell diameters to an intermediate cell and a further j = 1 cell diameter from that intermediate cell.

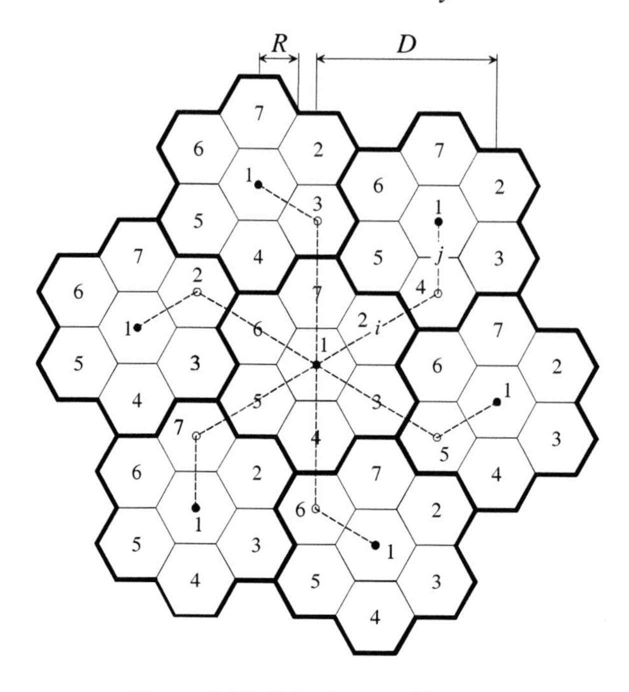

Figure 1.50 Cell cluster with $K = 7$

The cluster size is calculated from

$$K = i^2 + ij + j^2 \tag{1.32}$$

Common cluster sizes are 4 ($i = 2,\ j = 0$) and 7 ($i = 2,\ j = 1$) for city centres, and 12 ($i = 2,\ j = 2$) for rural areas. In the case of hexagonal cells the cluster size K is taken from a set of rhombic numbers $K \in \{1,3,4,7,9,12....\}$. A layout for cluster size 4 is given in Figure 1.51 and a layout for cluster size 12 is given in Figure 1.52

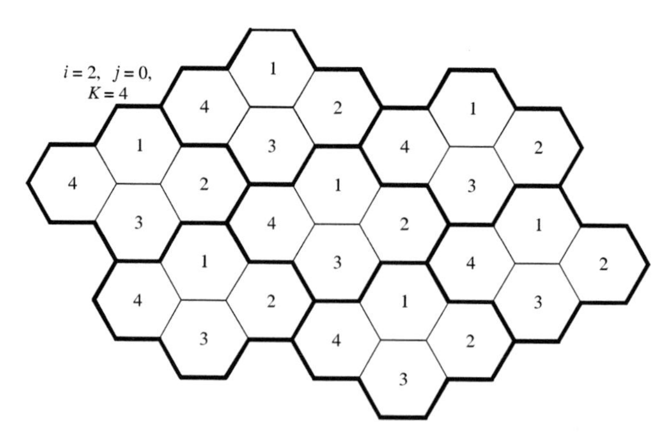

Figure 1.51 Cluster size $K = 4$

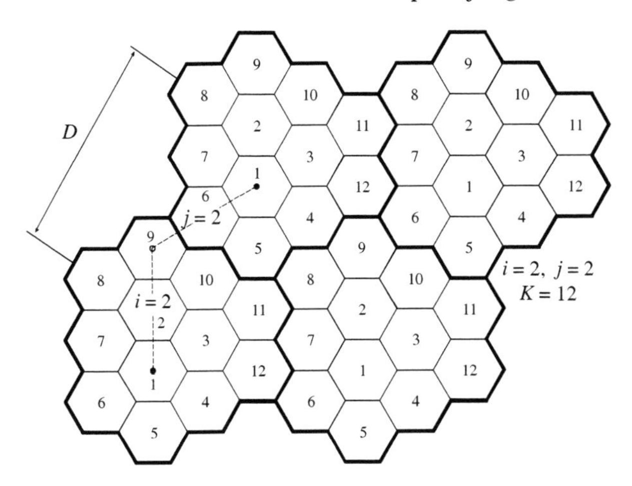

Figure 1.52 Cluster size $K = 12$

The frequency re-use distance D is shown in Figures 1.50 and 1.52. Assuming circular cells of radius R (based on the hexagon shape), the frequency re-use distance is related to cluster size and cell radius by

$$D = \sqrt{3K}.R \tag{1.33}$$

The corresponding re-use distances are given in Table 1.5. If all cells transmit the same power then as K increases the frequency re-use distance increases, thus increasing K reduces the probability of co-channel interference. However in order to maximise frequency re-use it is necessary to minimise the frequency re-use distance. Hence the design goal is to choose the smallest value of K which will meet the performance requirements in terms of capacity and interference.

Table 1.5 Re-use distances for cellular systems

K	D
4	3.46R
7	4.58R
12	6.00R

1.21 CELLULAR COVERAGE

Figure 1.53 shows two cells using the same set of frequencies at a re-use distance D. Assuming that the power by each base station is fixed the received power at a distance r from the base station is proportional to $r^{-\gamma}$. Using the plane earth propagation equation results in the substitution of $\gamma = 4$.

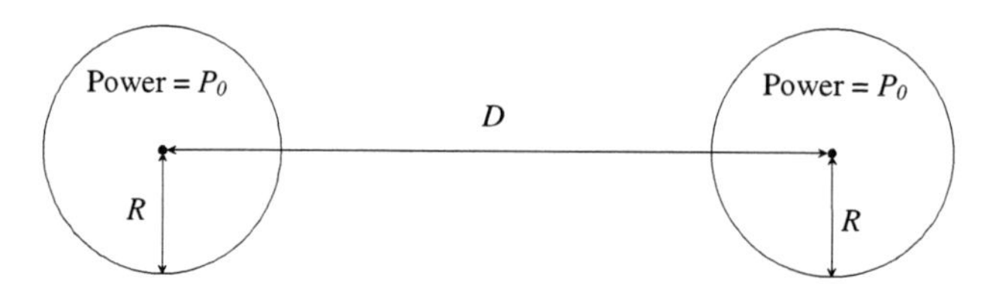

Figure 1.53 Signal to interference ratio

A mobile in one of the cells will receive a signal to interference ratio (SIR) which, on average, is a function of $q = D/R$. (On average the SIR at a mobile receiver will be the same as at the base station receiver.) It should be noted that the actual power level P_0 is assumed the same for all cells and q, *known as the co-channel interference reduction factor*, is independent of the value of P_0 . The signal to interference ratio within a cell depicted by Figure 1.53 is thus

$$\frac{S}{I} = \left(\frac{R}{D}\right)^{-\gamma} = \left(\frac{R}{D}\right)^{-4} \tag{1.34}$$

For a fully developed cellular system based on the hexagonal model there will be 6 interfering cells in the first tier of surrounding clusters as shown in Figure 1.54. If $\gamma = 4$ it can be assumed that the interference due to cells in the second tier can be ignored and the signal to interference ratio in one of the cells will thus be

$$\frac{S}{I} = \frac{S}{\sum_{k=1}^{6} I_k} \tag{1.35}$$

Assuming the local noise is much less than the interference level this can be written

$$\frac{S}{I} = \frac{R^{-\gamma}}{\sum_{k=1}^{6} D_k^{-\gamma}} \tag{1.36}$$

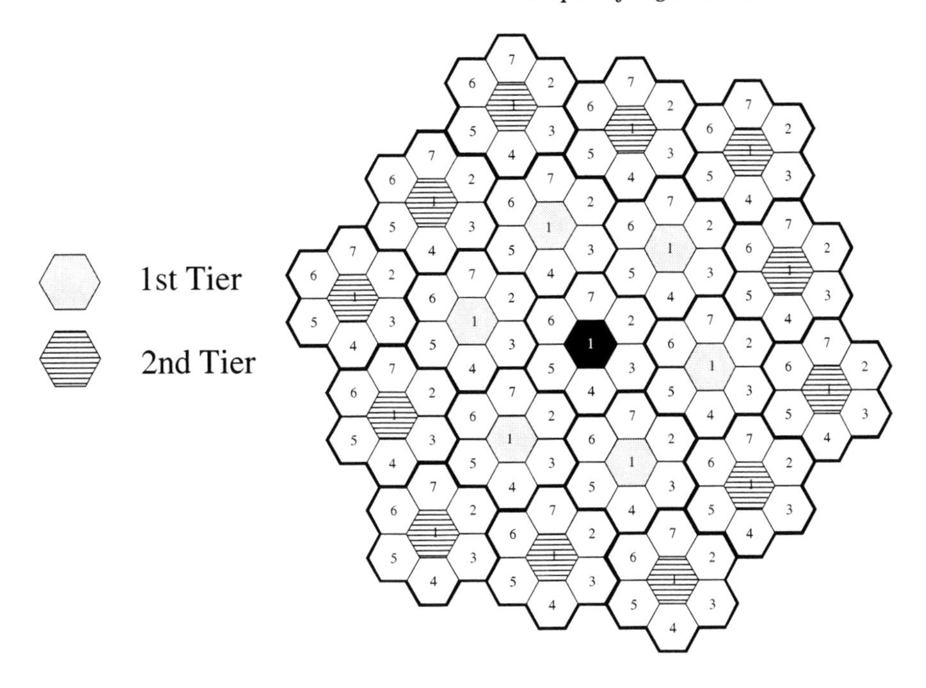

Figure 1.54 First and second tier interfering cells

Assuming that all values of D_k are equal (= D), this becomes

$$\frac{S}{I} = \frac{R^{-4}}{6D^{-4}} \tag{1.37}$$

The value of *S/I* necessary for acceptable performance depends on a number of factors relating to the modulation scheme etc. Typical values are shown in Table 1.6

Acceptable performance is clearly a subjective measure and is usually interpreted as the *S/I* at which 90% of users report that they have received an adequate service for 95% of the time.

Table 1.6 Minimum SIR values for mobile systems

System	*S/I*
TACS (analogue)	18 dB (63.1)
GSM (digital)	9 dB (7.9)
TETRA (digital)	19 dB (79.4)

Using the analogue SIR value from Table 1.6, as an example, equation 1.37 gives $q^4 = 6 \times 63.1$, hence $q = 4.41$.

For the hexagonal structure $q = \sqrt{3K}$, i.e. $K = 6.48$

Hence the cluster size required for this SIR is 7. It should be noted that this approximate analysis closely reflects the practical case, which is based on fixing a probability that a mobile will receive a SIR above a specified value.

Having established the cluster size it is then necessary to determine the cell radius. This is based on the number of available channels and the expected density of mobile subscribers (i.e. the average number of mobile subscribers/m^2). In order to derive this figure it is necessary to consider aspects of traffic theory developed for telephony. There are two major types of system used in telephony:

- blocked-calls-cleared system
- delay system.

Basically, in the blocked-calls-cleared system, when a subscriber requests a transmission channel (to set up a call) if all available channels are in use the call request is blocked. There is no mechanism for putting the call in a queue and the call is lost (or cleared). This is the situation encountered in public mobile communication systems. In the delay system a call request may be queued until a channel becomes free (there is usually a limit to the number of users which may be held in a queue). The blocked-calls-cleared system is described by the Erlang B formula which relates the traffic carrying capacity to the number of available channels for a specified blocking probability.

To illustrate this point it is assumed that the total number of channels available is 210. This means that, for a cluster size of 7, the number of channels per cell is 210/7 = 30. It is necessary to find the total offered traffic in the "busy hour". This is related to the average number of calls/hour and the mean duration of each call. To illustrate this relationship it is assumed that there are W subscribers per cell and that during the busy hour a fraction η_c of these subscribers make or receive a call of duration T minutes. Thus the total number of calls in the busy hour is $Q = \eta_c.W$.

The offered load is then $A = Q.T/60$ erlangs

To obtain the number of channels for this traffic it is necessary to attach a "blocking probability" for each call. A typical value for this is 2%. The relationship between offered traffic, blocking probability and number of channels is given by the Erlang B formula which is usually represented in tabular form and is given in Appendix 1. From the table it may be seen that 30 channels can support an offered traffic of 21.9 erlangs with a blocking probability of 2%. It is possible to relate this figure to the number of subscribers which the cell can support

i.e. $21.9 = Q.T/60 = (\eta_c.W.T)/60$

from which $W = 60 \times 21.9/(\eta_c.T)$

In this expression T is the mean call duration in minutes. The value of T for public systems is directly affected by the tariff levied by the cellular operator but is usually accepted to be 1.76 minutes. (The duration of calls in private mobile systems are of the order of tens of seconds.)

If it assumed that 60% of the total subscribers make a call during the busy hour then

$$W = 60 \times 21.9/(0.6 \times 1.76) = 1244.3$$

Hence with 30 channels available a cell could support 1244 subscribers. If the cell radius is R metres then the user density = $1244/\pi R^2$.

It is now possible to calculate the cell radius required for an average user density. For example, if it is assumed that the number of users/ km^2 is 1600, this represents a user density of $1.6 \times 10^{-3}/m^2$.

Thus $1244.3/\pi R^2 = 1.6 \times 10^{-3}$, which gives $R = 497.5$ m

i.e. the approximate cell diameter is 1 km.

Thus in dimensioning a cellular system it is necessary to determine the minimum signal to interference ratio which will produce an acceptable performance. This, in turn, determines the re-use distance from which the cluster size K is obtained. The objective is to use as small a value of K as possible, which maximises frequency re-use and also maximises the number of channels available per cell. Once K has been determined the number of channels per cell is calculated from the available bandwidth. If the total allocated bandwidth is B_{tot} and the channel bandwidth is B_{ch} the total number of channels available is $n_{tot} = \dfrac{B_{tot}}{B_{ch}}$ from which the number of channels/cell $n_{ch} = \dfrac{B_{tot}}{B_{ch}K}$ where K is the cluster size.

The value of n_{ch} determines the traffic load, in erlangs, which the cell can support. The cell radius is then based on the projected user density and the estimated mean call duration. There are clearly many approximations which are made in order to arrive at this result, the most important of which is that the mean SIR is determined by a function of $q = D/R$. In practice not all mobiles will experience a SIR equal to the mean value and in heavily loaded systems the actual SIR experienced by particular mobiles can be significantly less than the mean value. Under these circumstances it is necessary to adopt procedures which can increase SIR, one of which is to deploy base station antennas with directional characteristics, this is known as sectoring.

1.22 SECTORING

The approximate analysis presented in the previous section indicated that it is necessary to plan frequency re-use on a cluster size of 7, in order to achieve a mean SIR of 18 dB.

However, it is found that in areas of high traffic density this value can be inadequate. The worst case situation is illustrated in Figure 1.55 in which a mobile is on the boundary of its serving cell and interference is produced by all six interfering cells.

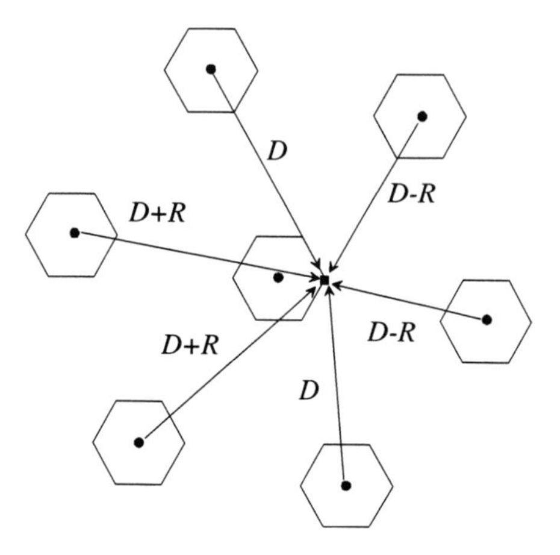

Figure 1.55 Worst case interference

In this case the distances from the mobile to interfering cells varies from *D–R* to *D+R*. The signal to interference ratio in this case is given by

$$\frac{S}{I}=\frac{R^{-4}}{2(D-R)^{-4}+2D^{-4}+2(D+R)^{-4}}=\frac{1}{2\{(q-1)^{-4}+q^{-4}+(q+1)^{-4}\}}$$

Substituting for q in this expression gives a SIR = 17 dB, which is less than the desired value. The situation can actually be worse than this, when the characteristics of the radio environment are taken into account, and a more conservative SIR estimate would be about 14 dB, which is 4 dB less than the acceptable value of 18dB used in the example.

Clearly, increasing the value of K to improve the SIR would reduce the efficiency of frequency re-use and this is not an attractive option. An alternative is to reduce the co-channel interference in a cell by using directional antennas at the base station and dividing the cell into a number of sectors. The 3 sector arrangement is shown in Figure 1.56. The original frequencies allocated to the cell are divided between the sectors. Reduction in interference is achieve by choosing a frequency re-use pattern such that the front lobe of any base station transmitter illuminates only the back lobe of its co-channel counterpart. What this means, in effect, is that the number of interfering base stations is reduced from 6 to 2 with a corresponding increase in SIR.

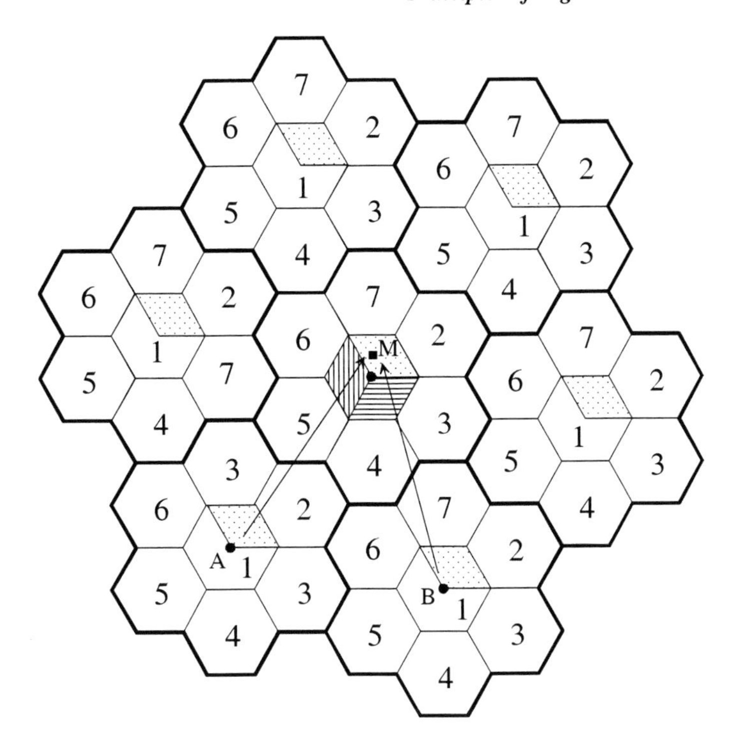

Figure 1.56 Cellular system with 120° sectored base station antennas

This effect is shown in Figure 1.56 in which it may be seen that only base stations **A** and **B** actually cause interference to mobile **M**. This produces a reduction in interference of approximately 5 dB over the omni-directional case. Hence by employing sectored antennas in areas of high traffic density it is possible to achieve the required SIR values. It should be noted that the sectored approach effectively increases the cluster size to 21 without any increase in the frequency re-use distance, however handovers may be required between sectors of the same cell.

1.23 STATISTICAL ISSUES

Equation 1.37 gives the mean value of SIR experienced when all six first tier interfering cells are contributing to the overall interference. In practice not all interfering cells will be active all the time and, in any case, mobiles are continually moving. The actual interference experience on a particular channel will therefore be subject to continual change and it is more appropriate to consider the interference experienced on a channel in terms of the probability that the SIR is less than some specified value. A typical characteristic is shown in Figure 1.57.

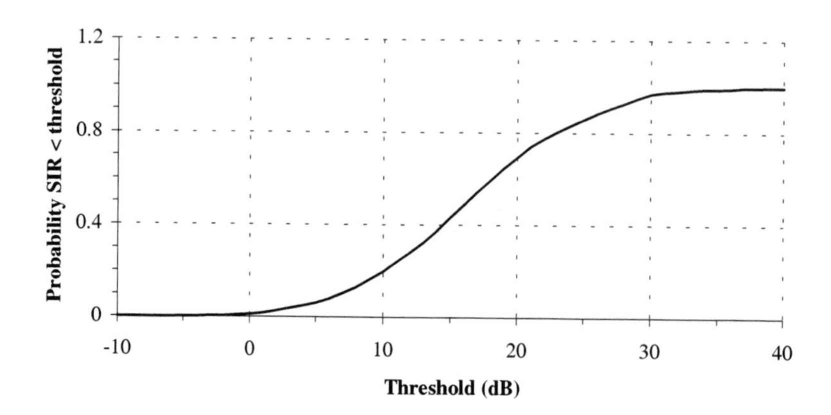

Figure 1.57 Cumulative distribution of SIR on a traffic channel

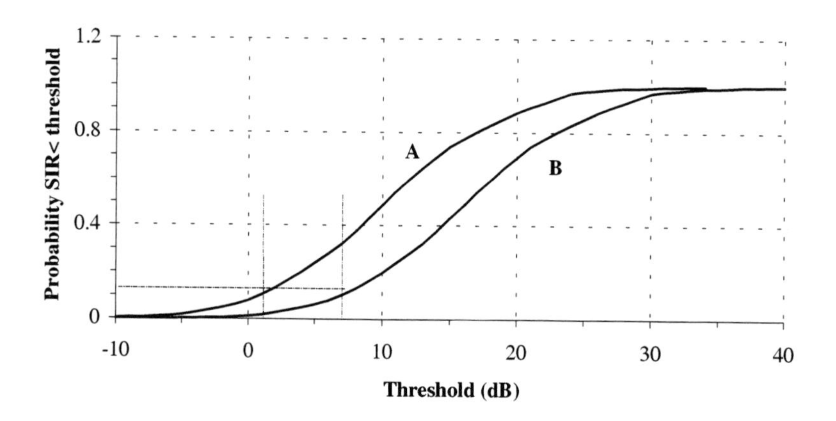

Figure 1.58 Effect of re-use distance on SIR distribution

From this figure it may be noted that the probability that the SIR is less than 10 dB is 0.2. An alternative interpretation of this graph is that on average 20% of mobiles will experience a SIR of less than 10 dB. In order to ensure an acceptable quality of service to users, it is necessary to place some limit on the minimum acceptable SIR. GSM, which is covered in Chapter 2, is designed to operate on a minimum SIR of about 9 dB hence a typical limit might be to specify that at least 90% of subscribers should have a SIR greater than 7 dB (alternatively 10% of users will have a SIR less than 7 dB).

The effect of changing re-use distance on the SIR is illustrated in Figure 1.58. In this figure curve A indicates that 10% of users have a SIR of less than 1 dB, which would clearly not meet the desired criterion for GSM. The SIR can be increased by increasing the re-use distance, which results in the shifting of the cumulative distribution function along the SIR axis. In this particular example it is necessary to increase the SIR by 6 dB. Thus to obtain a shift of 6 dB the required increase in re-use distance is given by

$40\log_{10}\left(\frac{d_B}{d_A}\right) = 6\text{dB}$. From which the increase in the re-use distance is $d_B/d_A = 1.41$ (i.e. a factor of 1.41).

1.24 CONCLUSIONS

This chapter has introduced the fundamentals of digital mobile radio and has indicated that the mobile radio channel presents a hostile, and far from ideal transmission medium for digital information, whether that be as a result of multipath propagation, or interference generated by other users in the system. The following chapters will indicate how mobile systems such as TETRA are designed to overcome the problems associated with the mobile environment and will emphasise the importance of adopting international standards in the design of such systems.

REFERENCES

[1] Dunlop, J. and Smith, D. G. 1998, *Telecommunications Engineering* (3rd Ed), Stanley Thornes, pp 62-63

[2] Dunlop, J. and Smith, D. G. 1998, *Telecommunications Engineering* (3rd Ed), Stanley Thornes, pp 156-158

[3] Bertoni, H. L. Honcharenko, W. Maciel, L. R. and Xia, H. H. 'UHF Propagation Prediction for Wireless Personal Communications', *Proceedings of the IEEE*, Vol 82, No 9, 1994, pp 1333-1359.

2

Public Digital Mobile Radio Systems and Environment

2.1 DIGITAL CELLULAR RADIO

The first public mobile systems (1st generation) were based on analogue techniques and employed frequency division multiplexing to divide the available spectrum between the active users. This chapter deals with digital mobile radio systems developed for public networks collectively known as 2nd generation systems. The overriding goal of public systems may be stated as the provision of a guaranteed quality of service to as large a user population as possible, using the minimum radio resource. Since the radio bandwidth allocated to public mobile communications will always be finite, large scale increases in capacity are possible only by utilising the allocated bandwidth more efficiently. The alternative is to make the cells smaller which means that problems with frequent handovers would then have to be addressed. It is stressed that there is no solution which is optimum in all respects. However, there are a number of significant advantages to be achieved with digital systems which has contributed to their adoption as second generation systems. The most significant advantage of digital systems is that digital signal processing techniques may can be used both to reduce transmission bandwidth requirements and to extract signals in poor SIR conditions. There are other advantages, for example digital transmissions may be encrypted to avoid unauthorised eavesdropping. This chapter considers 2nd generation public systems based on TDMA (GSM and DECT) and CDMA (cdmaOne^{TM}).

2.2 THE GLOBAL SYSTEM FOR MOBILE COMMUNICATIONS (GSM)

GSM was originally developed as a European Standard and was specified by a specially formed group of the "Conference Europene des Administrations des Postes et des Télécommunications" (CEPT) known as the Groupe Special Mobile (GSM). The standard which has been subsequently developed is known as "The Global System for Mobile Communications" or simply as GSM. A variation of this standard has been developed for low power terminals is known as DCS 1800 (PCS 1900 in the USA). GSM can handle both voice and data traffic, the voice waveform being digitally encoded before transmission. Individual users are given access to the radio channel for a limited period and transmit a burst of digital information. This chapter gives an outline of the GSM standard and begins with a description of the GSM network architecture.

2.3 THE GSM NETWORK ARCHITECTURE

The GSM network architecture is relatively complex and a simplified version of this is shown in Figure 2.1. It is composed of a number of functional entities which are interconnected by interfaces with detailed specifications. Each *location area* in the GSM system is served by a gateway mobile switching centre (GMSC) which acts as the interface with the public switched telephone network. In areas of high traffic additional mobile switching centres (MSC) may be provided. Each GMSC is provided with a home location register (HLR), which contains the details of all mobile subscribers belonging to the area related to the GMSC and a visitor location register (VLR) which contains details of all mobile subscribers currently present in the area related to the GMSC (or additional MSCs where these are provided).

The (G)MSC is connected to a number of base station controllers (BSC) by what is known as the *A* interface. Each BSC controls a number of base station transceivers (BST) which are connected to it by the *Abis* interface. The base station transceivers are the fixed network terminations for the radio link with the mobile terminals. The radio interface between the BSTs and mobile subscribers is known as the *Um* interface. When considering the detailed specification of GSM it is convenient to start with the radio (*Um*) interface.

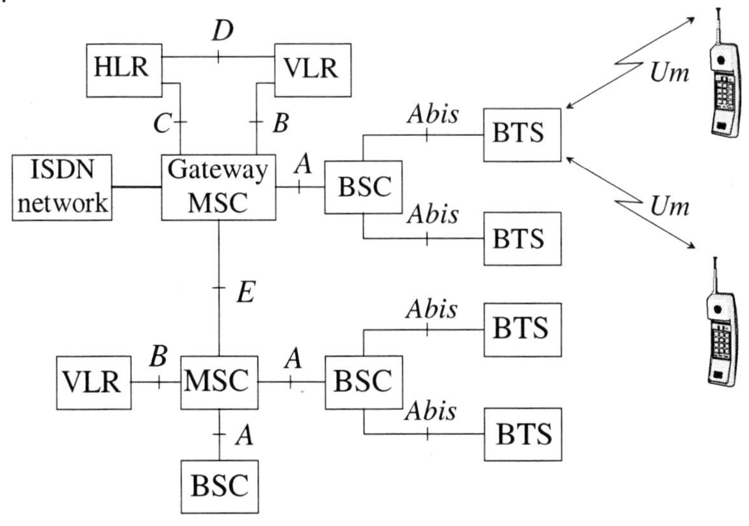

Figure 2.1 GSM network architecture

2.4 THE GSM RADIO INTERFACE

The radio subsystem constitutes the physical layer of the link between mobile and base station transceivers. In GSM different carrier frequencies are used for uplink and downlink, the carriers differing in frequency by 45 MHz. Thus GSM uses frequency division duplexing (FDD) to separate the uplink and downlink. The main attributes of the GSM interface are:

- time division multiple access (TDMA) with 8 channels/carrier.
- 124 radio carriers in a paired band (890 to 915 MHz mobile to base station, 935 to 960 MHz base to mobile, inter-carrier spacing 200 kHz).
- 270.833 kb/s per carrier.
- Gaussian minimum shift keying with a time bandwidth product $BT = 0.3$.
- Slow frequency hopping (217 hops/second).
- Synchronisation compensation for up to 233 μs absolute delay.
- Equalisation of up to 16 μs time dispersion.
- Downlink power control.
- Discontinuous transmission and reception.
- Block and convolutional channel coding coupled with interleaving to combat channel perturbations (overall channel rate of 22.8 kb/s).
- Full rate channel 13 kb/s voice coder rate using regular pulse excitation/linear predictive coding (RPE/LPC), half rate channel 6.5 kb/s voice coder rate using vector sum excited linear predictive coding (VSELP)
- Overall full rate channel bit rate of 22.8 kb/s

The radio sub system is the physical layer of the link between mobiles and base stations. Each cell can have from 1 to 16 pairs of carriers and each carrier is time multiplexed into 8 slots. The carrier frequency/slot combinations form the *physical channels* of the GSM system. The operation of the radio subsystem is divided into a number of *logical channels* each of which has a specific function in terms of handling the transmission of information over the radio sub system. The logical channels are mapped onto the available physical channels, as illustrated in Figure 2.2.

2.5 LOGICAL CHANNELS IN GSM

The logical channels in GSM may be divided into two categories:

1. Traffic CHannels (TCH), which carry voice and data;

2) Signalling Channels, which carry broadcast control information (Broadcast Control CHannel BCCH, Common Control CHannel CCCH) and call specific control information (Stand-alone Dedicated Control CHannel SDCCH, Fast Associated Control CHannel FACCH, and Slow Associated Control CHannel SACCH).

Some of the signalling channels are divided into sub channels which, conceptually, exist in parallel. However, the existence of one signalling channel may exclude the presence of another, for example the Paging CHannel (PCH) and Access Grant CHannel (AGCH) are never used in parallel. The radio subsystem requires two channels for its own purposes. These are the *Synchronisation CHannel* (SCH) and the *Frequency Correction CHannel* (FCCH).

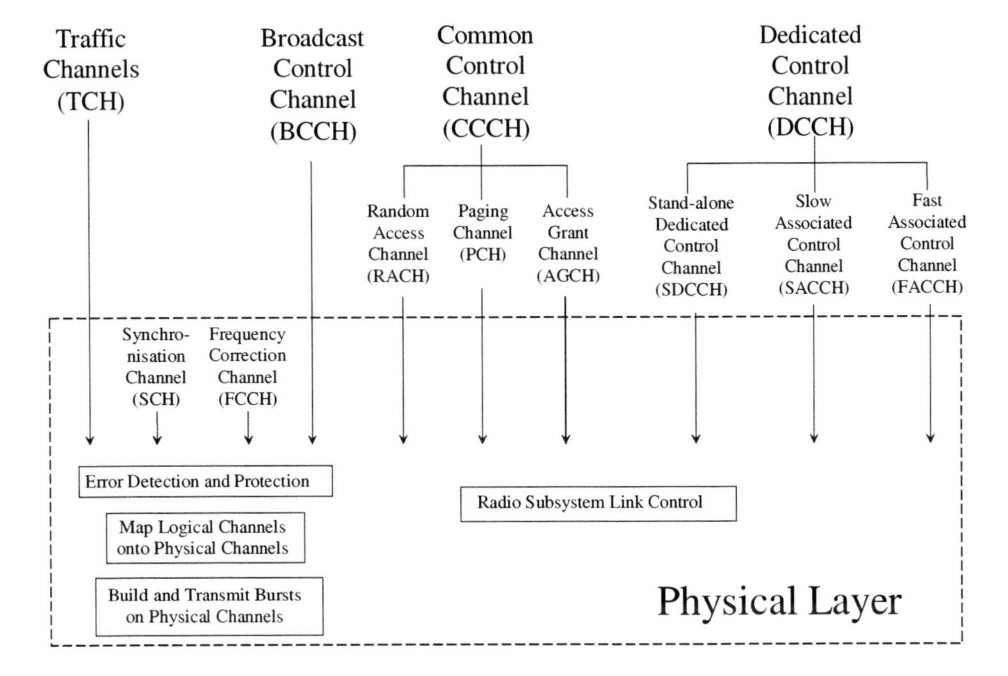

Figure 2.2 Mapping of physical channels in GSM

The logical channels defined for GSM are as follows:

- The *Broadcast Control CHannel* (BCCH) is a base to mobile channel which provides general information about the network, the cell in which the mobile is currently located and the adjacent cells.
- The *Access Grant CHannel* (AGCH) is part of the Common Control CHannel (CCCH) and is a base to mobile which is used to used to assign dedicated resources (i.e. a SDCCH or TCH) to a mobile which has previously requested them via the RACH.
- The *Paging CHannel* (PCH) is also part of the CCCH and is a base to mobile channel used to alert a mobile to a call originating from the network.

- The *Synchronisation CHannel* (SCH) is a base to mobile channel which carries information for frame synchronisation and identification of the base station transceiver.
- The *Frequency Correction CHannel* (FCCH) is a base to mobile channel which provides information for carrier synchronisation.
- The *Stand-alone Dedicated Control CHannel* (SDCCH) is a bi-directional channel allocated to a specific mobile for exchange of location update information and call set-up information.
- The *Slow Associated Control CHannel* (SACCH) is a bi-directional channel used for exchanging control information between base station and a specific mobile during the progress of a call or a call set up procedure. The SACCH is associated with a particular traffic channel or stand alone dedicated control channel.
- The *Fast Associated Control CHannel* (FACCH) is a bi-directional channel which is used for exchange of time critical information between mobile and base station during the progress of a call. The FACCH transmits control information by stealing capacity from the associated traffic channel.

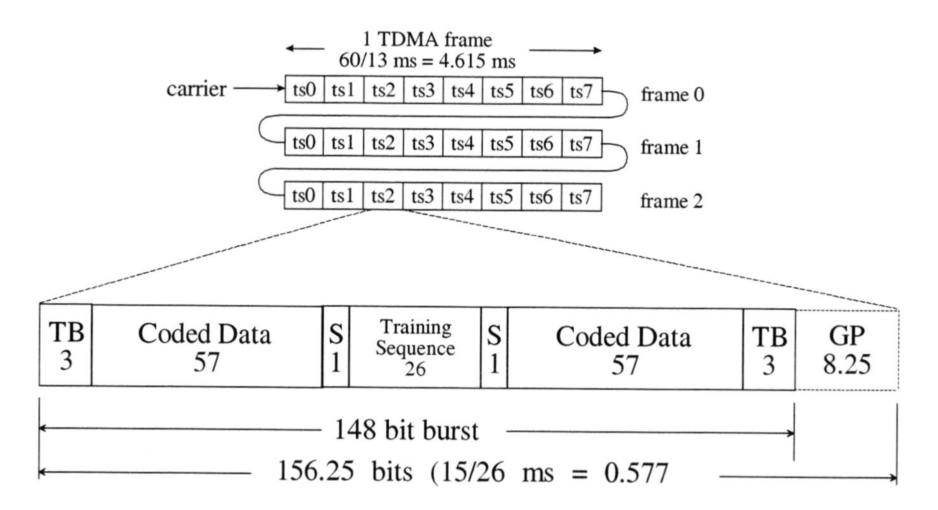

Figure 2.3 Normal burst in GSM

The logical channels are mapped onto the basic TDMA frame structure which is shown in Figure 2.3. Bursts of transmission from base station and mobile occur in the slots of the up and down carriers. The bit rate on the radio channel is 270.833 kb/s which gives a bit duration of 3.692 μs. A single time slot consists of 156.25 bits and therefore has a duration of approximately 0.577 ms. The recurrence of one particular time slot on each frame makes up one physical channel. This structure is applied to both up and down links. Data is transmitted in bursts which are placed in these time slots, the length of the bursts being slightly shorter than the duration of the time slots. This is to allow for burst alignment errors, time dispersion on the propagation path, and the time required for smooth switch on/off of the transmitter. The numbering scheme is staggered by 3 time slots to remove the necessity for the mobile station to transmit and receive at the same time. This is illustrated in Figure 2.4

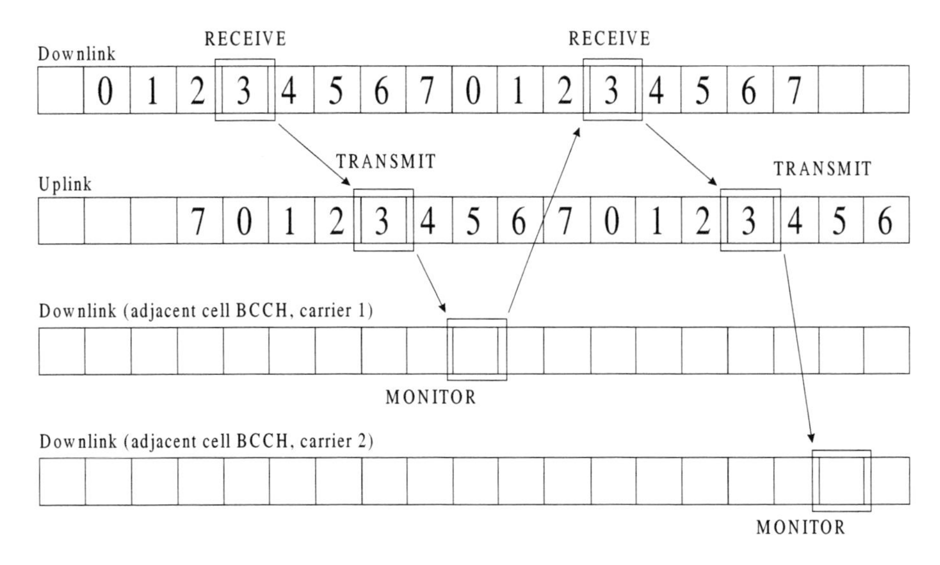

Figure 2.4 GSM slots and scanning structure

The purpose of the radio subsystem is to provide a "bit pipe" with a defined throughput, acceptable transmission delay and quality for each of the logical channels. To achieve this the physical layer (the reader is referred to Section 4.1 for a fuller description of the OSI layered model) performs a variety of tasks which can be grouped into the following categories:

- Create physical channels by building data bursts and transmitting them over the radio path.
- Map the logical channels onto the created physical channels, taking into account the throughput needs of particular logical channels.
- Apply error protection to each logical channel according to its particular needs.
- Monitor and control the radio environment to assign dedicated resources and to combat changes in propagation characteristics by functions such as power control.

The traffic and control information transmitted over the GSM radio interface is mapped onto standard burst structures which are transmitted in the individual time slots.

2.5.1 Burst Structure for GSM

There are four types of burst which can occupy a time slot in the GSM radio interface, these are:

- Normal burst (148 bits + 8.25 guard bits).
- Frequency correction burst (148 bits + 8.25 guard bits).

- Synchronising burst (148 bits + 8.25 guard bits).
- Access burst (88 bits + 68.25 bits), used to access a cell for the first time in case of call set up or handover.

The data structure within a normal burst is shown in Figure 2.3 and consists of 148 bits transmitted at a rate of 270.833 kb/s. Of these bits 114 bits are available for data transmission, the remaining bits are used to assist reception and detection. A training sequence (26 bits) in the middle of the burst is used by the receiver to synchronise and estimate the propagation characteristics. This allows the setting up of an equaliser to compensate for time dispersion produced by multipath propagation. Tail bits (3 bits) transmitted at either end of the burst enable the data bits near the edges of each burst to be equalised as well as those in the middle. Two stealing flags (one at each end of the training sequence) are used to indicate that a burst which had initially been assigned to a traffic channel has been "stolen" for signalling purposes in the FACCH. The frequency correction burst, synchronisation burst and random access burst, used in GSM, are shown in Figure 2.5.

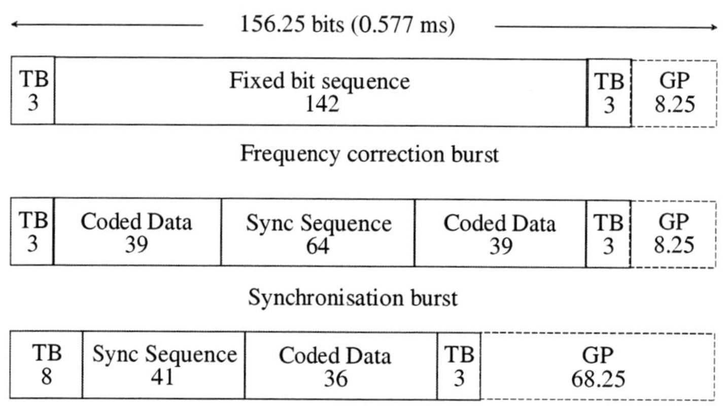

Figure 2.5 Other bursts used in GSM

Each burst in the GSM system modulates one of the carriers assigned to a particular cell using Gaussian minimum shift keying (GMSK). If frequency hopping is not employed, each burst belonging to one particular physical channel is transmitted using the same carrier frequency. A network operator can implement slow frequency hopping (SFH) which can be used to overcome the problems of fading. When SFH is implemented the carrier frequency is changed between transmitted bursts within a logical channel. The carrier frequency for each burst is selected from the set of carrier frequencies available within a cell according to a specified sequence.

2.5.2 Mapping of Logical Channels in GSM

Having described the burst types in GSM it is appropriate to consider the way in which the logical channels are mapped onto the physical channels. There are seven different cases of logical channel mapping in the GSM system, these are:

I. Mapping of a full rate speech channel (TCH/FS) and its slow associated control channel (SACCH) onto one physical channel.

II. Mapping of one half rate speech channel (TCH/HS) and its slow associated control channel (SACCH) on to one physical channel (in this case alternate frames are idle).

III. Mapping of two half rate speech channels (TCH/HS) and their slow associated control channels (SACCH) onto one physical channel.

IV. Mapping of the broadcast control channel (BCCH), the common control channel (CCCH), the frequency correction channel (FCCH) and the synchronisation channel (SCH) onto one physical channel and is assigned to slot 0.

V. Mapping of four stand-alone dedicated control channels (SDCCH) and their slow associated control channels (SACCH), the broadcast control channel (BCCH), the common control channel (CCCH), the frequency correction channel (FCCH) and the synchronisation channel (SCH) onto one physical channel and is assigned to slot 0. (This is for lower capacity than case III, i.e. when a cell is allocated 1 or 2 carriers.)

VI. Mapping of the broadcast control channel (BCCH), the common control channel (CCCH) onto one physical channel. This is used in combination with IV when a base station is required to handle a large number of mobiles and is assigned to slot 2, 4 or 6.

VII. Mapping eight stand-alone dedicated control channels (SDCCH) and their slow associated control channels (SACCH) onto one physical channel. This is used in combination with IV and VI for large capacity base stations to provide dedicated control channels for call set-up and registration

This list indicates that there are at least two logical channels to be mapped onto each physical channel. The time slots of the physical channel must therefore be assigned to the logical channels on a structured basis. For this purpose two multiframe structures have been defined:

1. A multiframe consisting of 26 TDM frames (resulting in a recurrence interval of 120 ms) for the TCH/SACCH cases I and II.

2. A multiframe consisting of 51 TDM frames (resulting in a recurrence interval of 235.38 ms) for signalling channels, cases III, IV and V.

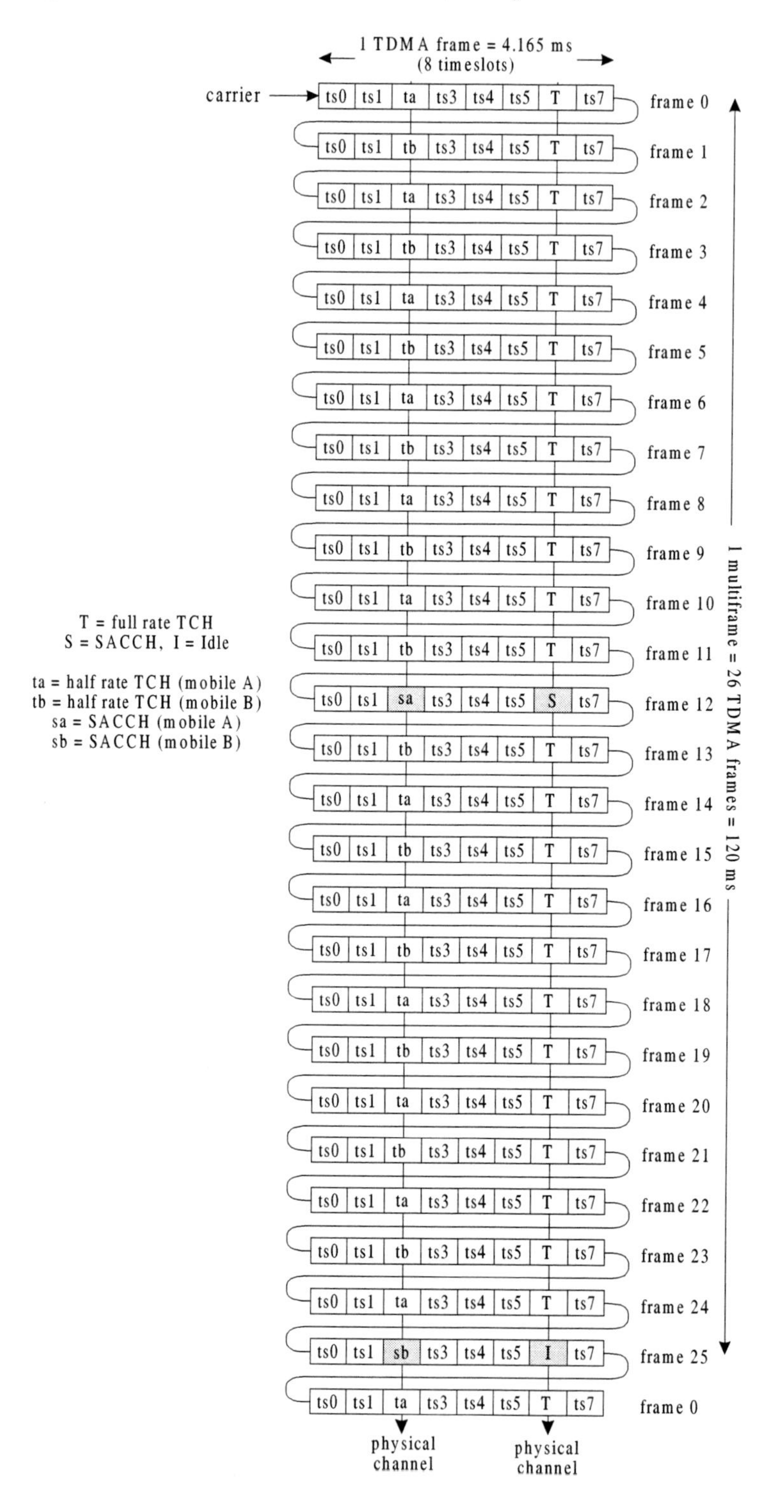

Figure 2.6 GSM full rate (case I) and half rate (case III) channels

2.5.3 Mapping of Traffic Channels and Associated Control Channels

The mapping of the traffic channel and its SACCH for cases I and III are shown in Figure 2.6. In this diagram it is assumed that a mobile has been allocated a particular time slot on a carrier. This time slot (of duration 15/26 ms) will occur at intervals of 60/13 ms and during frames 0 to 11 it is used for the transmission of traffic data. The SACCH is transmitted in frame 12 and frames 13 to 24 are used for traffic data. On the full rate channel the next frame 25 is idle. The full rate TCH thus uses 24 frames out of the 26 available in the multiframe. The duration of the multiframe is therefore $26 \times 60/13$ ms = 120 ms.

Provision is also made for the transmission of half rate speech channels with a coder rate of 11.4 kb/s. In this case it is possible for two mobiles to share a particular slot on a carrier (case II). Mobile A uses the even numbered frames and mobile B uses the odd numbered frames. The SACCH for mobile A is transmitted in frame 12 whilst that for mobile B is transmitted in frame 25. Case II is used when only a single half rate channel is required, and results in idle alternate frames.

The fast associated control channel (FACCH) is not shown explicitly in Figure 2.6 as this is substituted for all or part of the traffic channel, when the need arises, for transmission of high priority signalling. The presence of the FACCH is indicated to the receiver by setting the appropriate stealing flag.

The gross bit rate per full rate traffic channel is derived as follows:

Data bits per normal burst = 114 bits

Number of normal bursts per 120 ms multiframe = 24

$$\text{Gross bit rate} = \frac{24 \times 114}{0.12} = 22.8 \text{ kb/s}$$

The SACCH uses 114 bits per 120 ms, hence gross bit rate = 950 b/s

The throughput of the physical channel is 114 bits per 4.615 ms = 24.7 kb/s (including the idle frame). In the case of the half rate channel (case II) two half rate channels share one physical channel, the idle frame is then used to accommodate the SACCH for the second half rate channel. In this case each traffic channel occupies only 12 frames which results in a gross bit rate of 11.4 kb/s for each (each SACCH uses 950 b/s).

2.5.4 Mapping of the BCCH/CCCH

The mapping of the BCCH and the CCCH uses a multiframe of 51 TDM frames and is shared by all mobiles currently in a cell. In addition all sub channels transmitted on this structure exist in one direction only. The mapping of these sub channels onto a single physical channel (case IV) using a 51 multiframe is shown in Figure 2.7. This structure appears on time slot 0 of one of the allocated carriers in the cell, which is known as the BCCH carrier (or base channel). The uplink of the BCCH/CCCH structure carries only the random access channel as this is the only control channel which exists from mobile to base. A mobile may use any one of the 51 frames on time-slot 0 to access the network.

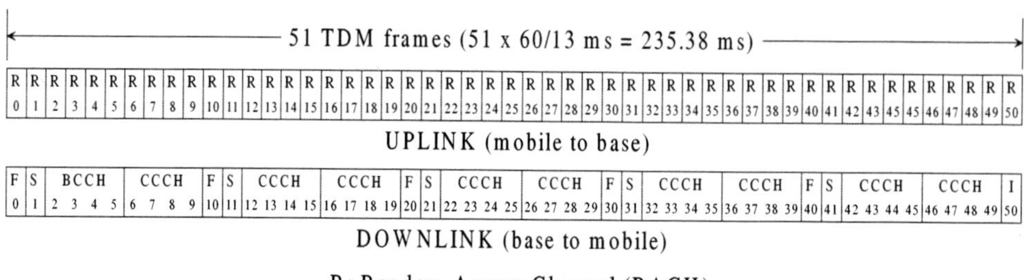

Figure 2.7 Multiframe structure for BCCH/CCCH (case IV)

On the downlink frames 0 to 49 are grouped into 5 sets of 10 frames and frame 50 remains idle. The gross bit rate for the BCCH is 4 frames of 114 bits per 235.38 ms = 1.94 kb/s.

2.5.5 Mapping of the SDCCH/SACCH

During registration, call se-up etc. considerable exchange of information occurs between the mobile and base station. This exchange is accomplished over a dedicated control channel allocated by the base station to a mobile for the duration of the operation. Eight stand-alone dedicated control channels (SDCCH) and their slow associated control channels (SACCH) are mapped onto one physical channel (case VII) as shown in Figure 2.8.

51 TDM frames (235.38 ms)

	SACCH 5	SACCH 6	SACCH 7	I	I	I	SDCCH 0	SDCCH 1	SDCCH 2	SDCCH 3	SDCCH 4	SDCCH 5	SDCCH 6	SDCCH 7	SACCH 0
1	0 1 2 3	4 5 6 7	8 9 10 11	12	13	14	15 16 17 18	19 20 21 22	23 24 25 26	27 28 29 30	31 32 33 34	35 36 37 38	39 40 41 42	43 44 45 46	47 48 49 50

	SACCH 1	SACCH 2	SACCH 3	I	I	I	SDCCH 0	SDCCH 1	SDCCH 2	SDCCH 3	SDCCH 4	SDCCH 5	SDCCH 6	SDCCH 7	SACCH 4
2	0 1 2 3	4 5 6 7	8 9 10 11	12	13	14	15 16 17 18	19 20 21 22	23 24 25 26	27 28 29 30	31 32 33 34	35 36 37 38	39 40 41 42	43 44 45 46	47 48 49 50

UPLINK (8 ×SDCCH + 8 ×SACCH)

	SDCCH 0	SDCCH 1	SDCCH 2	SDCCH 3	SDCCH 4	SDCCH 5	SDCCH 6	SDCCH 7	SACCH 4	SACCH 5	SACCH 6	SACCH 7	I	I	I
1	0 1 2 3	4 5 6 7	8 9 10 11	12 13 14 15	16 17 18 19	20 21 22 23	24 25 26 27	28 29 30 31	32 33 34 35	36 37 38 39	40 41 42 43	44 45 46 47	48	49	50

	SDCCH 0	SDCCH 1	SDCCH 2	SDCCH 3	SDCCH 4	SDCCH 5	SDCCH 6	SDCCH 7	SACCH 0	SACCH 1	SACCH 2	SACCH 3	I	I	I
2	0 1 2 3	4 5 6 7	8 9 10 11	12 13 14 15	16 17 18 19	20 21 22 23	24 25 26 27	28 29 30 31	32 33 34 35	36 37 38 39	40 41 42 43	44 45 46 47	48	49	50

DOWNLINK (8 ×SDCCH + 8 ×SACCH)

Figure 2.8 Multiframe structure for SDCCH/SACCH (case VII)

2.5.6 Mapping of Logical Channels in Low Capacity Systems

When the number of carriers allocated to a cell is limited to 1 or 2 it is necessary to multiplex the BCCH/CCCH and SDCCH/SACCH onto a single physical channel (case V). This mapping is shown in Figure 2.9.

51 TDM frames (235.38 ms)

	0 1 2 3	4	5	6 7 8 9	10 11 12 13	14	15	16	17	18	19	20	21	22	23	24	25	26	27	28	29	30	31	32	33	34	35	36	37 38 39 40	41 42 43 44	45	46	47 48 49 50
1	SDCCH 3	R	R	SACCH 2	SACCH 3	R	R	R	R	R	R	R	R	R	R	R	R	R	R	R	R	R	R	R	R	R	R	R	SDCCH 0	SDCCH 1	R	R	SDCCH 2
2	SDCCH 3	R	R	SACCH 0	SACCH 1	R	R	R	R	R	R	R	R	R	R	R	R	R	R	R	R	R	R	R	R	R	R	R	SDCCH 0	SDCCH 1	R	R	SDCCH 2

UPLINK (RACH + 4 × SDCCH + 4 × SACCH)

	0	1	2 3 4 5	6 7 8 9	10	11	12 13 14 15	16 17 18 19	20	21	22 23 24 25	26 27 28 29	30	31	32 33 34 35	36 37 38 39	40	41	42 43 44 45	46 47 48 49	50
1	F	S	BCCH	CCCH	F	S	CCCH	CCCH	F	S	SDCCH 0	SDCCH 1	F	S	SDCCH 2	SDCCH 3	F	S	SACCH 0	SACCH 1	I
2	F	S	BCCH	CCCH	F	S	CCCH	CCCH	F	S	SDCCH 0	SDCCH 1	F	S	SDCCH 2	SDCCH 3	F	S	SACCH 2	SACCH 3	I

DOWNLINK (BCCH + CCCH + 4 × SDCCH + 4 × SACCH)

Figure 2.9 Multiframe structure for low capacity cells (case V)

2.6 SECURITY IN GSM

The setting up of any call between mobile and base station, or vice versa, requires specific exchanges between mobile and base station in order that the mobile may obtain a channel. Initially mobiles use common channels, i.e. channels shared with other mobiles and listens to the broadcast channels from the base station. Under such circumstances the mobile is said to be in the *idle state*. The mobile moves into the *dedicated state* when it is allocated a bi-directional channel for its specific communication needs.

GSM is a public system operating on radio frequencies and hence it is necessary to build in security features which protect the network against fraudulent access and ensure subscriber privacy. The security functions include:

- *authentication* of the user, to prevent access by unregistered users;
- radio path *encryption*, to prevent unauthorised listening;
- user *identity protection*, to prevent subscriber location disclosure.

In GSM the mobile station is divided into two parts the first part contains hardware and software specific to the radio interface and the second part contains the user specific data and is known as the subscriber identity module (SIM). The SIM is a plug-in unit analogous to a smart card and may be regarded as a key to the use of the terminal. Once the SIM is removed the mobile unit cannot be used for any service which will be charged to the subscriber's bill. The SIM has several functions and a limited programming is possible by the user. However most of the information contained within the SIM is protected against alteration, and in some cases against reading. All security functions involve the SIM which is designed to be very difficult to duplicate. Hence the combination of the SIM and the security functions provides a high degree of protection against fraudulent access to the network.

2.6.1 Authentication

There are two methods used by GSM. The first uses a personal identity number (typically 4 digit) which is stored in the SIM. A user wishing to make a call enters the PIN which is checked by the SIM, without transmission on the radio interface. After this stage GSM uses a much more sophisticated system which interrogates the mobile unit. This is controlled from the MSC/VLR and occurs at call set-up, location updating, handover etc.

Each authorised user is provided with a secret parameter (key) known as K_i which is stored in a highly protected way in the SIM and is unknown to the user. This K_i is also known to the network operator and is stored in the mobile's HLR . The infrastructure transmits a 128 bit random number *RAND* to the mobile. This is used by the mobile, in conjunction with a secret algorithm known as *A3*, to produce a 32 bit response known as *SRES*. The infrastructure uses the same *RAND*, K_i and *A3* algorithm to produce a *SRES* which is then checked against the response from the mobile (in the public GSM network this is actually done in the mobile's HLR, which thus avoids the need to transmit K_i to the current MSC/VLR being used by the mobile).

In order to maintain the desired security level, *A3* is devised so that the computation of *SRES* from *RAND* and K_i is straightforward but the computation of K_i from *RAND* and *SRES* is extremely complex. It should be noted that *A3* is operator dependent, which is a further reason for performing the network *SRES* computation at the mobile HLR. The authentication process is illustrated in Figure 2.10.

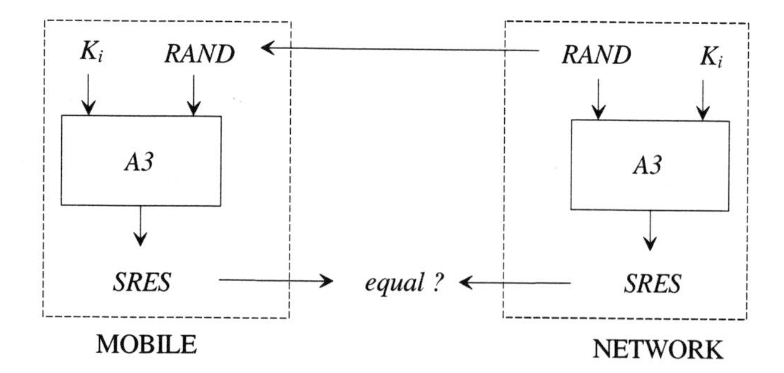

Figure 2.10 The GSM authentication procedure

2.6.2 Encryption

Encryption (or ciphering) is employed in GSM to prevent unauthorised listening. It is used for all data transmitted between mobile and base station in the *dedicated state*. This includes user information (voice, data), user related signalling (which includes the called numbers) and system related signalling (handover signalling etc.). There are actually two modes of transmission, one is the protected mode (encrypted) and the other is the open mode (clear text) and it is necessary to switch from one mode to the other as appropriate. When in the open mode it is still necessary to protect the actual user identity. This is achieved by using an identity alias known as the Temporary Mobile Subscriber Identity (TMSI) instead of the International Mobile Subscriber Identity (IMSI). This alias is agreed between the mobile and the network during protected signalling exchanges.

Ciphering is achieved by "exclusive OR-ing" the 114 data bits of each normal burst with a pseudo random sequence. Deciphering follows exactly the same operation and reproduces the original 114 bit data ("exclusive OR-ing" twice with the same pseudo random sequence reproduces the original data stream). The algorithm used to generate the pseudo random sequence is known as *A5*.

The *A5* algorithm generates the pseudo random sequence from two inputs one being the frame number (22 bits) and the other being a key K_c (64 bits) agreed between mobile and network. In fact two different pseudo random sequences are generated by *A5*, one being used on the uplink and the other on the downlink.

K_c is actually computed during the GSM authentication process and is stored in a non volatile memory in the SIM. It is therefore remembered after the mobile is switched off. This "dormant" key is also stored in the MSC/VLR last used by the mobile and is ready to be used at the start of encryption. This means that when the next transition from clear mode to cipher mode occurs the dormant key can be used for encryption and becomes the "active" key. During the next authentication a new value of K_c is computed and this becomes the dormant key. Hence the value of K_c is not constant but changes on each transition from clear mode to cipher mode. The value of K_c is generated from the same *RAND* used in the authentication process by a secret algorithm known as *A8*. This is also operator dependent. The algorithms *A3* and *A8* are always run together and in most cases are implemented as a single operator specific algorithm known as *A3/A8*.

2.6.3 User Identity Protection

This ensures that the location of a user cannot be devised from communications over the radio interface during clear mode operation. It avoids sending the IMSI on an open transmission. A TMSI is allocated to a mobile station the first time it registers with a location area and is released when the mobile leaves that area. This means that any request for the full IMSI can take place in ciphered mode and is thus fully protected.

2.6.4 Sequence of Events

It is clear from the previous descriptions that the VLR and HLR play an important role in the security aspects of GSM. Each location area is controlled by a MSC and currently each MSC has an associated VLR. In the *idle state* the mobile continually monitors the BCCH and can therefore determine when it enters a new location area from the information broadcast from the base station. When this happens the mobile initiates a registration update request to the new BS which includes the identity of the old registration area and the TMSI which it was using in the old area (this is in unprotected mode).

This request is passed to the new MSC/VLR over the "fixed network". The new VLR cannot translate the TMSI into the ISMI and hence makes a request to the old VLR to send the ISMI corresponding to the known TMSI. The old VLR responds with the ISMI and also the required authentication information. The new VLR then initiates an authentication procedure with the mobile. If this authentication succeeds the new VLR uses the IMSI to determine the address of the HLR of the terminal and sends the location update information. The HLR sends a registration confirmation to the VLR with all relevant subscriber profile information required for call handling. The new VLR then assigns a new TMSI to the mobile. Hence both the old VLR and the HLR are involved in transferring information to the new VLR over the fixed network. This sequence of events is shown in Figure 2.11.

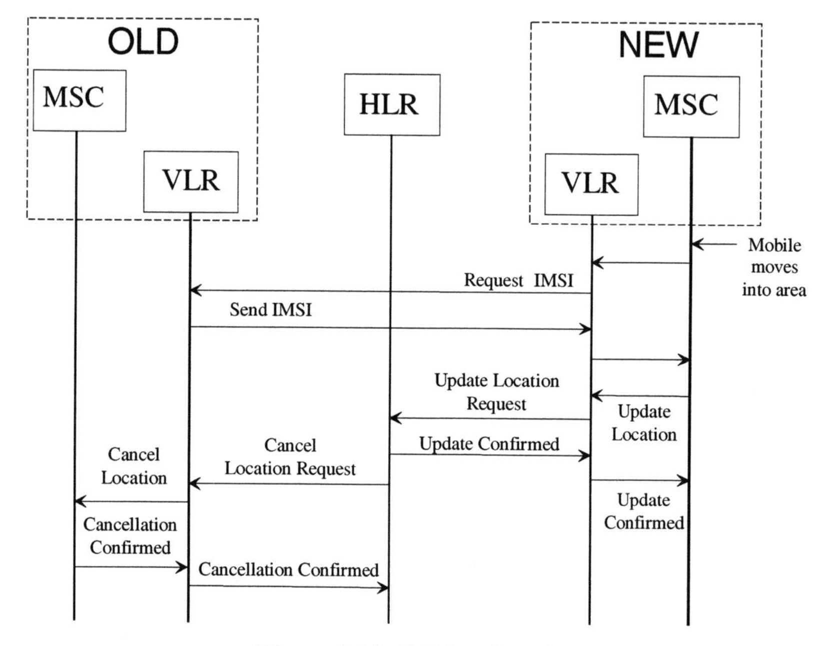

Figure 2.11 GSM registration

It should be noted that when in the idle state (unprotected) no user sensitive information is transmitted over the radio interface.

2.7 OPERATION OF THE GSM SYSTEM

When a mobile station is first switched on it is necessary to read the BCCH in order to determine its orientation within the network. Before this can be done the mobile must first synchronise in frequency and then in time. The FCCH, SCH and BCCH are all transmitted on the same carrier frequency (known as the base channel) which has a higher *power density* than any of the other channels in a cell because steps are taken to ensure that it is transmitting information at all times. The mobile scans around the available frequencies, picks the strongest and then selects the FCCH. The frequency correction channel is transmitted at regular intervals on the base channel. The modulation used in GSM is Gaussian minimum shift keying (GMSK) which is described in section 1.7.11. The carrier modulation is arranged in such a way that if all data bits in the frequency correction burst are zero, differential decoding produces a pure sine wave of frequency $f = f_c + 67.7$ kHz where f_c is the allocated carrier frequency. This principle is used to correct the transmission frequency of mobiles at regular intervals.

The frequency correction burst is always followed by a synchronisation burst which is similar to a normal burst . The frequency transmitted for the first information digit in a standard burst (i.e. excluding the initial 3 tail bits) is based on the convention that this digit follows an infinite series of 1s. Consequently if the first information digit is binary 1 this will be transmitted as a frequency $f_c + 67.7$ kHz (which is the frequency transmitted by the GSM frequency correction burst). If the first digit is binary 0 this will be transmitted as frequency $f_c - 67.7$ kHz. Subsequent digits, including the training sequence, are transmitted with frequencies according to Table 1.3.

The FCCH allows the mobile to synchronise its oscillator with the frequency of the base channel. The purpose of the synchronisation burst is to allow the mobile to synchronise with the transmitted bit interval and to receive information on the base station identity and which of the eight possible training sequences is being used. The synchronisation burst of Figure 2.5 represents the physical mapping of the Synchronisation CHannel (SCH).

The Synchronisation CHannel (SCH) always follows the FCCH in the base channel. The SCH contains information on the current frame number and the training sequence used within the cell. Once this information is available to the mobile it is able to read the BCCH. This contains information on the cell location and various other overhead messages. This 3 stage synchronisation process normally takes between 2 to 5 seconds but can take up to 20 seconds. When a mobile is switched off it retains information about the cell including the frequency of the base channel. When it is switched on again it looks for the base channel on the frequency stored in the subscriber identity module (SIM). If it is switched on again in the same cell the scanning process is avoided and the synchronisation process is much faster.

2.7.1 Location Updating

If a mobile is switched on in a different area from that which is stored in the SIM (where it was last switched off), or if it enters a new area, the mobile station initiates a location updating procedure to inform the network of its new location. Location updating can also be forced by the network.

Each location area has a mobile switching centre with a home location register (HLR) and a visitor location register (VLR). The HLR is a database of all mobiles normally resident in that location area. The VLR is a database containing a record of all mobiles in the area which are not normally resident within that area. If a mobile enters a new location area, location updating is executed via the fixed network. GSM then supports two alternatives:

1. The VLR immediately issues a Mobile Subscriber Roaming Number (MSRN) to be associated with the actual identity over the radio path (i.e. the International Mobile Subscriber Identity [IMSI]). The international mobile subscriber identity and the mobile subscriber roaming number are then conveyed to the home location register of the mobile over the fixed network. At the end of this procedure the home location register contains the unique directory number of the mobile coupled with the international mobile subscriber identity and the current mobile subscriber roaming number. Note that when a mobile first registers in a location area it is allocated a Temporary Mobile Subscriber Identity (TMSI). This is used in preference to the IMSI in the interests of security.

Table 2.1 Location updating

System activity	Channel	Mobile activity
System overhead parameters and other overhead messages	BCCH →	Mobile switched on, searches for base channel and synchronises. Monitor BCCH for current location.
Receive channel request	← RACH	If current location different from that stored in SIM generate a channel request.
Assign stand alone dedicated control channel	AGCH →	Receive stand alone dedicated control channel assignment and store in memory.
Receive location updating request.	← SDCCH	Request for location updating (registration).
Request authentication from mobile	SDCCH →	Receive authentication request.
Receive and check authentication	← SDCCH	Authentication response
Request mobile to transmit in ciphered mode	SDCCH →	Receive request and switch to ciphered mode.
Receive acknowledgement	← SDCCH	Acknowledge cipher mode request.
Confirm location updating including the optional assignment of a temporary identity (TMSI).	SDCCH →	Receive location updating including and (TMSI) and store in SIM.
Receive acknowledgement	← SDCCH	Acknowledge new location and TMSI.
Send channel release	SDCCH →	Switch to *idle update* mode, monitor BCCH and CCCH.

2. In this case the network identity of the VLR or MSC, rather than the mobile subscriber roaming number, is reported to the VLR. A call for a particular mobile is then routed to the appropriate home location register. In the first alternative the mobile subscriber roaming number is available at the HLR and the call is directed to the VLR and the mobile is subsequently paged by transmitting the international mobile subscriber identity over the appropriate paging channel. In the second alternative the HLR signals the designated MSC and transacts for a mobile subscribers roaming number which is assigned by the VLR. Subsequently the mobile is paged with the assigned mobile subscriber roaming number. It should be noted that whilst in the *idle update* mode the mobile compares the identity of the location area transmitted on the BCCH with that stored in its SIM. Should there be any change the mobile will initiate a *registration*.

2.7.2 Call Establishment from a Mobile

A channel is requested on the RACH and may be in contention with other mobiles. A slotted ALOHA protocol is used (see Section 4.7.2). If a request is received without a collision a dedicated control channel can be assigned by the network by a response on the Access Grant CHannel (AGCH). To minimise the probability of a collision during channel access a short access packet format is used which can be transmitted within one burst (see also Section 2.13.1 on adaptive frame alignment).

The access packet contains a 7 bit random number which is used by the network in conjunction with the access slot number to address the originating mobile station for channel allocation. The full mobile identification is delivered once a dedicated control channel has been allocated. These channels are used for various functions such as authentication etc. Detection of possible collision (or transmission errors) is performed within the network through a check of the received access burst. If a collision (or error) is detected the network aborts the procedure. If the mobile does not receive an access grant on the AGCH (which will be monitored 5 TDM slots later) a new access attempt will be made on the next slot with a given probability. (In effect this means that the mobile chooses a random number from 1 to n, which represents the next slot on which an access attempt will be made). It is possible that even when packets collide the FM capture effect will ensure that one packet is received without error.

When an *access grant* is received the mobile proceeds with the call set-up on the allocated dedicated control channel by sending a *set-up* message to the network. This contains addressing information and various network information. The network accepts the call establishment by returning a *call proceeding* message on the SDCCH. In the normal call set-up procedure the network will assign a dedicated traffic channel before it initiates the call establishment in the network. (The network may queue the traffic channel request up to a maximum queuing period.)

When called party alerting has been initiated an *alerting* message is sent to the mobile over the FACCH (on the allocated traffic channel) and a ringing tone may be generated by the network and sent to the mobile.

Table 2.2 Call establishment from a mobile

System activity	Channel	Mobile activity
System overhead parameters and other overhead messages	BCCH →	*(Idle Updated)* monitor BCCH and CCCH (PCH) for mobile control message.
Receive channel request	← RACH	Generate channel request.
Assign stand alone dedicated control channel	AGCH →	Receive stand alone dedicated control channel assignment and store in memory.
Receive call establishment request	← SDCCH	Send call establishment request.
Request authentication from mobile	SDCCH →	Receive authentication request.
Receive and check authentication	← SDCCH	Authentication response.
Request mobile to transmit in ciphered mode	SDCCH →	Receive cipher mode request.
Receive acknowledgement	← SDCCH	Acknowledge cipher mode request (all further transmissions in cipher mode).
Receive set-up request	← SDCCH	Send set-up message and desired number.
Send call proceeding indication to mobile and route call to desired number	SDCCH →	Receive call preceding indication.
Assign traffic channel to mobile	SDCCH →	Receive traffic channel assignment.
Receive traffic channel acknowledgement on allocated channel	← FACCH	Switch to traffic channel and send an acknowledgement.
Send alert signal that called number is available and the phone is ringing	FACCH →	Receive alert signal ringing sound.
Send connect message when called party accepts	FACCH →	Receive connect message.
Receive connect accept response	← FACCH	Acknowledge connect message and switch to traffic channel.
Exchange of user data	← TCH →	Conversation commences on TCH.

When the call has been accepted at the remote end a *connect* message is transferred to the mobile, indicating that the connection is established in the network. The mobile station responds by sending a *connect acknowledge* message and then enters the active state (further signalling takes place over the SACCH or FACCH).

Table 2.3 Call establishment to a mobile

System activity	Channel	Mobile activity
System overhead parameters and other overhead messages	BCCH →	*(Idle Updated)* monitor BCCH and CCCH (PCH) for mobile control message.
Receive incoming call, generate a paging message	PCH →	Receive paging message.
Receive channel request	← RACH	Generate channel request.
Assign stand alone dedicated control channel	AGCH →	Receive stand alone dedicated control channel assignment and store in memory.
Receive paging acknowledgement	← SDCCH	Answer paging message from network.
Request authentication from mobile	SDCCH →	Receive authentication request.
Receive and check authentication	← SDCCH	Authentication response.
Request mobile to transmit in ciphered mode	SDCCH →	Receive cipher mode request.
Receive acknowledgement	← SDCCH	Acknowledge cipher mode request, switch to cipher mode.
Send set-up message for incoming call	SDCCH →	Receive set-up message.
Assign traffic channel to mobile	SDCCH →	Receive traffic channel assignment.
Receive traffic channel acknowledgement on allocated channel	← FACCH	Switch to traffic channel and send an acknowledgement.
Send alert signal	FACCH →	Receive alert signal and generate ringing sound.
Send connect message when user off-hook	FACCH →	Receive connect message.
Receive connect accept response	← FACCH	Acknowledge connect message and switch to traffic channel.
Exchange of user data	← TCH →	Conversation commences on TCH.

2.7.3 Call Establishment to a Mobile

In this particular case a paging message is routed to the traffic area in which the mobile is registered and transmitted on the paging channel. In responding to the page the mobile must first request a channel as in the previous case. When access grant is received from the base station the mobile responds with a *call confirmed* message on the allocated dedicated control channel. A traffic channel is then allocated and the call proceeds (i.e. the mobile enters the active state).

2.7.4 Call Release

Call release can be initiated either by the mobile or the fixed network, via the SACCH by sending a DISCONNECT message. If the release is initiated by the mobile the network responds with a RELEASE message. The mobile responds with a RELEASE COMPLETE message and releases the TCH. The mobile then enters the idle state and monitors the CCCH (PAGCCH).

2.8 VOICE CODING IN GSM

A significant objective in GSM is to minimise the bandwidth required in each traffic channel. This is achieved by reducing the number of bits/s representing a voice waveform by capitalising on the inherent redundancy in such a waveform. GSM also makes use of the fact that during a normal conversation one of the participants is either listening or pausing for breath during approximately 60% of the time of the conversation and it is necessary to transmit a signal only during periods of activity. This gives rise to Discontinuous Transmission (DTX). Voice coding in GSM is based on a technique known as analysis by synthesis predictive coding which is based on the principle of transmitting information about the speech production process rather than the speech waveform. This principle is also used in the TETRA voice coder and it is described in detail in Section 4.9.

There are however significant differences between the GSM voice coder and the coder used in TETRA. In the case of GSM the full rate voice coder produces an output data rate of 13 kb/s using a technique known as Regular Pulse Excitation/Linear Predictive Coding (RPE/LPC). In effect, the excitation source (the lungs) is modelled by pulses which are assumed to be regularly spaced and whose amplitudes are computed in a feedback loop such as that shown in the encoder section of Figure 4.12. These pulses are passed through synthesis filters to produce a synthesised waveform. The coefficients of the filters are adjusted to minimise the error between the actual voice waveform and the synthesised version. The resulting pulse amplitudes and filter coefficients are then transmitted to the receiver where they are used to reproduce the synthesised speech.

The voice coder effectively produces a block of 260 bits every 20 ms and these are sorted into three different classes according to their function and importance. The most important are 50 class Ia bits which describe the filter coefficients, speech sub block amplitudes and LPC parameters. Next in importance are 132 class Ib bits which consist of the excitation sequence and further LTP parameters. The remaining 78 bits are known as class II bits. Channel coding is added to the 260 bits and results in 456 bits per 20 ms. The class Ia bits receive greatest channel coding, the class Ib bits receive less robust channel coding and the class II bits receive no channel coding. Some transmission errors may be compensated for in the speech coder by interpolation, e.g. by retaining the previous filter coefficients when new ones are not available. If uncorrected errors are detected in the class Ia bits the entire speech frame is discarded and the missing frame is extrapolated from the previous frame.

2.9 NON VOICE SERVICES

The basic transmission rate on the GSM traffic channels is 22.8 kb/s. When these channels are used for data transmission a significant overhead is required for error protection and the user bit rate has a much lower value. GSM provides data transmission at user rates of 9.6 kb/s (ful-rate channel), 4.8 kb/s (full rate and half rate channel) and 2.4 kb/s (full rate and half-rate channel). For services offering both full rate and half rate channels the error protection available on the full rate channels is much greater. When considering the data services offered by GSM it is convenient to make the comparison with similar services on the fixed network as shown in Figure 2.12.

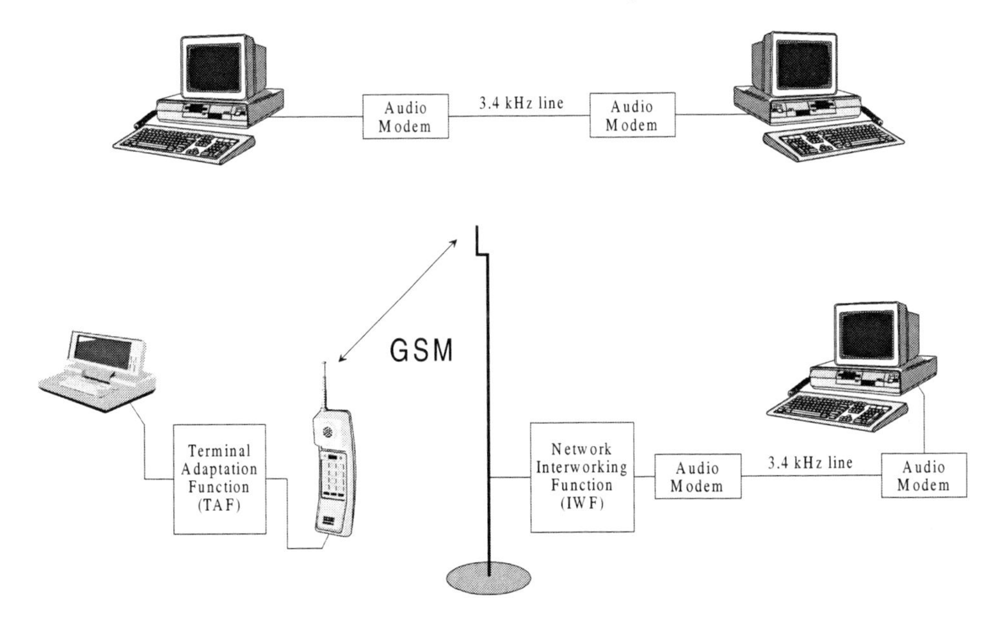

Figure 2.12 GSM data services

In this figure the terminal adaptation function (TAF) and Network Inter-working function (IWF) act as interfaces with the GSM bearer capabilities. In fact the IWF is able to interface with other networks, such as ISDN, and is not confined to the PSTN. Between the TAF and IWF the GSM network is required only to transport digitally provided data. The traditional interface between a computer and modem includes several wires on which several information flows are carried in parallel. Two of these wires contain the user data bit streams and the others contain flow control signals which allow the terminal to control the modem. Because in GSM the modem is remote from the terminal the user data and flow control signals must be transmitted over the GSM bearer and this accounts for the difference between the user bit rate (9.6 kb/s) and the capacity (12 kb/s) of the full rate data channel (additional error coding increases the actual bit rate to 22.8 kb/s).

2.9.1 Data Transmission Transparent Mode

There are two modes of transmission employed in GSM data services known as the Transparent (T) Mode and the Non-Transparent (NT) mode. In the T mode error correction is accomplished entirely by a forward error correction mechanism which is provided by the radio interface transmission scheme. In the NT mode a combination of forward error correction and repeat transmission is used. When the forward error correction is insufficient to correct all errors at the receiving end the information is repeated.

The T mode transmission is derived from ISDN specifications and the path between the TAF and IWF is regarded as a synchronous circuit and the information exchanged between the two ends is composed of user data at rates between 600 b/s and 9.6 kb/s plus some auxiliary information and three intermediate rates have been defined. The lower the intermediate rate, the higher the added error protection and the lower the residual error rate. The error protection scheme in the T mode is given in Table 2.5 and this is combined with an interleaving depth of 22 frames. The performance of the available T mode options, for a typical urban area, is shown in Table 2.4.

Table 2.4 Performance of T mode options

User rate (kb/s)	Intermediate rate (kb/s)	Channel type	Residual error rate (%)
9.6	12.0	full rate	0.3
4.8	6.0	full rate	0.01
		half rate	0.3
≤ 2.4	3.6	full rate	0.001
		half rate	0.01

2.9.2 Data Transmission (Non Transparent Mode)

In the NT mode the transmission on the GSM circuit connection is considered as a packet data flow, although the offered service, end to end, is a circuit switched service. The basic transmission rate is the same as the T mode with the same forward error correction. The NT transmission mode groups bits in successive frames of 240 bits which include redundancy to enable the receiver to detect residual errors and request retransmission, this is shown in Figure 2.13. Thus the throughput of the NT mode is not constant but varies with the quality of the channel. Because the NT mode includes this additional redundancy space must be found for this within the intermediate bit rate. This results in a reduced transcoding compared to the T mode which capitalises on the fact that the NT mode has its own built in flow control mechanism, known as the radio link protocol (RLP). GSM has adopted an Automatic Repeat reQuest (ARQ) system based upon a 20 ms RLP frame structure. The RLP frame has a fixed length of 240 bits consisting of a header (16 bits), an information field (200 bits), and a frame check sequence (FCS) field (24 bits) generated using a fire code which provides an error detecting capability of "at least" 14 erroneous bits. The information field is actually composed of 192 data bits and an additional 8 bits for modem control.

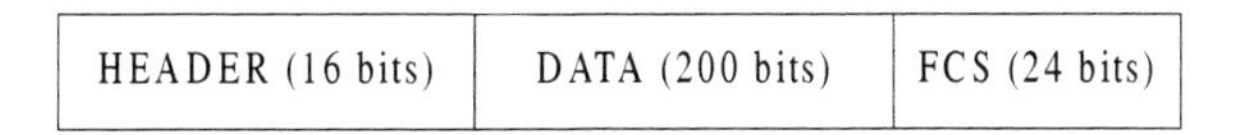

Figure 2.13 GSM NT data mode packet format

For reasons of efficiency the header carries supervisory information as well as identification for the information field the combined header and information fields forming what are termed Information + Supervisory (I + S) frames. The command exchange necessary to implement the RLP protocol of transmission is thus carried by the different fields of an I+S frame header. The flow control provided by the RLP is specially adapted to GSM transmission. Error free frames are acknowledged and a negative acknowledgement results in a retransmission. If there is a frame sequence that cannot be transmitted successfully after N2 repetitions, the RLP link will be reset or disconnected.

2.9.3 The GSM Short Message Service

There are two different kinds of short message service (SMS) specified in the GSM standards, these are point-to-point (PP/SMS) and cell broadcast (CB/SMS). From the GSM user's point of view, the PP/SMS delivers short text messages of length up to 160 characters from one GSM terminal to another. The network delivers the message from the sender to the selected recepient.

The short message service is provided by the subscriber's network operator. Without this subscription it is not possible to send or receive SMS messages. All the facilities provided by the operator are associated with the Subscriber Identity Module (SIM card), rather than with a particular terminal. The SMS is termed a connectionless service as there is no connection established between the sender and the actual receiver, as is the case for voice and data calls. The sender therefore cannot know when, or if, the message is received and the reception order of different messages is not guaranteed.

The delivery of a PP/SMS message from one mobile to another is the concatenation of two separate tasks: sending the message from a mobile to a special entity in the network called the short message service-service centre (SMS-SC), termed the Mobile Originating (MO) SMS, and then sending the message from the SMS-SC to the receiving mobile, termed the Mobile Terminating (MT) SMS. Once the network has accepted the message, it can be stored until delivered to the actual receiver. The maximum storage time is network operator dependent, but can be also affected with a special parameter of a short message. The SMS utilises six different protocol data unit (PDU) types as listed in Table

2.5. The SMS-COMMAND PDU contains a command to be executed relative to an earlier issued SMS-SUBMIT.

The SMS PDUs are sent via the SACCH, during a call, or via a SDCCH at other times. Sending messages from one mobile to another is the most basic SMS option. Messages can be sent to a SMS-SC and vice versa from a number of sources depending on the provisions of the network operator. For example short messages may be sent as e-mail, fax, voice or paging messages. The maximum message length of 160 characters with normal text messages (seven bit data) is achieved with a SMS specific packing.

Table 2.5 GSM short message service PDU types

PDU type	Direction	Function
SMS-SUBMIT	Mobile → SM-SC	Delivers a short message
SMS-DELIVER-REPORT	Mobile → SM-SC	Delivers a failure cause (if necessary)
SMS-COMMAND	Mobile → SM-SC	Delivers a command
SMS-DELIVER	SM-SC → Mobile	Delivers a short message
SMS-SUBMIT-REPORT	SM-SC → Mobile	Delivers a failure cause (if necessary)
SMS-STATUS-REPORT	SM-SC → Mobile	Delivers a status report

As the PP/SMS is a connectionless protocol, the timings are not exact and the delays are typically longer than in a connection oriented protocol. Sending a SMS message one way, Mobile → SM-SC or SM-SC → Mobile, produces a delay of between 3 to 5 seconds. As SMS messages are sent via control channels the throughput is affected by other signalling traffic.

The CB/SMS functionality permits a number of unacknowledged general messages to be broadcast to all receivers within a particular region. Cell Broadcast (CB) messages are broadcast to defined geographical areas, one or more cells. The CB service can be used for delivering weather, traffic, advertising or other local information.

2.10 ERROR PROTECTION

Error protection in GSM is based upon a combination of channel coding and interleaving. The coding schemes employed vary for traffic and signalling channels, and this is also the case for interleaving.

2.10.1 Channel Coding in GSM

The logical channels in GSM have different channel coding schemes and therefore different degrees of error protection. In general the coding used is a combination of block coding and convolutional coding, the convolutional code being the *outer* code. The inner code used in some channels is a block code consisting of information bits plus associated parity check digits (the signalling channels use a fire code). The combination of information and parity check digits are then further encoded using a form of half rate convolutional coding. The resulting bit stream is then interleaved over a given number of bursts (which is different for different logical channels). The received bit stream is de-interleaved and then decoded in a convolutional decoder. The resulting information + parity bits (which might contain errors) are then further decoded in a block decoder.

There are several stages of coding and decoding in the GSM traffic channels and these are shown in Figure 2.14.

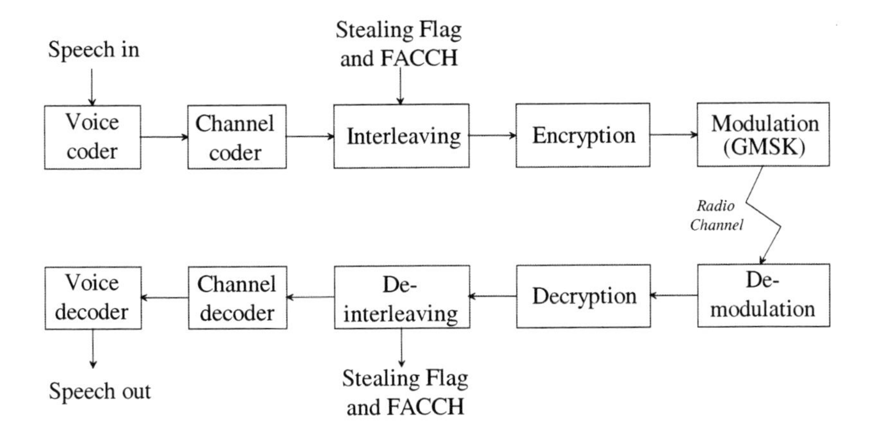

Figure 2.14 Coding and decoding in GSM

The speech coder used in GSM produces a net bit rate of 13 kb/s (260 bits per 20 ms) which is increased to a rate of 22.8 kb/s (456 bits per 20 ms) by the addition of channel coding bits. The bits produced by the voice coder are divided into class I and class II bits. As errors in the class I bits have a marked effect on the quality of the decoded voice waveform these bits are error protected and the class II bits are not. Of the 182 class I bits 50 (class Ia) are additionally protected using three redundancy bits resulting in a bit stream of 185 bits. To these bits a further 4 zero bits are added giving a total of 189 bits which are protected using a convolutional code with a half rate efficiency, producing 378 bits. These bits are then added to the remaining 78 class II bits to produce a coded block length of 456 bits. The actual coding employed for the full rate voice coder is shown in Table 2.5. It should be noted that $456 = 8 \times 57$, which is significant when considering interleaving.

The situation is slightly different in the NT data channels where the radio link protocol (RLP) has an influence. The complete NT packet of 240 bits is encoded using a punctured half rate convolutional code (1 bit out of 15 is punctured). This produces a gross block size of 456 bits which is then interleaved over 22 unequal bursts (actually called an interleaving depth of 19). At the receiver after de-interleaving and de-convolutional decoding the FCS can then detect (but not correct) up to 15 residual errors. Should there be up to 15 errors (as a result of the inability of the convolutional code + interleaving to correct the transmission errors) the RLP will request re-transmission of the 456 bit block. In this case the FCS is effectively the "inner code" so that the transmission is quite robust.

Table 2.6 Error coding in the GSM logical channels

Channel and transmission mode		Input rate (kb/s)	Input Block	Coding	Output block	Interleaving
Full rate voice	Ia	13.0	50	Parity (3 bits) + convolutional (1/2)	456	On 8 half-bursts
	Ib	13.0	132	Convolutional (1/2)		
	II	13.0	78	None		
Full rate data (9.6) Half rate data (4.8)		12.0 6.0	240	Convolutional (1/2) punctured 1 bit in 15	456	Complex, on 22 unequal burst portions
Full rate data (4.8)		6.0	120	32 null bits added + convolutional (1/2)	456	Complex, on 22 unequal burst portions
Full rate data (2.4)		3.6	72	Convolutional (1/6)	456	On 8 half-bursts
Half rate data (2.4)		3.6	144	Convolutional (1/3)	456	Complex, on 22 unequal burst portions
SCH			25	Parity (10 bits) + convolutional 1/2	78	On 1 sync burst
RACH (+ handover access)			8	Parity (6 bits) + convolutional 1/2	36	On 1 access burst
Fast associated signalling on half and full rate traffic channels			184	Fire code 224/184 + convolutional 1/2	456	On 8 half bursts
SDDCH, SACCH, BCCH, PAGH			184	Fire code 224/184 + convolutional 1/2	456	On 4 full bursts

2.10.2 Interleaving

The channel coding is most efficient when bit errors are uniformly distributed within the transmitted bit stream. However errors due to fading cause errors to occur in bursts. The problem is reduced by a technique known as bit interleaving which spreads out adjacent bits in a block over several normal bursts. On the traffic channels the sequence of 456 bits is split into eight groups of 57 bits, each one carried by a different burst and this is termed an interleaving depth of 8. The individual bits in a block are allocated to the 8 bursts as shown in Figure 2.15. Since each normal burst has 114 data bits each burst contains bits from two separate speech blocks. Clearly it is necessary to have two speech blocks available before a normal burst can be formed. This requires an interval of 40 ms and hence the coding delay for GSM is of the order of 40 ms. The de-interleaving process is simply the reverse of the interleaving process and results in any burst errors being uniformly distributed in the reconstituted speech blocks.

The normal bursts are also used for transmitting messages on the FACCH. The FACCH signalling also occurs in blocks of 456 bits and undergoes the same interleaving process. To distinguish between a normal TCH and a FACCH transmission the stealing flags are set to 1 when the channel contains FACCH transmissions. Because of the interleaving process it is clear that a normal burst can contain both 57 bits of a TCH and 57 bits of a FACCH. Hence there are two stealing flags in each normal burst, one for each half burst. The control channels use a different error protection scheme with an interleaving depth of 4.

The radio path is subject to multipath propagation which can produce a delay spread of several μs. This becomes apparent at the receiver as inter-symbol interference (ISI). The training sequence of each burst is used by the receiver to estimate the multipath delay spread being experienced by that burst. This information is used to set up the appropriate equaliser coefficients to minimise the ISI. The delay spread is a dynamic parameter which changes from burst to burst. Placing the training sequence in the middle of a burst reduces the time over which the delay spread can change relative to that during the transmission of the training sequence.

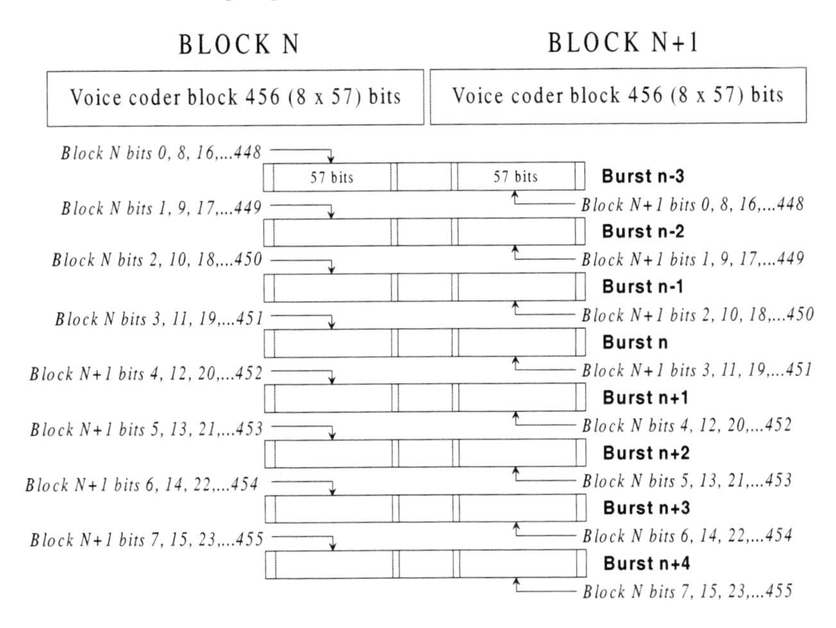

Figure 2.15 Interleaving in GSM

2.11 HANDOVER IN GSM

The possible types of handover defined for GSM are:

- Intra-Cell Handover - this occurs between traffic channels within the same cell.
- Inter-Cell Handover - this occurs between traffic channels on different cells.
- Inter-MSC Handover - this occurs between cells belonging to different MSCs.

Handover may be used in a number of different situations, these are:

- **To maintain link quality**. When the SIR falls below a given value the mobile will be required to handover to an adjacent cell which provides a stronger signal.
- **To minimise interference**. Even though a mobile has an acceptable SIR, situations can exist in which it may be causing unacceptable interference to a call in a co-channel cell. This interference may be avoided by initiating a handover to a different channel.
- **Traffic management**. In an urban environment where cell sizes are small a mobile can possibly be served adequately from a number of cells. In such circumstances the network can request a handover in order to evenly distribute traffic throughout the cells (thereby avoiding congestion within particular cells). In order to implement such a traffic management policy it is necessary for the network to have a detailed description of the area in which the mobile is operating. Measurements on signal levels, interference levels, distances, traffic loading etc. must therefore be collected and processed.

2.12 GSM HANDOVER MEASUREMENTS

The GSM system is able to assess the quality of both the up link and down link since these can be considerably different. The measurements performed in the GSM system are as follows:

- The received signal level (RxLev) and received signal quality (RxQual) on the uplink, measured by the serving base station.
- RxLev and RxQual on the down link, measured by the mobile and reported to the network every 0.5 seconds by means of the SACCH.
- The signal level of the BCCH of adjacent cells (Adjacent cells are identified by the mobile by reading the base station identification code and frequency of the carrier. The results for the six strongest cells are reported every 0.5 seconds via the SACCH).
- The distance of a particular mobile from its serving base station (this is determined from the adaptive frame alignment procedure, which is employed to cater for varying propagation delay within cells, and is directly available to the base station).
- The levels of interference on free traffic channels may be measured in the serving cell and possible target cells.
- Traffic loading on serving and adjacent cells (may be measured by operations and maintenance functions).

The data generated by the handover measurements must be processed before any handover is initiated. The processing involves the following stages:

- Averaging of measurements over several seconds to avoid the effects of fast fading.

- Comparison of serving cells with predetermined thresholds which trigger the handover requirement. Handover is only initiated if link quality cannot be improved by increasing transmitted power.
- If handover is required the best cell to handover to is determined from one of a number of algorithms (e.g. lowest path loss, strongest signal, acceptable signal level in a low traffic cell etc.).
- The resources are then allocated and the actual handover signalling is initiated.

The detailed algorithms for handover implementation have not been defined by GSM but have been left open for manufacturers and operators. There is however an optional recommendation which does contain the specification of a basic handover algorithm, the signalling sequence is shown in Figure 2.16. The air-interface handover signalling has been designed such that the break in traffic which occurs during handover is minimised. Under most conditions the break, essentially caused by the stealing of the traffic channel by the FACCH, is less than 100 ms which is only a barely perceptible break in speech.

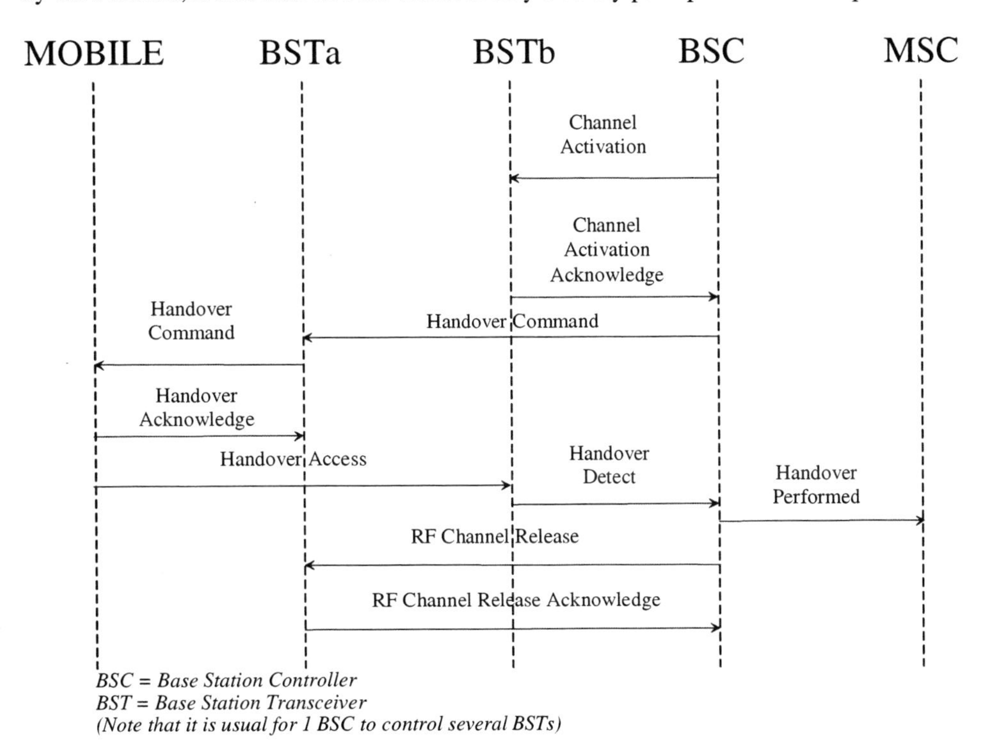

Figure 2.16 Handover sequence in GSM

To prevent the mobile station from exceeding the planned cell boundary whilst still using the same radio channel a strategy can be applied which leads to a handover whenever an adjacent cell is entered which allows communication with less power. This is possible with GSM since the mobile listens to other base stations and takes measurements during the periods when it is not receiving or transmitting on an assigned traffic channel. The current base station receives measurements from the mobile, via the SACCH, and decides when handover should be initiated. This is known as network initiated handover.

2.13 FEATURES OF THE GSM SYSTEM

The GSM specifications contain several features which are designed to accommodate the movement of mobile terminals and to minimise interference.

2.13.1 Adaptive Frame Alignment

Figure 2.4 illustrates that a mobile staggers its transmission by 3 time slots after a burst from the base station. Thus there is a nominal delay of 3 TDM slots between transmit and receive frames at the base station which must be added to the round-trip propagation delay between base station and mobile. The propagation delay varies with the distance between base station and it is possible for a burst from a mobile near the perimeter of a cell to overlap with a burst from a mobile close to the base station which is on an adjacent time slot. The timing advance required to ensure that bursts arrive at the base station at the beginning of their time slots is calculated at the base station and transmitted to mobiles on the corresponding SACCH. This avoids the need for a long guard interval, which would be an inefficient use of the radio resource.

The initial timing advance is obtained by monitoring the RACH from the mobile, which contains only access bursts with a long guard interval of 68.25 bit periods. This ensures that there will not be any overlap problems for a mobile separation from the base station of up to 37 km. The required timing advance is specified in terms of bit periods by a 6 bit number transmitted on the SACCH. This means that an advance between 0 and 63 bit periods can be requested. During normal operation, when the TCH has been established, the base station continually monitors the delay from the mobile. If this changes by more than 1 bit period the new advance will be signalled to the mobile on the SACCH. For cell radii greater than 35 km GSM specifies the use of every other time slot. This allows for cell radii up to 120 km but does reduce capacity.

2.13.2 Adaptive Power Control

It is indicated in Section 1.21 that if the transmit power is constant then the mean SIR is a function only of frequency re-use distance. However it is not necessarily desirable to work with constant power and the goal is rather to ensure that a minimum transmitted power is used on both the up and down links in order to maintain adequate speech quality. This also has the advantage of conserving battery power for hand held mobiles.

GSM specifies that mobiles must be able to control transmitted power on all bursts in response to commands from the base station. For a class 1 mobile (with a maximum power output of 20 W) there are 16 possible power levels separated by 2 dB (the minimum power level is 20 mW). For initial access on the RACH the mobile is constrained to use the maximum power specified for the cell (as broadcast on the BCCH). After initial access the mobile power level is determined by the base station and transmitted on each SACCH message block. The mobile will change by a 2 dB step every 60 ms until the desired value is reached. The mobile confirms its current power level by signalling this to the base station on the uplink SACCH. It is apparent that adaptive power control is an alternative to handover.

2.13.3 Slow Frequency Hopping

The radio environment is subject to frequency dependent fast fading. The effect of fast fading depends on the coherence bandwidth and how this is related to the overall signal bandwidth. If the transmitted bandwidth is less than the coherence bandwidth a deep fade will result in communication being lost completely. In such circumstances a mobile moving with reasonably high velocity will experience fades of short duration and the error correction procedures combined with interleaving will be sufficient to provide an acceptable service. However if the mobile is moving slowly (or is stationary) the fade duration becomes longer and can exceed the interval over which bit interleaving is effective. This will result in errors in the class I bits of the transmitted encoded voice signal and will give rise to bad frames and degraded speech quality.

To overcome the problem of long duration fades the sequence of bursts making up a traffic channel are cyclically assigned to different carrier frequencies defined by the base station. Timing signals are available at the base and mobile to keep transmitters and receivers in synchronism on the defined hopping sequence. The result is that the positions of nulls in the received waveform change physically from one burst to the next. Hence the bit interleaving can correct errors even when a mobile is stationary. This is known as frequency diversity. In fact GSM has a coherence bandwidth of the order of 53 kHz in a typical urban environment which should be compared with the bandwidth of a GSM carrier (> 400 kHz). Thus fast fading does not produce a complete loss of signal but will produce a loss over a band of frequencies equal to the coherence bandwidth.

Frequency hopping is an option for each individual cell and if implemented it is possible for the base station to instruct a mobile to switch from a non-hopping to a hopping mode should a drop in signal. The base channel which contains the CCCH and BCCH is not allowed to hop as this is the pilot channel for a particular cell In effect when frequency hopping is switched on the base station assigns a mobile a set of RF channels rather than a single RF channel. There are 63 different hopping sequences which may be assigned to a mobile and the mobile is informed of which hopping sequence and which carrier frequencies are to be used by the base station.

Another advantage of slow frequency hopping is that co-channel interference is more evenly spread between all the mobile stations and this is known as interference diversity. This essentially reduces the value of SIR required for acceptable quality from 11 dB to 9 dB.

2.13.4 Discontinuous Transmission and Reception (DTX)

During normal conversation a speaker is active for only about 44% of the time. The rest of the time the speaker is listening or pausing for breath. Measurements have shown that the percentage of time that both speakers talk at the same time is very low (typically 6% of the active period). This means that a traffic channel will only be used in one direction for approximately 50% of any conversation. Voice activity detectors (VAD) are employed to suppress TCH transmissions during silent periods which results in the following advantages:

- the level of co-channel interference is reduced, on average, by 3 dB;
- the battery life of the mobile can be significantly increased since it is not necessary to transmit a carrier during silent periods.

In practice it has been found that the silence periods are quite disturbing to the person at the other end of the link as the impression is given that the call has been disconnected. Hence a compromise is reached in which low level "comfort noise" is synthesised during periods of silence. This requires periodic transmission of the background noise parameters during silence periods.

Discontinuous reception may also be employed to conserve battery power when a mobile is in the stand-by mode. The paging channel on the downlink CCCH is organised in such a way that the mobile needs to listen only to a subset of all paging frames. Hence a mobile can be designed to make the receiver active only when needed.

2.13.5 GSM Evolution

GSM was designed essentially as a public digital cellular voice communication system and has been the most successful 2nd generation system world wide, by a large margin. There have been several enhancements to the GSM standard designed to accommodate the increasing use of mobile communication systems for multi-media traffic. In this respect the GSM specification is being evolved to accommodate a High Speed Circuit Switched Data (HSCSD) service (up to 64 kb/s) [1] and a General Packet Radio Service (GPRS) [2], both of which are aimed at non-voice services. The GSM specifications have also been evolved to accommodate the needs of European railway operators, resulting from the work of the Union International des Chemins de Fer (UIC) [3]. Further enhancements are underway to increase the available bit rate per user to more than 300 kb/s [4]. It is clearly outside the scope of this text to cover the full capabilities of GSM, however it is necessary to indicate that GSM will continue to evolve to meet the needs of the 3rd generation environment.

The next section considers a parallel development to GSM in public digital mobile communication systems based on cordless communication technology.

2.14 CORDLESS COMMUNICATION SYSTEMS

The GSM system is characterised by wide area coverage and the ability of the mobile unit to move freely over large distances. This clearly adds significant infrastructure and operating costs to cellular systems. A notable reduction in these costs can be achieved if the area of coverage is limited as low power terminals are then possible and mobility management functionality is drastically reduced. A fundamental feature of cordless communications is the Telepoint concept. This is essentially an extension of the fixed part (i.e. the base station) of the common domestic cordless telephone to handle a large number of mobile handsets using digital technology. The Telepoint service is a cordless payphone accessed by a portable terminal which is small enough to be carried at all times by the owner. The Telepoint base station is effectively an access point to the fixed network with the supporting administration and billing systems. The basic Telepoint concept is shown in Figure 2.17 and it is interesting to note that both up and down channels of a particular mobile use the same carrier frequency. This means that data is compressed and transmitted in what is termed time division duplex (TDD) mode.

2.15 THE DIGITAL ENHANCED CORDLESS TELECOMMUNICATIONS SYSTEM (DECT)

The main objective of the Digital Enhanced Cordless Telecommunications system standard (DECT), is to support a range of applications such as residential cordless telephone systems, business systems, public access networks (Telepoint) and radio local area networks. In addition DECT provides a system specification for both voice and non-voice applications and supports ISDN functions. DECT is a multi frequency, TDMA-TDD cordless telecommunication system which supports handover.

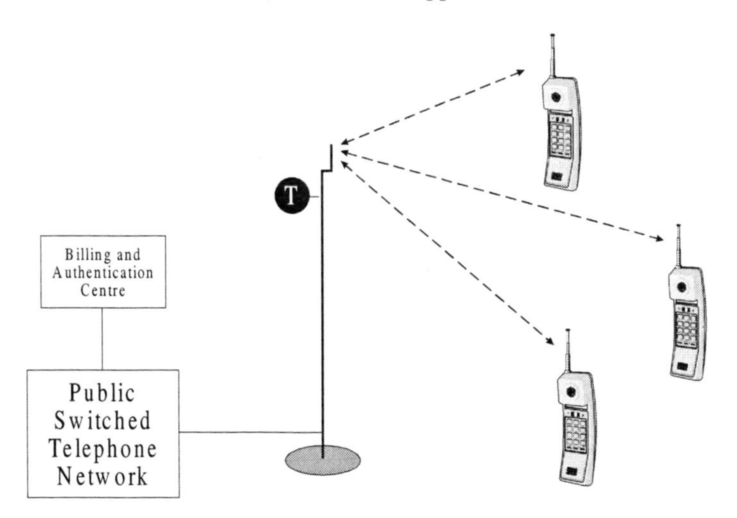

Figure 2.17 The Telepoint concept

DECT has a layered structure similar to that of the OSI model, which is described in Section 4.1 and, in common with TETRA, is divided into a control plane (for signalling data) and a user plane (for user data). The DECT structure uses four layers for communication between a DECT terminal and the DECT network, whereas the OSI model uses only three layers. The main reason for this discrepancy is that the OSI model does not adequately provide for multiple access to a particular transmission medium. The layered description of the DECT standard is shown in Figure 2.18.

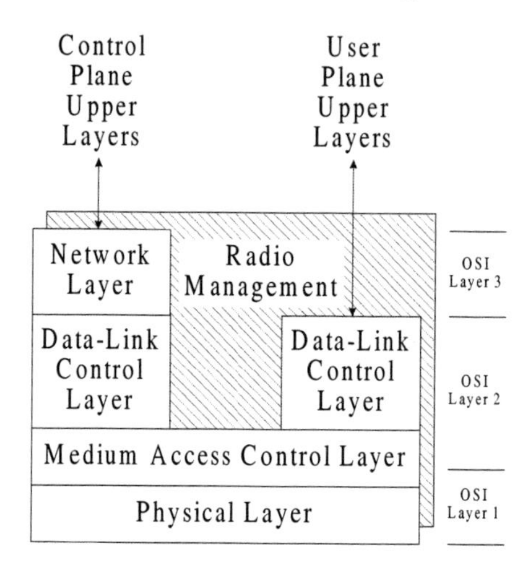

Figure 2.18 DECT layered structure

2.16 VOICE CODING IN DECT

The coder chosen for DECT is the 32 kb/s adaptive differential pulse code modulation (ADPCM) standard [5]. This clearly has a higher bandwidth than the 13 kb/s codec specified for GSM which uses a regular pulse excitation algorithm with a long-term predictor (RPE-LTP).

The two basic reasons for this choice are that the processing delay of the ADPCM coder is much less than the GSM equivalent and that such a coder is already widely available as a low cost integrated circuit. Furthermore, the speech quality is high and the coding is robust to radio-path variations. Processing delay is a major issue because DECT remains essentially a cordless extension to the fixed network and therefore must conform to line system standards which permit a maximum round trip delay of 5 ms in the speech path. The TDD transmission scheme introduces a delay of approximately 1 ms and the ADPCM codec can keep the speech processing element within the remaining 4 ms permitted. The issue of compatibility with line rather than cellular radio standards is important because of the fundamental difference between the Telepoint service and cellular radio.

2.17 THE DECT PHYSICAL LAYER

The physical layer deals with dividing the radio transmission into physical channels. Its functions are as follows:

- to modulate and demodulate carriers with a defined bit rate;
- to create physical channels with fixed throughput;
- to activate physical channels on request of the medium access control (MAC) layer;
- to recognise attempts to establish a physical channel;
- to acquire and maintain synchronisation between transmitters and receivers;
- to monitor the status of physical channels (field strength, quality etc.) for radio control.

In the DECT system ten carrier frequencies have been allocated in the band 1880 MHz to 1900 MHz the spacing between each carrier being 1.728 MHz. Each carrier is divided into 24 time slots occupying a period of 10 ms, the duration of each slot being approximately 416.7 μs. DECT employs time division duplex transmission with slots 0 to 11 being used for base station to handset and paired slots 12 to 23 being used for the reverse direction. Both slots in a pair can be used for transmission in one direction in response to unsymmetrical UP and DOWN traffic. Each slot transmits bursts of 420 bits in an interval of 364.6 μs which is 52.1μs shorter than the slot duration. This guard space allows for timing errors and propagation dispersion. The sequence of one burst every 10 ms constitutes one physical channel which represents a mean bit rate of 42 kb/s. The voice waveform is digitised at 32 kb/s using ADPCM and is compressed and buffered for transmission at 1152 kb/s. The structure of the DECT physical layer is shown in Figure 2.19.

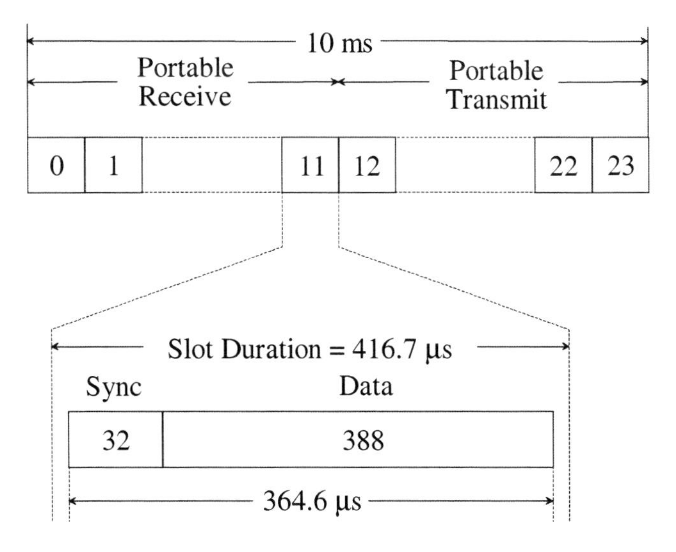

Figure 2.19 DECT physical layer

2.18 DECT MEDIUM ACCESS CONTROL

The medium access control (MAC) allocates radio resource by dynamically activating and deactivating the physical channels which must accommodate the signalling channel (C-channel), the user information channel (I-channel), the paging channel (P-channel), the handshake channel (N-channel) and the broadcast channel (Q-channel). In addition the MAC layer invokes whatever error protection is appropriate for the service (in the case of speech there is no error protection). It should be noted that DECT differs from the standard OSI model in assigning error protection to the MAC layer which is the most efficient way of treating individual radio links.

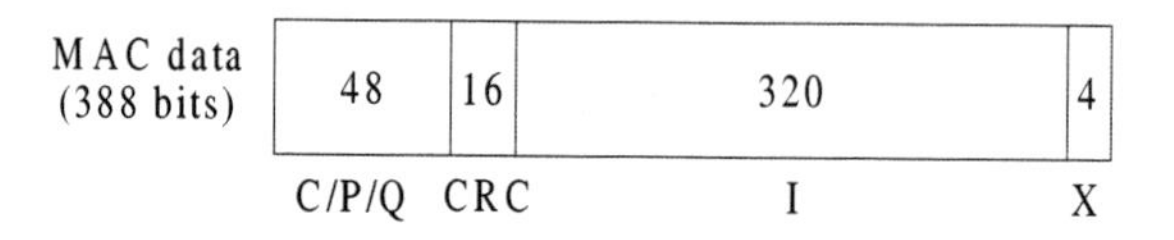

Figure 2.20 DECT MAC layer

The multiplexing scheme for logical channels used during a normal telephone conversation is shown in Figure 2.20. In order to lock on to a particular base station the portable must verify the identity of the base station and receive call set up parameters. This information is broadcast on the Q-channel. The paging channel is used by the base station to initiate network originated calls and is therefore broadcast to all mobiles in a cell. The signalling information on the C channel is for a specific mobile. The N-channel is used to exchange identities of the portable and base stations at regular intervals. The multiplexed C-, P-, N- and Q-channels are transmitted in 48 bits of each burst and capacity is allocated on demand whilst a minimum capacity for each channel is guaranteed. These 48 bits are protected by a 16 bit CRC and, if transmission errors result, an automatic repeat request (ARQ) procedure is used. The X bits in the packet are used to recognise partial interference in the I-channel independently of the user service.

2.19 CALL MANAGEMENT IN DECT

The DECT base station consists of one single radio transceiver that can change frequency from slot to slot. Hence each slot operates independently and can use any of the 10 DECT carriers. Thus unlike GSM, which uses a fixed frequency allocation, DECT uses a Dynamic Channel Selection (DCS) procedure.

A physical channel is a combination of any of the DECT time slots and any of the DECT carrier frequencies and every base station transmits on at least one channel known as the beacon (when several channels are active there are an equal number of beacons in the cell). All active channels broadcast system information and base station identification. When a portable (known as the cordless portable part CPP) is in the idle mode it scans for the beacon of a nearby base station (known as the radio fixed part RFP) and locks onto the strongest channel. In the idle state the portable listens at 160 ms intervals for a possible paging call from the system. If the signal level drops below a fixed threshold the mobile will scan for another beacon and will lock onto one of appropriate strength.

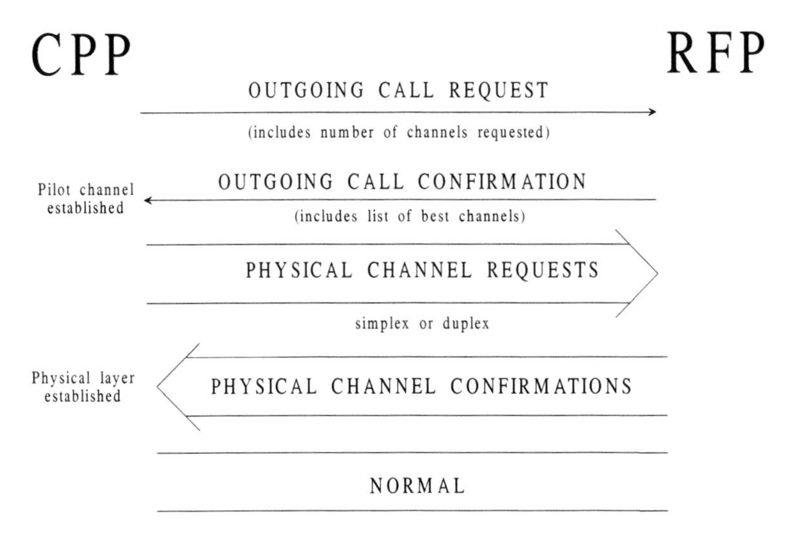

Figure 2.21 Mobile initiated call set-up sequence

In order for the portable to initiate a call set-up a number of exchanges are required between the CPP and the RFP and these are shown in Figure2.21.

The sequence of events is as follows:

- An OUTGOING CALL REQUEST is transmitted by the CPP on a single channel which has been selected on the criteria of minimum interference. This is effectively the dynamic channel selection (DCS) procedure in which the portable selects a free channel with the minimum interference level. The transmission includes a field identifying the number of physical channels that the CPP envisages the call will require.

- If the RFP receives this request and if the channel is free, half a TDM frame later (5 ms) the RFP transmits an OUTGOING CALL CONFIRMATION packet. A "pilot" link has now been established between the CPP and RFP which may occupy either a half or full rate physical channel. This link is sufficient for voice communications and is always duplex regardless of whether it is used for voice or data communications. Through this pilot link further physical channels can be activated so that the connection can support higher data rate services. If further physical channels are to be activated then the RFP will transmit a list of its available channels to the CPP.

- The CPP may generate further physical channel requests with a high probability of successful confirmation by combining the RFP's channel information with its own signal strength measurements into a map. With reference to this map the CPP will transmit PHYSICAL CHANNEL REQUEST packets on sufficient channels to satisfy the link capacity that is required. The requests may be on half or full rate physical channels and include information as to whether they will be used as simplex or duplex channels. When the channel is to be used as a *SIMPLEX* link then both up and downlink sections of the frame are used for transmission in one direction only e.g. CPP to RFP. The ability to activate simplex physical channels permits efficient spectrum allocation when assigning several physical channels for a high data rate asymmetric connection. Data calls are often asymmetric in their bearer capacity requirements and simplex physical channels permit the DECT radio interface to reflect this. When the channel is to be used as a *DUPLEX* link up and downlinks are used for CPP to RFP and RFP to CPP transmissions respectively, as with voice transmissions.

- The RFP will transmit PHYSICAL CHANNEL CONFIRMATIONs on all the channels that it has received a request on and are acceptable to the FRP.

The procedure for network originated calls is similar with the addition of a paging transmission to initially alert the CPP that a connection is required. This paging channel, which is a broadcast channel will be multiplexed onto the beacon to which portable is locked. A suitable communications channel has now been established between the CPP and RFP for the connection.

2.20 HANDOVER IN DECT

DECT is designed for relatively small cells and supports handover between cells. The emphasis in DECT is for the handover procedure to be rapid without any interruption of service. While the portable is communicating on a particular channel it scans the other channels and records the free channels and identities of base stations that are stronger than the one it is currently using. Handover is initiated as soon as another base station is stronger than the current one in use.

The current link is maintained in one time slot whilst the new link is set up, in parallel, in another time slot. When the new link is established the new base station requests the central control to make a switch from the old to new base station. As the old and new channels both exist in parallel, on different time slots, there is no break in service during the handover period. The DCS system is designed so that handover is completed before a significant loss of quality occurs.

2.21 DYNAMIC CHANNEL SELECTION (DCS)

The DCS strategy used in DECT is known as the Least Interfered Channel strategy. The DECT handsets scan the assigned channels at least every 30 seconds and list the 10 most preferred channels for call set-up and assignment. Calls may be setup on a channel when the SIR level is above 25 dB and may be re-assigned when the SIR drops below 20 dB. Channels are classified as free if the measured interference is less than –50 dBm. A

DECT system can use up to 120 channels but some of these channels may not be available if the number of transceivers per base station is limited.

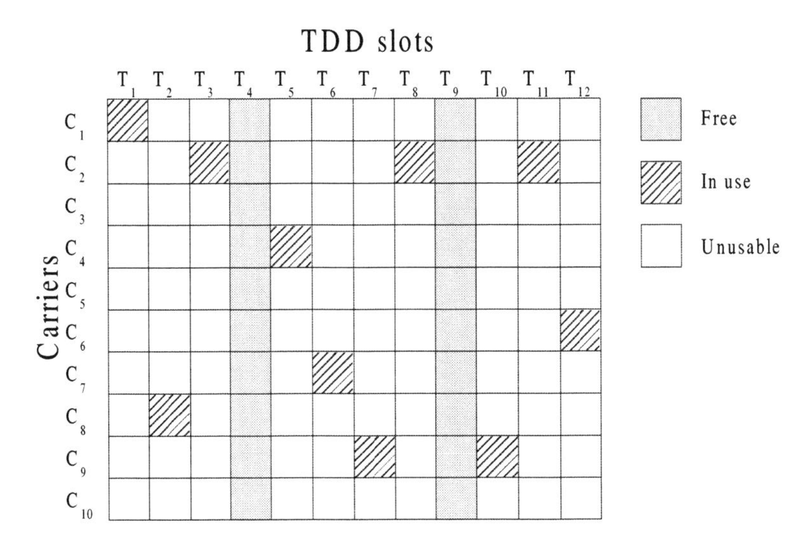

Figure 2.22 Time slot/frequency matrix for a DECT system

In single transceiver base stations a maximum of 12 calls may be in progress at any time. The time slot/frequency matrix for a DECT system is shown in Figure 2.22, but a limited number of transceivers per base station constrains the possible channel choice in the operation of DCS. The Least Interfered Channel (LIC) algorithm has three modes of operation, set-up, update and handover.

Set-up: Access is via the base station with the strongest signal. The interference is measured on the down-link and if greater than the free channel threshold this channel is discarded. This is repeated for all channels, the one with the highest SIR being stored. If the highest measured SIR is greater than the call set-up threshold, the call is set up on that channel otherwise it will be blocked.

Update: In this mode the preferred channel information within the portables is refreshed on a regular basis.

Handover: In this mode the least interfered channel algorithm tests the specified number of preferred channels which were stored at the last data update. Calls that cannot be assigned to a preferred channel are dropped. Once a call has been re-assigned the mobile's preferred channel information is updated.

2.22 SPREAD SPECTRUM SYSTEMS

In a code division multiple access (CDMA) system users are distinguished from each other by a code rather than by allotted time slot as in GSM. In a typical CDMA system a single physical channel (link) is 1.23 MHz wide and, typically, up to 12 subscribers share the same link simultaneously. An important property of CDMA is that neighbouring cells, or sectors in cells, can use the same physical channel. CDMA systems allow multiple use of the allocated radio spectrum by deliberately spreading the spectrum occupied by each user with a high-speed code word unique to that user. The spectrum spreading is achieved by multiplying the user data by the identifying code and modulating a carrier with the resultant waveform.

The technique is therefore called *spread spectrum radio.* At the receiver the original user data may be recovered by correlating the demodulated waveform by the original spreading code. All other signals remain fully spread and are not subject to demodulation. It could be concluded that deliberately spreading the spectrum of a signal is wasteful. However, spreading the signal bandwidth by some factor lowers the signal power spectral density by the same factor.

2.23 CODE DIVISION MULTIPLE ACCESS

CDMA is an example of a *power-limited system.* Such systems are not constrained by bandwidth and the more the information is spread with bandwidth the less is the effect of noise. There are several kinds of spread spectrum systems: direct sequence, frequency hopping, and chirping types. In a spread spectrum system, the bandwidth of the transmitted signal is much wider than that required for the information to be carried, and the transmitted signal is modulated a second time with a waveform which is not related to the transmitted information. This second modulating waveform determines the final bandwidth of the transmitted signal. The frequency hopping used in GSM is a form of spectrum spreading but GSM is not regarded as a spread spectrum system as the hopping rate is much less than the data rate.

The spectrum spreading technique used in CDMA is of the direct sequence variety. The spreading is accomplished by multiplying the narrowband information by a much wider spreading signal, which is usually a pseudo random noise sequence (PN). If each user has its own specific PN sequence then all received signals except the desired one appear as noise to a receiver in such a system.

An important characteristic of a spread spectrum system is its *processing gain,* G_p, which is proportional to the ratio of the spreading code rate to the data rate.

$$G_p = R_{chip}/R_{data} \qquad (2.1)$$

The PN spreading code is often called a *chipping code,* because it "chops up" or "chips" the much slower information bits. In effect each data bit is represented as a sequence of

N_c pulses, known as chips, within the bit period of the data signal T_{data}. Each chip has a duration $T_{chip} = T_{data}/N_c$ where N_c is known as the chip length, thus

$$G_p = T_{data}/T_{chip} \tag{2.2}$$

The performance of a spread spectrum system is most dramatic for G_p values greater than 1,000. In such circumstances, the energy per bit is spread out over $G_p = N_c$ chips. In the receiver, the energy in each of the G_p chips is accumulated over the much longer interval T_{data}. This means that the wanted received signal is de-spread back to the original bandwidth whereas the unwanted signals are not. The filter which follows the de-spreading action will thus pass only a very small amount of the signal energy present in the signals which have not been de-spread. Hence the processing gain and the conclusion that compensation for a low value of energy per bit/noise spectral density (E_b/N_o) is possible with a high bandwidth.

2.23.1 Power Control in CDMA

There is a need for strict power control of individual mobile transmitters in a spread spectrum system in order to ensure that one or more non de-spread mobiles do not contribute a substantial amount of power at the receiver. This is in contrast to bandwidth-limited systems where individual transmitters use a power level required in order to ensure good communications. Precise, accurate, and fast power control requires that the power in a CDMA channel be adjusted around *800 times per second.* Power control is the greatest challenge of a CDMA system and essentially replaces the frequency planning, which is an essential element of FDD and TDMA systems.

2.23.2 Diversity in CDMA

Slow frequency hopping is an example of frequency diversity used in GSM to overcome the problems of fading (for slow moving mobiles). CDMA systems make use of frequency diversity, spatial diversity, and time diversity.

2.23.3 Frequency Diversity

Since each signal in a CDMA system covers a relatively wide part of the spectrum, frequency diversity is an inherent feature of CDMA. Multipath fading is caused by different delays among the alternative paths between a mobile and a base station. A CDMA signal may be regarded as a noise signal with a bandwidth of 1.23 MHz as shown in Figure 2.23. Since fading is a frequency selective phenomenon a fade will be equivalent to filtering the noise spectrum with a notch filter. The width of the notch is effectively the coherence bandwidth and is approximately equal to the inverse of the delay between the paths arriving at the receiver. For example, if the delay spread is 5 μs, then the notch will be about 200 kHz wide. If the receiver moves, the notch will move across the spectrum occupied by the signal. As the delay between the paths becomes shorter in time, the cancelling of the signals from the alternative paths becomes more complete and the notch becomes wider. If the delay spread is less than 1 μs, the width of the notch becomes comparable with the CDMA signal bandwidth and the signal suffers a deep fade. Delays much longer than 1 μs cause only a power reduction of the received CDMA signal.

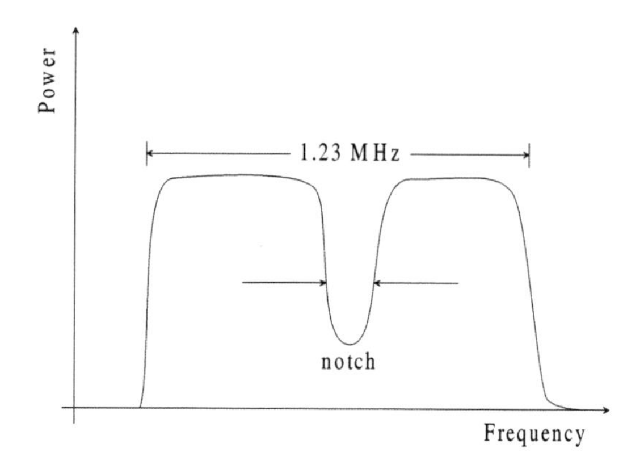

Figure 2.23 Effect of multipath on the CDMA spectrum

2.23.4 Spatial Diversity

Spatial diversity is possible in both TDMA and CDMA and corresponds to the use of multiple antennas at the base station. If two antennas at a receiver site are far enough apart, then the delays between the paths to these antennas from the mobile will be different, and it will be unlikely that a fade at one antenna will also be experienced at the other antenna at the same instant. An interesting feature of CDMA is that the system allows multiple base stations to transmit to a single mobile, and this allows soft handovers to be employed, i.e. handovers in which communication occurs with the target base station before communication with the serving base station is terminated. In GSM the channel from the serving base station is released before the channel in the target base station is activated.

The ability to allow multiple base stations to transmit to a single mobile in CDMA, with its inherent advantages against multipath fading, is made possible by a device known as a rake receiver.

2.23.5 Time Diversity

In GSM time diversity takes the form of channel coding (convolutional coding) and interleaving. Time diversity is also used in CDMA. For example on the IS-95 standard deployed in the USA (also known as cdmaOneTM), the forward channel (base to the mobile) employs half-rate convolutional coding, which doubles the number of bits representing the original information bits. The reverse channel (mobile to the base) employs one-third-rate convolutional coding. This means that the number of bits representing the original information is multiplied by three. All the bits, both information and redundancy bits are interleaved and both of these measures have the effect of separating adjacent bits and filling the space between them with redundant ones. At the receive end of the channel a Viterbi decoder, with soft decision points is used. Viterbi decoding is a fast way to decode a convolutional code. The decoder makes 1 and 0 bit decisions as it works its way through the received code toward the decoded output data. The demodulator ahead of the decoder can pass additional information to the decoder concerning the quality of each recovered symbol from the demodulator. The demodulator marks noisy or questionable symbols as they are passed to the channel decoder. This allows the decoder to ignore bad symbols rather than try to work through them in an effort to recover the original data and this is the basis of soft decision decoding.

2.24 THE FORWARD (BASE TO MOBILE) LINK IN cdmaOne™

CDMA is a mechanism for multiplexing many users by using different spreading codes. As with GSM there are a number of logical channels which must be accommodated for reliable system operation. In the cdmaOne™ system, for example, there are 64 logical channels available on each carrier. There is one pilot channel, one synchronisation channel and 62 other channels. All of the remaining 62 channels can be used as traffic channels but up to seven can be used for paging. On the forward channel use is made of the fact that each of the 64 logical channels are synchronised.

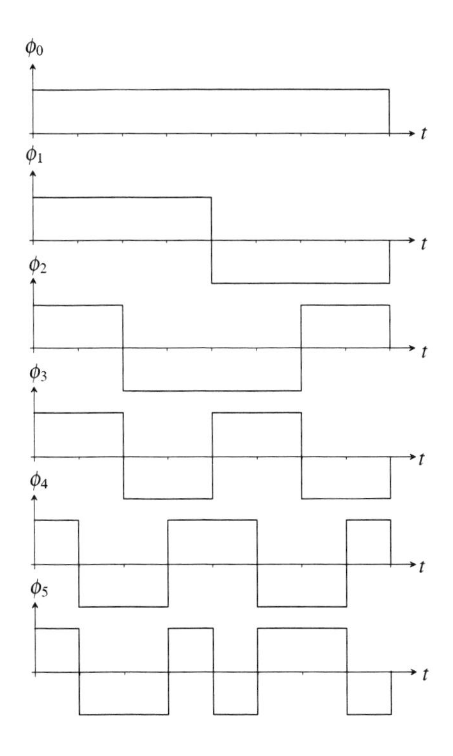

Figure 2.24 The first six Walsh functions

The basic user channel data rate is 9.6 kb/s and this is spread to a channel chip rate of 1.2288 Mchips/s using a combination of techniques. The data stream in each channel (except the pilot channel) is encoded using a 1/2 rate convolutional code interleaved and spread by one of 64 Walsh codes of length 64 bits, as shown in Figure 2.24, (this stage of spreading is known as *Walsh Cover*). Each channel in a particular cell uses a different Walsh code but to reduce interference from mobiles using the same Walsh code in a different cell all channels in a particular cell are scrambled using a pseudo random binary sequence of length 2^{15} chips. This sequence also provides the desired wideband spectral characteristics as not all of the Walsh functions yield a wideband spectrum (this stage of spreading is known as *Quad Cover*).

Orthogonality amongst users (channels) within a cell is preserved by the individual Walsh functions. The pilot channel is provided on the forward link for channel estimation and synchronisation which allows coherent detection to be used. The pilot channel is transmitted at a higher power level than the other channels and has a payload data which is a constant string of zeros. The pilot channel in each cell has its own identifying spreading code, called *W0*, for the mobiles to refer to.

2.24.1 Synchronisation Channel

The Synchronisation CHannel (SCH) is usually present, but may be omitted in very small cells. In the latter case, a mobile gets its required synchronisation information from a neighbouring cell. When the SCH is present, it is always spread with Walsh function number *W32*. The data in the SCH contains a large amount of information which all the mobiles in a CDMA system need, and which includes such things as the system identification number (SID), some access procedures, and some precise time-of-day information. The information in the SCH is created at rate of 1.2 kb/s and is half-rate channel-encoder before it is repeated only once to yield a 4.8 kb/s data rate entering a bit interleaver. The output of the interleaver is modulo-2-added with Walsh function *W32* of the SCH.

2.24.2 Paging Channels

Paging channels (PCH) are optional, and a maximum of seven paging channels are provided on a downlink. Paging channels are covered with Walsh functions *W1* to *W7* at 1.2288 Mb/s. PCH data occurs at one of three rates (9.6, 4.8 and 2.4 kb/s), which is half-rate-encoded and is repeated a number of times inversely proportional to the original data rate. The output of the interleaver is a constant 19.2 kb/s, which is modulo-2-added with a scrambling code. After Walsh covering, the 26.66-ms frames are quad covered in an identical fashion to all the other channels on the downlink.

2.24.3 Quad Cover

All 64 logical channels are finally modulated at the base station with a common PN code, which is $2^{15}-1$ bits long. This code is called a short code and has a rate of 1.2288 Mb/s. All base stations have the same PN spreading code, but each base station sets its own offset in the short code. There are 512 possible offsets for the short PN spreading code termed 0 to 511. A mobile can easily distinguish transmissions from two base stations by their distinguishing short-code offsets, and then the individual logical channels are distinguished from each other with their Walsh functions. The pilot, synchronisation, and paging channels area simplified versions of the forward traffic channels.

2.25 THE cdmaOne^{TM} REVERSE LINK

The reverse link is the sum of transmissions from all the mobiles, within a particular cell, at a frequency 45 MHz below the forward channel frequency. The spreading strategy used on the reverse link is different to that on the forward link because each received signal arrives at the base station at a different time due to the differing propagation delays associated with the individual mobiles. On the reverse link each channel is coded with a 1/3 rate convolutional code. The encoded data is then interleaved and split up into blocks of 6 symbols ($2^6 = 64$). Each block of symbols is then mapped onto one of the 64 Walsh functions producing a stream of 307.2 kchips/s [$64 \times (9.6k \times 3)/6$]. A final 4 fold spreading to 1.2288 Mchips/s is achieved with a user specific code of period $2^{42}-1$ (*known as a long code*) and a base station specific code of 2^{15} chips. The second process is gives the mobile a cell-specific code to augment the mobile-specific long code that preceded it and is equivalent to the quad cover (or *quadrature spreading*) used on the downlink and actually uses the same spreading code as the downlink.

The rate 3 coding and the mapping onto Walsh functions results in a greater tolerance to interference than would be realised from traditional spreading (by means of a repetitive code). This added robustness is important in the reverse link due to the use of non coherent detection and the presence of in-cell interference (there is no orthogonality amongst in-cell interferers).

2.25.1 Access Channel

There are only two types of channels on the reverse link, traffic channels and access channels. The access channels are almost identical to the traffic channels. Mobiles use the access channel to initiate communications in the system and to respond to PCH messages. The data access channel occurs at a fixed rate of 4,800 b/s in 20-ms frames and contains information required by the network to properly log the mobile terminal into service. The 20-ms frames are 1/3 rate channel-encoded and repeated once to yield 28.8 kb/s channel-encoded data which is then mapped onto Walsh functions and quad covered as in the traffic channels.

2.25.2 Payload Data

Voice transmission in cdmaOne^{TM} is based on a *variable-rate speech encoder* block which is sensitive to the amount of speech activity present on its input, and its output will appear at one of four rates: 9.6, 4.8, 2.4 or 1.2 kb/s. These rates change in proportion to how active the speech input may be at any time. The rate is subject to change every 20 ms. The output of the speech encoder is convolutional coded at a half rate, thus doubling the data rate to 19.2 kb/s when the input is 9.6 kb/s. When the speech encoder operates in one of its three reduced rates, the error coder repeats the 20-ms frames as often as required in order to maintain a full 19.2 kb/s output to the interleaver. Walsh covering and quad covering are then applied as previously described.

2.25.3 Signalling

Signalling can occur on the forward traffic channel. One example is the power control sub channel, which is always present in the forward traffic channel and effectively steals bits in the information stream. Other types of signalling must periodically occur in order to maintain the integrity of the channel. This type of signalling appears in either blank and burst mode or one of three dim and burst modes. In the blank and burst mode all of the speech data from the speech encoder are replaced by important signalling data at rates inversely proportional to the accompanying speech data rates.

2.26 THE RAKE RECEIVER

In a typical multipath environment the different rays from the base station arrive at the mobile antenna at slightly different times, and will have some arbitrary phase relationship with each other. This gives rise to the fast (Rayleigh) fading phenomenon described in Section 1.12. CDMA can take advantage of frequency diversity gain because of its wide bandwidth, however, considerable improvement in performance can be achieved by adding the power from several of the reflected rays in a *rake receiver* which is particular to CDMA systems.

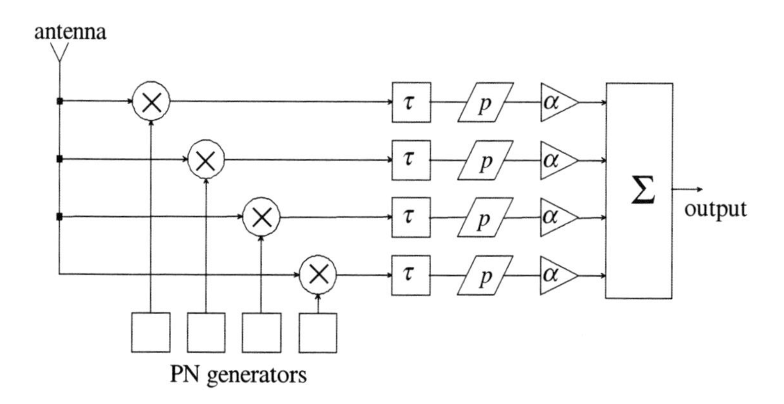

Figure 2.25 The RAKE receiver

Conceptually, a spread spectrum signal is usually recovered in two steps. In the first step the quad spreading sequence of the selected signal is removed by exclusive Oring with the appropriate Quad code. This de-spreads the spectrum of the required signal whilst the spectrum of the other signals remain spread. In the second step the de-spread signal is demodulated. Several rays appear at the receiver antenna, and each ray will have a slightly different phase from the others. The relative phase relationships will change in a random fashion with time. A selected channel from each ray is separately de-spread with the appropriate PN sequence (the same PN sequence that was used to spread the original channel at the base station transmitter). Since each ray arrives at the receiver antenna at a different time, each de-spreading PN generator in the rake receiver has its phase dynamically adjusted to correlate with different rays containing the same channel information. The recovered and de-spread data streams appear at the outputs of the mixers in Figure 2.25.

The de-spread signals no longer contain the PN chipping waveform and are directed to a linear optimal combiner with n branches ($n = 4$ in a CDMA mobile).

Each branch has three stages:

- The τ stages adaptively cancel delay spread among each of the de-spread rays and is called τ dither tracking.
- The p stages adaptively adjust the phase of each branch.
- The α stages adaptively equalise the output level of each branch.

Each of the four branches is called a *finger*, and three of the fingers are optimally combined. The fourth finger is called a *roving finger*, which is used to seek out the next ray to be assigned as one of the three combining fingers.

In addition to time diversity, the rake receiver is used to add spatial diversity and accomplish the soft handover function. As the mobile moves toward the edge of a cell equivalent of the MSC detects that the mobile is near the limit of its RF power range and assigns the PN sequence used by the mobile to a neighbouring cell which is assigned the same carrier frequency. Two base stations thus transmit traffic to the mobile with the same identifying spreading sequence on the same physical channel. The rake receiver in the mobile treats one of the rays coming from the newly assigned base station as another ray in need of optimal combining. As the mobile moves further into the coverage area of the new base station, all its rake fingers finally correlate with rays from the new site, and the old site drops the PN sequence and corresponding traffic from its transmitter. Thus there is no break in transmission during handover.

2.27 POWER CONTROL LOOPS IN cdmaOne™

For the CDMA system to operate correctly, RF power in the system must be controlled in two aspects. Firstly, all transmissions from mobiles must be received at the base station receiver within 1 dB of each other, even under fast multipath fading conditions. Secondly, in order to allow as many users as possible to share the use of a cell, only the minimum RF power required for reliable communications is allowed from any base station transmitter. To accomplish these goals:

- The base station creates a separate RF power control sub channel to adjust the RF output power of each mobile 800 times a second in small steps.

- The RF power of mobiles is switched off when transmissions are not absolutely necessary. This would be the case under reduced data rate conditions.

There are three major RF power control mechanisms in the CDMA system which are described below.

2.27.1 Open Loop Power Control

The mobile begins by estimating the loss in the path between itself and the base station, and then it makes an initial course adjustment in its RF power output based on this estimate. To do this, it de-spreads and then measures the RF power on one of the forward channels from the base station. The mobile combines this reading with some control information the base station sends to the mobile during some initial signalling transactions. This information is a set of operating parameters such as the actual output power of the base station, the current traffic load serviced by the base station, and some other details the mobile needs to make an initial RF output power adjustment.

2.27.2 Closed Loop Power Control

Path losses vary with frequency in the 800 MHz band. The initial path loss estimated by each mobile is based on the forward channel of the base station. Since the mobile responds to the base station on the reverse channel, which is 45 MHz lower in frequency, it is unlikely that the initial open loop path loss estimate will be accurate for the reverse channel. The base station therefore measures the level at which mobile transmissions are received and returns a 0 (increase power) or a 1 (decrease power) in the RF power control sub-channel at a rate of 800 per second. The mobile responds to the bits it receives on the forward power control sub channel by making small adjustments between 0.5 and 2.0 dB in its RF output power. The size of the adjustment depends on the data rate of the mobile at the time the adjustment is needed, as well as the time since an adjustment was last made. It is possible to make closed loop adjustments within 24 dB above or below the initial power setting of a mobile.

2.27.3 Base Station Transmission Power Control

In order to keep the total RF power on the forward channel low, which will allow a maximum number of mobiles to share the use of a cell and not disturb neighbouring cells, the power level of the base station, on a per-channel basis, is restricted to the lowest possible level that each individual mobile absolutely needs. The output of the base station, is continuously lowered in small steps until the responding mobile signals the base station for more power. This routine is performed individually for each of the CDMA channels on a single forward link.

2.28 MOBILE ACCESS IN cdmaOne™

For each forward PCH in the downlink, there will be one or more reverse access channels from the mobiles in a cell. Mobiles gain access to the CDMA resources by first listening for a PCH in order to decode a paging message from the network. When the mobile decodes a *page*, or if the mobile user initiates a call to the network, it is necessary for the mobile to start an access procedure. This is done by choosing an access channel and then building the proper logical channel structure for a reverse access channel. The access channel is modulated using OQPSK with a quadrature PN cover, which the mobile adopts from the pilot channel spreading sequence of the base station. The access channel is also orthogonally modulated, but the long code that selects the 64-bit Walsh codes one after the other is calculated from a mask specific only to access channels. The payload data occurs at a fixed rate of 4.8 kb/s in 20 ms frames, and is encoded in a 1/3 rate convolutional coder. The frames are always repeated once, yielding a 28.8 kb/s encoded data rate.

Once an acknowledgement is received on the PCH, the base station assigns a traffic channel (with a corresponding Walsh code), and the mobile responds by changing the mask, with which it calculates its long code, from a public access channel mask to a private mobile-specific channel mask.

2.29 CONCLUSION

This chapter has outlined the main digital public mobile radio systems and illustrated the techniques which are employed in order to provide reliable communication in typical environments in which public systems operate. The most important characterising parameters of public mobiles systems is the need to provide continuous coverage for high density user populations. This means that many of the design goals have been determined by the requirement of using allocated spectrum as efficiently as possible which means re-using available frequencies over quite short distances. The consequence of this is that public mobile systems are designed and dimensioned to work in interference limited conditions with high mobility management overheads. The next chapter considers the private mobile environment and the effects that this has on the development of private mobile radio (PMR) standards.

REFERENCES

[1] Dunlop, J. 'Potential for compressed video transmission over the GSM HSCSD service' *Electronics Letters* Vol 33, No 2, 1997, pp121–122.

[2] Brasche, G. and Walke, B. 'Concepts, Services and Protocols of the new GSM Phase 2+ General Packet Radio Service' *IEEE Communications Magazine*, Vol 35, No 8, 1997, pp 94–104

[3] Webb, W. T. and Shenton, R. D. 'Pan-European railway communications; where PMR and cellular meet' *Electronics and Communications Engineering Journal* Vol 6, No 4, 1994, pp 195–202

[4] Olofsson, H. and Furuskar, A. 'Aspects of Introducing EDGE in Existing GSM Networks' *Proceedings of 7th IEEE International Conference on Universal Personal Communications*, Florence, October 1998.

[5] Dunlop, J. and Smith, D. G. *Telecommunications Engineering* (3rd Ed), Stanley Thornes, 1998.

3

The Private Mobile Radio Environment

3.1 INTRODUCTION

Professional (or private) mobile radio (PMR) systems are systems set up by a company or group of users to provide mobile radio services for that group of users alone. In this way they differ from the public cellular mobile systems described in the previous chapter.

The simplest forms of PMR are 'walkie-talkie" systems where mobiles communicate with each other directly and there is no need for base stations or a controlling network. Mobiles are self contained, but while such systems are simple to set up and cheap to run, they are not very flexible as mobiles need to be within range of each other and calls to other networks or users are not possible. However, for communications on a single site – a building under construction, for example – such a system may be all that is required. A new service, called PMR446 [5] has recently been deployed. Eight 12.5kHz channels have been harmonised across Europe, in the Private Mobile Radio band at 446 MHz, supporting up to about 1 or 2 million users in the UK depending on their concentration [2]. The channels are unlicensed, which makes them unsuitable for users requiring guaranteed access, and PMR446 equipment is not permitted to be connected to the public telephone network, but it otherwise provide a cheap solution for on-site and short range use. PMR 446 replaces short range business radio (SRBR) services which were introduced in the UK [4] and France. The USA has a similar system called family radio service (FRS).

At the other end of the PMR spectrum, systems rival or even exceed the complexity of public cellular systems. Users may group together to run joint systems, or have such systems run for them, in so called public access mobile radio (PAMR). PAMR systems are shared by a number of different users.

PAMR systems, or PMR systems with a common standard and interworking arrangement, have the advantage of allowing users on different PMR systems to communicate with each other directly. For different branches of the emergency services, or for services in neighbouring countries, this is important. While such interworking can be arranged by routing calls from the calling PMR system through a fixed network to the called system, this adds delays and means direct mode operation, where mobiles communicate directly without using base stations, is not possible. These interworking requirements, as well as the advantages of scale shown by the GSM cellular radio standard, have caused a move towards open standards in the PMR field.

An important question facing potential PMR users is which option will provide them with the most efficient service: providing the service themselves with PMR, using a PAMR provider, or using a standard cellular service provider. Traditionally, users were driven to PMR systems for reasons such as to ensure adequate coverage, to reduce cost, or to provide supplementary services. With the development of good public cellular radio networks, the situation is changing. In particular, coverage in cellular systems is improving to the point where it can equal, and, for many users, exceed, the dedicated coverage provided by the PMR user. The cost of public cellular operation is reducing as well. While PMR is cheaper for large numbers of users or for operations in limited areas, the balance is swinging against PMR. In the past, PMR handsets operated over a limited range of frequencies, perhaps with manual frequency selection, and with very simple call control. This made them far less complex than cellular handsets. The cost advantage is being lost as PMR handsets become more complex with the move to digital and the drive to increase flexibility, capacity and security. The relative size of the PMR marketplace means cellular users have the advantages of economies of scale. However, in the third area, that of services, public cellular has yet to become a serious contender for the PMR provision outside fairly simple speech and data transmission requirements with relatively lax call set-up requirements and without requirements for very high reliability.

This chapter will examine the PMR user community and its requirements. The following section will look at some of the requirements for PMR services and systems, before going on to compare PMR and PAMR with cellular systems and to examine the PMR marketplace.

3.2 THE PMR USER COMMUNITY

There are a wide variety of users of PMR systems. The PMR user community can be divided into a number of different sectors:

- public safety: emergency services (police, fire, ambulance, mountain rescue, etc)
- non-safety national government: other governmental agencies, such as non-emergency health, customs, etc.
- non-safety local government
- transport: railways, buses, taxis, etc.
- other utilites: water, electricity, gas, coal
- on-site PMR: general purpose businesses operating in local areas or within their own premises
- other PMR: operating over larger areas
- PAMR.

The UK provides a good example of a mature market for PMR, having the second largest number of PMR users of any country in Europe and a medium PMR penetration of 1.5% (see Figure 3.1). The UK market is divided as shown in Figure 3.2.

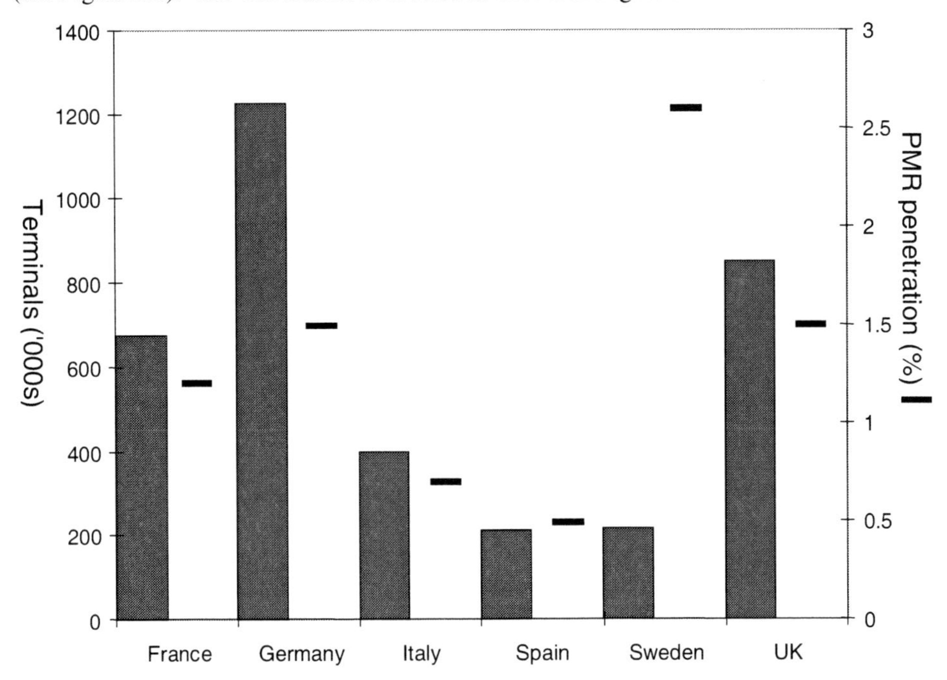

Figure 3.1 PMR terminals and PMR market penetration in the EU countries having the largest PMR markets

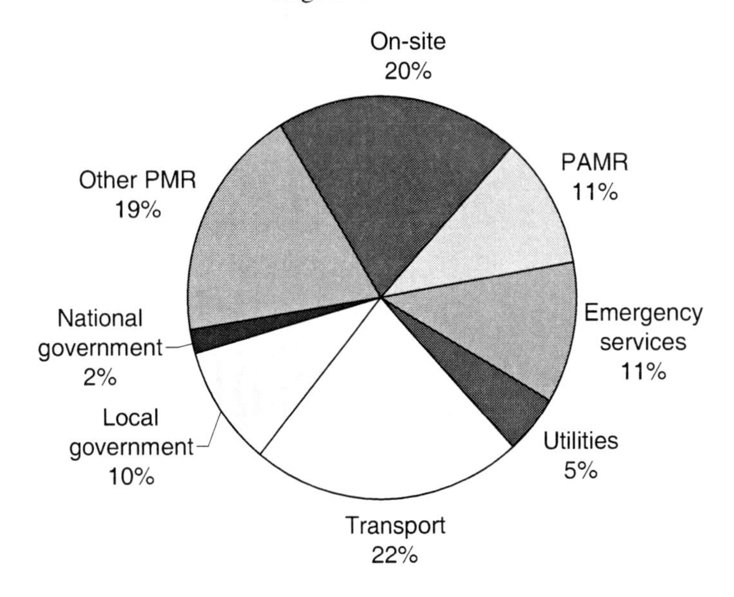

Figure 3.2 Breakdown of PMR users in the UK

The UK market is divided for licensing purposes between wide area private business radio and on-site private business radio, with about 20% of users falling into the latter category [2]. This is for general purpose businesses operating within a single site of up to about 3 km with one or two frequencies. Such systems will often just operate between mobile handsets without using a base station.

For the wide area PBR category, only a little over 10% (75,000) of users terminals have access to nationwide channels for their exclusive use. Most users operate with shared channels transmitting from a single fixed base station with a cell radius of up to 30km, although the emergency services, utilities and some transport users have larger regional networks.

3.2.1 Emergency Services

Emergency services form a large group of PMR users. Public safety operations account for about 15% of the EU PMR market (see Figure 3.3), and represent an even higher proportion of the market in other countries with less mature PMR markets.

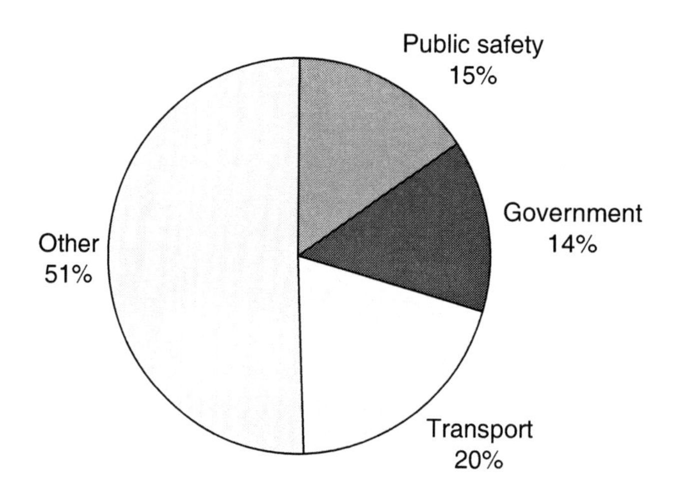

Figure 3.3 Breakdown of major PMR users in the EU

Many European countries, in particular those in the Schengen Group,[1] but also some outside the EU in Eastern Europe, are planning to update existing analogue technology with digital systems in the near future. Older systems are not able to meet current demands, in particular for security and reliability. Emergency services have severe demands, including good coverage and very resilient operation. This includes the ability to cope with large surges of traffic should an emergency occur. Public safety users also

[1] The Schengen Group is includes signatories to the Schengen Agreement on co-operation relating to cross border controls, and includes all EU countries except the UK and Ireland, which are observers.

require a wide range of call services and are prepared to pay a premium for them. In Europe, in particular, there is a wish to harmonise equipment between neighbouring authorities in order to simplify inter-agency operations.

3.2.2 Utilities

Utilities have to maintain large supply networks, and so require good coverage over large areas. They also require high reliability, and are likely to provide their own PMR requirements. In general, utilities are making more innovative use of PMR, pioneering the use of data services in areas such as work scheduling. Data communications costs are quite small in comparison with other maintenance costs, with the result that utilities are insensitive to pricing pressures. In this market segment, features are more important than price.

3.2.3 National Government

National governments normally require operation over large areas or country-wide. Depending on their communications needs, they may join in a national PAMR network or with an emergency network if call priorities can be suitably enforced to allow sharing of that network. These users usually face fewer regulatory difficulties and are relatively price insensitive. However, the size of national governments is such that communication needs may be devolved to separate departments and different government agencies may run different, perhaps incompatible, networks for different purposes.

3.2.4 Local Government

Local government usually face greater pricing pressures than national governments, but their communication requirements are more localised, reducing demands for a complex network. Co-operation between agencies at this level is more likely so that a common system may be used by many departments.

3.2.5 On-site

Some 20–25% of PMR users only wish to communicate within a single site and only require simple infrastructure. In some cases, such as a construction site, the system may only be required on a relatively temporary basis. In fact, unless dispatcher operation or PSTN access is required, direct communications between terminals is sufficient and the costs of infrastructure can be avoided. Such systems are usually voice only, and make an ideal application for the new PMR 446.

3.2.6 PAMR Operators

PAMR operators provide PMR services to many users with many different requirements. As a result, they are looking for a feature-rich system. In addition, it is in their interest to use a spectrally efficient system, as capacity constraints will limit the opportunity for selling on the service. Good coverage over a wide area is also required to market the system, although the operator may have the ability to disregard some areas if it would not be commercially viable to provide services there.

3.2.7 Transport

Transportation systems offer an ideal application for PMR. By their very nature, transport systems involve a large number of dispersed operations which have to be co-ordinated. However, the characteristics of the system requirements are highly dependant on the type of transportation. The market can be divided into three distinct areas: railways, buses and coaches, and taxis and dispatch/courier operators (Figure 3.4).

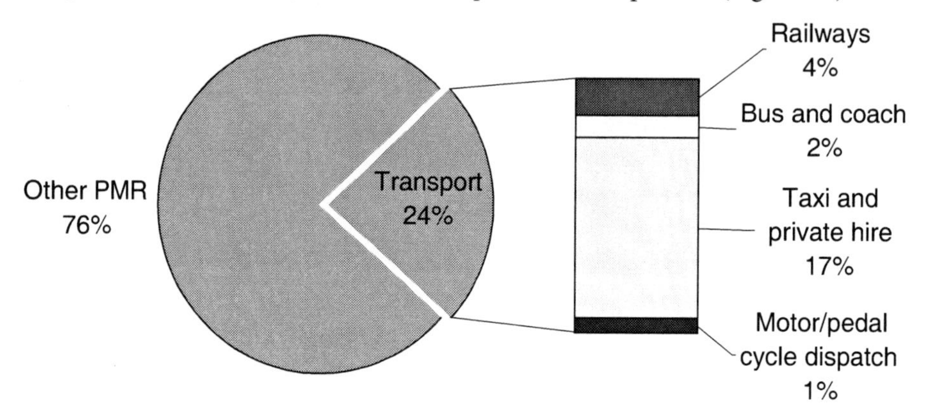

Figure 3.4 Breakdown of UK PMR usage in the transport sector

3.2.7.1 Railways

Railways are large operations, which can be compared to utilities, although from a PMR viewpoint they require some additional features. The railway network is fixed, the capital costs are high, and requirements change relatively infrequently. This means that infrastructure can be well planned and located on the companies own property, and the market is comparatively insensitive to price. Coverage is required over a wide area, and traffic may be regional or country-wide, requiring a sophisticated network infrastructure. UIC[2] service requirements [7] include a large number of supplementary services and data services.

3.2.7.2 Bus and coach operators

While bus and coach operations can be country-wide, or even international, they tend to have smaller geographical networks than railways, and the network is not fixed, which makes dedicated provision of coverage by a PMR scheme costly and inefficient. Unless the company operates in a relatively limited area, provision by PAMR or by a public cellular service may be more cost effective. This market is less sophisticated than

[2] Union Internationale des Chemins de Fer, the international union of railway companies.

railways in terms of requirements supplementary and data services, and the smaller, less capital-intensive businesses are more price sensitive.

3.2.7.3 Taxis, private hire, couriers

This forms the opposite end of the scale to railways in terms of the transport market. Businesses are usually locally based, and voice is the predominant service, although short data messages have large potential for this type of service industry. A cheap single site analogue system can be all that is required. Such users, the majority of which are small businesses, may be slow to switch to more expensive digital systems. However, they represent a very large part of the PMR market. In the UK in 1995, 170,000 licences were for the on-site market segment, representing 20% of the overall PMR market of 850,000 users [2].

3.2.7.4 Other off-site PMR

There are many PMR users who do not fall into the above categories. These include larger companies operating over several sites for which on-site operation is not sufficient, but which only require district-wide or regional coverage.

3.3 REQUIREMENTS OF PMR SERVICES

The requirements of a private mobile radio system can be summed up very simply as giving the ability for users to communicate with each other reliably. More specifically, it is possible to identify a number of key requirements of PMR users. In no particular order, these are:

- **Reliability**. Many PMR services are used in safety critical systems. One advantage to the user of being involved in the operation of the service is that they are in the position to ensure reliability and are not dependent on other operators. The lack of public cellular systems to guarantee quality of service or grade of service in all circumstances, or their unwillingness to take liability for safety critical services, may force the use of a PMR system. A survey that looked at the importance of PMR features [6] found that service availability was classed as "extremely important" by two-thirds of those questioned, the highest proportion of any requirement.

- **Speech and data transmission capability**. Mobile data services are increasingly being used for tracking, telemetry or information updating services. Examples of innovative data service use include BT, a national telecommunication operator, which sends daily work orders direct to repair technicians so that the working day can start at the first job rather than with a trip to the depot. Simoco is conducting trials with Langdale Ambleside Mountain Rescue Team in the UK on transmitting medical telemetry, including still images, video, text messages and GPS data to assist in rescue operations. As data services develop, so will the applications which make use of them. A flexible data service provision is therefore essential. In the survey [6], almost 80% of users classed data communications as important, with over half of these saying that it was "very important". 10% of respondents classed data calls as "not important", but all respondents classed speech calls as important to some degree.

- **Centralised and decentralised operation**. In many businesses, PMR is used to organise users, and a central dispatch point is therefore required. However, it may also be important that users are able to contact each other in the absence of a central control point or even any infrastructure at all. Again the survey [6] found almost 80% of users classed direct mode operation as important, with over half of these saying that it was "very important".

- **Point-to-point, group calls and broadcast calls**. If PMR systems are used, a flexible group call structure is essential so that users can share information directly rather than having to relay it via others. Therefore, group calls, calls involving a number of defined users, and broadcast calls, where the call includes all terminals, are required in addition to point-to-point (single terminal to single terminal) calls.

- **Fast call set-up**. Rather than dialling a number to set up a call, with the called party answering a phone, PMR systems usually have a pressel or "push-to-talk" to activate a call to the dispatcher or user group, with the receiving terminal annunciating the message without an answering procedure. Calls may therefore consist of a sentence or two, and users expect to be connected to the called terminal without delay. This is particularly important in the emergency services where the radio may be used to give urgent commands and the dropping of the first few words of the message due to delaying in setting up the call might have serious consequences.

- **Good coverage**. Professional mobile radio users usually have less choice as to where to make a call than a cellular user. The call location is often stipulated by the location of the work the user is undertaking. In the case of a utility this may mean having good coverage over a wide area, and for public safety users, constraints can be even more severe. For example, a mountain rescue service may require coverage in areas where public cellular systems would not be provided, but even in more benign radio environments, as well as overall area coverage, the absence of blackspots within a covered area is also very important.

- **Long battery life**. User maintenance costs money in terms of lost work time in PMR systems, and reliability of service is also important. This compares with public cellular systems where the users are responsible for battery charging.

- **Flexibility**. Flexibility takes many forms. Flexibility with regard to services has already been covered, but another input aspect of flexibility is the ability of the system to change with the developing needs of the operator. In particular, the system should be scalable so that growth can be handled, and sufficiently adaptable to allow new services, which were not anticipated when the system was installed, to be added later. Businesses will not want to invest large sums of money in a system which cannot be modified easily once installed.

- **Low total cost of ownership**. Companies using PMR systems will consider costs over the entire life of the equipment, including capital costs for the infrastructure and maintenance costs for the equipment in addition to the "headline" cost of the terminals themselves. Unsurprisingly, no respondents to the survey in [6] classified costs as unimportant, with 95% classifying them as "important" or "extremely important".

A number of other requirements may not be necessary in all cases but will be needed by a large number of users. Any PMR system will therefore have to take them into account.

- **Security**. Many PMR users have a requirement for high levels of security. Security takes a number of different forms, both in terms of reliability of operation and protection of the transmitted information from tampering and interception.

- **Call priorities**. PMR operators may wish to be able to differentiate between users to give different call priorities or qualities of service to different user or call types. For example, an emergency call may be able to pre-empt other call types to gain access to the network.

- **Communication between networks**. Many companies operate over large areas or with several sites. They may not want to provide the complete network themselves, or they may use different networks on different sites due to equipment replacement cycles or regulatory restrictions. Their PMR networks may therefore have to communicate with each other. Also, in many circumstances, communication with general telephone or data networks is a useful feature.

- **Ease of licensing**. This issue involves not just the bureaucratic process of obtaining permission to use a radio channel, but also the issues of the availability of channels and any co-ordination which may be required with other users in the same area. The problem of licensing hundreds of different users operating in numerous different areas is much more complex than that of organising a small number of national cellular operators. It is only possible if the PMR radio channels are as self-contained as possible from the point of view of interference with other users.

- **In-house control of system**. In [6], almost two thirds of PMR users preferred to control their own network, with less than one sixth thinking that control was not an issue. An obvious reason for this is to ensure security, but other advantages relate to cost control and service guarantees.

One requirement not included above is that of capacity, or the efficient use of radio resource. In fact, capacity is not normally an issue to PMR users due to the length of the call, and since licences have been relatively cheap in most countries. PMR users have only been concerned at a more general level as channels become scare and have to be reused more frequently in high traffic areas such as cities. Of more general concern is the fact that the division of the available spectrum between different PMR users means there is little trunking efficiency. This results in there being fewer channels than required in most large urban areas. The regulatory authorities are therefore likely in insist that PMR systems are spectrally efficient and use a narrow carrier spacing.

3.4 PMR CONFIGURATIONS

The simplest PMR configuration is point-to-point direct terminal communication. This is used by PMR 446. The system has no infrastructure, and in most cases all terminals within range receive messages. It is possible, however, to conduct private conversations through the use of signalling tones or messages which mute terminals which the message is not intended for so that it is not heard. Either a single common frequency is used, or different frequencies can be used for different call groups. Communication is only possible between terminals when they are in range of each other, and given the power limitations on battery operated portable devices, this may be a significant restriction.

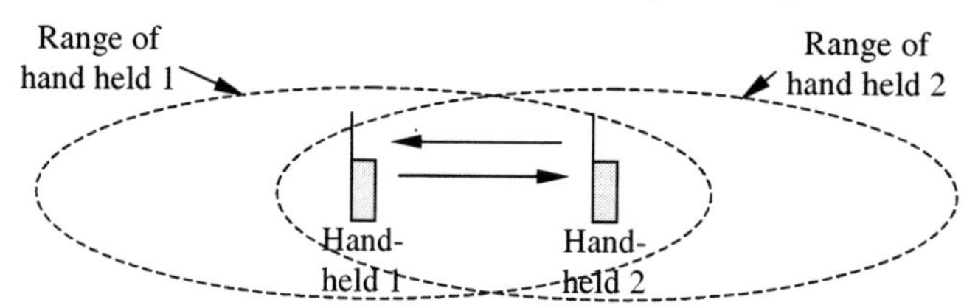

Figure 3.5 Simple direct mode PMR configuration

One of the most common PMR configurations is dispatch operation. At least two channels are used, one for uplink communications between terminals and the base station, and one for the downlink to the terminals. Messages from the dispatcher on the downlink can be received by all terminals (although again individual addressing is possible), whereas messages from the terminals can only be received by the dispatcher. Mobile to mobile communication is possible via the dispatcher. Links with the public switched telephone or data networks are possible, again via the dispatcher.

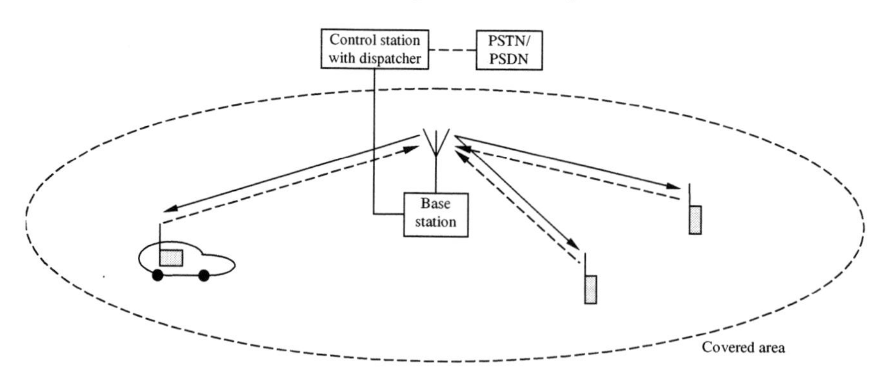

Figure 3.6 Dispatch mode PMR configuration

A number of refinements to this basic system are possible. If extended coverage is required, but central dispatch or PSTN network access is not necessary, the base station can be connected as a repeater. This is called "talkthrough" mode where any uplink messages are retransmitted on the downlink, effectively extending the range of mobiles to that of the base station. In Figure 3.7, the transmission from mobile 1 is received by mobiles 2 and 3, even though they would not have been in range if the message had been transmitted directly.

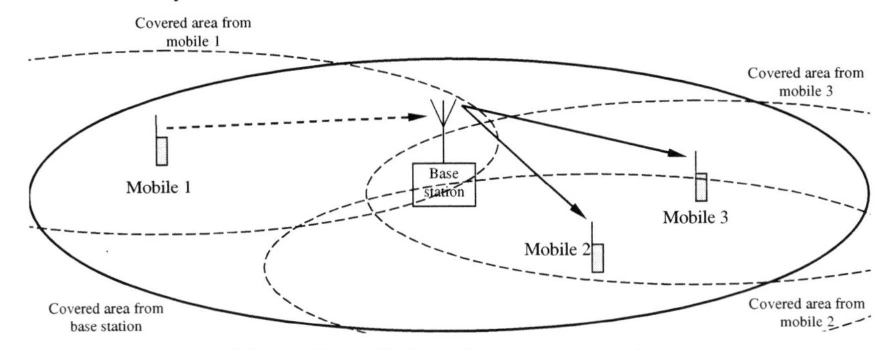

Figure 3.7 Talkthrough repeater operation

Different organisations can share repeaters (so-called "community base stations" or "community repeaters") if the different users have signalling to identify their messages. The signalling is retransmitted by the base station so that mobiles in other groups are muted and privacy maintained. Since users in groups do not hear all the messages it is necessary to keep usage low to ensure access. Such systems therefore include time outs to ensure that users to not hog a channel.

A better option, although one which requires more complexity, is trunked operation. In this case, several channels are available, pooled between different PMR operators. This allows trunking efficiency and makes it more likely that a free channel will be available.

In many cases, a single base station will not be able to cover the entire service area. If the uncovered area is limited to relatively small areas, such as in the shadow of a building, a remote radio port can be provided to illuminate this area.

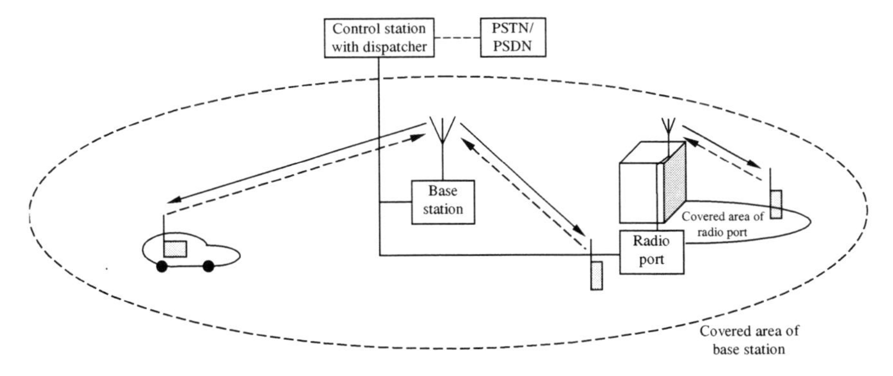

Figure 3.8 Using a radio port to fill a coverage blackspot

Since hand-held terminals usually have lower power than mobile terminals mounted in vehicles (due to battery and safety restrictions), mobiles can receive signals at greater ranges than hand-helds. Portable vehicle-mounted repeaters can therefore be used to provide hand-held coverage to users working near to their vehicles. This mode of operation is commonly used by the emergency services.

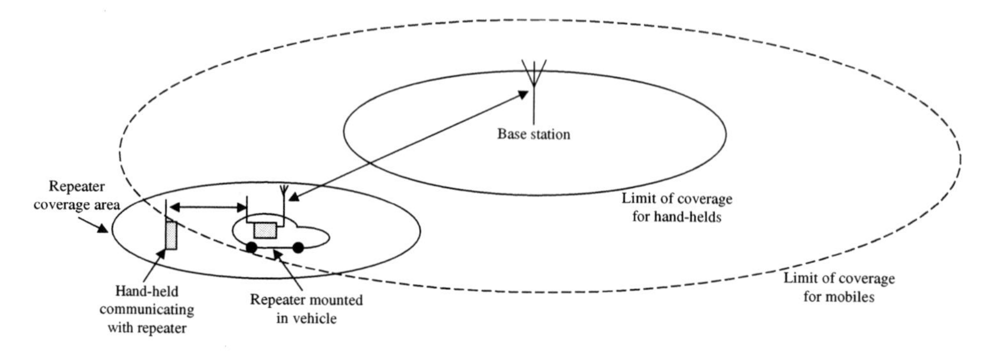

Figure 3.9 Vehicle-mounted repeater for local hand-held coverage

If larger areas have to be covered, several base stations must be used. If only a relatively low capacity is required, these can all transmit the same signal in a system known as simulcasting, and the system acts in the same way as one large cell. In analogue systems the frequencies used in the different cells vary by a few hertz which reduces problems in the overlap regions that receive signals from two or more cells. In digital systems, this is not possible, and systems have to be carefully designed to ensure that terminals can receive an adequate signal in the overlap region.

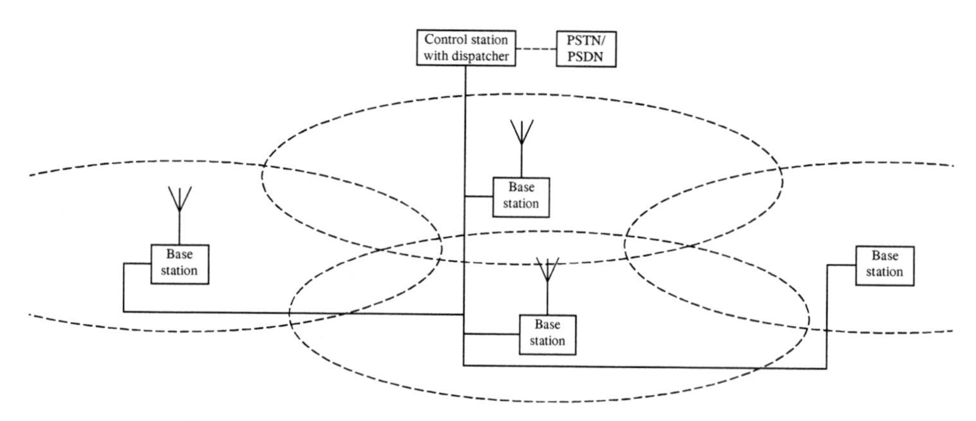

Figure 3.10 Wide area coverage using several base sites

Larger capacity systems require the use of cellular re-use schemes as described in Section 1.17. Such systems are considerably more complex than other configurations, requiring switching between the base stations and handover of mobiles between cells. However, large PMR operators, and PAMR operators, need to use cellular configurations to give them sufficient capacity. Even large PMR or PAMR systems do not have as much traffic as public cellular systems, and so will have a relatively flat architecture compared to the complex hierarchical network architecture of GSM, for example (see Section 2.3). The next section compares PMR and cellular operation more generally.

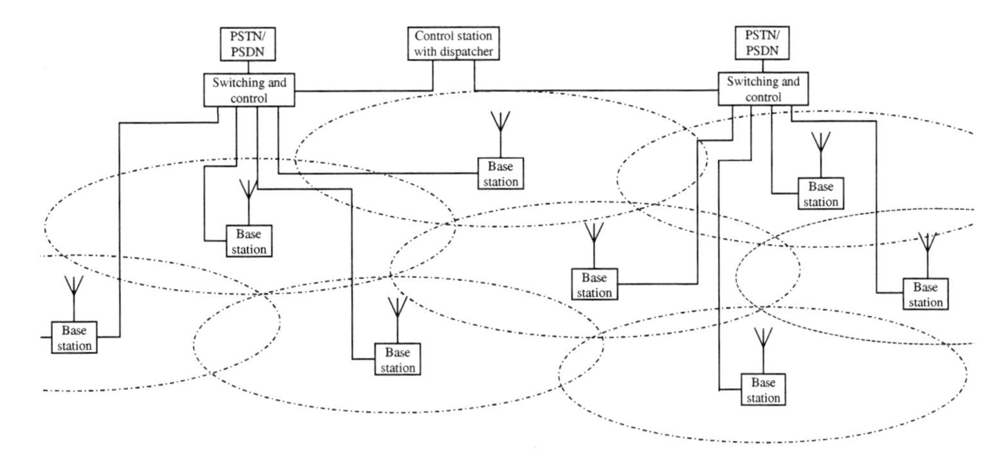

Figure 3.11 Cellular PMR configuration

3.5 COMPARISON BETWEEN PMR AND CELLULAR

PMR systems are in many ways similar to the public cellular systems described in Chapter 2. However, there are some significant differences between the two types of system which means that their design requirements are very different. The main differences between PMR requirements and the requirements of cellular systems are as follows.

Group calls. Cellular users have a much lower requirement for group calls than PMR users, and such requirements can usually be covered by having some sort of conferencing facility to link calls. PMR systems, on the other hand, must have flexible group call facilities, including allowing parties to enter and leave groups, and the ability to contact all users in a particular area.

Dispatcher operation. Many PMR systems have a centralised dispatcher controlling and monitoring the system. This facility is not required in a cellular system.

Decentralised operation. PMR systems are often required to work in a direct mode, where mobiles contact each other directly rather than via fixed base stations and network infrastructure. This allows operation outside the coverage of the fixed infrastructure and also in an emergency. Cellular systems must route all communications through the fixed infrastructure to allow for control and billing.

Fast call set-up. Cellular users dial a number, and wait for their call to be connected. This may take tens of seconds depending on the call's destination and call handling issues such as billing. In contrast, PMR users with a push-to-talk expect to do exactly that – press and talk – without delay.

Supplementary services. Supplementary services are additional call services over and above the basic communication service. Examples include call forwarding for a voice call. PMR users are more likely to want supplementary services tailored to their particular needs, such as variable priorities for different users, the ability to break into or monitor conversations, and so on. Cellular operators, on the other hand, have a much broader user community and will wish to offer a fixed range of simpler services which they can be confident will be commercially viable.

Traffic patterns. With a PMR system which operates without dialling (i.e. a push-to-talk to contact the dispatcher or other users), calls are very short, consisting of a sentence or two. Usage regulations request a limit on shared PMR channels of 15–20 seconds, and the system may include a time out limiting the length of activity periods to 30 seconds or one minute so that one user is not able to hog a channel [3]. In contrast, cellular calls will consist of a conversation, and so be longer. The average length of cellular calls is just under two minutes [1].

Another difference between PMR and cellular is the destination of calls. Most cellular calls originate or terminate outside the mobile network, with only a small proportion of mobile to mobile calls within the operator's network. On the other hand, PMR calls are usually intended for other users on that network, and the facility to route calls to other networks may even be absent.

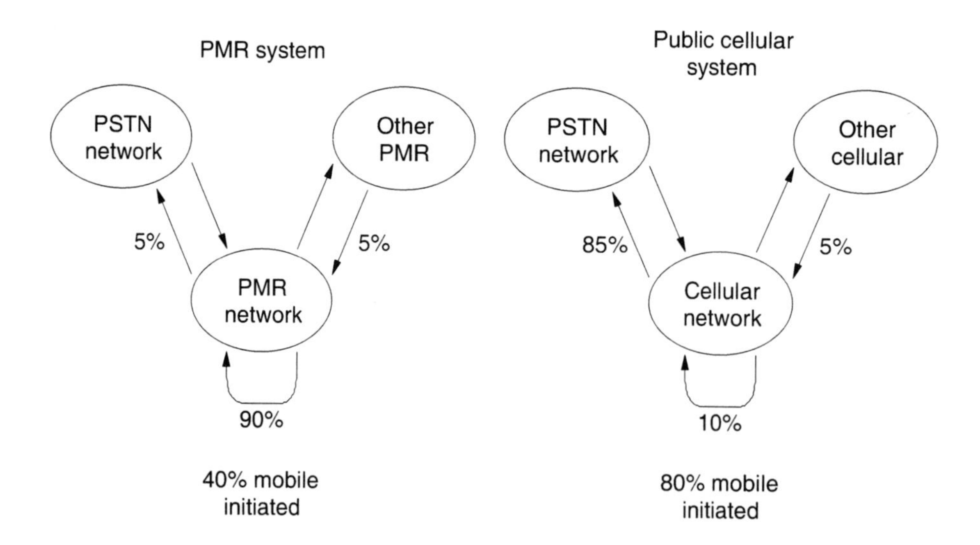

Figure 3.12 Sources and destinations of calls in PMR and cellular networks

Capacity. Cellular operators have a fixed allocation of frequency and they wish to maximise the number of users on the system to maximise their revenue. Their user base is large compared with their spectrum allocation and there is therefore an incentive to provide a large number of base stations, small cells, high re-use and efficient air-interface techniques to increase the number of simultaneous users supported. A PMR system is likely to have a much lower user base and the traffic is lower due to shorter calls, so capacity for the PMR operator is not likely to be an issue. PMR operators with their lower capacity requirements will want to minimise infrastructure costs, and so will have much larger cells, in the order of tens of kilometres. Cellular operators usually have cells limited to a few kilometres in radius at most.

Capacity does affect PMR operators in another way when it comes to obtaining licences. In many urban areas there are so many PMR operators that there are no spare frequencies and channels have to be shared. A PAMR system allows trunking of calls and results in a more efficient use of the spectrum.

Frequency planning. In a cellular system, frequencies are planned throughout the whole system. This is not the case in PMR, where frequencies will be allocated to users for specific areas and there may be no co-ordination between users in a particular area. This means that a PMR system must obey strict interference limits with regard to neighbouring carriers, whereas cellular systems can tolerate adjacent carrier interference because neighbouring cells can be planned with this in mind.

Control, billing and authentication. In a cellular system the user is authenticated and billed for each call. This is in contrast to a PMR system where permitted users may use the system at will. The PMR operator has to pay for the infrastructure but this is effectively a standing charge and there is no per call charge.

Relationship between the service provider and the user. In a PMR system the users are providing their own service, or will employ someone to provide the service on their behalf. The quality of the service is therefore directly within the control of the user. A cellular system provides a standard service, though perhaps with some amendments. The user therefore has much less control. A PAMR system falls somewhere between these two extremes.

Coverage. A cellular operator will provide coverage where it is economic to do so, which is where people who want to make mobile phone calls are. Cellular operators normally quote coverage in terms of the percentage of the population rather than the land mass, as complete coverage of the land mass of a country would be extremely expensive, and unless external factors such as government support are involved, may not be undertaken.

On the other hand, while cellular users may be able to operate their system at capacity in some areas, judging that the extra infrastructure costs would not be recovered by the additional traffic served. PMR operators may not have the option of dropping or queuing high priority calls, and will therefore have to provide additional capacity to meet worst-case, rather than average, traffic load.

PMR operators usually require coverage over predefined areas of operation. Cellular users must be able to use their phones over as wide an area as possible, including internationally. PMR users may not be interested in use outside their specific location, although in the case of police services or truck drivers this might still be a considerable area requiring roaming between mobile networks.

3.6 PMR STANDARDS

3.6.1 The Need for and Development of Standards

With the move towards digital PMR systems, there has been a trend away from proprietary systems toward a public standard with which equipment must conform, thus allowing equipment from different manufacturers to be used together. Moves towards public standards have come from manufacturers and operators, as in the case of TETRA, or the user community, as in the case of APCO25. The attitudes of governments to the standardisation process are quite varied. A hands off approach has been taken in the United States of America, where it has been decided not to insist on the APCO25 system but to allow the market to dictate which system is used. In contrast, in Europe the European Commission is far more proactive in setting standards and even defining them at a technological level through ETSI. This insistence that for certain government contracts systems conforming to ETSI proposals must be used is one of the reasons cited for the opening of the Matra PMR system resulting in the TETRAPOL standard.

Public standards have a number of advantages, as was shown by the success of GSM in the mobile radio sector. Public standards enlarge the market, allowing an economy of scale as well as opening the market for niche players in more specialised areas. All the various standards include defined interface points allowing users to source different parts of the system from different suppliers. As well as forcing more competition between providers, it means that suppliers are no longer required to produce all the components of the system, although turnkey solutions are still certain to be required by some users.

Public standards also allow users more freedom to move equipment between networks. This is less of an advantage than it would be in the case of public cellular systems, where some users want a high degree of mobility and roaming between networks. However, it can still be seen to be an advantage to many PMR users, especially those, such as the emergency services, who co-ordinate with each other.

3.6.2 Analogue PMR

Early PMR systems were analogue and proprietary. However, a wish to share infrastructure costs and a need to share spectrum led to the development of trunked radio systems, and with this development came the need for standards so that equipment could be sourced from different suppliers. The earliest such system, which is still available from a number of different suppliers, is LTR (logic trunked radio), developed by E F Johnson. A number of these systems are in operation worldwide.

Another major analogue PMR standard is MPT1327 (Ministry of Post and Telecommunications), which developed in the UK in the late 1980s, but has been adopted by manufacturers and implemented worldwide. It is the most widely used PMR standard, common everywhere except the United States of America, where proprietary systems by Motorola, and to a lesser extent Ericsson, dominate. Although MPT1327 is more complex and expensive than some simpler analogue systems, it is relatively efficient in terms of spectrum use (digital systems are still better by a factor of two), and offers some data capabilities as well as the normal PMR voice call features such as group calls, fast call set-up, and priorities.

PMR networks are expensive, and decisions to replace or upgrade are not taken lightly. Existing analogue systems are likely to remain until capacity, maintenance or required features force a replacement. At the time of writing in 1999, manufacturers of LTR and MPT1327 equipment were still promoting their analogue systems, in particular for use in countries or areas without spectrum capacity constraints.

3.6.3 Digital PMR

Digital systems[3] offer a large number of advantages over analogue systems. A major advantage is the ability to recover the signal completely as long as the noise level is below a particular threshold. This compares with the analogue case, where noise is always cumulative and degrades the quality of the signal. There is a disadvantage in that when the noise level approaches the threshold of a digital system, the system performance falls off very rapidly, whereas in an analogue system the quality falls off steadily, giving clear warning of the system's limits.

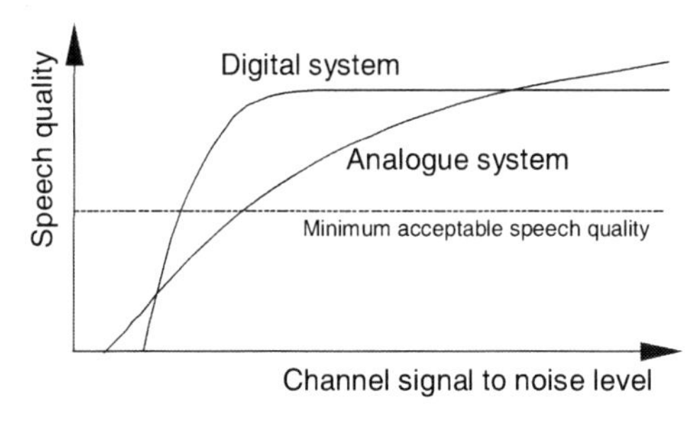

Figure 3.13 Comparison of analogue and digital speech quality with differing signal to noise levels

Additional advantages relate to the sending of data, which can be sent directly in a digital system without the requirement for a modem, and for trunking, as a digital signal can be manipulated more easily than an analogue one. While digital modulation is more complex than analogue systems, the transformation of a speech signal into digital form allows the use of very efficient compression techniques so that the spectrum required for a speech signal is lower with digital modulation and good speech coder than with an analogue modulation scheme.

Digital systems are more complex, and therefore more expensive. However, the increased flexibility, availability of services, and efficiency, in combination with the increased

[3] Digital systems are defined as those where the modulation scheme is digital. Some providers advertise analogue systems such as LTR and MPT1327 as digital since they have digital signalling and data services, but the basic systems are in fact analogue.

quality of service, means that the PMR market is now moving towards digital in the same way as the cellular market five years ago.

3.6.3.1 EDACS

EDACS (Enhanced Digital Access Communications System) is a proprietary digital PMR system from Ericsson. The first systems were installed in the late 1980s, and the system has found application in the military field, as well as its principle use in public safety. When the system was launched, a major selling point was its data services, which were unusual for a mobile radio system of that time, and the system has achieved considerable success, particularly in the USA.

3.6.3.2 Geotek-FHMA

Geotek-FHMA is a digital system which uses slow frequency hopping on an FDMA structure. The technology employed is novel for the civil mobile radio environment, being more common for secure military communications. As well as developing the system through its Israeli subsidiary, Geotek operated a limited number of digital networks itself in the USA, but the system suffered from limited take-up. A link up with IBM in 1997 failed to raise fortunes, and Geotek withdrew from digital network provision in 1998. The system itself, which has been installed in about half a dozen countries, is promoted as the PowerNet system for public safety applications along with Rafael, the Israeli defence firm which was a partner in its development. National Band 3, a UK PAMR operator owned by Geotek, was going to adopt Geotek-FHMA, but this network is now owned by Telesystem International Wireless of Canada, which through its Dolphin Telecom subsidiary is using TETRA for new digital PAMR operations in the UK and France.

3.6.3.3 APCO25

APCO25 was an initiative by the Association of Public-safety Communications Officials – International, Inc. (APCO) to try to create a standard PMR system for public safety applications. While the emphasis in Europe has been to create a cross-border standard for such systems, the United States has a more market-orientated culture and different countries may have incompatible systems from different suppliers. The idea of a federally required standard has since been watered down, but the system has continued to be developed with Motorola, which owns rights to much of the key technology, licensing this to other manufacturers. Motorola's Astra system was the first APCO25 system, launched in 1996.

Recently APCO25 and TETRA agreed to cooperate on future developments. In its current form, APCO25 is an FDMA system with 12.5 kHz carrier spacing, which is compatible with existing analogue channel spacing. However, future plans foresee halving the bandwidth requirements for speech channels to make more efficient use of spectrum. A narrow-band FDMA approach with 6.25 kHz carrier spacing has been proposed to allow this, but this is being reconsidered in the light of TETRA developments which may see a TDMA approach being employed for this development.

3.6.3.4 iDEN

iDEN is a proprietary digital radio system from Motorola. It has a high level modulation scheme (16-QAM) which allows very efficient use of the radio spectrum, allowing it to support 6 speech channels into a 25 kHz radio channel. There are many iDEN systems operating worldwide, but none are in operation in Europe so far.

iDEN is efficient, and provides a full suite of PMR features. The system is more appropriate for large users or PAMR systems, since it shares a TDMA system's disadvantage of having a minimum number of channels, six in this case compared to the four of TETRA. However, since the system is proprietary, even from an industry leader like Motorola, potential for frequency allocation and interoperability is limited. Without significant market penetration costs are likely to remain high.

3.6.3.5 GSM derivatives

There have been a number of attempts to develop GSM as a PMR system, as opposed to users using a GSM cellular service instead of PMR. An early attempt was GSM-R, which is a railway operator initiative to extend GSM to meet their requirements. These include a number of PMR features such as fast call set-up, priorities and group calls, as well as railway-specific features such as operate at very high speeds (500 kph). GSM-R was adopted by the railway operators in preference to TETRA since TETRA was not then available, but industry support for the relatively small railway market has been lukewarm and since its recommendation, some railways have decided to install TETRA or TETRAPOL systems.

A related development is GSM-ASCI (Advanced Speech Call Items) which adds some PMR features to GSM. Also, GSM phase 2, although aimed at the cellular market, adds packet switched and high rate data services to the range of available services. Economies of scale will mean GSM-based PMR equipment would be relatively cheap, but even with ASCI, GSM will still lack essential PMR features like sufficiently fast call set-up to allow push-to-talk operation, and direct mode. Users who do not require these features may well simply purchase dedicated capacity on a cellular network.

A very significant disadvantage for GSM in a PMR environment is its minimum spectrum requirement of 600 kHz for a single carrier system, since GSM assumes frequency planning and free adjacent carriers are available. Only very large users, willing in effect to construct their own cellular system, could make use of such large spectrum allocations efficiently.

3.6.3.6 TETRAPOL

TETRAPOL is a digital FDMA PMR system. It has its origins in the opening of a proprietary standard developed by the French company Matra. By combining digital modulation with proven FDMA technology, TETRAPOL offers a digital PMR solution at lower cost than trunked systems like TETRA, albeit with slightly lower efficiency and a more restricted range of services. TETRAPOL has had a significant advantage in that it has had a presence in the market for a couple of years before the roll out of TETRA systems, and has been adopted by users in 15 countries, mainly in the public safety area.

TETRAPOL faces a problem in terms of take-up in the EU, since although it is recognised by the ITU, it is not a formal standard approved by ETSI, and moves to convert the TETRAPOL PAS (Publicly Available Specification) to an ETSI standard have recently stopped. Most European Union countries are planning to use TETRA, although TETRAPOL has been recognised by Schengen Group along with TETRA, and it is in use by security forces in France, Spain and Austria, as well as other European countries such as Romania, Slovakia and the Czech Republic.

3.6.3.7 TETRA

The main focus of this book is the TETRA standard. TETRA was developed from the start as an open harmonised digital PMR standard within ETSI. As an ETSI approved system, it has a significant commercial advantage within Europe, both from the point of view of manufacturer and operator support, and from the point of view of governments, which in Europe will specify ETSI approved systems for their contracts. The wide user and producer base should provide significant economies of scale. However, the PMR market has a number of existing 2nd generation digital systems in operation already, and with 3rd generation cellular systems only a few years away it is important that TETRA gets off to a good start if it is to establish a dominant position in the marketplace.

TETRA is a feature-rich system, providing everything from specialised safety services to cellular operating modes. It also has a wide selection of data services. The trunked TDMA access technique allows more efficient use of the radio spectrum, but means that an operator must be assigned a minimum of at least four voice channels. However, the system has a number of operating modes, which allow wide area coverage with a single radio carrier without resorting to cellular frequency reuse schemes that would increase radio carrier demands still further. Of more potential concern is that the complexity of the system will make the infrastructure and terminals relatively expensive, which should be offset by the economies of scale if the system becomes popular.

3.7 PMR MARKET EVOLUTION

As was discussed in Section 3.3, the vast majority of existing PMR users require relatively simple speech services. With this background it is interesting to see how developments in feature-rich digital PMR will affect this marketplace.

For the traditional reasons for employing PMR – availability, features and cost – for the majority of users only the latter still applies. For users who require speech services alone, existing analogue PMR provides an acceptable service, the only difficulty being the provision of capacity in densely populated areas. Its principal advantage is one of low cost. The difficulty for digital PMR is that as PMR is, relative to cellular, a niche market, and that should capacity issues force users to make a switch it may be to digital cellular rather than digital PMR. The relative size of the PMR market compared with that of public cellular is in fact decreasing, as PMR is expected to grow slowly if at all in most European countries, while the cellular market achieves growth rates of, in some cases, more than 60% per annum.

New services such as data are likely to attract a large number of users, as they can make more efficient use of company resources. Examples include more efficient scheduling of workers, and multimedia facilities to support workers in the field from exports at the company base. However, in this area digital PMR faces competition from new data services such as GSM's High Speed Circuit Switched Data (HSCSD) and General Packet Radio Service (GPRS).

Traditional PMR also faces competition at the small scale end of the market. More than 20% of UK PMR use is for small cell single site use (up to 3 km), much of which consists of mobile to mobile communication without infrastructure. This market could be attacked by the PMR 446. Also, for single site use, some radio equipment provides are suggesting the use of systems based on the coreless telephone technology DECT, which while having the advantage of connection to the fixed telephone network, would have a much smaller range (of the order of 300 m). A survey of PMR users [6] showed that while 25% only required on-site use, only less than a fifth of those, or 4% of the overall market, had requirements limited to within 200 m of the base site (see Figure 3.14). Although PMR 446 and DECT use unlicensed channels and so do not offer protection from interference in heavily loaded conditions, a new digital FDMA PMR system called DIIS is being developed to provide flexible direct mode speech and data communication within standard 12.5 kHz PMR channels. Such licensed systems could offer greater range.

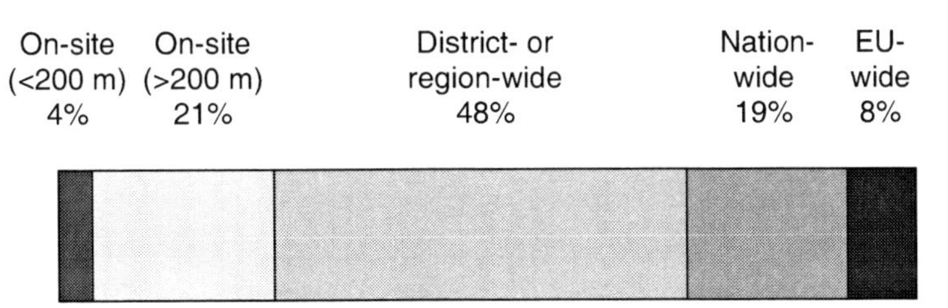

Figure 3.14 PMR area requirements

One external factor which may change the shape of PMR would be spectrum pricing. This could raise the cost of analogue PMR with its relatively wide band and speed its replacement. However, whether the market would then move to digital PMR or digital public cellular is unclear.

A further factor is the move across many business sectors of outsourcing operation not related to the core business activity of the firm. In many cases this would suggest a trend from providing PMR operations within the company to more specialist PAMR operators where suitable guarantees of service could be obtained. However, PAMR services have recorded slow growth in the marketplace so far, perhaps because of the long replacement cycles on mobile equipment. Also, as noted in Section 3.3, there is considerable resistance to losing control of the PMR operation.

The two clear applications for digital PMR are the emergency services, who need the range of features and the confidence of being involved in the service provision, and PAMR operators and major PMR users (government, the utilities, railways, etc.), who may also benefit from the additional features but must also receive service guarantees cellular operators may be unwilling to give. The popularity of these services beyond major users will depend in a large part on the pricing structure relative to public cellular. In this regard it is interesting that TETRAPOL, one of the most mature digital PMR systems, has made most of its sales to the emergency services and is currently being installed on the French railways. TETRA, with its efficient trunked operation, is particularly well placed to serve PAMR operators.

REFERENCES

[1] Hess, G. C., 1993, *Land-Mobile Radio System Engineering*, Artech House, Boston.

[2] Quotient Communications, *The Demand for Private Business Radio: A Study for The Radiocommunications Agency*, The Radiocommunications Agency, London, December 1997.

[3] RA 53, 'Sharing of Private Mobile Radio Channels', Revision 5, The Radiocommunications Agency, London, March 1996.

[4] RA 282, 'Short Range Business Radio - Notes for Radio Suppliers and Manufacturers', Revision 2, The Radiocommunications Agency, London, July 1997.

[5] RA 357, 'PMR 446 Information Sheet', The Radiocommunications Agency, London, March 1999.

[6] Richter, J A, 'TETRA – A Golden Opportunity', Tetra Workshop, Madrid, Spain, January 1996.

[7] Webb, W. T. and Shenton, R. D., Pan-European Railway Communications: Where PMR and Cellular Meet, *Electronics and Communications Engineering Journal*, August 1994, pp 195–202.

4

An Overview of the TETRA System

4.1 INTRODUCTION

Chapters 2 and 3 have considered the public and private mobile environments and have illustrated the essential differences between the operation of mobile systems within these environments. The remaining chapters of this book are devoted to a very significant development in the private mobile domain, the specification by the European Telecommunications Standards Institute (ETSI) of the Terrestrial Trunked Radio System (TETRA). This chapter gives an overview of the TETRA system and serves as an introduction to the layering concepts upon which TETRA is based, and which are used for the in-depth coverage presented in Chapters 5 to 9.

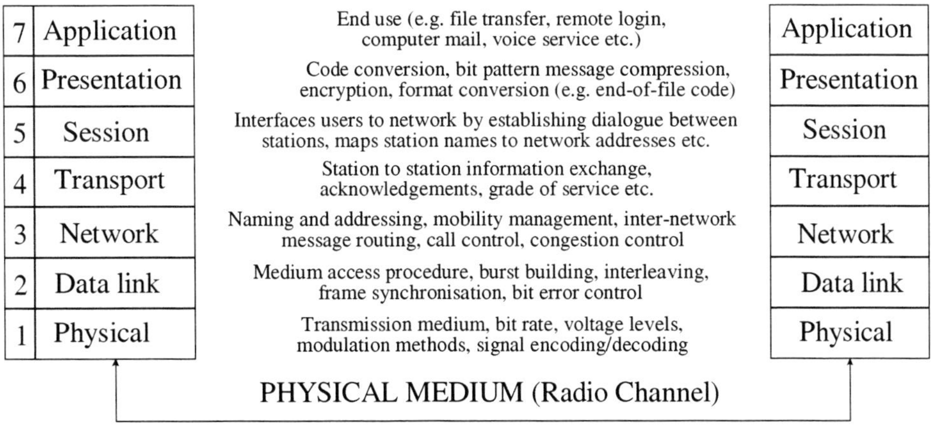

	Layer	Function	Layer
7	Application	End use (e.g. file transfer, remote login, computer mail, voice service etc.)	Application
6	Presentation	Code conversion, bit pattern message compression, encryption, format conversion (e.g. end-of-file code)	Presentation
5	Session	Interfaces users to network by establishing dialogue between stations, maps station names to network addresses etc.	Session
4	Transport	Station to station information exchange, acknowledgements, grade of service etc.	Transport
3	Network	Naming and addressing, mobility management, inter-network message routing, call control, congestion control	Network
2	Data link	Medium access procedure, burst building, interleaving, frame synchronisation, bit error control	Data link
1	Physical	Transmission medium, bit rate, voltage levels, modulation methods, signal encoding/decoding	Physical

PHYSICAL MEDIUM (Radio Channel)

Figure 4.1 Layered architecture of the OSI model

The Open Systems Interconnection (OSI) model was designed primarily to facilitate global communications between computers, irrespective of the characteristics of the particular networks of which they are part. The model, which is illustrated in Figure 4.1, does not specify how systems are implemented, but rather how they communicate. This means that many different networks, using products of different manufacturers, may be coupled together by mapping the OSI model onto the complete communications path. This is particularly important in the TETRA system, as will become apparent from the following chapters. The definition of the TETRA system is essentially confined to layers 1 to 3 of the OSI model and the main parameters are shown in Table 4.1

Table 4.1 Main parameters for TETRA

Parameter	Value
Carrier spacing	25 kHz
Modulation	$\pi/4$–DQPSK
Carrier data rate	36 kb/s
Voice coder rate	ACELP (4.56 kb/s net, 7.2 kb/s gross)
Access method	TDMA with 4 time slots/carrier
User data rate	7.2 kb/s per time slot
Maximum data rate	28.8 kb/s
Protected data rate	Up to 19.2 kb/s

The TETRA system has been designed to allow migration from analogue PMR systems and therefore the radio parameters have been adopted with this in mind and are therefore somewhat different to the radio parameters adopted for public systems such as GSM. Operation of the TETRA system is intended for existing VHF and UHF PMR frequencies. The spacing between TETRA carriers is 25 kHz, which allows direct replacement of two 12.5 kHz analogue FM channels or a single 25 kHz analogue FM channel. The key air interface parameters adopted for TETRA are shown in Table 4.1 and are indicative that the specific technology adopted for the TETRA system has been driven by the needs of PMR user groups. In particular TETRA provides features not catered for by current public digital cellular systems such as fast call set-up, broader range of services, scaleable architecture allowing a wide range of system configurations and a high level of user control over resource management.

Each carrier in the TETRA system accommodates four time slots which represent the *physical* channels available. These physical channels are shared between a number of logical channels which carry both traffic and signalling information.

4.2 THE BASIC SERVICES OF THE TETRA SYSTEM

TETRA supports a number of services which are characterised as bearer services and teleservices. A bearer service is defined as a service which provides information transfer between user network interfaces involving only low layer functions (OSI layers 1 to 3) and excluding the functions of the terminal. A teleservice provides the complete capability for communication between users including the terminal functions. A teleservice will therefore include attributes of the higher layers (4 to 7) of the OSI stack. This concept is illustrated in Figure 4.2.

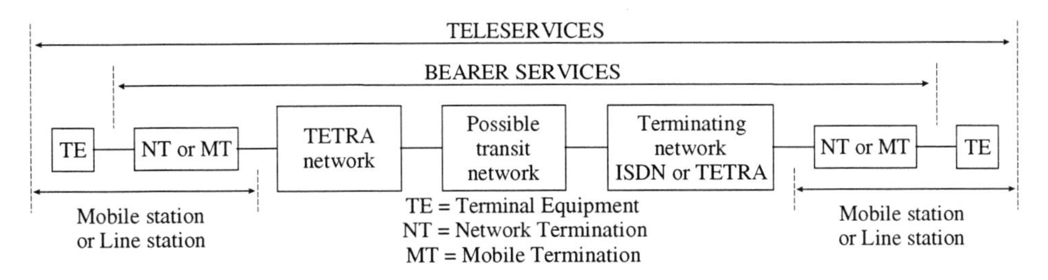

Figure 4.2 Bearer services and teleservices supported by TETRA

The bearer services in TETRA are individual call, group call, acknowledged group call and broadcast call for each of the following:

- Circuit Mode (Voice plus Data)
- Packet Connection Oriented Mode
- Packet Connectionless Mode

The teleservices supported by TETRA are clear speech or encrypted speech in each of the following:

- individual call (point-to-point)
- group call (point-to-multipoint)
- acknowledged group call
- broadcast call

TETRA also supports a number of supplementary services which modify or supplement a bearer service or teleservice. These supplementary services may divided into PMR type and telephone type services. A typical PMR type supplementary service is allocation of access priority, a typical telephone type supplementary service is call forwarding, for example.

4.3 TETRA NETWORK ARCHITECTURE

The TETRA network architecture is illustrated in Figure 4.3 and, unlike the GSM network architecture of Figure 2.1, is defined only in terms of six specified interfaces. The defined interfaces are those required to ensure interoperability, interworking and network management.

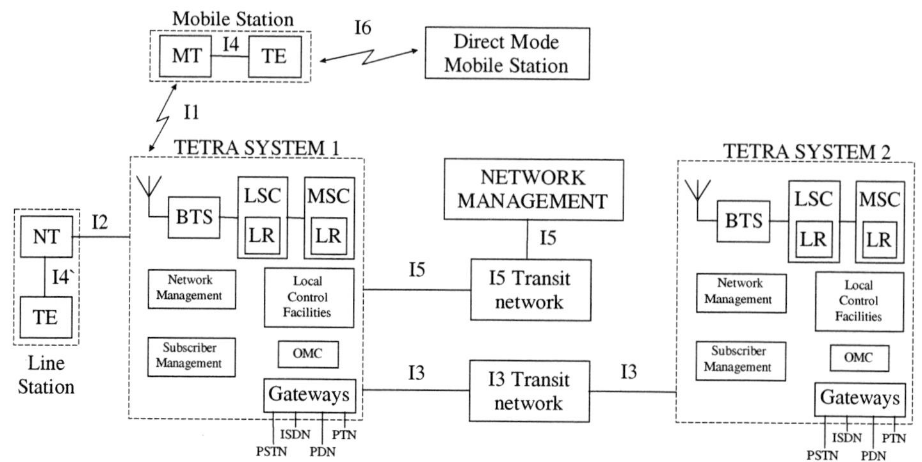

BTS = Base Transceiver Station, LSC = Local Switching Centre, MSC = Main Switching Centre,
LR = Location Register, OMC = Operations & Maintenance Centre,
PDN = Packet Data Network, PSTN = Public Switched Telephone Network, PTN = Private Telephone Network
ISDN = Integrated Services Digital Network

Figure 4.3 The TETRA network architecture with defined interfaces

The defined interfaces are:

- I1 = Radio air interface
- I2 = Line station interface
- I3 = Inter-system interface
- I4 = Terminal equipment (TE) interface for a mobile station (MS)
  (I4` = Terminal equipment (TE) interface for a line station (LS))
- I5 = Network management interface
- I6 = Direct mode interface

The functional sub-entities within an individual TETRA system, e.g. base transceiver station (BTS), main switching centre (MSC) etc., are *not* defined by the TETRA standard and are essentially propriety in nature. The interfaces and examples of TETRA sub-entities will be covered in more detail in later chapters.

4.4 TETRA CIRCUIT MODE (V+D)

This mode of operation, which is also known as the *trunked mode*, allows the simultaneous transmission of voice and data in a circuit switched mode. In this mode each source is allocated a traffic channel for the duration of a call, irrespective of whether that source is active. A traffic channel (TCH) is one of a number of logical channels specified in the TETRA system. The transmission mechanism for logical channels is provided by a physical channel (e.g. a specific carrier frequency/slot). A particular physical channel may be used for several logical channels on a shared basis and therefore involves the concept of multiplexing.

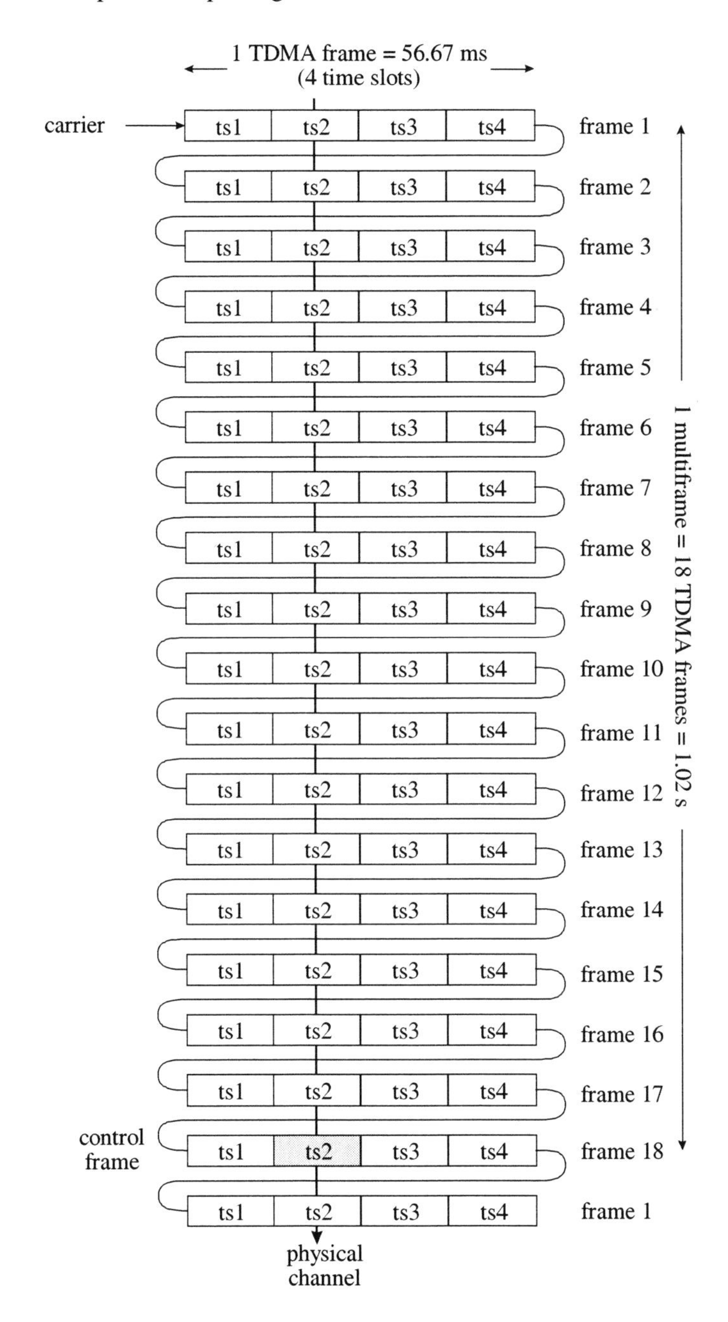

Figure 4.4 TETRA frame structure

Physical channels are identified in TETRA as Control Physical channels (CP) and Traffic Physical channels (TP). The logical channels are mapped onto both Control Physical channels and Traffic Physical channels depending upon the mode of operation.

TETRA is designed to operate in the frequency range from 150 MHz to 900 MHz and each cell is allocated one or more pairs of carriers (uplink and downlink). The separation between uplink and downlink frequencies is 10 MHz (in the VHF band) or 45 MHz (in the UHF band). Each carrier provides four physical channels by use of Time Division Multiple Access (TDMA) which divides the carrier into four slots of duration 14.167 ms. The TETRA TDMA frame (four timeslots) has a period of 56.67 ms. This frame is repeated 18 times in order to produce one multiframe of duration 1.02 s. The multiframe is repeated 60 times in order to produce a hyperframe of duration 61.2 s which is related to encryption and synchronisation. The frame structure is shown in Figure 4.4. The uplink and downlink transmission channels are offset in time by two slots to allow half duplex operation to be supported by low cost mobile terminals which do not require a duplexer.

One pair of carriers per site (cell) is designated to carry the main control channel MCCH. In normal operation time slot 1 of every frame (both uplink and downlink) is allocated for control purposes. This is known as the Control Physical channel (CP). The remaining three time-slots (channels) are used for traffic and represent the Traffic Physical channels (TP). Considering slot 2, Figure 4.4 indicates that traffic may be transmitted in the first 17 occurrences of this slot and that the 18th occurrence is used for signalling purposes. Thus the Traffic Physical channel may be regarded as 17 consecutive TDMA frames followed by a control frame. Data must therefore be compressed in the ratio 17:18 for transmission on a Traffic Physical channel.

4.4.1 Burst Structure

Each timeslot in the TDMA frame has a duration of 14.167 ms which corresponds to 510 modulating bit periods or 255 modulating symbol periods. As TETRA uses $\pi/4$ –DQPSK modulation, a modulation symbol has a duration of two bit periods. Within each slot data is transmitted in the form of bursts. There are eight burst types defined for TETRA, examples of which are given in Figure 4.5. It will be noted from this figure that the uplink bursts (mobile to base station) differ from the downlink bursts in that an allowance is made for power ramp-up (and possible linearisation of the power amplifiers) in the mobile terminals at the start of each burst. The ramp-up interval is equivalent to 34 modulation bit periods and therefore the capacity of the downlink burst, which does not have such an allowance, is greater than that of the uplink burst. This extra capacity is used for the transmission of 30 bits of low layer information on every downlink slot which forms the Broadcast BlocK (*BBK*).

The normal downlink burst contains three independent fields Block 1 (*BKN1*), Block 2 (*BKN2*) and the Broadcast BlocK. The normal uplink burst contains 2 independent fields called *SSN1* and *SSN2*. A separate *logical channel* may be mapped onto each of these fields. The Broadcast BlocK contains 30 scrambled bits gross (14 bits net) and is used exclusively for the Access Assignment CHannel (AACH). In order to appreciate the way in which the logical channels in TETRA are mapped onto (or multiplexed within) the available physical channels it is appropriate to consider the logical channels in more detail.

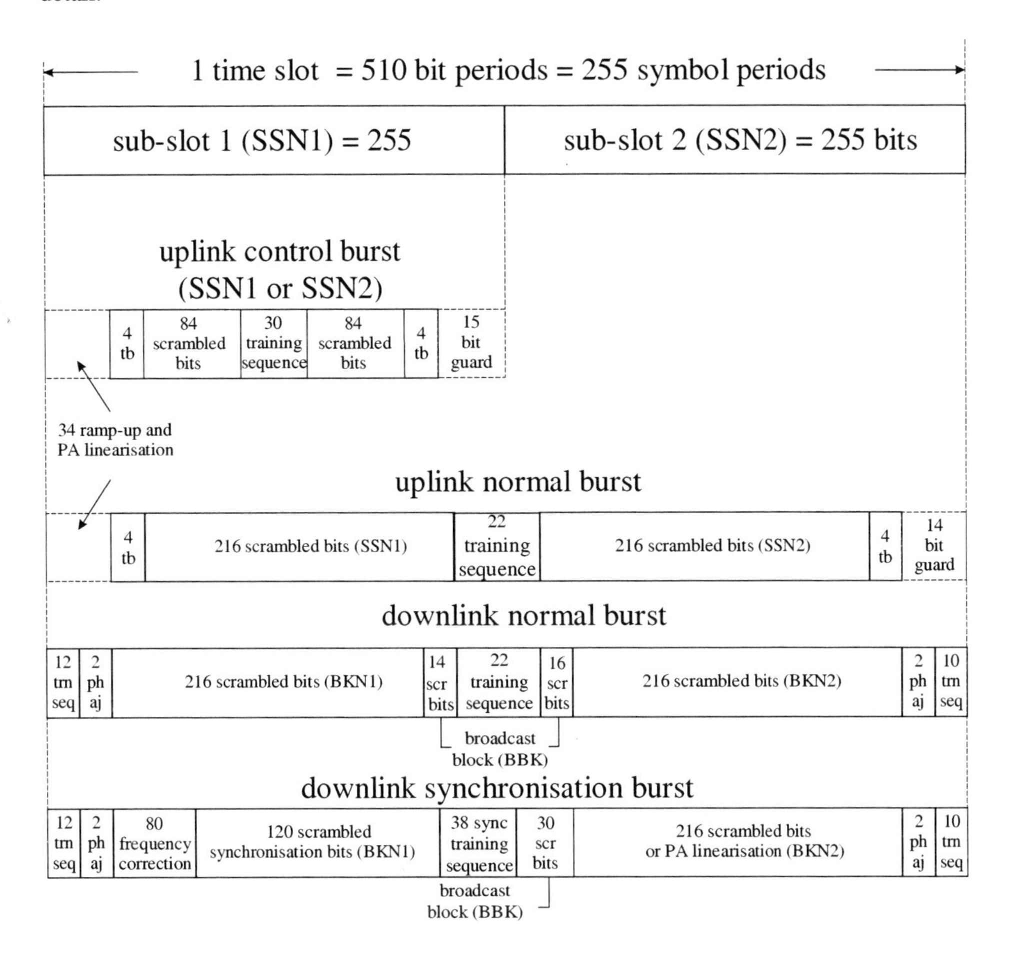

Figure 4.5 TETRA burst types

4.5 LOGICAL CHANNELS HIERARCHY

A limited number of logical channels have been mentioned in the previous section. The concept of logical channels is crucial to the operation of the TETRA system and it is appropriate in this chapter to give a brief overview of the logical channel hierarchy. This hierarchy is quite complex and is mapped onto the physical layer in a variety of ways. In order to understand the need for this hierarchy it is necessary to expand on the OSI model of Figure 4.1. The function of layer 2 (data link layer) is to add coding etc. to provide an error free link, via layer 1 (the physical layer) between layer 3 (the network layer) at transmitter and receiver. However, in the case of mobile radio dedicated communication links between the layer 3 peer entities do not exist. Therefore layer 2 must also provide and control the necessary links, when they are required, and consequently is divided into a Logical Link Control sub-layer (LLC), which interacts with layer 3 and provides reliable communication, and a medium access control sub-layer (MAC) which interacts with layer 1 and provides the necessary communication resources. The medium access and link control data are separate from the user data but they must be combined in some way for transmission over the physical medium via layer 1. It is necessary therefore to separate the information handled by layer 2 into a control plane and a user plane. This leads to the division shown in Figure 4.6

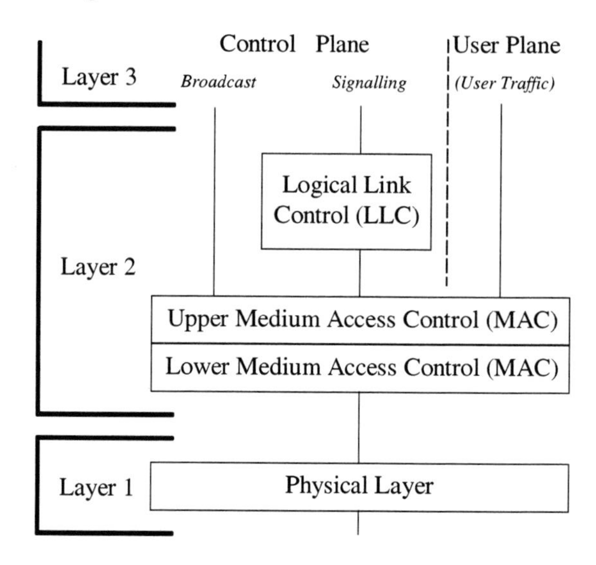

Figure 4.6 Control plane and user plane division in the OSI model.

It will be noted from Figure 4.6 that the medium access control sub-layer is itself divided into an upper MAC and a lower MAC. In TETRA the communication between layers of the OSI model takes place via formalised interfaces known as service access points (SAP). This concept is illustrated in Figure 4.7 which also indicates the way in which the functionality between upper MAC and lower MAC is partitioned. A detailed functional description of the functionality of the upper and lower MAC are given in Chapter 5.

Information is passed to the MAC from the upper layers and the MAC also generates its own information for peer to peer communication. Figure 4.7 illustrates that the information passed to the upper MAC is composed of

- C-Plane broadcast information, via the TSB – SAP, destined for all mobiles (base to mobile only).
- C-Plane signalling information comprising two-way control information and packet data, via the TSA – SAP, destined for specific mobiles.
- U-Plane user traffic comprising circuit mode voice and data plus end-to-end user signalling information, via the TMD – SAP.

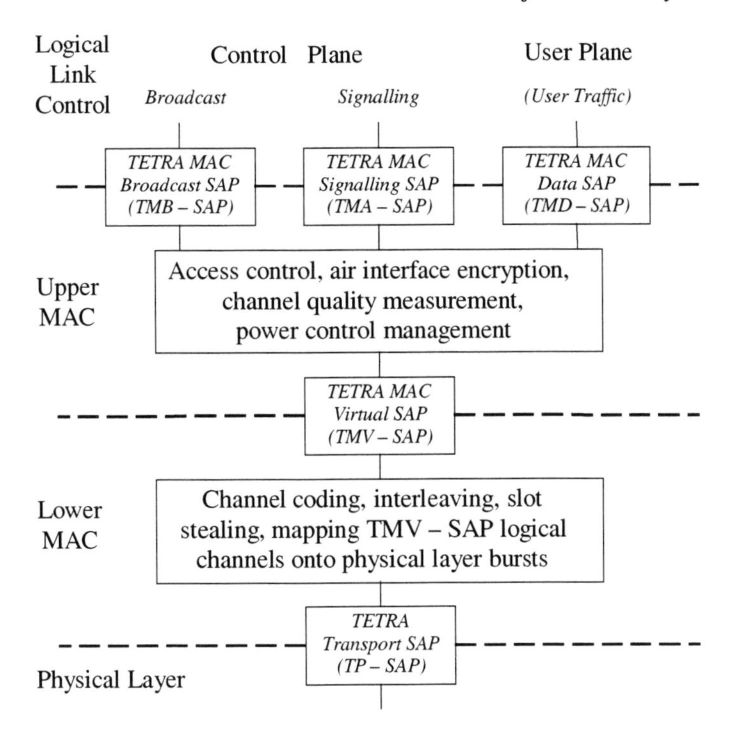

Figure 4.7 Service access points (SAP)s in the TETRA OSI layered architecture

Each service access point is provided with logical channels for transmission of this information.

4.5.1 Broadcast Control CHannel (BCCH)

The Broadcast Control CHannel (BCCH) is passed via the upper MAC (TMB – SAP). The BCCH is a uni-directional channel for common use by all mobiles, its purpose being to broadcast general information. The BCCH is actually divided into two categories, the Broadcast Network CHannel (BNCH) and the Broadcast Synchronisation CHannel (BSCH). The BNCH broadcasts network information to mobile stations (downlink only) and is passed to the lower MAC (via the TMV – SAP) as part of the Signalling CHannel (SCH). The BSCH broadcasts information to mobile stations used for time and scrambling synchronisation (also downlink only), and is passed to the lower MAC as the BSCH. The BCCH is normally broadcast in slot 18 (see Tables 4.2 and 4.3).

4.5.2 Common Control CHannel (CCCH)

The Common Control CHannel (CCCH) is passed via the upper MAC (TMA – SAP). The CCCH is a bi-directional channel for transmitting control information to or receiving control information from mobiles not actively engaged in a circuit mode call. The CCCH comprises the Main Control CHannel (MCCH) and the Extended Control CHannel

(ECCH). The CCCH is passed between upper and lower MAC (via the TMV – SAP) as part of the Signalling CHannel (SCH).

4.5.3 Associated Control CHannel (ACCH)

The Associated Control CHannel (ACCH) is also passed via the upper MAC (TMA – SAP). The ACCH is a bi-directional dedicated signalling channel associated with a channel which has been allocated for circuit mode traffic (i.e. a TP). It comprises the Fast Associated Control CHannel (FACCH), the Slow Associated Control CHannel (SACCH) and the Stealing CHannel STCH. The FACCH uses frames 1 to 17 when they are not used for traffic and the Slow Associated Control CHannel always uses slot 18. When a mobile is first assigned a physical channel (i.e. a carrier timeslot combination) the channel is provided as the fast associated signalling channel for the control signalling which occurs at the start of a call (a typical example is the BS allocating transmit permission to a particular mobile). When this signalling phase is complete the assigned channel becomes a traffic channel (TCH). At the end of a call the assigned channel reverts to a FACCH to pass control signalling to the system. Thus it is clear that the FACCH and TCH are mutually exclusive and cannot exist at the same time.

It is also necessary for signalling information to pass between the mobile and base station during the progress of a call. There are two mechanisms provided for this, the SACCH uses frame 18 of the physical channel which has been assigned to the mobile engaged in a call. However frame 18 is not used exclusively for the SACCH associated with a particular call but also carries broadcast messages which are used for system control. Hence it is necessary to indicate to mobiles which logical channel will occupy the next occurrence of frame 18 and this is done by means of the Broadcast Block, which is transmitted on every burst. The other method available for signalling during a call makes use of the stealing channel. Capacity is stolen on the uplink or the downlink for urgent messages, such as handover signalling. In this case part of the assigned channel is allocated to the stealing channel (STCH) and thus the TCH and STCH can *nominally* exist at the same time. Whenever part of a slot is allocated to the STCH this is signalled by a change in the training sequence transmitted within the burst.

The FACH and SACH are passed between upper and lower MAC (via the TMV – SAP) as part of the Signalling CHannel (SCH). The STCH is passed between upper and lower MAC as the STCH.

4.5.4 Access Assignment CHannel (AACH)

The Access Assignment CHannel (AACH) is generated within the upper MAC and is not therefore passed to the upper layers. It is a uni-directional channel (downlink only) the purpose of which is to indicate access rights on control channels and to indicate the assignment of the uplink and downlink slots on each physical channel. The AACH information is carried exclusively in the Broadcast BlocK (BBK) and is present in every downlink slot.

4.5.5 Common Linearisation CHannel (CLCH)

The CLCH is also 'generated' within the upper MAC, and is an opportunity for mobiles to linearise their transmitters. The CLCH is an uplink channel which is shared by all mobiles. A corresponding BS Linearisation CHannel (BLCH) is defined which is a downlink channel used by a BS operating in a discontinuous transmission mode. The CLCH is passed to the Lower MAC (via the TMV – SAP) as the CLCH. The way in which the Linearisation CHannel is multiplexed is illustrated in Tables 4.2 and 4.3. Although the Linearisation CHannel is regarded as a logical channel no useful information is actually transmitted over the radio interface. The CLCH may be regarded as a time interval during which mobile stations may transmit 'on air' for the purpose of monitoring and linearising their own power amplifiers after switching frequency (e.g. at a channel assignment). It is necessary to maintain a linear characteristic in order to avoid degrading the advantages of pulse shaping, as described in Section 1.8.

4.5.6 Traffic CHannels (TCH)

The Traffic CHannel is passed via the upper MAC (TMD – SAP). The TCH is a bi-directional for carrying user information. In TETRA different traffic channels are defined for speech or data applications and for different data message speeds:

The Speech Traffic CHannel (TCH/S) carries digitised voice information produced by an Algebraic Code Excited Linear Predictive (ACELP) coder at a net rate of 4.56 kb/s. This rate is increased to a 7.2 kb/s by the addition of error protection bits, the detail of which is given in Section 4.9.1.

Three data traffic channels are provided depending on the amount of inherent error protection. These are:

- (TCH/7.2) offering unprotected data at 7.2 kb/s net rate
- (TCH/4.8) offering low protected data at 4.8 kb/s net rate
- (TCH/2.4) offering high protected data at 2.4 kb/s net rate

Higher net rates up to 28.8 kb/s, 19.2 kb/s or 9.6 kb/s may be provided by allocating up to four traffic physical channels to the same communication, which must consecutive slots on the same frequency. TETRA accommodates three different depths of interleaving (with $N = 1, 4$, or 8) which may be applied to the data traffic channels TCH/4.8 and TCH/2.4.

4.5.7 Signalling CHannel (SCH)

The Signalling CHannel SCH is one of the logical channels passed by the lower MAC to the physical layer via the TP–SAP. The SCH is shared by all mobiles, but may carry messages specific to one mobile or one group of mobiles. System operation requires the establishment of at least one SCH per base station. The SCH may be divided into three categories, depending on the size of the message:

- Full slot Signalling Channel (SCH/F) which is a bi-directional channel used for full size messages.
- Half slot Downlink Signalling CHannel (SCH/HD) which is a downlink only channel, used for half size messages.
- Half slot Uplink Signalling CHannel (SCH/HU) which is an up-link only channel, used for half size messages.

4.6 MAPPING OF LOGICAL CHANNELS IN TETRA

The way in which the logical channels are accommodated (or mapped) on the physical channels in TETRA depends on whether the physical channel is a Control Physical channel (CP) or a Traffic Physical channel (TP). It will be recalled that each cell has a main control channel (MCCH) carried on slot 1 of the designated carrier, which is essentially the CP. The way in which the logical channels are mapped onto the CP is shown in Table 4.2.

Table 4.2 Logical channel mapping on a CP channel

Frame	DOWNLINK		UPLINK	
FN	Block *BKN1*	Block *BKN2*	Subslot *SSN1*	Subslot *SSN2*
1 to 17	SCH/F		SCH/F	
	SCH/HD SCH/HD	SCH/HD BNCH	SCH/HU CLCH	SCH/HU SCH/HU
18	BSCH	SCH/HD		

Table 4.3 Logical channel mapping on a TP channel

Frame	DOWNLINK		UPLINK	
FN	Block *BKN1*	Block *BKN2*	Subslot *SSN1*	Subslot *SSN2*
1 to 17	TCH STCH + TCH STCH + STCH		TCH STCH + TCH STCH + STCH	
18	SCH/F		SCH/F	
	SCH/HD BSCH SCH/HD	SCH/HD SCH/HD BNCH	SCH/HU CLCH	SCH/HU SCH/HU

The BS indicates on the AACH the type of logical channel(s) to be used on the next uplink subslot (SCH/HU or CLCH) or slot (SCH/F). This indication is valid within one frame and for one physical channel. A similar arrangement occurs for mapping of logical channels on a traffic physical channel, in this case the type of logical channel to be used on the next up-link sub slot or full slot is indicated by the BS on the AACH which is carried in the Broadcast BlocK of each burst. The channel mapping for a Traffic Physical channel is shown in Table 4.3.

It will be apparent from this section that there are a number of logical channels present in the MAC sub-layer the purpose of which is simply to support correct operation of the physical layer. The AACH, which indicates access rights on control channels and usage markers on traffic channels, and CLCH fall into this category. A further parameter exists for the support of correct operation of the physical layer known as the Slot Flag (SF). This corresponds to one of two synchronisation training sequences used at the physical level, in a normal uplink or downlink burst, to indicate whether one or two logical channels are present within the burst. The operation of the TETRA V+D mode is closely related to the logical channel structure outlined in this section and is described in Section 4.7.

4.7 OPERATION OF THE TETRA V+D MODE

In the TETRA V+D system channel acquisition is performed automatically when the mobile is powered up. The relevant channel is contained within the MS memory or a search is performed to find a channel. When a mobile is activated it must first acquire synchronisation before it can decode any of the messages broadcast by the base station. This is done by synchronising with the training sequence of the SYNC burst (Figure 4.4) of any frequency in the cell. The SYNC burst appears in the Broadcast Synchronisation CHannel (BSCH), which is always transmitted in subslot *BKN1* of frame 18 of the Control Physical channel and also shares subslot *BKN1* of frame 18 of the Traffic Physical channel (Table 4.3). Once synchronisation is achieved the mobile decodes the rest of the SYNC burst which identifies the cell, slot and frame and mode of operation. This gives the mobile full frame synchronisation with a particular base station.

The mobile then searches for the Broadcast Network CHannel (BNCH) on the current frequency and decodes information on the frequency of the main carrier, the number of secondary control channels (SCCH) in operation on the main carrier, power control information and some random access parameters. Having decoded this information the mobile then locates the MCCH on slot 1 of the main carrier, or a relevant SCCH. The mobile, at this point, has all the information needed to communicate with the system and may receive downlink messages and may transmit uplink messages.

In normal operation there are typically 4 or 5 pairs of carriers provided (16 to 20 physical channels) and the MCCH is present in time slot 1 of all frames 1 to 18. Mobiles which are not involved in a call listen to the downlink transmissions on the MCCH. The base station transmits on all downlink slots during normal mode and mobiles may be paged by an appropriate message on the MCCH.

4.7.1 Location Registration

The coverage area of a TETRA V+D system is divided into a number of location areas, which may correspond to a single cell or a group of cells. In order for a mobile unit to be paged within a particular location area it must be registered within that location area. Implicit registration can be performed by any system message that conveys the identity of the mobile, e.g. a cell change request, call request etc. The MS MAC sub-layer uses a random access procedure to initiate information transfer (unsolicited) to the BS.

4.7.2 Random Access

When a mobile wishes to make contact with the system this is done my making a random access, generally on a subslot on the MCCH, using the slotted ALOHA access mechanism. The slotted ALOHA access is a refinement of the ALOHA mechanism which allows a user to transmit a fixed access burst, as soon as it is formed. ALOHA relies on a positive acknowledgement to indicate that the burst was received without error. Assuming that each access burst requires a transmission time of P seconds then if such an acknowledgement is not received within a time $P + 2t_p$, where t_p is the propagation delay between transmitter and receiver, the access burst is retransmitted.
An access burst will be successfully transmitted at a particular instant t only if no other burst is transmitted P seconds before or after t (i.e. the vulnerable period is $2P$), this is illustrated in Figure 4.8.

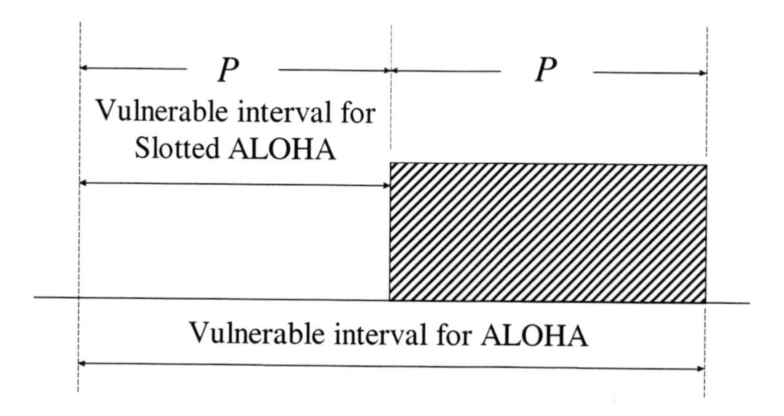

Figure 4.8 Vulnerable interval for ALOHA and slotted ALOHA

The normalised channel throughput S (the average number of successful transmissions per interval P) is related to the normalised traffic offered G (the number of attempted transmissions per interval P, including new and retransmitted bursts) by

$$S = Gp_0 \tag{4.1}$$

Where p_0 is the probability that no additional transmissions are attempted in the vulnerable interval $2P$.

If it is further assumed that burst arrival times are independent and exponentially distributed with a mean arrival rate of λ per second, the probability of k arrivals in an interval of duration t is then a Poisson process given by $P_t(k)$ where

$$P(k) = \frac{(\lambda t)^k e^{-\lambda t}}{k!}$$

The probability of no arrivals in time t is $P(0) = e^{-\lambda t}$. In P seconds there will be $\lambda.P = G$ arrivals, hence $\lambda = G/P$. Letting $t = 2P$ gives the throughput as

$$S = Ge^{-2G} \tag{4.2}$$

Equation 4.2 is based on the assumption that the propagation delay for all mobiles is the same. This expression must be modified to account for the fact that some mobiles will be closer to the base station than others. In such cases there will be a maximum *difference* in propagation delay of t_{dm} which has a normalised value of $a = t_{dm}/P$. equation 4.2 thus becomes

$$S = Ge^{-2(1+a)G} \tag{4.3}$$

The maximum throughput for pure ALOHA channel is found by differentiating equation 4.2 with respect to G and occurs at $G = 0.5$, thus

$$S_{\max} = 1/2e = 0.184 \tag{4.4}$$

The throughput characteristic for pure ALOHA is shown in Figure 4.9.

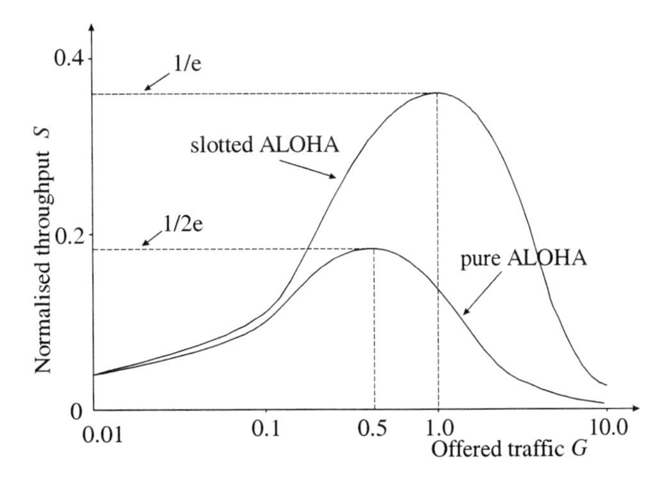

Figure 4.9 Throughput characteristic for ALOHA and slotted ALOHA

The figure shows that when the offered load G is low there are very few collisions and virtually all transmissions are successful and $S \cong G$. At higher values of offered traffic the number of collisions increases, which increases the number of retransmissions causing still more collisions and so on. The characteristic is thus unstable and S actually begins to drop. The maximum throughput of ALOHA is thus limited to 18.4% of the system capacity.

The throughput characteristic of pure ALOHA can be improved significantly if all transmissions are synchronised. The time axis is divided into intervals of time (or slots) of duration P and each user may transmit only at the start of a slot. Under these circumstances if collisions do occur then bursts will overlap completely and the vulnerable period is reduced to P. The throughput of slotted ALOHA is then given by

$$S = Ge^{-G} \tag{4.5}$$

This has a maximum value when $G = 1$ given by

$$S_{max} = 1/e = 0.368$$

The maximum throughput of slotted ALOHA is thus increased to 36.8% of the system capacity. The characteristic for slotted ALOHA is also shown in Figure 4.9.

It is clear from this figure that slotted ALOHA has a similar instability when the offered load exceeds the value at which the throughput is a maximum. This is avoided in TETRA by allowing a mobile to attempt an access only if it receives an access code on the AACH. There are 4 different access groups designated A, B, C and D and a mobile will belong to one of these groups by a process called binding. The binding process also includes the subscriber class and priority. A mobile may use a subslot, designated for a particular access code, only if the message priority and subscriber class, transmitted on the AACH, correspond to its current binding.

When a mobile does have permission to make a random access it randomises its response within a specified number of "access frames". This process is necessary to reduce the probability of collision. Once the initial random access has been performed all subsequent exchanges between BS and MS are by means of reserved slots. The way in which the slotted Aloha access mechanism is implemented in the TETRA system is covered in more detail in Section 6.7.

4.7.3 Call Set-up Procedure

This section provides an overview of the sequence of events for individual (as opposed to group) call set-up. The call set-up procedure is shown schematically in Figure 4.10. The message sequence starts by the calling mobile *M1* making a random access (known as a *u-setup*) in slot 1. This message is sent in a single burst and contains sufficient information to establish the service. The BS acknowledges with a message (known as *d-call proceeding*) and at the same time pages the called mobile *M2* with a message (known as a *d-setup*) in its next slot 1. The called mobile *M2* responds with a *u-connect* message in the subsequent slot 1. The BS then assigns the calling mobile a channel using the *d-connect* message and the called mobile using the *d-connect ack* message.

In principle the assigned channel can be slot 2, 3 or 4 on the same carrier or any slot on another frequency duplex pair. In practice if slot 2 is assigned on the same carrier it could not be used until the next frame due to limitations on the switching time of the synthesiser. In Figure 4.10 it is assumed that the assigned channel is slot 2 on a different carrier. In this case the first action of the mobile terminals involved is to linearise their transmitters. It should be noted that under these circumstances the BS gives immediate CLCH permission, on the assigned channel, by means of the CLCH permission flag. This flag is included to allow fast call set-up when a change of carrier is necessary for channel allocation. The normal provision of a CLCH occurs in frame 18 of all four physical channels at least once every four multiframe periods.

If both mobiles are on the same site the called mobile can be instructed by the BS to indicate its presence on the half slot following linearisation by sending a layer 2 message MAC access. If the mobiles are on different sites both can be instructed to indicate their presence in the half slot following linearisation. The BS then announces to the initiating mobile on the assigned channel that a connection has been established and grants transmit permission etc. (the assigned channel effectively exists as the FACCH at this stage). The initiating MS then proceeds to transfer coded voice frames.

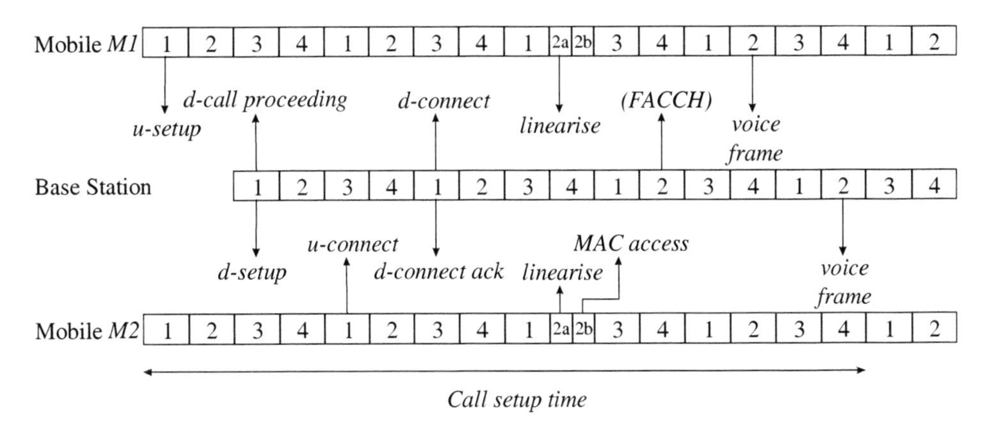

Figure 4.10 Call set-up procedure

The call set-up time is defined as the time between the initial access request (*u-setup*) sent by mobile *M1* and the first voice frame transferred to the called party. Under ideal conditions this is around 230 ms depending on which channel has been allocated (e.g. 16 $\times$ 14.167 ms = 226.67 ms), but may be longer if several attempts are required due to propagation conditions. The full call set-up time for TETRA is specified as less than 300 ms.

4.8 SPECIAL FEATURES OF THE V+D MODE

The basic call set up procedure for a V+D mode individual voice call was described in the previous section. The group call set up procedure is similar except that the BS immediately assigns a traffic channel (by issuing a *d-connect* message). In response to the *u-setup* message. The V+D mode has a number of special features associated with voice and data calls which may be listed as follows:

- Independent multiple concurrent tele/bearer services can be supported up to the physical capability of the air interface (19.2 kb/s net or 28.8 kb/s gross duplex), or to the limits of the class of mobile. The simplest class of mobile station is intended to work on alternate slots in the uplink and downlink, i.e. half duplex. A more elaborate MS will support full duplex operation over four contiguous time slots to give the data transfer at the full rate quoted above.

- Transmitter pre-emption is supported by the V+D mode. The transmitting MS must continuously monitor the downlink AACH to verify that transmit permission has been granted in the next uplink slot. This facility allows rapid re-assignment of the radio resources.

- Several grades of handover are supported by the V+D mode. During a call, or whilst monitoring the MCCH, a mobile may scan adjacent cells for better signal quality. The decision to choose an alternative cell is made by the MS. Depending on the MS capability and the call re-establishment option supported by the infrastructure, the MS re-establish request may be made to the present serving cell (seamless) or to the preferred new cell (fast). It should be noted that these options are equivalent to backward handover and forward handover as defined in Chapter 2.

- The TETRA V+D mode uses an event label, on the air interface, to minimise the occurrence of crossed calls. In the traffic mode every down-link slot has an identifier which is carried in the Broadcast Block. This ensures that if a mobile, engaged in a call, temporarily loses the signal and regains it a short time later, the event label can be used to determine whether the mobile has rejoined the correct call.

- Slot stealing during a voice, or circuit switched data, transmission can be performed for user-to-user signalling, for user-to-system signalling or for system-to-user signalling. This is done on a speech frame basis and, in principle, one or both speech frames within a timeslot may be stolen. The term speech frame refers to the 216 bits including channel coding, produced per 30 ms by the TETRA voice coder. When a speech frame is stolen for signalling the corresponding half slot made available is designated as the stealing channel. When two speech frames are stolen there are no speech parameters present and therefore the speech channel specifications no longer apply. When one speech frame is stolen it is assumed to be the first one, in which case the two half slots are encoded and interleaved separately (this is covered in more detail in Section 4.9.1).

- Different simultaneous access priorities are supported by the V+D mode. Mobile station groupings, call type and call queuing priorities allow multiple grades of service to be offered.

- High quality speech transmission at a net rate of 4.56 kb/s.

This coding rate is considerably less than that used in GSM and the principles involved are discussed in detail in Section 4.9.

4.9 VOICE CODING IN TETRA

Voice coding is accomplished in TETRA by modelling the actual speech production process, using a technique known as *analysis-by-synthesis* predictive coding. This is illustrated in Figure 4.11, in which the speech production mechanism is shown.

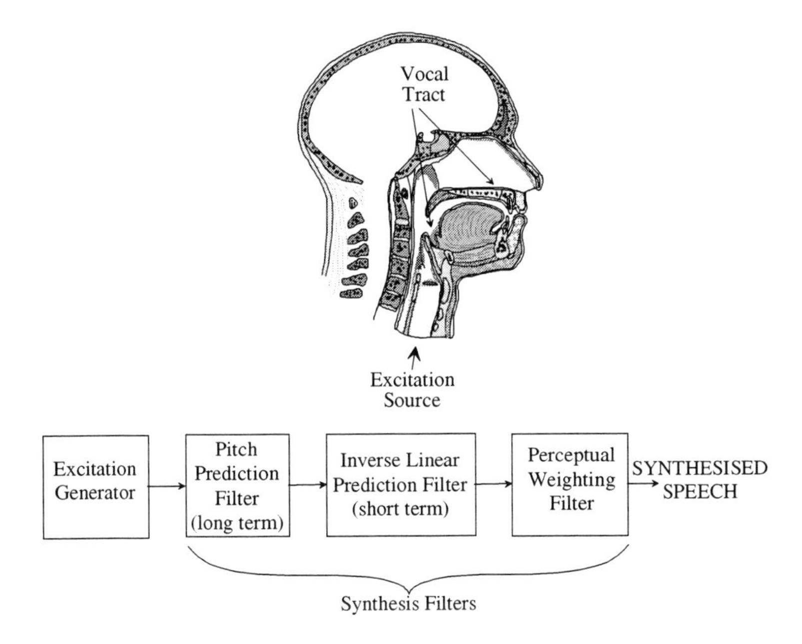

Figure 4.11 Human speech production mechanism and model

This mechanism can be represented as a source of excitation (the lungs), with a given pitch and loudness, which is filtered by the characteristics of the vocal tract (teeth, tongue, pharynx etc.). The filtering operation can be emulated by a series of digital filters with specific (time varying) parameters. Analysis-by-synthesis voice coders are based on the principle of minimising the mean squared error between the actual speech waveform and the synthesised speech by adjusting the excitation and the filter characteristics of the synthesiser. A schematic diagram of this technique is shown in figure 4.12 in which $\hat{v}(n)$ represents samples of the synthesised speech.

4.9.1 Algebraic Code Excited Liner Predictive Coding (ACELP)

The coder defined for TETRA uses an excitation source based upon a codebook of Gaussian sequences and belongs to a class of coders known as codebook excited linear predictive (CELP) coders. The particular codebook used in TETRA is based on highly structured algebraic codes which results in a form of coder known as an Algebraic Code Excited Linear Predictive (ACELP) coder. The reader is referred to the ETSI standard [1] for a full description of the TETRA ACELP voice coder and to Steele [2] for a full theoretical description of the ACELP process. This section will present an overview of the principles of operation of the ACELP coder and the associated channel coding which is applied.

The encoding procedure outlined in Figure 4.11 includes two steps, firstly the synthesis filter parameters are calculated from 30 ms of speech outside the feedback loop of Figure 4.12. Secondly the excitation sequence is calculated for this filter by dividing the 30 ms speech frames into 7.5 ms subframes, the excitation parameters being determined individually for each subframe.

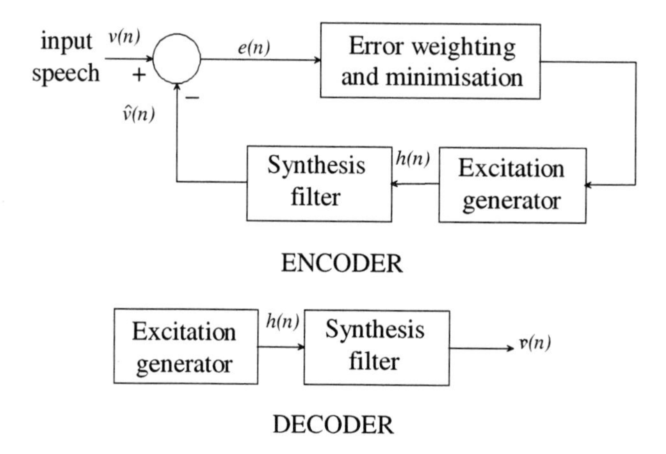

Figure 4.12 Analysis-by-synthesis voice coding

In this model a block of N speech samples is synthesised by filtering an appropriate innovation sequence from a codebook, scaled by a gain factor g_c. The quantised filter parameters and the address of the innovation sequence are then transmitted to the receiver with appropriate channel coding applied. At the receiver the excitation sequence is recovered from an identical codebook and passed through the synthesis filters to produce the synthesised speech samples $\hat{v}(n)$ which are an accurate representation of the original speech samples $v(n)$.

The pitch predictor filter estimates pitch parameters of the speech waveform over periods up to 160 samples (long term predictor). The pitch prediction filter is given by

$$\frac{1}{B(z)} = \frac{1}{1 - g_p z^{-T}} \tag{4.6}$$

Where T is the pitch delay and g_p is the pitch gain.

The short-term predictor models the short-term spectral envelope of a block of speech samples, of length N, by the transmission function of an all-pole digital filter

$$H(z) = \frac{1}{A(z)} = \frac{1}{1 - \sum_{i=1}^{p} a_i z^{-i}} \tag{4.7}$$

The coefficients a_i are known as the Linear Prediction (LP) parameters and p is known as the predictor order. In the case of TETRA $p = 10$. If $s(n)$ is the current speech sample the predicted value $\tilde{s}(n)$ may be calculated from a linear combination of previous samples.

$$\tilde{s}(n) = \sum_{i=1}^{p} a_i s(n-i) \tag{4.8}$$

The values of a_i are chosen to minimise the mean square error between $s(n)$ and $\tilde{s}(n)$ over short segments (30 ms) of speech. $A(z)$ is the *inverse filter* of Figure 4.11.

In the analysis-by-synthesis procedure synthetic speech is computed for all candidate innovation sequences. The sequence which produces the output closest to the original signal, according to a perceptually weighted distortion measure, is then selected as the sequence to be reproduced at the receiver. This is actually a computationally complex operation and a great deal of effort has been expended to reduce the complexity of the search operation by using specially tailored innovation sequences [2].

The perceptual weighting filter operates on the principle that noise in the formant regions of the speech is perceived less than in other regions. Thus the noise spectrum is modified to have a higher value around the formant regions. The filter characteristic used is given by

$$W(z) = \frac{A(z)}{A(z/\gamma)} = \frac{\sum_{i=1}^{p} a_i z^{-i}}{1 - \sum_{i=1}^{p} a_i \gamma^i z^{-i}} \tag{4.9}$$

A(z) is the inverse filter of equation 4.7 and $0 < \gamma < 1$is a coefficient which determines the amount of weighting applied to the noise spectrum at the formant frequencies, typically $0.8 > \gamma > 0.9$. The linear prediction parameters of both *A(z)* and *W(z)* are quantised before transmission. The pitch synthesis filter of equation 4.9 is actually implemented as an adaptive codebook in the TETRA system. The short-term prediction parameters are transmitted each 30 ms speech frame. The pitch and algebraic codebook parameters are transmitted every subframe.

The ACELP process implemented in TETRA produces 137 bits per 30 ms of speech (which is equivalent to a bit rate of 4.567k kb/s). These bits are assigned to one of three sensitivity classes based upon a bit sensitivity study. 30 bits (the most sensitive to error) are assigned to sensitivity class 2, 56 bits are assigned to sensitivity class 1 and 51 bits (the least sensitive to error) are assigned to sensitivity class 0. Channel coding is added to these bits to produce 216 bits per 30 ms (which is equivalent to a bit rate of 7.2 kb/s). It will be noted from Figure 4.5 that each normal burst has a payload of $2 \times 216 = 432$ bits, corresponding to two speech frames with associated channel coding. In fact two speech frames from the voice coder are encoded together. An 8-bit cyclic redundancy check (CRC) and a further 4 tail bits are added to the 2×30 sensitivity class 2 bits and a convolutional code of rate 8/18 is applied to the resulting 72 bits (see Chapter 6). A convolutional code of rate 2/3 is applied to the 2×56 sensitivity class 1 bits and the 2×51 sensitivity class 0 bits receive no channel coding. The 432 payload bits of each normal burst are then re-ordered (referred to as an interleaving depth of 1).

4.9.2 Error Concealment

The purpose of channel coding is to minimise the effect of errors in the received data stream on the decoded speech. Thus the sensitivity class 2 bits receive the greatest error protection and the sensitivity class 0 bits receive the least. There are three measures of performance quoted for the speech channel decoder. These are defined as follows:

- Bit Error Rate (BER) is the percentage of bits in error after channel decoding;
- Message Error Rate (MER) is the rate at which the CRC indicates an error in sensitivity class 2 bits;
- Probability of Undetected Erroneous Message (PUEM) is the rate at which the CRC fails to detect erroneous sensitivity class 2 bits.

In addition when a Bad Frame Indicator (BFI) is received (indicating that the frame is badly corrupted or lost), the decoder performs an error concealment procedure utilising the parameters of the last received "correct" frame. The error concealment procedure consists of retaining the previous filter coefficients and interpolating the bad speech frames.

4.10 DATA SERVICES IN V+D MODE

TETRA offers three types of data services to the user in the trunked (V+D) mode of operation (V+D) which may be listed as:

- short data service (up to 254 characters)
- circuit mode data
- packet mode data

The short data service provides a point-to-point and point-to-multipoint capability for sending short messages that comprise a limited number of data bits with the meaning being user defined. The SDS entity supports the following mobile originated and mobile terminated services:

- user defined and predefined reception and transmission for individual message;
- group message.

In the circuit mode data services an end-to-end circuit is established. This end-to-end circuit can be used unprotected, or have low or high forward error protection added. Optionally, the data may be encrypted by the standardised mechanisms of TETRA on the air interface or end-to-end.

The data rates offered are:

- Unprotected data: 7.2, 14.4, 21.6, 28.8 kb/s
- Protected data low: 4.8, 9.6, 14.4, 19.2 kb/s
- Protected data high: 2.4, 4.8, 7.2, 9.6 kb/s

The packet mode services offered to the user are divided in two categories:

- Connection oriented packet data services;
- Connectionless packet data services.

A *connection oriented* packet data service is a service which transfers X.25 packets of data from one source node to one destination node using a multiphase protocol that establishes and releases logical connections or virtual circuits between end users.

A *connectionless* packet data service is a service which transfers a single packet of data from one source node to one or more destination nodes in a single phase without establishing a virtual circuit.

In addition to the voice and data mode packet services TETRA also provides distinct Packet Data Optimised (PDO) services.

Equipment conforming to the PDO specification will support only packet data services which will be superior to similar services on the V+D system.

4.11 DIRECT MODE

The trunk mode of operation specified for TETRA is based on the radio interface definition known as I1. In addition to this mode of operation TETRA also provides for direct mobile-to-mobile communications. Direct Mode Operation (DMO) may also be used when the mobile station is outside the coverage of the network or it can be used as a more secure communication channel within the coverage of the network. The services provided by direct mode are:

- individual/group circuit mode call transmission and reception in simplex mode;
- call set-up with and without presence check;
- clear and encrypted circuit mode operation;
- pre-emption capability;
- user defined short message transmission and reception;
- pre-defined short message transmission and reception.

The direct mode of operation uses a radio interface definition known as I6 and there are two variations on this basic interface depending on the actual operational situation. The radio interface types are listed in Table 4.4.

Table 4.4 Trunked and direct mode radio interfaces

Interface	Application
I1	Trunked mode radio interface
I6	Direct mode: mobile to mobile radio interface
I6`	Direct mode: radio interface gateway from trunked mode
I6``	Direct mode: radio interface via a repeater

The mobile-to-mobile mode of operation (walkie-talkie) is illustrated in Figure 4.13. In this mode mobiles communicate directly with each other using a the direct mode radio air interface I6. In this particular case the TETRA trunked mode infrastructure is not involved and therefore the I6 interface does not support features such as mobility and radio resource management, which are supported by the interface I1.

The TETRA specifications are written such that a mobile station (MS) operating in the trunked mode can be contacted by a DMO mobile station that is within (mobile-to-mobile) range, and vice versa, a mobile station operating on direct mode can be contacted by the TETRA system (if it is within network coverage). This facility, which requires the

MS to monitor operation on the alternative transmission system is called 'dual watch' mode. The dual watch mode of operation is illustrated in figure 4.14.

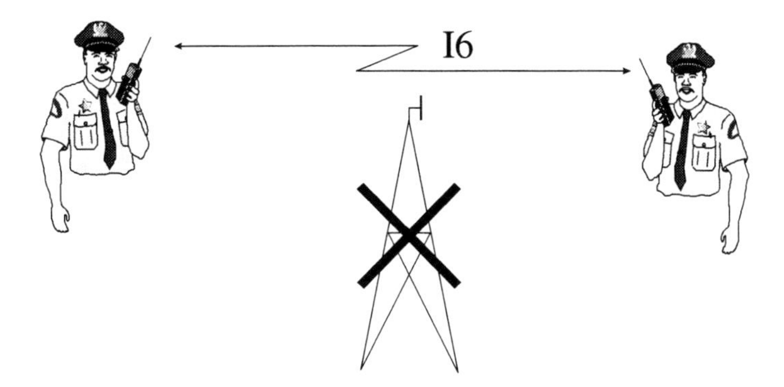

Figure 4.13 Direct mode mobile-to-mobile operation

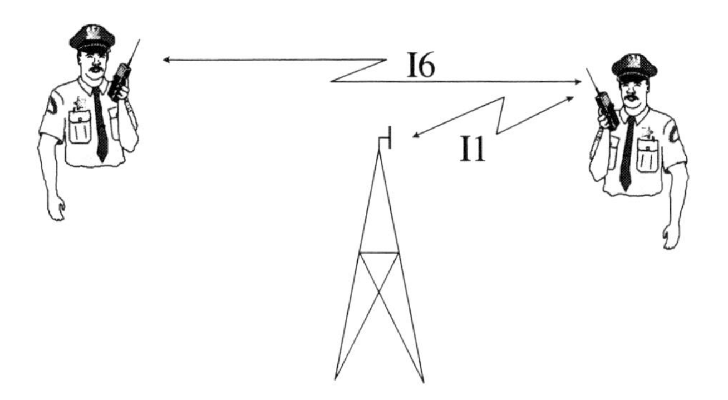

Figure 4.14 Direct mode dual watch operation

Direct mode terminals may also be contacted by the trunked system by means of a gateway mobile station. The gateway terminal essentially provides coverage extension and communicates with the trunked network via the I1 interface. The gateway terminal provides trunked TETRA/DMO interoperability at OSI layer 3, if the terminal is within radio coverage.

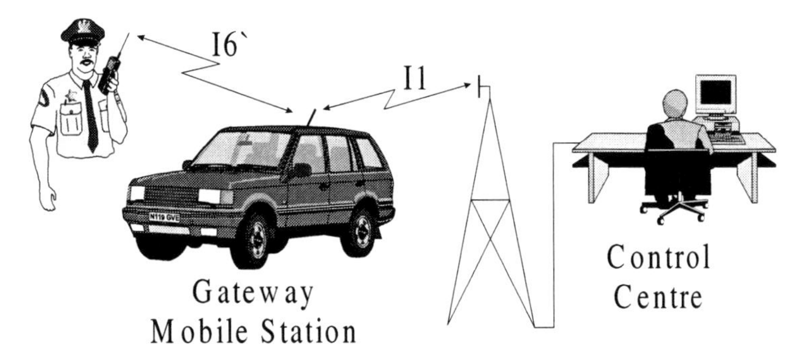

Figure 4.15 Direct mode coverage enhancement with a gateway station

Direct mode terminals communicate with a participating gateway mobile station using a modified I6` radio interface. This mode of operation is illustrated in Figure 4.15.

In many circumstances it may be necessary for personnel to operate in environments with poor coverage due to severe shadowing of the signal from a DMO terminal. Under such circumstances TETRA allows the use of repeater stations which are independent of the trunked network. The DMO terminals use a modified I6`` radio air interface. This mode of operation is illustrated in Figure 4.16.

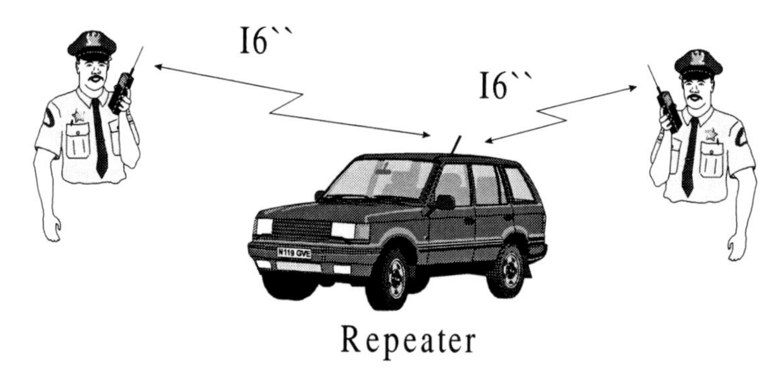

Figure 4.16 Direct mode coverage using an independent repeater

4.11.1 The Direct Mode Channel

Direct mode operation allows users to communicate when they are outside the coverage of a TETRA network. Communication takes place using specific Direct Mode channels which are distinct and separate from the standard TETRA trunked mode channels. The DM channels are programmed into the mobile terminal and a user selects one of these channels for communication. An individual call can be set up between two terminals that have selected the same DM channel, the normal mode of operation being simplex. The calling mobile may directly start transmission without checking for the availability of the called mobile which is the default method (known as the basic mode). Alternatively, the calling mobile may first perform a presence check on the called mobile within its coverage area and start transmission only after receiving a positive indication of presence. A typical coverage range is up to 400 m, in urban areas, and up to 2 km in rural areas. The maximum call set-up time for DMO clear (non encrypted) speech transmission is 150 ms.

When a channel is free it may be used by any direct mode MS which may tune to that channel. In DMO operation the absence of a base station requires that special procedures must be followed in order to achieve synchronisation between mobiles participating in a call [3]. When a channel is in active use in a group or individual call, a direct mode MS acting as "master" provides synchronisation using Direct Mode Synchronisation Bursts (DSB) in frames 6, 12 and 18 and transmits traffic in Direct Mode Normal Bursts (DNB) in frames 1 to 17. A direct mode mobile may transmit three types of bursts as shown in Figure 4.17.

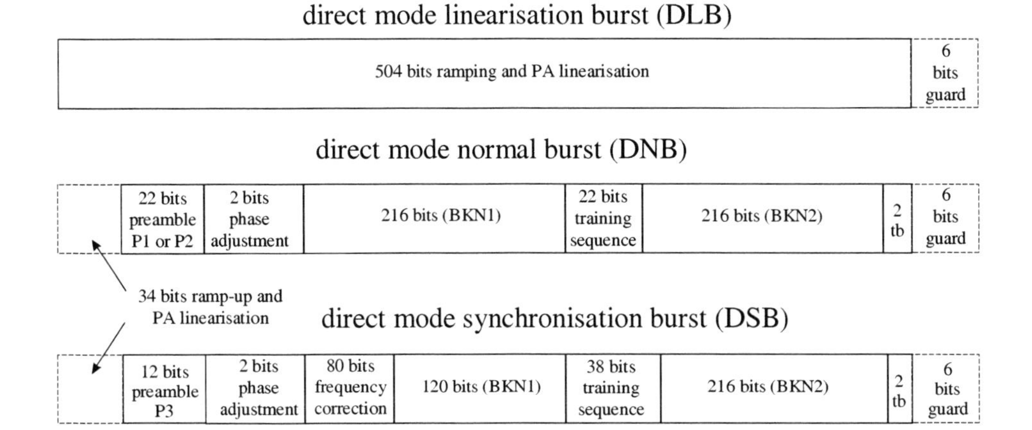

Figure 4.17 Direct mode bursts

4.11.2 Call Set-up in Direct Mode

The call set-up procedure (basic mode) will be illustrated by reference to Figure 4.18. It should be noted that the following constraints are placed on direct mode operation:

- frame 18 is always used for synchronisation purposes and carries a DSB in both slots 1 and 3;
- frames 6 and 12 carry reservation information in a DSB in slots 1 and 3;
- pre-emption is permitted in slot 3 of frames 2, 5, 8, 11, 14 and 17;
- linearisation may be permitted by means of a Direct mode Linearisation Burst (DLB) in slot 3 of frame 3, during a call;
- frames 1 to 17 usually carry traffic in slot 1 in a DNB, during a call.

The calling mobile will first determine the state of the channel. If the channel is occupied this can be detected by the presence of synchronisation bursts in slot 3 of frames 6, 12 and 18. If the channel is free the DM mobile may first linearise its transmitter and then establish channel synchronisation and its role as "master" by transmitting a sequence of synchronisation bursts, shown as *(su)* in Figure 4.18, using the DSB. These synchronisation bursts contain frame count information which defines their position in the 18 frame cyclic multiframe structure. The master DM mobile may then immediately transmit traffic (*tc*) using the DNB structure in the next available frame, which is frame 1 in Figure 4.18.

Frame	17				18				1				2				3				4			
Slot	1	2	3	4	1	2	3	4	1	2	3	4	1	2	3	4	1	2	3	4	1	2	3	4
	su	su	su	su	su	su	su	su	tc				tc		p?		tc		lc		tc			

Frame	5				6				7				8				9				10			
Slot	1	2	3	4	1	2	3	4	1	2	3	4	1	2	3	4	1	2	3	4	1	2	3	4
	tc		p?		tc		oc		tc				tc		p?		tc				tc			

Figure 4.18 Call set-up sequence for DMO (basic mode)

Also shown in figure 4.18 are the positions of the slots which are allocated to allow pre-emption requests to be made *(p?)*, the position of the slots available for linearisation *(lc)* and the synchronisation burst which are used to indicate occupation of the channel *(oc)*.

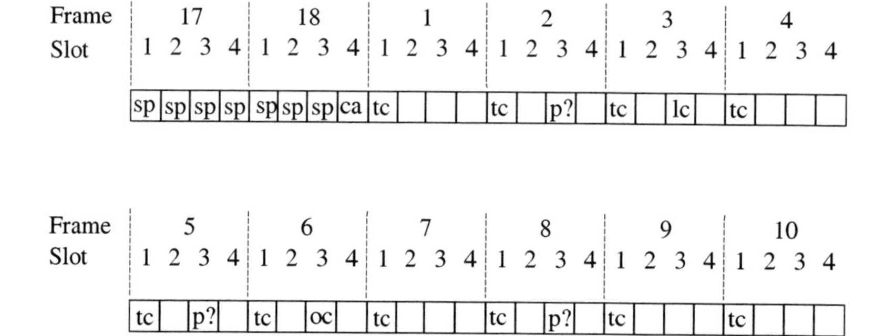

Figure 4.19 Call set-up sequence for DMO with presence check

The call set-up procedure with presence check is illustrated in Figure 4.19, the synchronisation bursts in this case *(sp)* contain a parameter which requests a response from the mobile which has been addressed as the recipient of the call. This mobile is defined as the '*slave*' for the duration of the transaction and responds with a connect message *(cm)*, indicating its wish to receive the call. The *(cm)* message may be repeated several times. When the *master* receives this message it responds with a connect acknowledgement message *(ca)* after which the *master* commences transmission of traffic.

In a DM call each transaction constitutes a separate transmission with designated *master* and *slave(s)*. The procedure for terminating one transaction and starting another is termed changeover. In order to change the *master* (sender) in a call, the *master* indicates that its transaction has finished. Recipients of the call can issue a changeover request in order to become *master* of a subsequent transaction. During a DM call provision is made for a mobile to access the DM channel for priority reasons such as an emergency call. In this case a pre-emption mechanism is provided. A full description of the pre-emption procedure, call termination and short data transmission procedures is given in [3].

4.11.3 Control Channels in Direct Mode Operation

The previous section has given an indication of the signalling procedures which occur during call set-up. The logical channels defined for direct mode are:

- linearisation (LCH);
- signalling (SCH);
- stealing channel (STCH).

There are three categories of SCH in defined DMO, and these are:

- Synchronisation Signalling CHannel (SCH/S), which is used for synchronisation messages (*su* and *sp* in Figures 4.18 and 4.19);
- Half slot Signalling CHannel (SCH/H), which is used for half slot messages;
- Full slot Signalling CHannel (SCH/F) which is used for short data messages.

4.12 SUMMARY

This chapter has provided an overview of the TETRA system and has indicated the major elements involved in V+D operation in both trunked mode and direct mode. The relationship of the architecture of the TETRA system to the OSI model has also been introduced. This is developed in more detail in subsequent chapters.

4.13 SPECTRAL EFFICIENCY

Having described the elements of both GSM and TETRA it is of interest at this point to illustrate the main parameter upon which cellular systems are compared. An important issue in cellular systems is the number of cells required to cover a specific subscriber density for a given allocated spectrum. This influences cost because the smaller the number of cells the smaller the number of base stations required. In comparing different systems a useful measure is the number of subscribers/cell/MHz which is the definition of spectral efficiency used in this text. The influence of traffic theory on cellular coverage was briefly considered in Section 1.21 and it should be noted that the number of subscribers/cell/MHz is not actually the same as the number of channels/cell/MHz because of the non linear relationship between traffic capacity and number of channels, specified by the Erlang B formula, as shown in Figure 4.20.

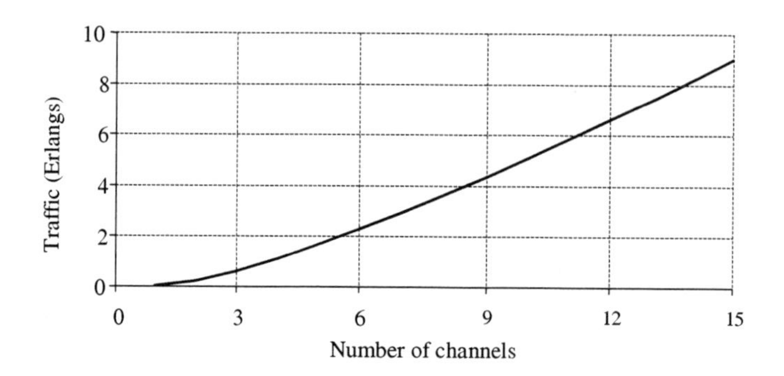

Figure 4.20 Erlang B for 2% blocking probability

As indicated in Section 1.21, if the total allocated bandwidth is B_{tot} and the channel bandwidth is B_{ch} the total number of channels available is $n_{tot} = \dfrac{B_{tot}}{B_{ch}}$ from which the number of channels/cell $n_{ch} = \dfrac{B_{tot}}{B_{ch}K}$ where K is the cluster size.

It is possible to compare TETRA and GSM in terms of spectral efficiency by making some simplifying assumptions. It should be noted that GSM has no facility for queuing blocked calls and the Erlang B formula applies. TETRA, on the other hand, does have a limited queuing ability and is described by the Erlang C formula. It is common practice to compare mobile communications systems on the basis of block-calls-cleared (Erlang B) and this is the procedure which will be adopted for comparing TETRA and GSM, however, the reader is referred to Section A4 for an explanation of the application of the Erlang C formula. A further assumption is that each base station radiates the same power. The SIR experienced by a mobile can thus be given as

$$\frac{S}{I} = \frac{R^{-4}}{6D^{-4}} \tag{4.10}$$

Assuming a hexagonal cell structure then

$$D = \sqrt{3K}.R$$

Thus when all base stations transmit the same power the SIR experienced depends only on the value of K. Letting $\gamma = 4$ (for illustration) gives

$$\frac{C}{I} = \frac{(3K)^2}{6} \text{ or } K = \sqrt{\frac{2}{3}\frac{C}{I}}$$

The number of channels/cell n_{ch} is given by

$$n_{ch} = \frac{B_{tot}}{B_{ch}\sqrt{\frac{2}{3}\frac{C}{I}}} \tag{4.11}$$

For TETRA B_{ch} = 25/4 kHz and the typical SIR = 19 dB (79.4)

For GSM B_{ch} = 200/8 kHz and the typical SIR = 9 dB (7.94), thus

$$\frac{n_{ch(TETRA)}}{n_{ch(GSM)}} = 1.3$$

Hence in theory a TETRA cell will have 1.3 × number of channels in a GSM cell. It is clear therefore that fewer TETRA cells will be required to support a given population density. The gain in capacity of TETRA relative to GSM depends on the traffic that the differing number of channels can support for a specified blocking probability. The actual spectral efficiency achieved will depend on the bandwidth allocated (as the possible cluster sizes are restricted to specific values).

To illustrate this point it is assumed that the total bandwidth available (uplink) is equivalent to 21 GSM carriers = 4.2 MHz. This is equivalent to 168 GSM channels and 672 TETRA channels (the seemingly high figure for TETRA is retained for comparison purposes).

For TETRA a SIR of 19 dB gives $K = 7$ hence $n_{ch(TETRA)} = 96$.

For GSM a SIR of 9 dB gives $K = 3$, hence $n_{ch(GSM)} = 56$.

$$\frac{n_{ch(TETRA)}}{n_{ch(GSM)}} = 1.7$$

With a 2% blocking probability 56 channels can carry 45.9 erlangs and 96 channels can carry 84.1 erlangs. The number of erlangs is directly related to the number of mobiles hence:

$$\frac{Spectral\ Efficiency\,(TETRA)}{Spectral\ Efficiency\,(GSM)} = \frac{84.1}{45.9} = 1.83$$

As pointed out earlier this result is derived for illustrative purposes and makes many simplifying assumptions. A further issue which must be taken into account in a real comparison is that the average call duration in a PMR environment is a few seconds, whereas in the public environment the call duration is typically 2 minutes.

Spectral efficiency can be increased further when account is taken of the characteristics of normal conversations. In effect, when two people are engaged in a telephone conversation, both are inactive for approximately 55% of the time, either listening to the other person or pausing for breath. GSM and TETRA make use of this fact by operating in a mode known as discontinuous transmission. Each terminal is equipped with a voice activity detector. When the user is silent the transmitter is switched off, which conserves battery power and also reduces the overall interference. However, this is an inefficient use of valuable spectrum and 3rd generation of digital mobile cellular systems [4] are already working on the principle of re-allocating slots to other users during periods of silence. This is known as statistical multiplexing and its effect is to increase the spectral efficiency.

In theory, it is possible to accommodate 2.2 separate users on each channel as each will be active, on average, for less than 45% of the time. This has two effects:

- it reduces the channel bandwidth to 45% of its nominal value;
- it increases the interference by a factor of 2.2, as each slot will be occupied, on average, for the 2.2 × the previous interval. To maintain the original SIR value K^2 must be increased by the factor 2.2

The overall effect may be obtained from equation 4.11. The value of B_{ch} is reduced by a factor 2.2, but K must be increased by a factor $\sqrt{2.2}$. This increases the number of channels/cell by $\sqrt{2.2} = 1.48$. Applying this to GSM would result in $1.48 \times 168 = 248$ channels/cell which could carry 233.8 erlangs. Hence the capacity of GSM could, in theory, be increased by a factor of 233.8/154.5 = 1.51 as a result of statistical multiplexing. It should be noted that, in practice, K would be an integer value from a set of rhombic numbers

REFERENCES

[1] European Telecommunication Standard ETS 300 395-2, 'Terrestrial Trunked Radio TETRA; Speech codec for full rate traffic channel; Part 2 TETRA codec', February 1998

[2] Steele R (Ed) *Mobile Radio Communications*, Pentech Press, London, 1994, ISBN 0-7273-1406-8, ch 3.

[3] European Telecommunication Standard ETS 300 396-3, 'Technical requirements for Direct Mode Operation; Part 3', March 1998

[4] Dunlop, J., Irvine, J. Robertson, D. and Cosimini, P. 'Performance of a Statistically Multiplexed Access Mechanism for a TDMA Radio Interface', *IEEE Personal Commun.* Vol 2 No 3 June 1995 pp 56–61

5

TETRA System Architecture, Components and Services

5.1 INTRODUCTION

The aim of this chapter is to introduce the TETRA system architecture and components in some detail. Most system descriptions are based on a conceptual representation of *functional configurations* and *reference points,* which is a widely adopted methodology for describing the interconnection of components within a network system. In the context of TETRA, this approach can be applied for identifying the various system interfaces and possible user equipment configurations. The network layer protocols associated with these interfaces will be the subject of Chapter 8 and the emphasis here will be on high level descriptions. Subsequent to system component descriptions, an overview of peripheral equipment interface (PEI) is presented. TETRA addressing and identity schemes, specified to uniquely identify subscribers, terminals, and networks are also summarised.

TETRA provides a range of services capable of supporting data-centric as well as multimedia-oriented applications. These are summarised by highlighting their salient features. To provide more insight into the features supported by TETRA, at the end of the chapter some comparisons are made with GSM system architecture and components.

The TETRA technical specification is driven by a comprehensive set of end user requirements. As such, most of the potential user groups of this network technology have been identified from the inception of the standard. Communication features such as group call, Direct Mode Operation, and numerous emergency-oriented supplementary services are all driven by end user demands during the standardisation process. As a background to this chapter, it will therefore be quite useful to be familiar with the user groups behind the TETRA technology as described in the section below.

5.2 THE TETRA USER GROUPS

5.2.1 Potential User Groups

Table 5.1 lists some of the potential TETRA user groups associated with the use of traditional PMR/PAMR technology. It is possible that new user groups will be added to this list with the realisation of the features TETRA can offer. The most important features of TETRA include group call, Direct Mode Operation, multimedia terminal capability, support for end-to-end encryption, and call priority schemes.

To highlight some of these points, the multi-slot operation of the TETRA V+D system, which is capable of offering a bit rate of up to 28.8 kb/s may be used as an example. A number of enhanced non-voice applications could be envisaged with such wireless transmission capacity. Examples abound for applications that rely on remote image database access, for maps, fingerprints, parts and maintenance schematics, etc. Table 5.2 lists some of non-voice applications which may be supported with multi-slot operation on a V+D platform.

For data-oriented applications, the packet data optimised (PDO) platform can provide even more enhanced services due to its *statistical multiplexing* scheme which is tailored to data traffic. It is also of interest to note that an extension to the PDO specification is currently under development by ETSI within the TETRA Project. An enhanced PDO system, named Digital Advanced Wireless System (DAWS), is to provide much higher data rates (in the order of 155 Mbps) and wide area mobility for wireless Internet access and other multimedia applications. This would mean higher frequencies of operation (current proposal around 5.3 GHz) and wider bandwidth than the current narrowband PDO system.

Table 5.1 TETRA User Groups

User groups	Business organisations
Public safety & public security	Ambulance, police, fire, rescue services (mountain, sea, etc), customs, etc.
Transport	Airports, buses, taxis, railways, special transport, river transport, etc.
Utilities	Electricity, gas, water, oil, etc.
Industrial	Manufacturing, distribution, construction, etc.
Public Services	Government departments, public health, environment protection, postal services, etc.
Support services	Private telecom operators, user services, maintenance, workforce management, etc.
Others	Possibly a new class of users, especially small businesses and with a narrow coverage.

Table 5.2 Examples of non-voice applications

Application types	Application examples
Data-oriented	Remote database access
	Internet access
	Computer file transfers
	Electronic mail and short message services
	Fleet management and despatch facilities
	Road informatics
	Remote data acquisition for monitoring or telemetry
	Image transmission for identification
	Access to image database for maps, fingerprints, etc
Video and multimedia	Remote database access with audio visual content
	Low bit rate videoconferencing
	Slow scan surveillance video camera

5.2.2 The TETRA MoU

As a background to the ETSI standardisation activities, a strong backing for the TETRA technology has been demonstrated with the creation of a forum that represents a significant number of end users, manufacturers, network operators, and other interested agencies. The forum was established in December 1994 as the TETRA Memorandum of Understanding (MoU), which, as of May 1999, represented 58 organisations from 18 countries, including 8 observers. Appendix A.2 provides the listing of the TETRA MoU members.

The TETRA MoU was created with the recognition of the importance of co-operation for operational and commercial aspects of TETRA. This undoubtedly will have a significant effect on the accessibility and expansion of this new technology. Numerous working parties under TETRA MoU are responsible for defining system and equipment inter-operability and inter-working requirements in various ways. Current activities encompass equipment type approval and validation, multi-vendor equipment procurement, market development, and aspects of business application and potential network operators. These activities can promote the acceptance of TETRA even wider since potential users would be better informed. For instance, at the time of writing, work was in progress for validating the inter-operability of TETRA equipment between different manufacturers as one of the objectives for establishing the credibility of a multi-vendor environment.

For more information on the activities of TETRA MoU, the reader is referred to their web site [1] which provides current developments on TETRA. The site also provides a wealth of diverse information on applications, marketing and business developments, and various technical issues.

5.3 SYSTEM ARCHITECTURE AND COMPONENTS

5.3.1 TETRA System Architecture

The TETRA network architecture consists of a number of system entities and defined interfaces. However, unlike the GSM network architecture, which was introduced in Section 2.3, the specification for TETRA is concerned only with the periphery of a TETRA system, referred to in the TETRA standards as *Switching and Management Infrastructure* (SwMI). That is, the internal interfaces of a TETRA network are not standardised. This was to allow manufacturers to implement the most cost-effective network solutions without constraints arising from standardisation [2].

Figure 5.1 shows a possible configuration for a TETRA SwMI. This representation is intended to show all the possible standard interfaces and system entities. In a particular implementation only a subset of these interfaces and entities may be present.

The definition of the SwMI could be articulated by identifying the following features:

- It comprises of up to six major system components. These have been introduced in Section 4.3.
- Its system components are interconnected with six defined interfaces which are required to ensure inter-operability, inter-networking, and network management between the various system components and networks. These components are described in Section 5.4.
- It provides a common network domain for the entities contained in it. That is, all the components within the SwMI are in the same address space. In fact, a *TETRA Network Domain* may be regarded as synonymous with SwMI. System addressing and identities are discussed in Section 5.7.

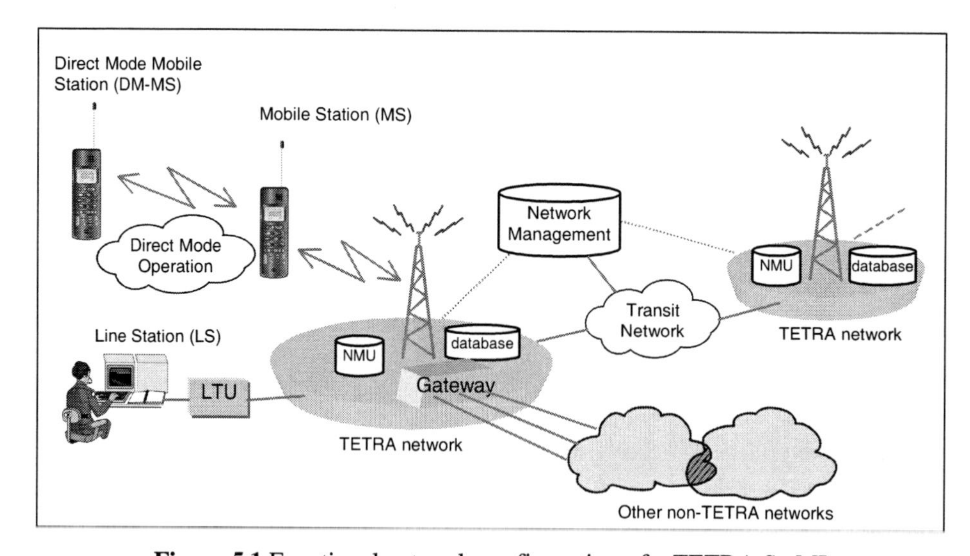

Figure 5.1 Functional network configuration of a TETRA SwMI

5.3.2 System Components

Six major system components can be identified with the network configuration shown in Figure 5.1.

- Individual TETRA network
- Mobile station (MS)
- Line station (LS)
- Direct Mode Mobile Station (DM-MS)
- Gateway
- Network management unit

All of the above components and the interfaces between them are the subject of TETRA standardisation with an exception of the internal architecture of an individual TETRA network. Each of the above components is briefly described below.

5.3.3 TETRA Network

This is an individual TETRA network system comprising of local switching centre, mobile switching centre (MSC), base transceiver station (BTS), gateways, switches, operations and management centre (OMC) and the associated control and management facilities. This is somewhat equivalent to GSM's base station subsystem and network subsystem put together. The aforementioned components contained within the individual TETRA system are not covered by the TETRA standard, except for the gateways that interconnect a TETRA network to the outside world.

It is important to note that a TETRA network is one component of the TETRA SwMI. Where the context causes no confusion with an individual TETRA network, the term *infrastructure*, or even network, will be used.

5.3.4 Mobile Station

The mobile station (MS) functionally comprises the mobile termination unit (MTU) and the associated terminal equipment (TE). TETRA mobiles may be categorised in terms of equipment portability, as *hand-portable* (or simply portable) and *vehicle-mounted mobile* (or simply mobile). Examples of both equipment types are depicted in Figures 5.2 and 5.3. The term *mobile station* will be used throughout to encompass both portable and vehicle-mounted mobile stations. Sections 5.6.2 and 5.6.3 describe the functional split of the TETRA MS.

5.3.5 Classes of MS

TETRA MSs are classified based on attributes such as power class and their capability for a given mode of operation. These are summarised below.

MS Transmit power class – Mobile terminals are rated in one of four transmit power classes. However, irrespective of the transmit power rating, they must be able to adjust their transmit power according to the power control instructions received from the network. Such power control can be used effectively for reducing co-channel interference and for conserving battery life. The MS transmit power classes are as follows.

- Class 1: 30 watts (45 dBm)
- Class 2: 10 watts (40 dBm)
- Class 3: 3 watts (35 dBm)
- Class 4: 1 watt (30 dBm)

Vehicle-mounted mobiles are usually rated as Class 1 or Class 2. Hand-portable mobiles are rated either as Class 3 or Class 4.

Nokia THR420
Ruggedly designed for demanding work environments; supports group call and dual watch for DMO.

Nokia THR600
Cellular like design with features like the THR420

Figure 5. 2 TETRA hand-portable mobiles (*Courtesy of Nokia*)

The Nokia TMR400 mobile, for a vehicle as well as desktop installation in an office. The TMR400 has similar capability as TMR420 described below.

The Nokia TMR420 mobile unit for car mounting. The TMR420 supports all the voice and data services of TETRA networks including trunked and direct mode and dual-watch function and has adjustable output power of up to 10 W. It provides TETRA air interface encryption for call privacy.

Figure 5.3 TETRA vehicle-mounted mobiles (*Courtesy of Nokia*)

MS receiver class – Three receiver classes are specified with their intended operating environments as shown below.

- Class A: optimised for use in urban areas and in areas with hilly or mountainous terrain.
- Class B: optimised for use in built-up and urban areas.

- Class E: intended to meet the more stringent requirements of quasi-synchronous systems. Quasi-synchronous systems are described in Section 5.14.3.

MS duplex capability – All MSs are required to support frequency half-duplex operation. An MS may optionally also support frequency full duplex operation. A frequency half-duplex MS can either transmit on an uplink frequency or receive on a downlink frequency at any time but is not able to transmit and receive at the same time. This type of MS requires time to switch from its transmit to receive frequency, which for TETRA must be less than one time slot duration. A frequency duplex MS, on the other hand, has the ability to transmit on an uplink frequency and receive on a downlink frequency at the same time. This type of MS can use all four uplink time slots and all four downlink time slots in a TDMA frame. A full duplex mobile may be used to support multiple or concurrent calls.

Support of concurrent calls – Concurrent calls may be supported by a full duplex MS as outlined above. Thus, there could be up to four simultaneous duplex voice circuits to different remote users, each with a different vocoder and a different encryption algorithm. This would also mean that a combination of voice and data teleservices to the same or different remote end points could be established.

Support of end-to-end encryption – MS support of end to end encryption is optional. Any end-to-end circuit switched user traffic encryption is in addition to air interface encryption which is standard.

5.3.6 Line Station

The line station (LS) functionally comprises the line termination unit (LTU) and the associated terminal equipment (TE). This is typical of a control room console terminal or dispatcher unit connected to a TETRA SwMI over an ISDN network. The essential difference between the MS and LS is the transmission media used, which is taken care of by the MTU for MS and by LTU for LS. Sections 5.6.2 and 5.6.3 describe the functional split of the TETRA LS.

5.3.7 Direct Mode Mobile Station

This consists of mobiles that communicate with each other directly without using the infrastructure in the trunked mode of operation. A Direct Mode Operation (DMO) mobile provides a point-to-point or point-to-multipoint communication by using the direct mode (DM) air interface. The TETRA standard specifies a number of options to extend the basic mode of operation:

- Direct mode repeater MS – to extend the range beyond two DMO mobiles;
- Dual mode switchable MS – to support both TETRA DMO and trunked TETRA V+D in dual-watch mode; and
- Direct mode gateway – to provide a link between TETRA DMO and TETRA V+D mode.

The different operating possibilities based on the above options can be described with DMO reference models [3]. Five DMO reference models have been defined and these are briefly described below.

- DM-MS: Direct Mode MS (DM-MS) is the basic DMO model where two mobiles communicate over the DM air interface, designated as U_d, in a "walkie-talkie" fashion. A DM-MS that initiates a call provides the air interface synchronisation reference and becomes the master DM-MS.
- DW-MS: Dual Watch-MS (DW-MS) model applies to a DM-MS, which is capable of both DMO and trunked V+D operation. A DW-MS can communicate with DM-MS or TETRA SwMI while at the same time monitoring for a V+D channel or DM channel. When idle, a DW-MS can monitor for both V+D and DM channels.
- DM-REP: Direct Mode Repeater (DM-REP) receives information from a transmitting DMO mobile and retransmits to another DMO mobile. A DM-REP mobile is regenerative in that it decodes the received information bursts and re-encodes them for retransmission in order to improve the overall link performance.
- DM-GATE: This model represents the link between TETRA DMO and TETRA V+D mode. A DM-GATE caters for the differences in protocol between the DM and trunked V+D air interfaces.
- DM-REP/GATE: This model is a special case of a combined repeater and gateway functionality possibly in single equipment. Such a combination can be easily catered for by a vehicle-based DM repeater and with additional gateway functionality for establishing a link to a TETRA V+D network.

5.3.8 Gateway

A gateway[1] enables calls to be set-up between users of a TETRA network and a non-TETRA network such as a public switched telephone network (PSTN). The need for a gateway specification [4] arises from the fact that other networks connected to a TETRA network use incompatible information formats and communication protocols for which some translation or conversion will be necessary.

Three types of gateways are envisaged: gateways for PSTN, ISDN and Public Data Network (PDN). The most important function of a TETRA gateway is in the role of interconnecting a TETRA infrastructure to an analogue PSTN since such interface involves signal coding between analogue and digital formats. For a digital ISDN gateway, on the other hand, no digital/analogue signal conversion would be necessary but digital speech transcoding will be required. Noting the analogue/digital signal conversion necessary for a PSTN gateway, the following are some of the main functions performed by a PSTN or ISDN gateway:

- Voice transcoding – TETRA uses a low bit rate voice coder at 4.567 kb/s (refer to Section 4.9), in contrast to a relatively high speech coder rate used in public networks

[1] This should not be confused with a DMO gateway which connects a DMO mobile to a trunked V+D.

(ranging from 32 kb/s ADPCM to 64 kb/s PCM). Voice transcoding would therefore be necessary to account for the codec incompatibility.

- Control signalling translation – The gateway allows signalling information to be passed between a TETRA infrastructure and an external network with appropriate translation so that the command and control signals will be interpreted correctly in each network. Some of the examples include translation of numbering schemes, and conversion between digital TETRA call control signals and analogue PSTN signalling which is mainly based on DTMF (dual tone multi frequency) signals.
- Echo cancellation – A PSTN gateway supporting full duplex calls may be required to incorporate echo cancellation for connection to a PSTN. Echo is caused by a combination of an excessive delay and a 2-wire subscriber circuit commonly used in the PSTN.

5.3.9 Network Management Unit

This provides local and remote network management functionality which is now becoming the norm in complex network systems like TETRA. This functions generally encompass system management for fault, configuration, accounting, performance, and planning.

The TETRA network management specifications are only concerned with the definition of standard management interface [5] and general requirements for interworking between different systems [6], [7], with the implementation of management functions left to network operators and equipment manufacturers. Chapter 9 will address some of the important aspects of network management.

5.4 SYSTEM INTERFACES

Section 4.3 has introduced TETRA's defined system interfaces. These interfaces are specified to ensure inter-operability between different network components regardless of the type of network implementation as summarised below.

- **Trunked Mode Air Interface (I1)**. This is the most important of the defined interfaces and ensures the inter-operability of mobile terminal equipment over the air interface.
- **Line Station Interface (I2)**. This interface is for terminals connected over the wireline connection (e.g. ISDN) as opposed to the I1 air interface specification.
- **Inter-System Interface (I3)**. This interface allows for interconnection of TETRA networks from different manufacturers.
- **Terminal Equipment Interface (I4 and I4')**. This interface standard facilitates the independent development of mobile to terminal equipment (interface I4), and line station (LS) to terminal equipment (interface I4').

- **Network Management Interface (I5)**. This interface standard caters for network management equipment inter-working with a TETRA network.
- **Direct Mode Air Interface (I6)**. This interface standard allows direct interoperability between mobile terminals that employ the TETRA DMO air interface.

5.5 TETRA REFERENCE CONFIGURATION

5.5.1 Basic Concepts

A reference configuration is a conceptual representation of a network system for the purpose of identifying various possible physical access arrangements. This conceptual representation has been in wide use since its formal introduction with ISDN standards as ITU-T Recommendation I.411[8].

Two concepts are used in defining a reference configuration:

- **Reference points**. These are conceptual points dividing functional groups. In a specific arrangement, a reference point may correspond to a physical interface between pieces of equipment, or in some cases there need not be any physical interface corresponding to the reference point.
- **Functional group**. This represents a set of functions which may be needed in a network access arrangements. Functional grouping can be used to specify the possible combinations or arrangements of a physical attachment to a network.

In the context of describing a TETRA system, the above conceptual representations can serve two important purposes.

- On the user interface side: for identifying the various possible physicals access arrangements.
- Both on the user interface side and network side: for identifying the various protocols that operate at various points within a network. All interface points within a network are identified by a reference point.

The user side of reference configuration is used for identifying the different subscriber terminals envisaged for connection to the network. For instance, mobile terminals with different interface capabilities can be easily specified with appropriate reference configurations without resorting to implementation details. User interface reference configurations highlight important system aspects such as user terminal types, support for peripheral equipment interface and external data services. This topic will be discussed in Section 5.6.

The TETRA standard specifies distinct reference points for connections to terminals as well as other networks. The reference points are used to designate the various network protocols (and protocol stacks) for packet mode services in V+D and PDO systems. This point will be clarified further in Chapter 8 but for now it will suffice to summarise the reference points with brief descriptions as shown below.

5.5.2 Reference Points in TETRA

Reference points in a TETRA system represent the physical interface points and they are marked R0 to R10. These reference points are specific to TETRA, i.e. although the conceptual representation is based on ITU-T I.411 principles, the reference points described below are outside the ITU-T I.411 specification.

- **R0:** This is the reference point within the mobile termination unit (MTU) corresponding to the network service boundary.
- **R1:** The reference point between packet mode terminal equipment and the MTU; there may be several alternative interface protocols at R1, including existing packet mode standards.
- **R2:** The reference point at the TETRA air interface. The main TETRA protocol architectures exist at this point.
- **R3:** The reference point between type TE2 LS and TETRA network.
- **R4:** The reference point for character mode terminal equipment connected to a packet assembler/disassembler (PAD). There may be several alternative interface protocols at R4, including existing PDN standards.
- **R5:** The reference point between the network management unit (NMU) and the TETRA network.
- **R6:** The reference point between one TETRA network and another TETRA network.
- **R7:** The reference point between one TETRA network and a non-TETRA packet data network.
- **R10:** A logical reference point equivalent to R0.

The reference points within the TETRA reference configuration specification will be described with brief remarks in the sections to follow. Their significance will be apparent as the functions and related protocols are elaborated at appropriate junctures. Most protocol related issues will be discussed in Chapter 8.

5.6 SUBSCRIBER ACCESS INTERFACES

The TETRA infrastructure supports various speech and data services that are accessible to the subscriber either through the MS (over the air interface), or through the LS (over the ISDN line). The subscriber side of interfaces is of major importance both to equipment manufacturers and end users if compatible terminals and flexible customer choices have to be realised. This suggests that the subscriber-access interface standardisation should take into account the anticipated conceptual configuration of customer premises equipment and the necessary interfaces.

Such a conceptual configuration is already in existence for ISDN under subscriber-access network *reference points and functional groupings* [8]. This principle is used as the basis for defining the TETRA (as well as GSM) subscriber access interfaces. An overview of

ISDN subscriber-access network reference configuration is first presented followed by the TETRA-specific definitions for MS and LS interfaces. Although the configuration for TETRA MS is expected to be somewhat different (due to its link over the air interface) a reference configuration functionally similar to that of TETRA LS is used.

5.6.1 The ISDN Reference Configuration

The ISDN approach to the subscriber-access network configuration is depicted in Figure 5.4. This is essentially a decomposition of the subscriber-access network to aid the development of standard ISDN implementations and also provide a clear demarcation between the end user and the network provider.

- The functional groupings define the possible combinations and arrangements of the physical user equipment. These include
  - Network termination (NT) functional group: Three types of NT on the network side: NT1, NT2 and NT12;
  - Terminal equipment (TE) functional group: Two types of TE physical interfaces identified on the subscriber side as: TE1 and TE2;
  - Terminal adapting (TA) functional group.
- The reference points designate the interfaces that separate the functional groups. These are *reference point R* (R for rate adaptation, provides a non-ISDN interface), *reference point S* (S for system, separates user equipment from the NT2 interface), and *reference point T* (T for terminal, separates user NT2 equipment from NT1 interface).

Network termination 1 (NT1) is the physical termination of the ISDN on the customer's premises and providing OSI layer 1 functionality. It forms the network boundary and is under the control of the network provider.

Network termination 2 (NT2) performs additional functions to the NT1 interface in order to provide full capabilities of an ISDN interface. These additional functions include multiplexing, concentration, and switching which may correspond to OSI layers 2 and 3. The NT2 resides at the customer side and can be an ISDN PBX serving several terminals over *reference point S*, or part of each customer equipment for interface with the NT1 over *reference point T*.

Network termination 12 (NT12) is a single termination with the combined functions of NT1 and NT2. Such termination may be used where the ISDN network provider also provides full service to the subscriber. In contrast to this approach, it should be clear that the splitting of the subscriber-network interface into NT1 and NT2 would allow competitive service provision (by other service providers) through the NT2 interface.

Terminal equipment type 1 (TE1) corresponds to customer equipment that supports standard ISDN interface. Examples include ISDN telephones, ISDN data terminals, digital fax machines, and various ISDN-compatible ancillary ports (e.g. for PC, fire/intruder alarm, etc).

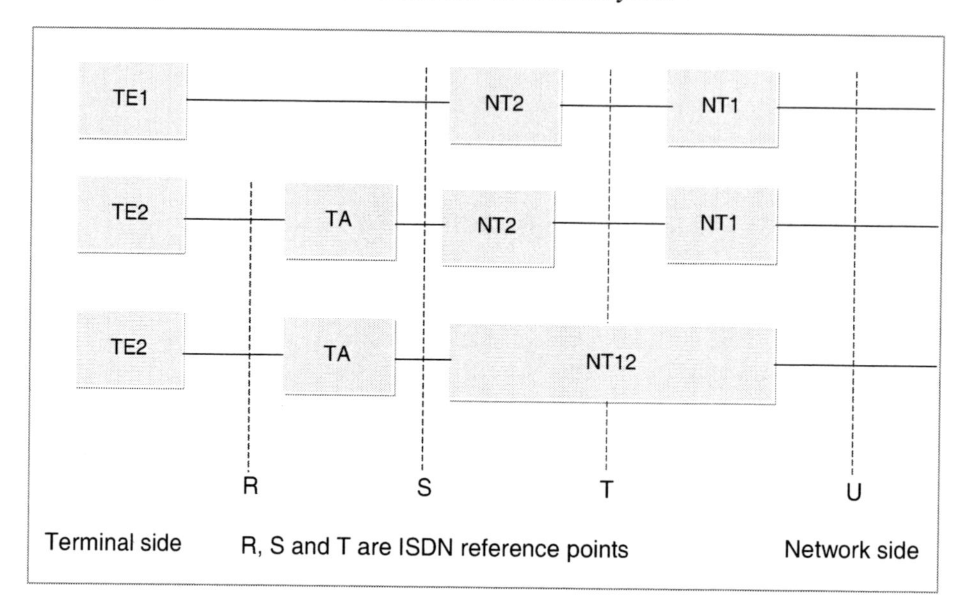

Figure 5.4 ISDN user-network interface reference configurations

Terminal adapting (TA) corresponds to customer equipment that provides the interface between non-ISDN TE2 functional group and NT2 functional group. The main functions of the TA are asynchronous (non-ISDN) to synchronous (ISDN) conversion, rate adaptation, and flow control. These TA functions are defined under the V.110 standard [9].

Terminal equipment type 2 (TE2) corresponds to customer equipment which is based on non-ISDN physical interface such as V.24/RS-232 asynchronous terminals, X.25 data terminals (on X.21 physical link), as well as various others based on non-standard or proprietary protocols. Such equipment requires a TA to plug into an ISDN interface as depicted in Figure 5.4.

5.6.2 Circuit Mode TETRA Reference Configurations

5.6.2.1 V+D LS reference configuration

A LS is connected to the network through the ISDN line and its user interface is therefore defined in terms of the ISDN reference configuration. The LS supports the following functionalities:

- NT functional groups NT1 and NT2
- NT functional group NT1 + NT2 (equivalent to ISDN NT12)
- TE functional group TE1 and TE2
- TA functional group

The above are identical to the ISDN configuration described under Section 5.6.1 and will not be repeated here.

5.6.2.2 V+D MS reference configuration

The access configuration for V+D TETRA MS comprises the following:

- Functional groupings:
  - mobile termination (MT);
  - terminal equipment TE2 (similar to TE2 for LS).
- Reference Point: R_T as a boundary between MT and TE2.
- Access Points:
  - AP2 as the access point for TETRA bearer services;
  - AP3 as the access point for man machine interface (MMI) to the user.

The MT supports the following general functions:

- air interface termination (Um);
- radio channel management;
- mobility management (MM);
- speech and data encoding/decoding;
- error protection/correction across the radio path for all information (speech, signalling, user data);
- flow control and mapping of signalling and user data;
- rate adaptation between user data and radio channel rate;
- support of TE.

There are two types of MT defined:

- MT0 with support of non-standard terminal interfaces that provide TE functionality;
- MT2 with the TE2 terminal interface, i.e. an asynchronous serial interface. The TE2 interface is used to manage the V.24 physical layer between an external terminal (e.g. a PC) and an MS over the R_T reference point.

Also refer to the definition of the peripheral equipment interface in Section 5.7 below.

A conceptual representation TETRA MS reference configuration is depicted Figure 5.5. In short, two types of TETRA MS are envisaged: one with a standard serial interface which can be used for interface to an external peripheral equipment (see also Section 5.7 on peripheral equipment interface), and the other without a serial interface. Both types support MMI (man machine interface) typically in the form of an LCD display and a keypad.

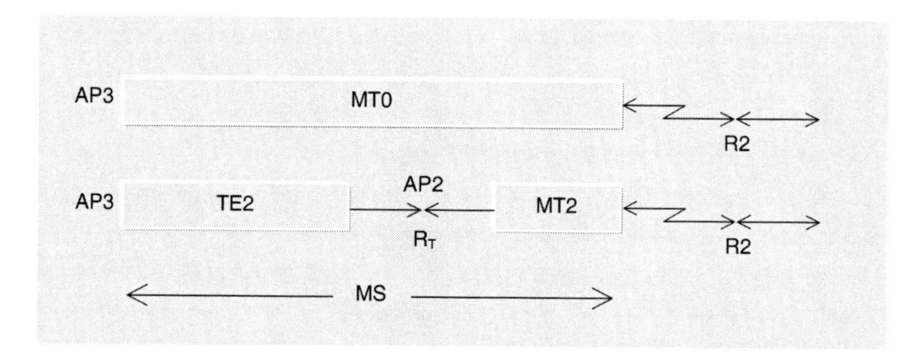

Figure 5.5 TETRA MS reference points and functional groupings

5.6.2.3 Direct mode TETRA MS reference

The DMO mobile supports only circuit mode services and it can be described with reference to the V+D TETRA MS configuration shown in Figure 5.5, with the following minor variations:

- The V+D air interface is replaced by the DMO U_d air interface.
- Unlike the V+D system, the user interface to the TETRA-specific equipment TE2 in is not specified.

The TE2 implementation for DMO may therefore be either proprietary or standardised as for V+D system. For a dual mode mobile supporting both trunked V+D and DMO, there is good reason to implement the same terminal equipment functionality.

5.6.3 Packet Mode Reference Configurations

The packet mode reference configurations described below apply both to V+D and PDO systems. Figure 5.6 shows packet mode TETRA MS reference points. TE3 is a character mode TE connected to a PAD. Note that MTU3 integrates the functionality of the PAD into the mobile termination unit. Packet mode LS reference configurations are shown in Figure 5.7 and in this case only TE2 and TE3 terminals are supported.

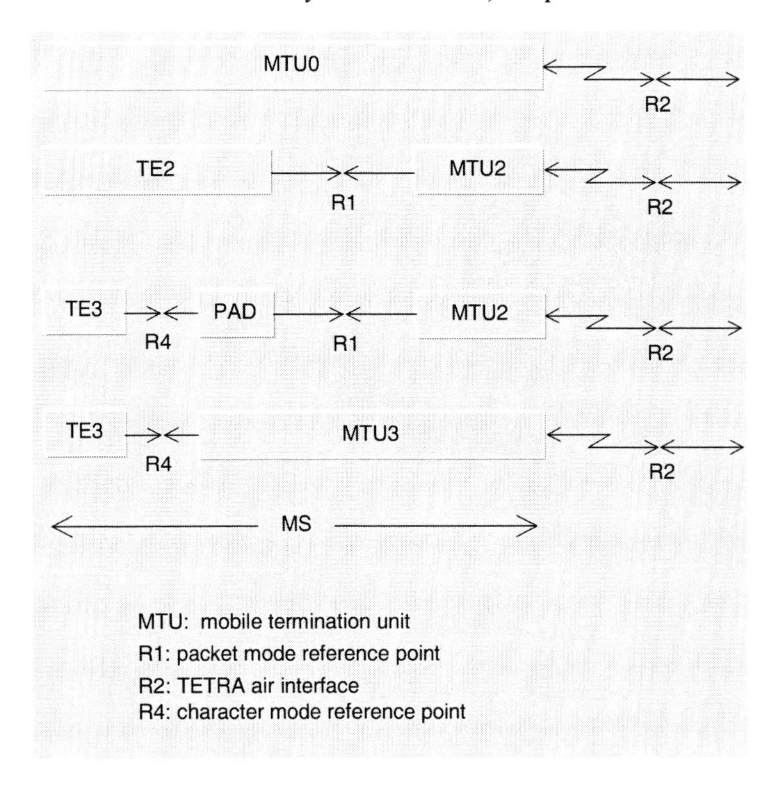

Figure 5.6 Packet Mode TETRA MS reference points

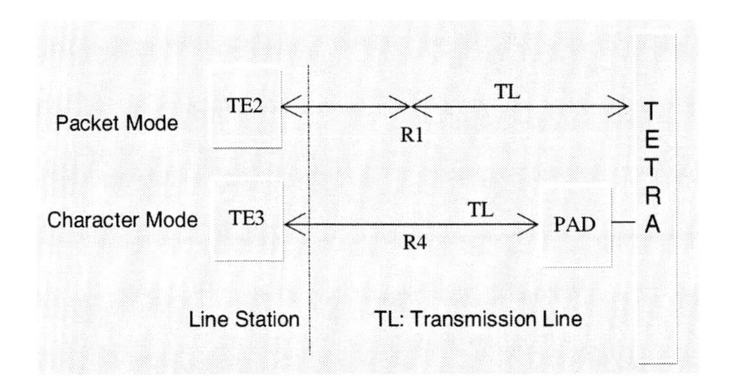

Figure 5.7 Packet mode TETRA LS reference points

5.7 PERIPHERAL EQUIPMENT INTERFACE

5.7.1 Overview

For the majority of the user population the telephone line provides the gateway to external information services. The dependence on the telephone line, be it at home or on the move, is set to continue with the ever increasing demand for Internet information access and personal communication needs such as fax, email, and soon with multimedia services such as videoconferencing.

The possibilities will be even greater with telephone terminals that are digital, mobile, and capable of multimedia support. All of these attributes apply to the V+D TETRA mobiles. With the data rate of up to 28.8 kb/s per carrier, TETRA mobiles offer the potential for supporting access to numerous network services. Therefore, it comes as no surprise that the TETRA standards embrace the peripheral equipment interface (PEI) specification [10] in order to realise the full potential of mobile to peripheral equipment interface from the outset.

5.7.2 The TETRA PEI

Unlike the days of GSM standardisation in the early 1980s, hardware and software support for access to data services through the telephone terminal are now relatively mature and stable enough to recommend specifications beyond the physical interface. In this context, three important developments should be noted from the experience of the IT industry over the recent past:

- The widespread use of modems for access to information services, together with the continual improvement of modem technology for higher transmission rate over an analogue voice channel (currently up to 56 kb/s);
- The significant convergence and improvement of the data link protocols for dial-up networks through the point-to-point protocol (PPP) [11], which is now a common standard for dial-up access to the Internet; and
- The development of wireless data systems such as the General Packet Radio System (GPRS), Cellular Digital Packet Data (CDPD), Mobitex and ARDIS.

The TETRA PEI specification mainly attempts to build on the above experiences and with the objective of minimising additional hardware and software requirements for the interface between TETRA mobile and peripheral equipment. The PEI specification is concerned with a link between a TETRA MT2, which is designated as reference point R_T, and a terminal equipment TE2, which is provided as a connection to a peripheral device such as a portable PC as shown in Figure 5.8. It should be noted, therefore, that only type MT2 mobile is expected to support the TETRA PEI. On the other hand, type MT0 mobile, at least from its reference configuration standpoint, is not expected to support the PEI functionality as it lacks a standard TE2 interface.

The key requirements of the TETRA PEI specification are [10]:

- a standard physical interface to all MT2s which is widely adopted in the IT world;
- minimal extra software in the TE2, at most only a device driver should be required;
- maximum compatibility with other wireless data systems;
- access to the full range of MT2 functionality.

The last point refers to user interaction with mobile-oriented functionalities, such as:

- set-up and control of speech calls, including call parameters;
- access to general information of MT2 and network, e.g. battery status, received signal strength;
- access to user applications located in MT2, e.g. short data service (SDS).

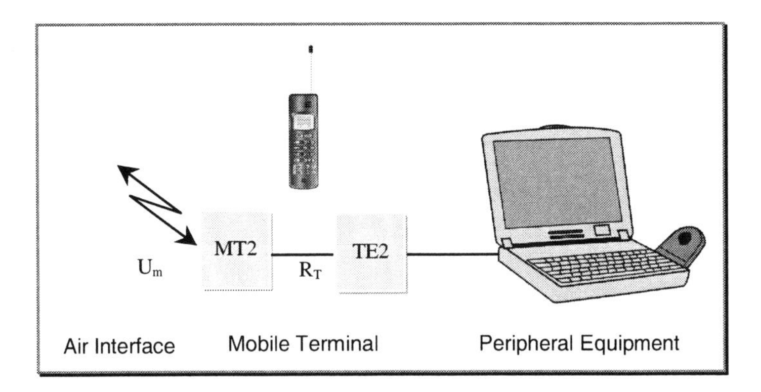

Figure 5.8 Components of the TETRA peripheral equipment interface

5.7.3 PEI Components

The TETRA PEI specification comprises three categories of service access as outlined below.

5.7.3.1 Modem functionality with AT commands

The AT command set (AT for *Attention*) is now a de facto standard for controlling modems (for configuring, testing and data communication) from a PC or other intelligent terminals. The TETRA PEI specification supports a number of AT commands for basic modem functionality that permits access to circuit data services, short data services (SDSs) and MT parameters and information. These include:

- Circuit mode data: with circuit mode connection, AT commands can be used as in conventional modem including dial-up applications;
- Short data service: AT commands can be devised to interrogate an SDS message handler to check if a message is received (or sent), read/unread status of a received message;

- Radio configuration: commands relating to network service and call control may be issued with AT commands. Examples include enabling/disabling encryption, selection of circuit mode error protection type, and enable/disable network registration;
- Radio status/configuration information: numerous radio and terminal status information can be retrieved with AT commands. Examples include radio signal quality, system parameters transmitted over the broadcast channel (cell selection, power control, etc.), battery status, SIM card related status, etc.

5.7.3.2 PPP and packet data

Packet data services at the PEI is supported by the PPP data-link protocol. The PEI also provides access to the following connections:

- IP version 4 (IPv4);
- IP version 6 (Ipv6);
- ISO connectionless data service; and
- X.25 services.

More explanations on the last two items can be found in Chapter 8. References [12] and [13] provide good coverage on the subjects of IPv4 and IPv6 and beyond.

5.7.3.3 PPP and radio remote control

This component of the TETRA PEI specifies a protocol that allows access to Circuit Mode Control Entity functions (call control, SDS and supplementary services), and mobility management functions from a TE2 peripheral device such as a PC. Some of the services supported in this category are also accessible with AT commands as described above. The specific services accessed include:

- circuit mode data
- short data service
- radio configuration
- speech call control
- radio status/configuration information
- supplementary services

TETRA PEI is set to offer flexible interface functionalities, which no other system has offered before at the standardisation level. With the functionalities such as AT commands and PPP connections built into a TETRA terminal, peripheral equipment interface should be a lot easier to configure and use since the task of setting up communication services can be handled by application set-up with minimal involvement by the user.

5.8 ADDRESSING AND IDENTITIES

Addressing is an important aspect of any communication system. In mobile communication systems in particular, where equipment and subscriber mobility is intrinsic to the system operation, it is also essential that equipment and subscriber identification schemes be defined so that they can be located on the move. TETRA uses a scheme similar to GSM for identification and addressing, with one notable difference in group-call identity, which is unique to TETRA.

The TETRA standard defines and specifies the addresses and identities, and their organisation in groups corresponding to the different functions. The TETRA identities are used to distinguish between equipment, subscribers and networks in the international context, while the TETRA addresses provide compatibility with other non-TETRA networks.

TETRA addresses and identities are designed to meet the following objectives [14]:

- to allow a large number of networks and network operators to co-exist, and for each network to support a large number of subscribers;
- to be able to uniquely identify any subscriber in any network.;
- to allow the use of shortened identities for intra-TETRA calls to reduce the signalling information in the set-up messages;
- to support efficient roaming and migration of subscribers.

TETRA addresses and identities are organised into the following groups:

- TETRA Equipment Identity (TEI)
- Mobile Network Identity(MNI)
- TETRA Management Identities (TMI)
- TETRA Subscriber Identities (TSI) and Short Subscriber Identities (SSI)
- Network-layer SAP addresses (NSAP)

The most important aspects of the above will be outlined in the following subsections.

5.8.1 TETRA Equipment Identity (TEI)

The TEI uniquely identifies one piece of TETRA equipment (MT or NT) internationally. This corresponds to the *International Mobile Equipment Identity* (IMEI) of GSM. The TEI it is allocated by the equipment manufacturer and used by the network operator to support various security and operational functions. As an example, a mobile terminal may be disabled and denied of service by identifying with its TEI if reported stolen or known to be obsolete or non-functional. It is also possible that a mobile terminal may be monitored with some operational or security checks over an extended period before it is barred. Likewise, terminals with valid TEI can be enabled over the air by the infrastructure to have access to services. The security of this mechanism of course relies on the difficulty of reprogramming the TEI by unauthorised individuals. The contents of the TEI are formed from three fields and a spare field as shown below.

6 digits	2 digits	6 digits	1 digit
Type Approval Code (TAC)	Final Assembly Code (FAC)	Electronic Serial No. (ESN)	Spare (SPR)

Figure 5.9 Contents of TETRA Equipment Identity

The connection of a terminal to a telecommunications network requires that the terminal be type-approved by an accredited test laboratory, independent of manufacturers and operators. For mobile communications, in particular, the need for type-approval, ideally to a common standard, is of critical importance to ensure that the equipment is functioning correctly when roaming different networks.

The type approval code (TAC) in Figure 5.9 identifies the test laboratory that carried out and granted the type approval. The final assembly code (FAC) identifies the manufacturer and place of equipment assembly. Both the TAC and FAC are allocated by a central body. The electronic serial number (ESN) is allocated by manufacturers in sequential order and it uniquely identifies each equipment within the high order numbers represented by TAC+FAC.

5.8.2 Mobile Network Identity (MNI)

The MNI is broadcast by all TETRA base stations to uniquely identify a network. As shown in Figure 5.10, the MNI is composed of the Mobile Country Code (MCC), identifying a country, and the Mobile Network Code (MNC), identifying a network.

10 bits	14 bits
MCC	MNC

Figure 5.10 Contents of TETRA MNI

The MCC uses a 10-bit field[2] to encode a 3-digit decimal country code as specified in the CCITT Recommendation for International Numbering Plan [15]. The MNC, on the other hand, is allocated to each operator (possibly by the national body for each country) based on a 14-bit binary field.

In the system broadcast of MNI, it is possible that additional operator information may be included as part of the broadcast information. The MNI is also used as part of the network management and subscriber identities are described below.

[2] Since the MCC is a 3-digit decimal, the binary values corresponding to decimal 1000 to 1023 are not used.

5.8.3 TETRA Management Identities (TMI)

The TMI is a non-transferable network layer 3 identity that should be allocated to a termination before it can be used. The TMI is allocated by the network operator and is used as an address only by the internal network management functions and therefore not accessible to normal network users. This also means that a visitor's TMI cannot be allocated to a migrating station, hence restricting the allocation only to the home network. It is also possible that secure networks may allow no TMI functions at all.

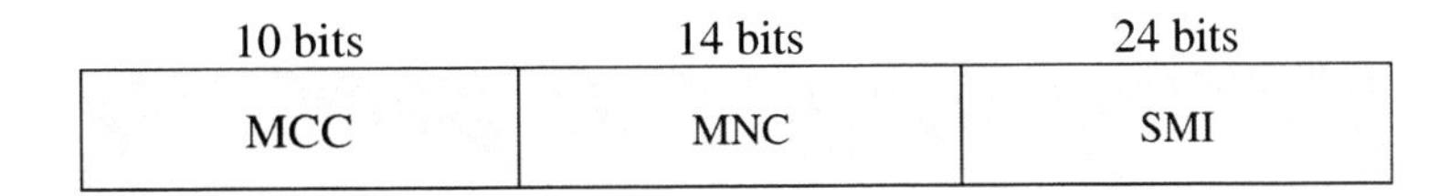

Figure 5.11 Contents of TETRA TMI

Contents of TETRA TMI, shown in Figure 5.11, are formed by concatenating a 24-bit field that identifies a network specific Short Management Identity (SMI), to the MNI described in Section 5.8.2.

5.8.4 TETRA Subscriber Identity (TSI)

The TSI, with its contents as shown in Figure 5.12, is an identity by which subscribers are known to mobile operators for subscribed services (possibly personalised according to individual needs), and for billing purposes. This information is usually stored in the subscriber identity module (SIM)[3] which is inserted into a mobile terminal to personalise the terminal. The TSI is also used as a network routing address with its network-specific part used as a MAC address.

One key aspect of the TETRA mobile system is the existence of group identities to facilitate group calls. Group identities within TETRA are treated in an identical way to individual subscriber identities and with the same allocation structure of Figure 5.12. The two types of TSI are distinguished as individual TSI (ITSI) and group TSI (GTSI). Unless indicated otherwise, any description of TSI will apply to both ITSI and GTSI.

A terminal (MS or LS) should have at least one valid ITSI before it can be used. To support secure network operations, the TETRA standard also defines a subscriber identity known as the Alias TETRA Subscriber Identity (ATSI) that can be associated with, and used in lieu of ITSI. The concept of secure communication with ATSI also extends to supporting subscriber roaming in visited networks where a temporary ATSI would be required. The ITSI-ATSI pairing is only known to the network operator, and subscribers will still be known to others only by their ITSI. At any given time, only one ATSI per ITSI will be in operation, although the network operator can change the definition of ATSI from time to time. For secure communication, the TETRA standard stipulates that

[3] The physical size of a SIM is commonly that of a credit card although smaller sizes are also used.

the ATSI should not be derived from a knowledge of the ITSI. There is no ATSI equivalent for group addressing.

In short, there are therefore three defined subscriber identities: the ITSI, ATSI and GTSI, all with identical TSI structure but only different in the way they are used and interpreted. Both ATSI and GTSI are expected to be assigned dynamically whereas the ITSI assignment is clearly a long term one. The partitioning of the address space between ITSIs, ATSIs and GTSIs is also non-distinguishable outside a given network.

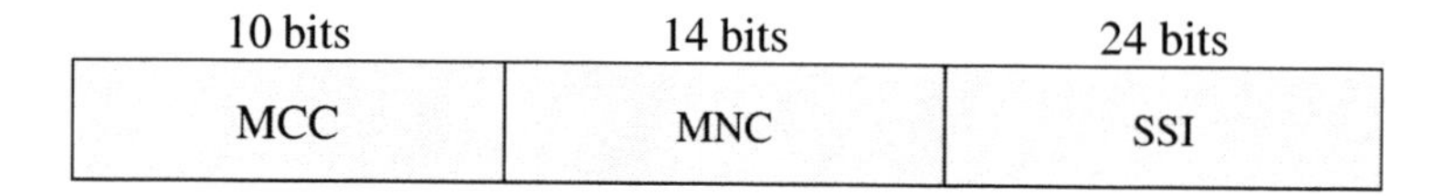

Figure 5.12 Contents of TETRA TSI

The TSI is composed of the Mobile Network Identity (MNI) part described in Section 5.8.2, and the short subscriber identity (SSI) as depicted in Figure 5.12. The TSI will be unique across all TETRA networks whereas the SSI will only be unique in one TETRA network. To take advantage of a short intra-network addressing with the SSI part, subscriber identities actually exist in two sizes:

- the full 48-bit long TSI, corresponding to ITSI, ATSI or GTSI, and
- the network-specific 24-bit truncation of the above: individual SSI (ISSI), alias SSI (ASSI), or group SSI (GSSI).

5.8.5 Network-Layer SAP Addresses (NSAP)

NSAP addresses are an additional method of addressing intended to provide direct compatibility with non-TETRA networks such as ISDN or public data networks. These addresses conform to one of the existing international standards (PSTN, ISDN, or data networks) and support direct inter-working with existing fixed networks by associating or *binding* them with ITSI numbers. The binding can be static, if an NSAP is allocated to a particular SIM card or terminal, or dynamic, if the association with the SIM card or terminal can be changed. With this approach, external data communications protocols such as X.25 and Internet IP can be easily accommodated, with prevalent use of this expected in the packet mode services.

NSAP addresses may also be used with V+D system for routing calls from external users (e.g. on call diversion). For the V+D system, however, the use of NSAP addresses is an operator option. For added flexibility, it is possible that the dynamic binding of NSAP address is carried out, not only by the network operator, but also by the user.

Since existing numbering plan standards do not support group calls, NSAP addressing is not bound to a GTSI.

5.9 TETRA NETWORK SERVICES

5.9.1 Overview

Telecommunication services in TETRA, as well as GSM, follow the ISDN model of grouping as bearer services, teleservices, and supplementary services. An overview of this has been presented in Section 4.2. Bearer services are basic communication facilities provided by the network at layers 1 to 3 of the OSI protocol stack, while teleservices make use of the 7 OSI layers to provide end-to-end services (Figure 4.2).

Bearer services are concerned with *how* data is transported from point to point regardless of the type of application or content of data being transported. They provide the basic "digital pipes" that are used as building blocks for more sophisticated network applications. TETRA supports three bearer services:

- circuit mode data;
- packet mode data, and
- short data service (SDS).

These are summarised in Table 5.3 for the three types of TETRA systems: V+D, PDO and DMO. The three TETRA systems can be identified in terms of services they support:

- V+D system – provides circuit mode speech and data, packet mode data, and SDS over four time slots.
- PDO system – provides only packet mode data over the equivalent of all four time slots in the V+D system.
- DMO system – provides circuit mode speech and data services, and SDS over a single time slot.

Table 5.3 Summary of TETRA bearer services

	circuit mode data (CMD) (kb/s)				packet mode data		short data service (SDS)				
CMD error protection high⇨ low ⇨ none ⇨	1-slot 2.4 4.8 7.2	2-slot 4.8 9.4 14.4	3-slot 7.2 14.4 21.6	4-slot 9.6 19.2 28.8	CONP	S-CNLS	pre-defined status (bits)	Type 1 16-bit	Type 2 32-bit	Type 3 64-bit	Type 4 ≤ 2047 bits
V+D	✓	✓	✓	✓	✓	✓	32,767	✓	✓	✓	✓
PDO					✓	✓					
DMO	✓						16	✓	✓	✓	✓

Note: CONP: connection-oriented data service S-CNLS: specific-connectionless data service

5.9.2 Circuit Mode Data

As shown in Table 5.3, circuit-mode data services are provided by V+D and DMO systems. The V+D system supports a multi-slot operation, that is, in addition to a single channel operation, channels can be aggregated as 2, 3 or 4 for increased capacity. For the DMO system, only a single slot operation is supported.

The circuit-mode data bearer services provide end-to-end circuit-switched connections over the air interface. Since no network routing is necessary, the bearer services are essentially made available at the data link or layer 2 of the OSI model.

A number of circuit-switched services are supported depending on the level of data error protection provided and the number of channels aggregated under multi-slot operation (Table 5.3). The level of data protection chosen for a connection will vary according to the nature of the data being transported. For applications that can tolerate some degradation, it may be appropriate to transmit low-protected or unprotected and accept any errors that may occur. Speech and image transmissions fall into this category of applications. On the other hand, data that must be transferred intact or with very low bit error rate transmission (signalling messages, user data, etc.) would be provided with the highest level of protection. The connection capacity, primarily determined by the protection level and the number of channels aggregated, is also dependent on a number of factors such as the volume of data to be sent, the required response time, and the cost of service.

5.9.3 Packet Mode Data

Packet mode data services are supported both by V+D and PDO systems. The V+D system uses a TDMA scheme which is also used for transmitting digitised speech. The PDO system, on the other hand, employs a statistical multiplexing scheme at the gross rate of 36 kb/s over the whole channel provided by a carrier (equivalent to 4 V+D time slots). Due to the bursty nature of data traffic (as opposed to a relatively regular traffic pattern for real-time speech, for instance) statistical multiplexing schemes generally perform better for data-oriented applications. This is indeed the reason why the PDO emerged as a standard.

TETRA supports two types of packet mode data services: connection-oriented and connectionless. The connection-oriented service establishes virtual connections between the sending and receiving terminals in order to transfer data packets as needed. The connectionless service, on the other hand, transfers a single packet of data without establishing the virtual connection. The functionality it offers is similar to that of the Internet Protocol (IP). More details on packet mode data services in connection with network protocols will be presented in Chapter 8.

5.9.4 Short Data Service

The TETRA short data service (SDS) is a datagram service, which is an unconfirmed service optimised for the exchange of predefined status messages or user-defined messages. Both point-to-point (individual messages) and point-to-multipoint (group or broadcast messages) are supported. SDS messages use the spare capacity of the signalling channel which is available from time to time. As such, SDS messages do not require an established channel of their own and exist as part of signalling. SDS messages can therefore be sent or received in parallel with an ongoing speech call. On the other hand, if the destination terminal is unavailable, the SDS message can be stored for later delivery.

Predefined status messages provide the most efficient coding of user messages where possible. Messages are simply designated by a 16-bit integer value, thus capable of representing up to 65535 distinct messages. For instance, predefined message number 0 designates an emergency call. Status messages are further categorised into two types by dividing the range as follows:

- 0 to 32767 reserved for system use.
- 32768 to 65535 available for TETRA network and user specific definitions.

User messages may be one of three fixed lengths in size of 16, 32, or 64 bits, or variable in length up to 2047 bits. User SDS types are defined as follows:

- Type 1 = 16 bits (fixed)
- Type 2 = 32 bits (fixed)
- Type 3 = 64 bits (fixed)
- Type 4 = 2047 bits maximum (variable)

5.9.5 Teleservices

A teleservice is a system service as seen by the user via the man machine interface of the terminal. It provides the complete capability, including all terminal functions, for communication between users.

TETRA teleservices include clear or encrypted speech communication with the following categories:

- **Individual call.** This is a point-to-point call established between two parties. The calling party receives acknowledgement of the call progress (answered, unanswered, rejected etc).
- **Group call.** This is a point-to-multipoint call established from one user to more than one individual. Call is established immediately and caller does not receive acknowledgement from any of the called individuals as to whether or not they are ready to communicate. This is the case because in this type of call a fast call set up time is considered to be the main objective.
- **Acknowledged group calls.** This is the same as the group call just described, except that, a call is not established unless acknowledgement is received from a defined number of called parties that they are ready to communicate. The group members may

be polled for presence and if the number of group members is deemed insufficient then the call may be cleared.

- **Broadcast calls.** This is a point to multipoint call from one party to more than one individual. With a broadcast communication, a one-way call is established immediately and called parties are not permitted to respond.

A group call is a semi-duplex communication between a calling user (commonly a dispatcher at a control centre) and one or more called users. Members of a group have a common number by which they will be addressed, and this group number can be dynamically modified and assigned at any stage. The calling user selects a group, and activates an equivalent of the PTT (press-to-talk) button to broadcast his or her voice to group members. The call ownership can be transferred to another member of the group by invoking the appropriate supplementary service which is described in the following section.

5.9.6 Supplementary Services

The TETRA standard defines supplementary services, which can modify or supplement the basic services. There are 30 supplementary services in total. These include PMR style capabilities such as call priority control and dynamic group number assignment; as well as telephone style services such as call diversion and call waiting. The following summarises the supplementary services under related grouping.

5.9.6.1 Call identification services

These are services that, when activated, can provide information regarding the identification of the users. Several services are provided in this category as listed below.

- *Calling Line Identification Presentation.* This is a service offered to the called user and identifies the calling user's number and, if applicable, the calling user's subaddress.
- *Connected Line Identification Presentation.* This identifies the connected user's number and, if applicable, the connected user's subaddress. Notice the difference between "calling user" and "connected user".
- *Calling/Connected Line Identification Restriction.* Prevents display of calling user's identification to others, i.e. it prevents calling or connected line identification from working.
- *Call Report.* Indicates identity to the called user who is not contactable, busy, or does not answer.
- *Talking Party Identification.* Informs all connected parties of the identity of the talking user.

5.9.6.2 Call forwarding services

With call forwarding services, calls incoming to the called user are, subject to conditions, diverted to another destination as defined by the called user at the time of activating the service. Three call forwarding conditions as well as unconditional forwarding are defined as summarised below.

- *Call Forwarding on Busy.* Instructs the network to send all incoming calls that encounter a busy condition to another number.
- *Call Forwarding on Not Reachable.* Instructs the network to send all incoming calls to another number when the called number is not reachable.
- *Call Forwarding on No Reply.* Instructs the network to send all incoming calls to another number on no reply.
- *Call Forwarding Unconditional.* Instructs the network to send all incoming calls to another number.

5.9.6.3 Call offering services

- *List Search Call.* Allows the user to define one or more alternative numbers to which the infrastructure will attempt to route unsuccessful incoming calls.
- *Call Authorised by Dispatcher.* Requires the dispatcher to verify and approve a call request before the call is allowed to proceed.
- *Short Number Addressing.* Allows the use of predefined short numbers (abbreviated addresses) instead of full addresses.
- *Area Selection.* This allows the user to choose a geographic area for outgoing individual calls. For incoming call area selection, this restricts the reception of incoming group calls to only those received when the user is in the defined area.
- *Access Priority.* This allows the user to gain access to the network when the radio link is congested.
- *Priority Call:* The infrastructure will give priority access to network resources to calls that have been sent with priority status.
- *Call Retention.* This allows the user to set a level of protection for calls against the probability of having the network connection resources pre-empted.

5.9.6.4 Call completion services

This category provides several services to do with the manner how a call connection is completed.

- *Call Waiting.* This alerts the user of an incoming call whilst busy on another call. The waiting call can be accepted, rejected, or ignored.

- *Call Hold.* Allows the user to put individual calls on hold, or multipoint calls if he or she is the call owner.
- *Call Completion on Busy Subscriber.* Automatically completes the call when calling user encounters a busy user. This is also referred to as *call back when free.*
- *Late Entry.* During a multipoint speech call, the infrastructure will send indications related to this call to allow group members to join the ongoing speech call.
- *Transfer of Control.* Allows the originator of a multipoint call to transfer ownership of the call to another user and then leave the call without it clearing down.
- *Pre-emptive Priority Call.* Will allocate resources to a user, even if this means that other calls with lower priority are disconnected.
- *Call Completion on No Reply.* This service automatically completes the call when the calling user encounters a user who does not reply. The caller essentially requests that the call be automatically completed when the called destination is next used. This is also referred to as *call back when next used.*

5.9.6.5 Multi-party services

- *Include Call.* Enables the user, while being involved in an active call with another user, to call a third user and have them included in the original call.
- *Dynamic Group Number Assignment.* Permits the user to create, modify and delete group(s).

5.9.6.6 Call restriction services

- *Barring of Outgoing Call:* Allows the user to bar selected categories of outgoing calls from being set-up by other users.
- *Barring of Incoming Calls:* Allows the user to bar selected categories of incoming calls from being received by other users.

5.9.6.7 Call intrusion services

- *Discreet Listening.* Allows a user to listen to one or more communications between TETRA subscribers without any indication that the communication is being monitored. Identification of the talking parties is given to the monitoring user.
- *Ambience Listening.* This service allows the calling user to place a TETRA MS or LS into a special type of individual voice call so that the called terminal transmits without any action from, or indication to, the called user. With this service enabled, the TETRA terminal essentially becomes a remote microphone for monitoring some emergency situation.

5.9.6.8 Charging services

- *Advice of Charge.* This service allows the user to receive information concerning charges for a call.

Table 5.4 summarises supplementary services that are intended for public safety applications. These should be self explanatory from the brief descriptions given above.

Table 5.4 Supplementary services essential for public safety applications

Call Offering Services:
Access Priority
Area Selection
Call Authorised by Dispatcher
Call Retention
Priority Call
Call Intrusion Services:
Ambience Listening
Discreet Listening
Call Completion Services:
Late Entry
Pre-emptive Priority Call
Multi-Party Services:
Dynamic Group Number Assignment

5.10 MOBILITY MANAGEMENT

5.10.1 Basic Principles

Mobility management is central to the concept of cellular mobile communications and is one of the key features that distinguishes a mobile network from a fixed network. The task of mobility management is essentially to track down the whereabouts of a mobile user. For this, a network database subsystem is generally used for maintaining the record of MS locations through registration procedures. Before describing a network database subsystem, the following paragraphs introduce some of the underlying concepts in mobility management.

5.10.1.1 Location area

Cell by cell registration is usually not desirable as this can increase the signalling traffic that could be wasteful of the communication resources. The increased signalling traffic could be even more pronounced if the cell sizes are small, thus increasing the cell crossing rate and consequent registrations; or if the mobile migrates into other networks, hence generating cross-region or cross-country signalling that could take up vital communication resources. This problem could be considerably reduced by dividing the coverage area into a number of location areas to facilitate an optimum registration procedure. A *location area* (LA) thus represents the minimum area in which an MS may be registered. This could comprise one or several cells usually depending on cell physical sizes (or based on some criteria that the network operator wishes to implement).

It should be straightforward to organise location areas by using addressing schemes. Cells that belong to a given location area (LA) are assigned with identical LA code (LAC). The LAC can be decoded by an MS from the network information which is broadcast in every cell. In a TETRA network, LAC decoding is performed by the radio link management protocol as described in Sections 8.4.2.3 and 8.4.5.1. Whenever an MS, upon selecting a cell, detects a different LAC, it will initiate registration with the infrastructure through a *location update* procedure.

5.10.1.2 Paging

Paging is an indication sent by the network to inform an MS (or a group of MSs) that there is an incoming call. An MS is paged only in the LA(s) in which it is registered and in response to the paging signal, the MS first needs to acquire a communication channel for the call to be connected. As a downside, it should be clear that increasing a location area over two or more cells would necessitate an increased paging effort by the network. That is, if an LA consists of several cells, then a few cells may have to be paged before locating the MS. Hence the choice of an LA will depend on an optimum balance between registration signalling load and the paging delay for locating an MS.

5.10.1.3 Registration procedure

Since the end result of registration is location update, registration procedures are carried out under a generic command for location update. The variants of registration such as registration with or without authentication can be specified with additional parameters to the location update command.

Registration procedure can be triggered by one of the following:

- when an MS demands – this could be initiated by a number of factors such as user's explicit demand (through a user application) for registration, which is also a means of authenticating the network, or when an MS is activated from power down or upon SIM insertion;
- when an MS changes an LA which requires a location update to be made; and
- when the infrastructure so demands, which is also a means to authenticate an MS.

There are of course a number issues that should be considered during registration. These include authentication of services the user has subscribed to, or whether roaming is allowed in a visited network, and so on.

5.10.1.4 Multiple registration

As network operator's option, it is possible for an MS to register in two or more LAs so that the registration area can be enlarged. This can be time-limited or for an indefinite period of time. From an MS standpoint this in effect enables the LA to be reconfigured dynamically and could be employed to achieve some desired performance target. In general, multiple registration (and others discussed below) could serve as additional techniques for optimising system performance within an LA-based mobility management.

5.10.1.5 Periodic registration

In certain situations it may be necessary to maintain a more accurate status of an MS within a registration area. TETRA supports this feature and can be enabled on MSs with a programmable timer. Although in principle the network could command the MS to perform location area updates periodically, in TETRA this control is performed by the MS. Section 8.5.8 demonstrates how periodic registration and other related functions are implemented within the MM protocol.

5.10.1.6 Implicit registration

The above discussion refers to explicit registration where the registration procedure is initiated either by the MS or the network. Implicit registration can also be performed based on a system message that conveys the identity of the MS such as response to paging or cell change request. It is a network operator option whether the system or part of the system will accept call requests from an MS that is not registered. In secure radio implementations an implicit registration will activate the authentication function in the same way as an explicit registration.

5.10.1.7 Deregistration

In mobile communication, negating a registration (or deregistration) when appropriate is as important as the registration procedure. Deregistration can avoid unnecessary control signalling and radio resource usage if the mobile cannot receive a call. This could be when a mobile is switched off or its SIM card is removed. Such events normally invoke a procedure that causes the mobile to *detach* itself from the network, in effect deregistering the MS so that the network will not attempt paging in vain. With the knowledge of the MS detachment, for instance, the network can direct the call to a voicemail or to an alternative number that may have been arranged with the service provider.

5.10.1.8 Paging last registered LA

Although in theory the system would not page a deregistered MS, in certain situations it would be sensible or even desirable to page an MS in its last registered LA even if it has not yet sent a registration message. The need for such cases can be easily illustrated with an example [2]; In situations where a large fleet of vehicles leaving their depot in the morning could cause control-signalling overload, it is possible that some MSs are unable to register immediately. By paging the last registered LA, it is possible that such unregistered MSs could still be able to receive calls.

5.10.2 Basic Mobility Procedures

Mobility management in TETRA is accomplished much like GSM with various signalling procedures for user/mobile identity registration and location updates. A simplified functional subsystem for the TETRA mobility management is shown in Figure 5.13. The reader should be reminded that this representation is only conceptual and not stipulated by the TETRA standard as the internals of a TETRA network are not specified (see also Section 5.3.1). The essential system entities and components involved in mobility management are the MS identity or ITSI, user identity (through the SIM card), subscriber home database (HDB), and if the user is migrating to another network not subscribed to, the visitor database (VDB).

As an aside on terminology, the terms roaming and migration have specific interpretations in the TETRA standard. An MS is said to *migrate* when it changes to a new LA in a network, either with different MNC and/or MCC, where the user does not have subscription for that network. This is known as roaming in GSM. In TETRA, an MS is said to be *roaming* when it changes its LA within a network of the same MNC/MCC, and for which the user has a valid registration. This in GSM simply represents change of a location area.

The key element of the subsystem shown in Figure 5.13 is the home database (HDB) which, in addition to the record of locations for MSs, also holds information of all network users. The information typically contains records such as user identity, cipher (or encryption) keys and subscribed services. Within the home network, the MS is authenticated and registered by the infrastructure based on the information available in the HDB. In a visited network, a visitor database (VDB) is instead used which in the first instance involves authentication through the HDB and then downloading of essential user information into the VDB for subsequent authentication within the visited network with minimal or no interaction with the HDB.

Note that in Figure 5.13 the HDB and the VDB are shown as individual and group, hence I-HDB, G-HDB, and so on. The databases (I-HDB, G-HDB, etc.) could be distributed database (DDB) rather than centralised in order to facilitate fast call set-up time, but as remarked above, such an implementation requirement is outside the TETRA standard and will remain an operator option.

The authentication functions may exist as separate entities from registration as depicted in Figure 5.13. The home authentication centre (HAC) provides the authentication and Over The Air Re-keying (OTAR) parameters. OTAR [16] is a mechanism for distributing or updating encryption keys from the SwMI to an MS directly over the air interface. The visitor authentication centre (VAC) contains the authentication and OTAR parameters provided by the home SwMI for the migrated user.

Chapter 8 provides more detail on TETRA mobility management in terms of network layer protocols. The protocol descriptions will clearly demonstrate the interactions between the various network entities which in this section are only briefly overviewed at a subsystem level.

In the wider sense of mobility management concepts, it is also worth noting that other mobility management schemes are possible. Of particular interest are techniques with the application of Intelligent Network (IN) concepts whereby the mobile databases could be handled within the IN architecture. An example of IN technique in a mobile system is described in [17].

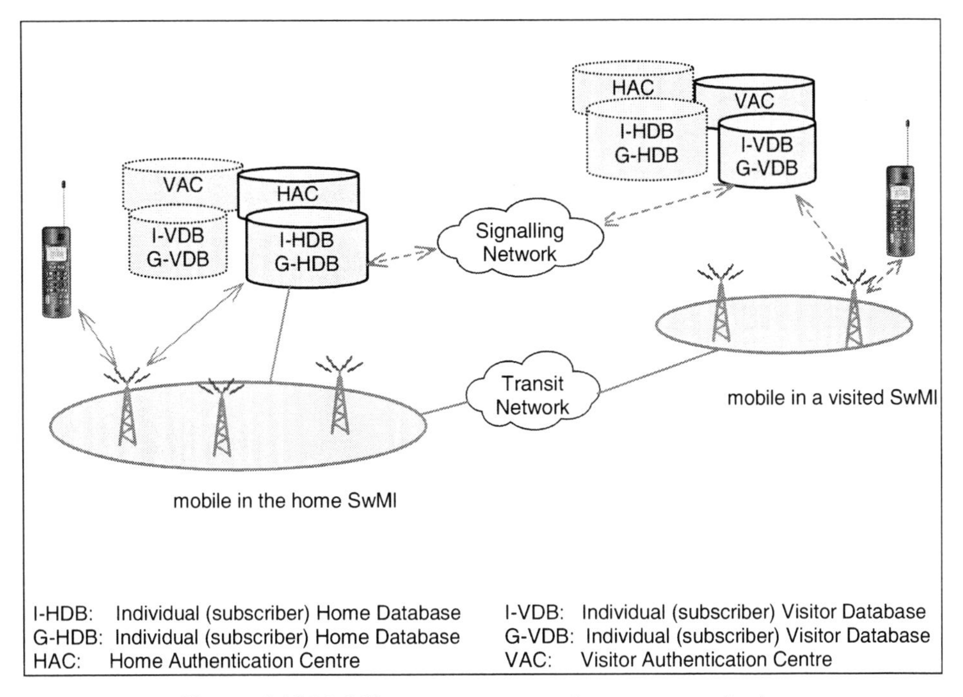

Figure 5.13 Mobility management subsystem organisation

5.11 TETRA INTER-SYSTEM INTERFACE

5.11.1 ISI Overview

The coverage requirement for a TETRA network is expected to be different from public networks in a number of ways. In PMR-oriented applications a TETRA network could be installed simply to satisfy a specific organisational need, such as serving a manufacturing site or an airport. In such cases it is possible that a TETRA network exists as an island in the radio environment sense, but with possible connection to a public network such as the ISDN. In other cases, the requirement could be for a wider coverage where the application could be, for instance, to facilitate regional emergency services.

The evolution of TETRA network installation (by an organisation, say) is therefore likely to be driven by the need to service specific sites, rather than providing a contiguous and complete coverage as in public networks. Given this scenario, one could envisage the advantages of interconnecting the separate "TETRA islands" in order to form one large logical network as shown in Figure 5.14. The networks could be owned by a single organisation or by different organisations. Such an interconnection can allow network resource sharing at various levels and can also provide an integrated network operation.

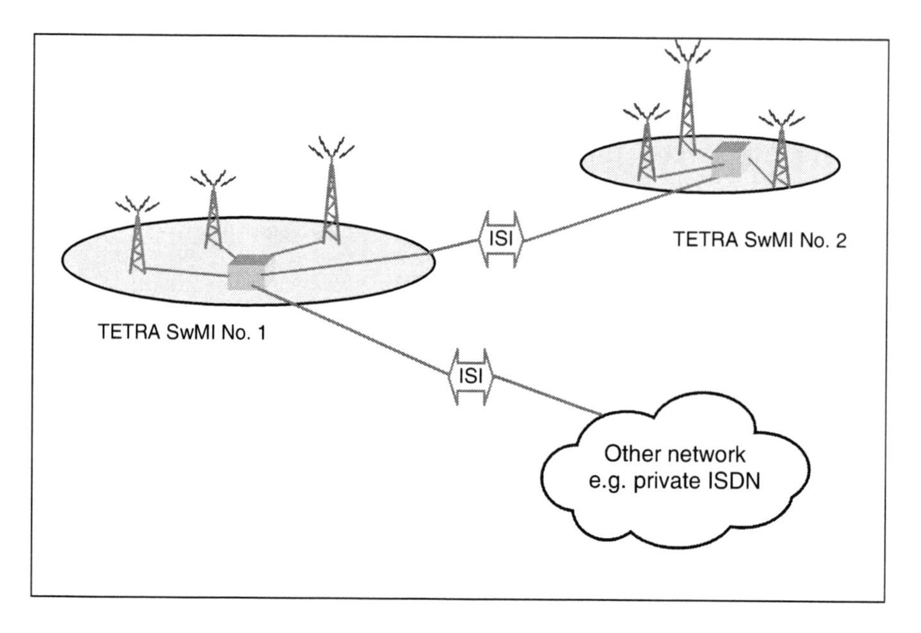

Figure 5.14 Inter-system interface between two TETRA networks

Interconnection between different networks is covered by the TETRA standard which specifies the various requirements for inter-working over the inter-system interface (ISI). The ISI specification is concerned with the inter-working of different TETRA networks (as well as non-TETRA private networks), possibly provided by different manufacturers. This specification complements the inter-working of TETRA networks with public networks, which is catered for by gateway interfaces as described in Section 5.3.8 above.

5.11.2 PISN Architecture

The TETRA ISI is based on the private integrated services network (PISN) configuration [18] which nowadays is also referred to as corporate network, which is basically the fixed network counterpart of the PMR. A PISN is much more than a private ISDN, or a PBX, and can comprise various types of networking equipments which, in the PISN parlance, are called Private Integrated Network EXchange (PINX). Examples of PINX include PBX, CENTREX, multiplexors, LAN connection, etc. A TETRA network that is provided as part of an organisational PISN can be regarded as a PINX, and for many organisations this is the most likely route for the fixed and mobile networks to evolve. This is depicted in Figure 5.15.

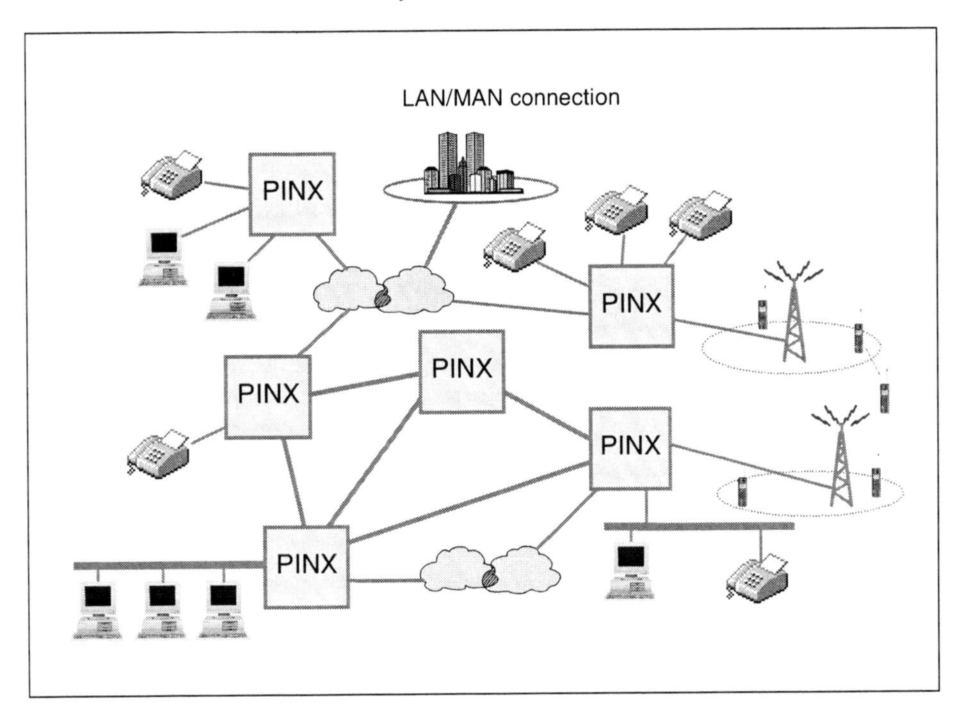

Figure 5.15 An integrated fixed/mobile corporate network architecture

Considerable efforts have gone into the standardisation of the PISN to provide enhanced networking capabilities over and above those provided by the public ISDN. The standardisation is carried out under the auspices of an international body comprising 11 of the world's leading PBX manufacturers, which are also signatories to the 1994 QSIG MoU[4]. The PISN standards are produced by the European Computer Manufacturers Association (ECMA) and submitted to ISO for adoption as global standards. The PISN is therefore mature and flexible enough to create a synergy between the fixed and mobile private networks and indeed this is of direct relevance to the development of TETRA.

5.11.3 TETRA ISI Signalling

The TETRA ISI presented here is only intended to provide a system level overview, consistent with the theme of this Chapter. There are protocol mechanisms and security requirements for the ISI which will be deferred until Chapter 9 after the various TETRA network protocols have been described in Chapter 8.

[4] QSIG stands for *Q*-reference point *sig*nalling, which is of the essence for the MoU as the inter-PINX signalling standard used for interconnecting multi-vendor equipment. QSIG is discussed in Chapter 9.

5.12 TETRA COMPARISONS WITH GSM

To highlight some of the notable features of TETRA, the following summarises comparisons of TETRA with the GSM system architecture.

5.12.1 System Level Comparison

1. One key difference between TETRA and GSM is that of the network architecture. GSM network architecture is described in Section 2.3. Unlike the hierarchical GSM network architecture, which is defined in terms of two internal interfaces, *A* and A_{bis}, TETRA uses a non-constrained (possibly flat) structure. This choice was made not to constrain the range of the architecture that manufacturers might want to choose to implement the system.

2. Only interfaces on the periphery of a TETRA system are defined. These are interfaces I1, I2, I3, I5 (Section 5.4), and the gateway interface (Section 5.3.8), which is covered by the TETRA standard. The interfaces within a TETRA network are not defined.

3. TETRA supports Direct Mode Operation (DMO) which is perhaps the most significant feature that distinguishes TETRA from GSM. Direct Mode Operation allows terminals to communicate directly with each other, an essential feature where network coverage is not available (or where network infrastructure need to be avoided for further security). In addition to DMO, mobile terminals may also be operated as a repeater for extending the coverage of the network.

4. TETRA provides three types of terminals:

   - trunked mobile station (MS), like GSM
   - trunked line-connected station (LS)
   - direct mode mobile station.

5. The trunked LS is another new feature of TETRA. The LS can be used to support dispatcher functionalities (central supervisor/control centre of traditional PMR systems) over an ISDN connection instead of the air interface. With such an arrangement more reliable connection can be had for the dispatcher and it also means that the LS traffic does not compete for mobile radio resources (at least between the TETRA SwMI and LS).

6. GSM is designed to support high subscriber density with coverage supported both with small and medium cell dimensions. Due to the anticipated PMR user density, TETRA, on the other hand, is intended for low to medium subscriber density with wide area coverage that can extend up to 60 km. The wide area coverage can provide a significant cost saving in terms of the necessary radio infrastructure.

5.12.2 System Parameters and Others

Table 5.5 summarises some of the notable system parameters that differentiate TETRA from GSM. Features available or identical to both systems, such as support for air interface encryption, are not listed.

Table 5.5 Notable TETRA/GSM system features and parameters

System parameters	TETRA	GSM
Frequency	around 400 MHz	around 900 MHz
Bandwidth	6.25 kHz	25 kHz (full rate codec); 12.5 kHz (half-rate codec)
Call set-up times	< 1 sec (300 ms typical)	< 10 sec (1-3 s typical)
Gross data rate per Hz	1.44 b/Hz (36 kb/s per 25 kHz)	1.35 b/Hz (270 kb/s per 200 kHz)
End-to-end encryption	yes	no
Cell coverage	wide area	small to large

The reader who is interested in the detail of standards description will notice differences even of a lexical nature between TETRA and GSM. These differences arise mainly from the fact that the architectures for the two systems are different and therefore some variations of system component implementations will be expected in TETRA. For instance, since the internal of a TETRA network is not specified, some variation of GSM's HLR (Home Location Register) can be expected, and TETRA uses a more generic term HDB (Home Database) instead. Some of the terminological differences are noted in Table 5.6.

Table 5.6 Summary of some TETRA/GSM technical terms

TETRA	GSM
HDB (home database)	HLR (home location register)
VDB (visitor database)	VLR (visitor location register)
Short data service (SDS)	SMS (short message service)
Migration	Roaming
Roaming	Change of location area
SwMI	LMN (land mobile network)
Cell selection (reselection)	Handover (or handoff)

5.13 SUMMARY OF TETRA PARAMETERS

The following table summarises the TETRA features and system parameters.

Table 5.7 Summary of TETRA features and parameters

System parameters/features	Parameter values and/or descriptions
Carrier spacing	25 kHz
Multiple access	TDMA
Channels (time slots) per carrier	4
Traffic channel bandwidth	25 kHz/4 = 6.25 kHz for V+D and DMO; 25 kHz for PDO
Duplexing and duplex offset	FDD, 10 MHz (for current frequency allocation)
Frequency allocation (present)	380-400 MHz (public safety); 410-430 MHz (commercial)
Symbol rate (gross bit rate)	18 k symbols/s (36 kb/s)
Digital modulation format	$\pi/4$–DQPSK
Net bit rate	Up to 19.2 kb/s
Random access	Slotted Aloha
PDO statistical multiplexing	Fast packet transfer (<100 ms for octets)
Call set-up (re-establishment time)	< 0.3 sec (< 1 sec)
Call connection types	Point-to-point (individual); point-to-multipoint (group call, acknowledged group call); point-to-multipoint (broadcast call)
Call connection modes	Normal (blocked calls cleared) and queued calls (for call completion on busy or on no reply)
Cell handover decision	Mobile-based
Encryption	Air interface encryption standard; End-to-end optional
Speech coding	Low bit rate ACELP codec at 4.6 kb/s; lossy decoding
Data services	Circuit mode, Packet mode, SDS
Supplementary services	30 in total; about 15 of these are tailored for emergency and safety use
Circuit mode user terminal equipment	TE1-ISDN (V+D LS); TE2 (V+D MS and LS, and DMO)
Packet mode user terminal	TE2 and TE3 (V+D and PDO data terminals)
Mobile power classes	1W (30dBm), 3W (35dBm), 10W (40dBm), 30W (45dBm)
Uplink power levels from mobiles	0.03W (15 dBm) to 30W (45 dBm) in 5 dB steps depending on mobile power class
Receiver sensitivity for 2% BER	-104 dBm (hand portable); -107 dBm (vehicle-mounted)
Base station power classes	0.6W (28 dBm) to 40 W (46 dBm) in 10 steps
Downlink power levels	28 dBm (0.6W) to 46 dBm (40W) in 2 dB steps depending on base station power class
Vehicle speed	Up to 200 km/h
Cell sizes	Up to 60 km for wide area coverage

5.14 TETRA CONFORMANCE TESTING

5.14.1 Scope of Conformance Testing

TETRA is an open standard and to ensure inter-operability between multi-vendor equipment there should be systematic and clearly specified rules for validating capabilities and conformance to the standards. This is ensured through tight specifications of conformance testing that lays down well defined type approval and operational test procedures. These are addressed under different categories of the conformance testing specification [19] as outlined below.

- Radio conformance testing specification – this deals with transmit and receive mode operations such as unwanted output power, modulation distortion, receiver sensitivity and spurious rejection performance, unwanted emissions, etc.
- Protocol testing specification for Voice plus Data (V+D) – this deals with the testing of conformance and capabilities of the various protocol entities, for instance call control protocols for call set-up, group call control.
- Protocol testing specification (PDO) – conformance test for PDO.
- Protocol testing specification (DMO) – conformance test for DMO.
- Protocol testing specification for TETRA security – this deals with TETRA security protocols including the end-to-end encryption protocols.

The radio conformance tests are generally carried out by certified test laboratories that specialise in tests for the relevant equipment electromagnetic compatibility (EMC) approval as well as those specific to mobile communications. Protocol tests may be carried out in-house with the use of specialised protocol tester equipment or by test houses specialising in such tasks. The TETRA standard makes provision for the protocol implementor to complete a documentation of the protocol implementation conformance statement (PICS) proforma as summarised below.

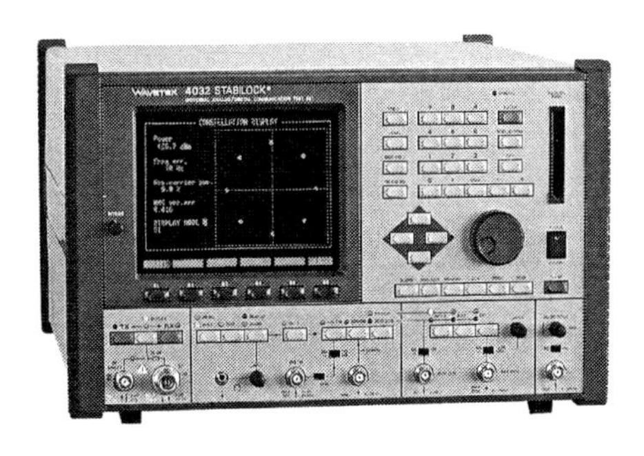

Some measurement capabilities of the Wavetek STABILOCK 4032:

- Transmitted power
- Frequency error
- Residual carrier power
- RMS and peak vector error
- Timing error
- Modulation spectrum
- Constellation display
- TX tests on MS in test mode
- TX tests on MS in with call set-up
- TX tests on base stations

The *Wavetek* STABILOCK 4032 measures TETRA terminal performance on the traffic channel

Figure 5.16 Test equipment for TETRA mobile terminals (*Courtesy of Wavetek Ltd.*)

5.14.2 Conformance Statement

The TETRA PICS proforma specification concerns the implementation statement by the protocol implementor. Its purpose [20] is to provide a mechanism whereby a supplier of an implementation of the requirements defined in the standard (e.g. V+D protocols or TETRA security mechanisms) may provide information about the implementation in a standardised manner. The PICS proforma provides a detailed checklist that can be used at various stages of the implementation and test phases including the use by the following [21]:

- the protocol implementor – as a mechanism for reducing the risk of failure to conform to the standard through oversight;
- the (potential) user – as a basis for initial check of inter-working with another implementation;
- the supplier and acquirer – as an indication of the capabilities of the implementation;
- the protocol tester – as the basis for selecting appropriate tests against which to assess the claim for conformance of the implementation.

As an example, the following table shows the protocols that are subject to the PICS declaration for V+D and TETRA security protocols.

Table 5.8 Example of protocols subject to PICS declaration

V+D Protocols:	
	Circuit Mode Control Entity (CMCE)
	Mobility Management (MM)
	Mobile Link Entity (MLE)
	Logical Link Control (LLC)
	Medium Access Control (MAC)
	Connection Oriented Network Protocol (CONP)
	Specific Connectionless Network Protocol (SCLNP)
Security Protocols:	
	Authentication
	Over The Air Rekeying (OTAR)
	Secure enable/disable
	Air interface encryption
	Encrypted short identities
	TEI (TETRA Equipment Identity) delivery
	MM extended functions
	PDU (protocol data unit) support
	End-to-end encryption

The above V+D protocols will be the subject of Chapter 8, except for the MAC and LLC, which will be covered in Chapter 6 under data link layer. Security protocols will be introduced in Chapter 9.

5.15 CONCLUSIONS

This chapter has presented TETRA's system architecture and described the main system components, interfaces and services. The subscriber-access network architecture of TETRA is defined on the basis of ISDN reference configuration and the various user equipment arrangements possible have been outlined.

The TETRA peripheral equipment interface (PEI) specification is the first of its kind to consider the development of mobile data applications with such great detail. It integrates AT commands and IP connectivity with full access to resources on the network side. An overview of the PEI specification has been presented with an outline of specific services available through the PEI.

TETRA employs comprehensive address and identity schemes for flexible and efficient addressing of equipment, networks, and subscribers, and important aspects of these have been outlined. Finally, the salient features of TETRA have been highlighted based on a comparison with the GSM architecture and system components.

The TETRA network standard is defined by the lower three layers of the open system interconnection (OSI) architecture, namely the physical layer, the data link layer and the network layer. The following three chapters address each of these layers in depth, beginning with the physical layer in Chapter 6 next.

REFERENCES

[1] TETRA MoU Web Site, http://www.tetramou.com

[2] ETSI Technical Report ETR 300-1: "Terrestrial Trunked Radio (TETRA); Voice plus Data (V+D); Designers' Guide; Part 1: Overview, Technical Description and Radio Aspects", May 1997.

[3] ETSI Technical Standard pr ETS 300 396-1: "Technical Requirements for Direct Mode Operation (DMO); Part 1: General Network Design", July 1996.

[4] ETSI Technical Standard ETS 300 392-4: 'TETRA Voice plus Data, Part 4: Gateways', Dec. 1997.

[5] ETSI Technical Report ETR 300-4: "Terrestrial Trunked Radio (TETRA); Voice plus Data (V + D); Designers' guide; Part 4: Network management", Dec. 1997.

[6] ETSI Technical Standard ETS 300 392-3: 'TETRA Voice plus Data, Part 3: Inter-working', Feb. 1998.

[7] ETSI Technical Report ETR 292: "TETRA voice plus data, Technical requirements Specifications for Network Management", July 1997.

[8] CCITT Recommendation I.411: "ISDN User-Network Interfaces – Reference Configurations".

[9] CCITT Recommendation V.110: "Support of Data Terminal Equipments (DTEs) with V-Series type Interfaces by an Integrated Services Digital Network (ISDN)".

[10] ETSI Technical Standard ETS 300 392-5: "Terrestrial Trunked Radio (TETRA); Voice plus Data (V+D); Part 5: Peripheral Equipment Interface (PEI)", May 1998.

[11] Sun, A., *Using & Managing PPP*, O'Reilly & Associates, Inc., Sebastopol, CA, 1999.

[12] Comer, D., *Internetworking with TCP/IP, Volume I: Principles, Protocols, and Architecture*, Prentice-Hall, Inc., Englewood Cliffs, N. J., 1995.

[13] Stallings, W., *Data and Computer Communications*, (5th Edn.), Prentice-Hall, Inc., Upper Saddle River, N. J., 1997.

[14] ETSI Technical Standard ETS 300 392-1: "Radio Equipment and Systems (RES); Terrestrial Trunked radio (TETRA); Voice Plus Data (V+D); Part 1: General Network Design, Clause 7", Feb. 1996.

[15] CCITT Recommendation X.121: "International Numbering Plan for Public Data Networks", 1988.

[16] ETSI Technical Standard ETS 300 392-7: "Radio Equipment and Systems (RES); Terrestrial Trunked Radio (TETRA); Voice plus Data (V+D); Part 7: Security", December 1996.

[17] Laintinen, M. and Rantala, J., "Integration of intelligent network services into future GSM network", in *IEEE Communications Magazine*, vol. 33, no. 6, June 1995.

[18] The ISDN PBX Network Specification (IPNS) Forum, *The QSIG Handbook* (Available at *http://www.qsig.ie/qsig/index.htm*), 1999.

[19] ETSI Technical Standard ETS 300 394 series of standards documents on Conformance Testing Specifications (Part 1 to Part 5).

[20] ETSI Technical Standard ETS 300 392-14: "Terrestrial Trunked Radio (TETRA); Voice plus Data (V+D); Part 14: Protocol Implementation Conformance Statement (PICS) proforma specification", Dec. 1997.

[21] ECMA Standard 165: "Private Integrated Services Network (PISN) – Generic Functional Protocol for the Support of Supplementary Services – Inter-Exchange Signalling Procedures and Protocol", Annex A, June 1997.

6

The Physical Layer

6.1 OVERVIEW, FUNCTION AND REQUIREMENTS

The physical layer is responsible for the transmission of raw data from one end of the link to the other. The physical layer is not responsible for ensuring the data arrives without error, or even in the correct order, as this is the responsibility of the upper layers. Nor will the physical layer have to concern itself with the choice of resources to use for the transmission, since this will also be controlled by the upper layers. The responsibility of the physical layer extends to the physical transmission of the data at the correct point in time, modulating it and receiving it at the other end. An overview of the different functions is shown in Figure 6.1.

A number of requirements were identified in Chapter 3 for PMR radio systems. The physical layer of the TETRA system must meet all the relevant requirements if the system as a whole is going to perform well. Those requirements that affect the physical layer include:

- **Reliability**. The system must work well under a variety of operating conditions, including poor radio channels and with interference.
- **Flexibility**. The system must provide users and operators with a wide choice of services and modes of operation.
- **Narrow bandwidth**. The system must be capable of sharing capacity between a large number of operators, which requires a narrow modulated carrier bandwidth. In addition, co-existence with existing PMR would be advantageous as it would simplify phased implementation and spectrum refarming (i.e. re-allocation of spectrum from superseded systems to new systems).
- **Good coverage and the capability to have large cells**. Many PMR operators provide their own infrastructure and since traffic may be low, they will want large cells in order to reduce infrastructure costs, while still being able to contact mobiles wherever they may be within the coverage area.

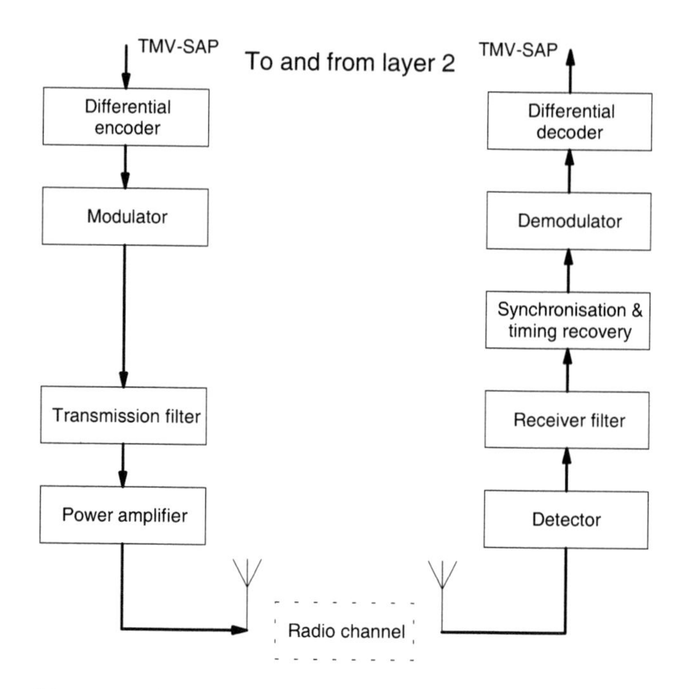

Figure 6.1 Overview of data processing functions in layer 1

- **Efficient use of radio resources**. With use of the radio spectrum increasing, and system should make efficient use of the radio resource.

This chapter will examine the TETRA physical layer, and the options chosen for the various parts of the transmission chain, comparing the TETRA solutions with those of other PMR systems.

6.2 FREQUENCY ALLOCATION

PMR systems have to work within a crowded radio spectrum, and as a result TETRA is designed to operate over a wide range of frequencies, from 150 MHz to 900 MHz. As TETRA is an FDD system, it has different bands for uplink and downlink transmission. Harmonised bands across Europe have been agreed in the range of 380–390 paired with 390–400 MHz for Public Safety users, and 410–420 paired with 420–430MHz is being assigned in some countries for commercial use. In addition, CEPT[1] has made recommendations for use in the following frequency bands:

450–460 paired with 460–470MHz
870–888 paired with 915–993MHz

The duplex spacing (i.e. spacing between uplink and downlink carriers) is 10 MHz, except in the proposed 900 MHz band, where it is 45MHz.

[1] Conférence Européenne Postes des et Télécommunication.

The propagation loss is higher at higher frequencies, which restricts the coverage which can be achieved. Figure 6.2 shows the effect of different frequencies on the pathloss in a rural environment (using the Hata open area model with a 10 dB allowance for clutter). Also shown for comparison is the propagation at a frequency of 150 MHz. Cell sizes are described in Section 6.12.6.

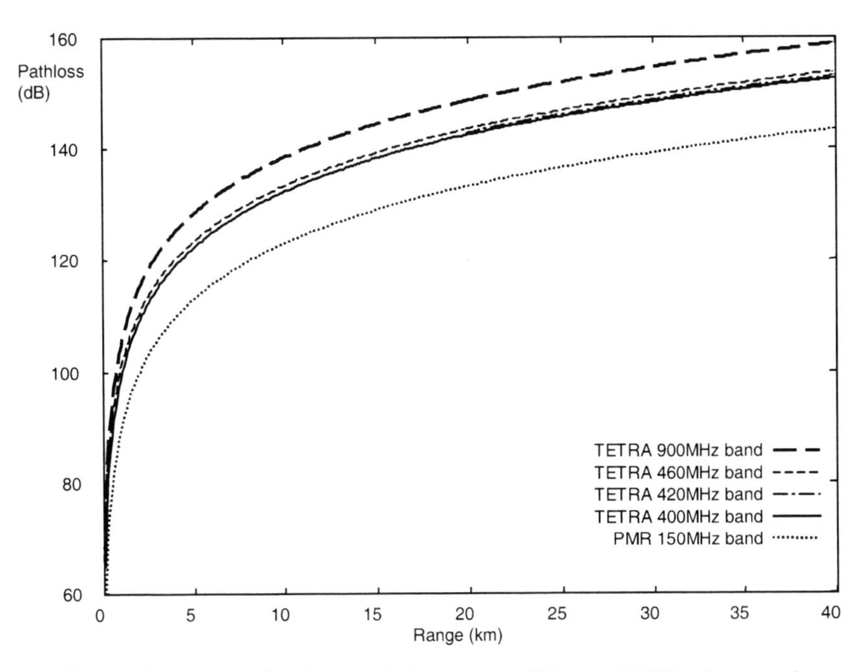

Figure 6.2 Propagation loss with distance at different TETRA frequencies

6.3 CHOICE OF MULTIPLEXING METHOD

A PMR system will be assigned a particular band of frequencies in the radio spectrum within which to operate. Its use of these frequencies must conform to the requirements set for the system, which include having options for voice and data services. Also, the bandwidth used by a particular PMR system must be as small as possible because there will be many PMR operators.

There are a number of other requirements that must be taken into account for a PMR system, which may be different from other types of radio system. The PMR spectrum will be used by a large number of operators, so frequency planning is difficult, if not impossible. There is also a requirement for a number of different services, including data. There will therefore be a number of different service data rates.

As discussed in Section 1.3, the principal methods of dividing up the radio spectrum between many users are CDMA, FDMA and TDMA. The following sections discuss the issues related to each of them for PMR operation, culminating in the TDMA system, which is employed in TETRA.

6.3.1 CDMA Operation of PMR Systems

CDMA, or Code Division Multiple Access, is described in Section 2.23. The modulation method used is varied so that different modulated signals can use the same frequencies at the same time and are distinguished by their modulating code. CDMA is very efficient in its use of spectrum, since the same pool of frequencies can be used by many different transmitters at the same time. However, the terminals are complex, with good control algorithms such as power control being essential for efficient operation. Also, CDMA requires wide bands of frequencies within which to spread individual users signals so that they can be differentiated. This means that it would not be possible to migrate from an analogue system to a digital one by introducing the digital service on individual carriers – the analogue spectrum would have to be converted as a block. This makes the option less attractive to current PMR operators seeking improved use of their systems with a conversion to digital.

6.3.2 FDMA Operation of PMR Systems

FDMA, where each communication channel is given its own frequency band, is the traditional method of dividing up radio resource. It is detailed in Section 1.3.

The guard band, which is present between each carrier, means that in practice it is not possible to use the whole range of each frequency band. For a PMR system, frequency planning is not possible, so the guard bands have to be relatively wide to minimise adjacent channel interference. For example, in an FM PMR system with a 25 kHz carrier separation, the carrier deviation is less than about 5 kHz, resulting in a bandwidth of about 12 kHz so that the carrier power is reduced by at least 60 dB at the edge of its allocated band [9]. This guard interval reduces the efficiency of the system, because as the number of bands increases, a larger proportion of the frequency band is taken up by the guard intervals.

FDMA is used for analogue PMR systems (using FM) with either 25kHz or, more recently, 12.5 kHz carrier spacing. Two digital PMR systems which use FDMA are TETRAPOL and APCO25. Both have options for 12.5 kHz carriers, which can be used in place of analogue carriers, and TETRAPOL also allows 10 kHz carrier spacing.

The frequency of the modulating signal affects the bandwidth of the resulting modulated signal. The band allocated to the carrier has to be as wide or wider than the bandwidth of the modulated carrier to avoid interference with adjacent carriers. This means that as the width of each frequency band is reduced, so is the amount of information that can be transmitted in each band. A limit is reached for speech transmission at about 5 kHz. The second stage of the American APCO25 system proposes using FDMA bands with a spacing of 6.25 kHz in order to transmit speech signals. In an effort to increase capacity, the UK Radiocommunications Agency has allocated some spectra for use with 6.25 kHz carrier spacing in order to encourage the deployment of such systems prior to the widespread introduction of more spectrally efficient digital schemes.

FDMA is a conceptually simple system, but it has a number of limitations. Each link takes up one frequency band, and requires one transmitter and one receiver. In order to have a simultaneous bi-directional (full duplex) link, it is necessary to be able to transmit and receive at the same time and have a transmitter and receiver active at the same time, which requires quite complex radio frequency circuitry. The alternative is to have a system that is only half-duplex, and have a pressel or similar device so that when one party is talking the other party is silent. A further problem is that the transmitted signal is continuous in time, so it is not possible to transmit signalling information in order to control the call without interrupting the transmitted data through a process called stealing, or alternatively using a different frequency for control information. This additional frequency could either be a tone which is modulated with the speech signal, increasing the bandwidth required, or alternatively an additional radio carrier which would necessitate a second pair of transmitters and receivers. Such complexity adds cost and power demand, while stealing has the disadvantage of discarding the information that would have been transmitted during the stolen period, which reduces the service quality.

A more significant problem, which does not affect systems such as APCO25, which are primarily voice based, is the transmission of data. If data transmission is required at the same rate as speech, this can be achieved by replacing the digital speech signal with the digital data signal and transmitting in the normal fashion. However, there is a problem with multi-rate data or with packet data, two requirements of a flexible PMR system.

The difficulty in the case of multi-rate data is that in order to increase the data rate above that of the speech signal in an FDMA system, it is necessary to either increase the width of each frequency band, or to use several bands. If the width of each band is increased, this means that those bands which do not require the higher data rates, such as those used for speech calls, waste resources. This is not a very efficient scenario. It would be possible to have different bands, some narrow and some wide, but this would mean having two (or more) different types of RF circuitry to cope with the different bandwidths. This would greatly increase the costs of the system while reducing the operational flexibility (because certain services would be constrained to use certain bands).

The alternative solution would be to use more than one frequency band to increase the data rate. However, this would mean using several transmitter/receiver pairs simultaneously, which would increase power requirements as well as the complexity of the handset. In addition, using several frequency bands is not as efficient as a single wide one (due to the effects of interference and the guard intervals).

6.3.3 TDMA Operation for PMR Systems

In a TDMA system, each burst of information is transmitted on a slot, and as with the guard bands which were required in frequency in an FDMA system, guard times are required between slots to allow the transmitter to switch on and off and to allow for propagation delays from different transmitters. The period over which burst transmissions from the various information sources repeat is called the frame.

Note that an inherent issue with TDMA systems is one of delay. As the data is not transmitted continuously, it must be buffered until enough data is available to form a burst, which is then sent. Usually the data is encoded and the whole burst must be received before it can be decoded and read out again.

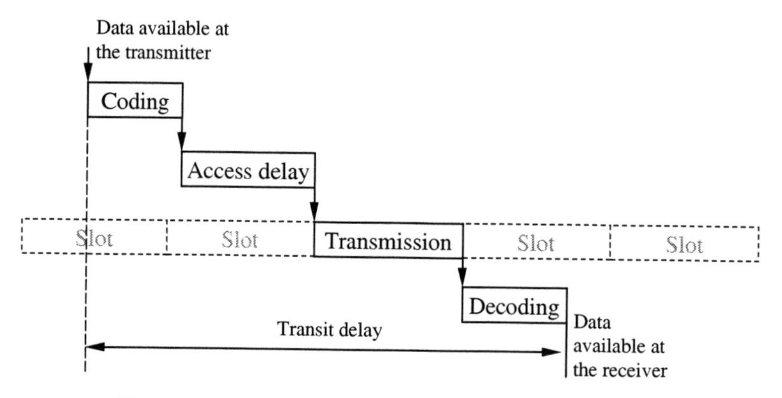

Figure 6.3 Transit delay for data in a TDMA system

Figure 6.3 shows the delay to a packet of data. The data is first encoded, and then there is a wait for the next available time slot (the *access delay*). The transmission of the data itself takes a time equal to the length of the timeslot, and finally there is a decoding delay. The total time taken is the *transit delay*.

This example assumes that the data is immediately available. For a speech service, there must be a sampling interval over which the speech is recorded. Some source and channel coding may be able to take place while this sampling is undertaken, but there is still likely to be a delay between the end of the sampling time and the time when the data is encoded ready for transmission. The access delay can be removed if the speech coding is synchronised in time to the channel allocated for its transmission. Figure 6.4 shows the delay to a speech service assuming a system, like TETRA, where the speech packet is transmitted in a single burst. In GSM, where speech is interleaved, or for interleaved data channels in TETRA, an additional delay is caused by the fact that the interleaving process spreads data over several bursts, all of which must be received before the data can be decoded.

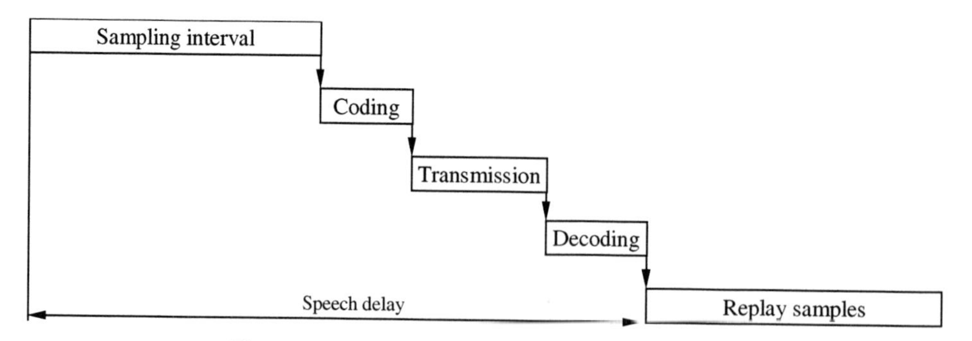

Figure 6.4 Delay for a speech service in TETRA

A TDMA system has a number of significant advantages in the support of high rate data services. In a TDMA system this can be achieved by allocating more slots. For example, TETRA, which uses TDMA, can support a data service of 7.2 kbit/s on a single slot, and of up to 14.4 kbit/s, 21.6 kbit/s or 28.8 kbit/s simply by allocating 2, 3 or 4 contiguous slots to a mobile. It is possible in an FDMA system to allocate more than one carrier to increase the information capacity, but this would now require the mobile to have several transceivers, increasing its cost. There is no increase in the number of transceivers required in the TDMA system, unless the service requires more slots than are available in the frame.

It is also possible to make innovative use of the TDMA system to allow the user to have full duplex operation with a transceiver that is only half-duplex, reducing costs. A mobile using a single slot service will be idle for the other 3 slots (although it may access control information). This time can be used for transmitting in the other direction. TETRA time slots are numbered such that the uplink slot x is two slot times after downlink slot x. If a mobile wishes to set up a bi-directional link (i.e. transmit and receive at the same time) and is allocated slot x, it receives the downlink message and then switches to transmit mode to transmit on the uplink slot two time slots later. This can be compared to a pure FDMA system that would require a separate transmitter and receiver to work at the same time in order to achieve full duplex operation. Note that this only applies to a single slot service – multi-slot operation would have to be restricted to half-duplex operation or more transceivers would be required.

The following diagram shows the TETRA frame structure, with the uplink slot numbering two slots behind the down link. A mobile has been allocated the channel in slot 1 for communicating in both the uplink and downlink directions to the base station.

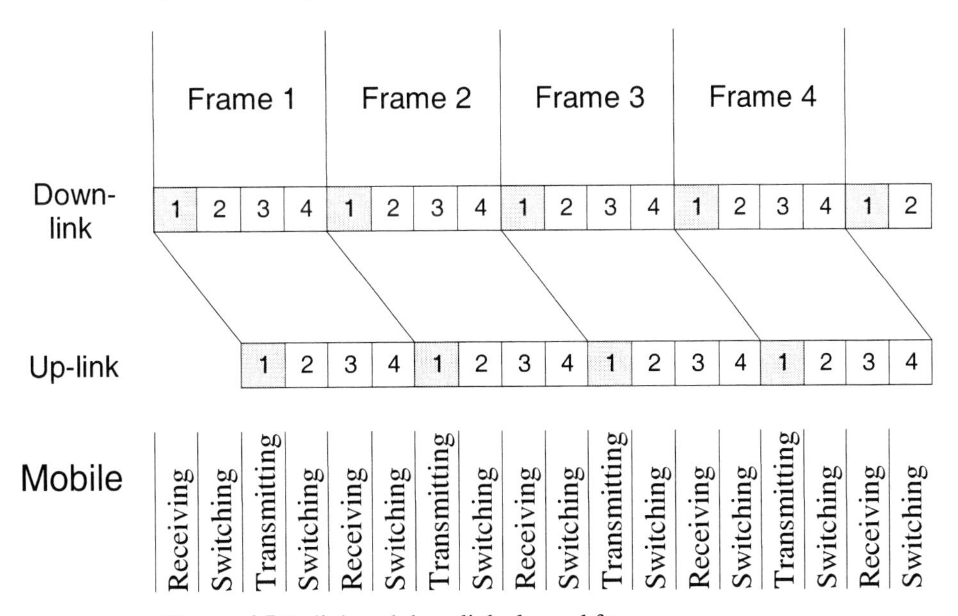

Figure 6.5 Uplink and downlink slot and frame arrangements

Another more general advantage of TDMA operation is that it allows frequency hopping (see Section 2.13.3). Since each burst is self-contained, i.e. they start and end with the transmitter switched off, the transmitter can move to carrier on a different frequency between bursts, avoiding interference or fading specific to a particular frequency. If mobiles hop in different patterns, there is a further advantage that the interference a mobile receives changes from burst to burst. This means that a mobile is not prevented from transmitting by a single dominant interferer. However, frequency hopping introduces further complexity and is not proposed for TETRA.

TDMA does have some drawbacks. Since the carrier rate is increased, multipath fading causes more problems and quasi-synchronous operation is more difficult. Also, since TDMA groups a number of channels on one carrier, it makes the system less suitable for small PMR operators. For example, TETRA uses a four-slot system, so the minimum quantity of TETRA channels an operator can have is four, although there are various operating modes to allow slots to be shared amongst different cells (see Section 6.14.4).

6.4 CHOICE OF TDMA PARAMETERS

A major question in the design of a TDMA system is the number of bursts that make up a frame and the rate of information transfer on the carrier. If the frame length is increased, the time between bursts increases and with it the amount of information which must be transmitted on each burst. In addition, the delay suffered by the data stream rises. If the data rate is increased, more bursts, and therefore more users, can fit in each frame, but the bandwidth required increases. While this is generally useful from the point of view of the use of the radio spectrum, as wider carriers mean fewer guard bands, in a PMR system it is useful to have relatively narrow bands of frequency due to the restricted availability of radio spectrum and the number of private networks. Most analogue PMR systems use a carrier spacing of 25 or 12.5 kHz. By keeping to a similar bandwidth, digital PMR systems can more easily co-exist with such systems. This is discussed in more detail below.

Due to the problem of timing, a short guard space is required between each burst. In addition, it takes time to switch on and off the transmitter, and during this time the channel cannot be used (this does not affect the base station to mobile link, as the base station is always transmitting – it is just the intended receiver that changes). This wasted resource between bursts would tend to suggest that slots should be relatively long to reduce the number of slot intervals. However, slots should not be too long, or there will not be enough data available to fill them in relatively low rate services like speech, unless the delay is excessive. Most mobile systems are designed around voice, which can be coded at a rate of around 5 to 13 kbits/s, depending on the quality which is required. If it is possible to transmit a speech signal in 6.25 kHz using FDMA (no framing), two users can be accommodated in a two slot system in 12.5 kHz, or four in a four slot system in 25 kHz. The reduction in guard bands means the four slot 25 kHz system should be able to support higher quality than the no-slot 6.25 kHz system. In fact, Motorola's iDEN system does this and squeezes six speech channels into 25 kHz, but at the expense of using higher order modulation.

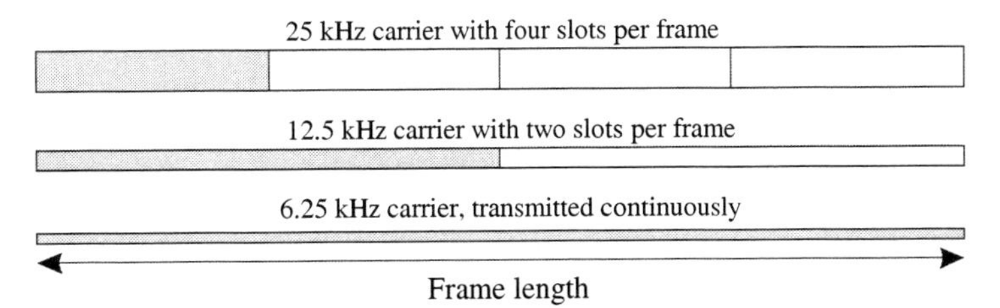

Figure 6.6 Trade-off between carrier wide and slot length

A bandwidth greater than the 25 kHz analogue standard would be a disadvantage from the point of view of migration to the TDMA service, which suggests a four-slot per frame TDMA system. The frame length, and slot size, should be as large as possible within the constraints of delay, which for a speech service should not exceed about 200 ms if it is not to be too noticeable. Section 6.8 shows that a frame length of just over 50 ms results in a delay of about 200 ms in the worst case for a single site PMR service.

In the case of TETRA the parameters chosen are a carrier rate of 36 kbits/s (18 ksymbols/s with the chosen modulation method) which can be used with a carrier spacing of 25 kHz, and a frame with a length of 56.67 ms made up of four slots. One carrier can therefore cater for four users with either a speech service or a data service of up to 7.2 kbits/s, allowing for framing and guard times.

The number of slots per frame, carrier rate and carrier spacing of TETRA is much lower than that of public cellular systems. For example, the GSM system has a carrier spacing of 200 kHz, and has eight slots capable of handling speech services or 9.6 kbit/s data services. One reason for this is to keep the delay as small as possible as GSM speech blocks may undergo considerable delay in the fixed network, especially if, for example, satellite links are involved. Figure 6.7 shows the bandwidth time arrangements for GSM, both full and half-rate, TETRA and TETRAPOL.

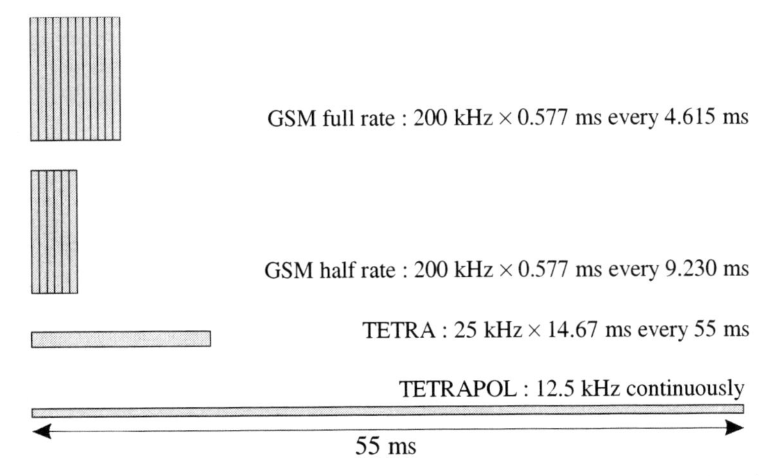

Figure 6.7 Bandwidth and time product for speech transmission over a 55 ms interval using GSM, TETRA and TETRAPOL

At first sight it would seem that TETRA is more efficient than GSM since it could fit eight four-slot carriers into the same spectrum that GSM could fit a single carrier with eight slots into. However, the picture is more complex since GSM can cope with higher levels of interference and so can re-use the same frequency more often. This reflects the different design requirements for a public operator providing a service to a large number of users over a wide area and private systems where the frequency has to be shared amongst a much larger number of operators.

A more relevant comparison would be with other PMR systems like TETRAPOL, iDEN and APCO25. All these systems are digital like TETRA. iDEN is a TDMA system like TETRA, and has the same carrier spacing of 25 kHz. However, iDEN supports six voice calls on a 25 kHz carrier, compared to TETRA's four voice calls. In order to do this, it uses a higher order of modulation, which is more susceptible to interference, and therefore needs smaller cells and larger reuse distances (see Section 1.19). On the other hand, TETRAPOL and APCO25 use FDMA rather than TDMA. They therefore have a separate carrier for each active user. TETRAPOL has carrier spacing of 10 kHz or 12.5 kHz, and can therefore fit two users into the same spectrum as a single TETRA carrier. In the absence of co-channel interference, TETRA would therefore be able to support twice as many users with a given spectrum allocation. However, if frequencies are re-used, TETRAPOL is able to cope with approximately double the amount of interference that TETRA will operate with, so in such circumstances the systems would work out about the same.

APCO25 has two transmission proposals, initially using a carrier spacing of 12.5 kHz, with a proposal to use a more advanced form of modulation similar to that used by TETRA to allow a carrier spacing of 6.25 kHz. When this advanced modulation becomes available, APCO25 will be able to support the same number of users as TETRA, although only between terminals supporting this modulation.

Table 6.1 Comparison of PMR multiple access methods and spectrum requirements

How TETRA compares...

	TETRA	TETRAPOL	iDEN	APCO25	GSM (full rate)	GSM (half rate)
Access method	TDMA	FDMA	TDMA	FDMA	TDMA	TDMA
Carrier spacing	25 kHz	12.5 kHz	25 kHz	12.5kHz/ (6.25 kHz projected)	200 kHz	200 kHz
Spectrum per speech user	6.25 kHz	12.5 kHz	4.167 kHz	12.5 kHz/ (6.25 kHz projected)	25 kHz	12.5 kHz

6.5 MODULATION

Given the various disturbances to the radio signal, a robust modulation system is required. Chapter 1 covers a number of different types of modulation that can be used in a mobile system. The criteria for a good modulation technique is that it should be:

- robust and reliable (i.e. that it should transmit information with as few errors as possible)
- simple to implement
- efficient in terms of the use of radio spectrum

The different types of modulation which have been proposed for mobile radio were described in Section 1.7 and are as follows:

- BPSK
- QPSK
- Offset QPSK
- π/4–QPSK
- 8PSK
- GMSK
- QAM
- Offset QAM

Errors will occur if noise in the radio channel is so great as to change one symbol into another. QPSK (or 4PSK as it is sometimes known) has the same performance in terms of received Bit Error Rate (BER) as BPSK (2PSK) because the in phase and quadrature components can be considered separately. Similarly, the various offset forms simply make it easier to decode the signal but do not increase the spacings between symbol points in the signal constellation and so have the same performance as BPSK/QPSK. Given these points, π/4–QPSK is the best of this family of techniques because it is less prone to non-linearities in the power amplifier.

A similar argument applies between QAM and Offset QAM, giving three main choices of QAM, GMSK or π/4–QPSK.

QAM can be very efficient. QAM with 16 symbols (16 QAM) can transmit 4 bits per symbol and has a BER performance about the same as QPSK in a mobile fading channel. For this reason it is proposed for two new PMR systems, Motorola's iDEN system and the Japanese digital PMR standard (along with π/4–QPSK). Although a higher carrier bandwidth is required, bandwidth efficiencies higher than 2 bits/s/Hz can be obtained. However, 16 QAM is unproven technology in mobile radio, and is more complex than GMSK or π/4–DQPSK.

GMSK is a popular choice for mobile radio, and is used for GSM and DCS1800. In PMR it is used by TETRAPOL. It is a constant envelope scheme, and so does not require linear amplification, but has a poorer bandwidth efficiency. Figure 6.8 compares the frequency spectra of a π/4–DQPSK signal with that of a GMSK signal. Both the π/4-DQPSK signal and the GMSK signal have the same bit rate (36 kbit/s), as in TETRA. The BT of the GMSK signal is 0.3, as in GSM. The GMSK signal can be seen to spread out more. For a public system such as GSM, where frequencies can be planned so that adjacent channels are not used in adjacent cells, this spreading is not so much of a problem. However, in a PMR system such planning is not possible and the only option is to constrain the signal within a narrow band by reducing the modulation rate. The public GSM system has a carrier rate of 271 kbit/s with a 200 kHz carrier spacing, giving 1.35 bit/s/Hz, but adjacent carriers will interfere with each other.

Also shown in Figure 6.8 is the spectrum of TETRAPOL, which uses GMSK. The absence of frequency planning means that the TETRAPOL signal must adhere to strict limits on its bandwidth. In order to do this, the carrier rate is limited to 8 kbit/s, giving either 0.8 bit/s/Hz or 0.64 bit/s/Hz. It has a BT of 0.25, giving it a smaller bandwidth than a comparable GSM signal at the expense of a slightly higher BER [13]. The TETRAPOL signal fits into a 12.5 kHz band completely (i.e., with adjacent channel interference less than –60 dB relative to the carrier). The TETRAPOL signal also fits into a 10 kHz band sufficiently well to allow adjacent TETRAPOL carriers to operate (adjacent channel interference less than –42 dB to the carrier). The TETRA signal fits into a 25 kHz band.

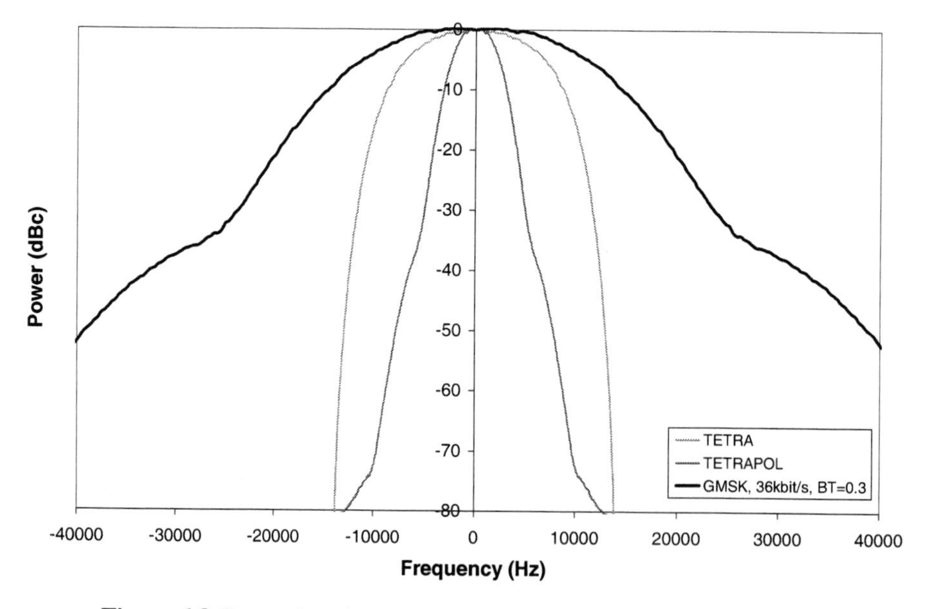

Figure 6.8 Comparison between the spectra of π/4–DQPSK and GMSK

When demodulating the signal, there are two main design options: coherent detection, which uses a copy of the carrier signal, and non-coherent detection, which uses the received signal alone. Coherent detection is conceptually simpler, because the received signal can be compared to the carrier in order to find the modulated signal, but it is difficult to obtain a reliable copy of the carrier in a mobile radio system which makes non-coherent detection easier.

In the case of non-coherent detection, differential encoding can be used. This reduces error performance because if a mistake is make in decoding it will affect more bits. Differential encoding reduces BER performance by about 3 dB, but this is worthwhile in terms of reducing decoding complexity.

This leaves π/4–DQPSK, which is the modulation used by TETRA. When TETRA was first proposed, this was a new technology, but it has matured, and is now one of the most popular modulation schemes for mobile radio, being used, for example, by the American IS-54 and Japanese PDC cellular systems and proposed for APCO25.

π/4–DQPSK is described in Section 1.1.7. While the modulation scheme is a PSK system with eight possible symbols, only four of the eight symbols are possible at each symbol interval, with the symbol positions shifting by π/4 after each symbol. This means that each symbol can transmit two bits. The symbol rate is therefore half the bit rate, and as modulation rate is reduced the bandwidth is also reduced. In addition, the differential nature of the scheme means that the first transmitted symbol provides a reference and transmits no information bits.

The lines show the possible signal transitions. Odd symbols have phases of π/4, 3π/4, –3π/4 and –π/4, whereas even symbols have phases 0, π/2, π, and –π/2.

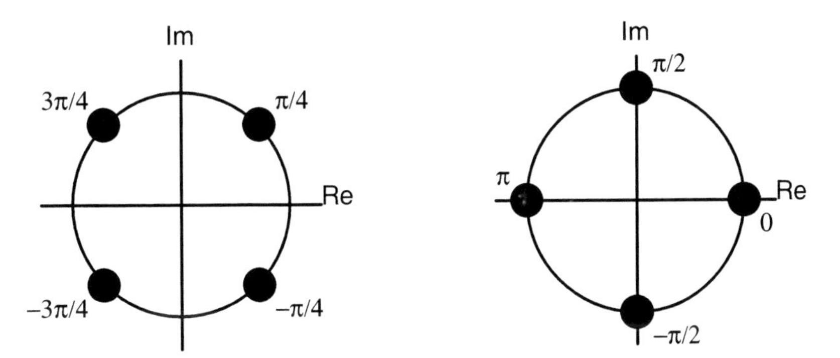

Figure 6.9 Constellation for odd symbols **Figure 6.10** Constellation for even symbols

The disadvantage of having two bits for each symbol is that if a symbol is received incorrectly, two bits may be corrupted. To minimise the chance of this happening, a careful mapping is chosen between the bits to be transmitted and the modulation symbols. If an error does occur it is more likely that a symbol will be mistaken for one of its neighbouring symbols in phase rather than the one directly out of phase with it. Therefore, a symbol with phase 0 is more likely to be mistaken for a phase of π/2 or –π/2, than for a symbol with phase π. By ensuring that neighbouring symbols represent bit pairs with only one change of bit between pairs in the sequence, bit errors can be minimised. Such a sequence is called a Gray code, and would be, for example, 00, 01, 11, 10, cycling back to 00. Compare this to a normal binary sequence 00, 01, 10, 11, cycling back to 00, which has two bit changes 01 to 10 and 11 to 00. The mapping between bits and symbols in TETRA is as follows.

Bit 1	Bit 2	Change in phase from previous symbol
1	1	$-3\pi/4$
0	1	$+3\pi/4$
0	0	$+\pi/4$
1	0	$-\pi/4$

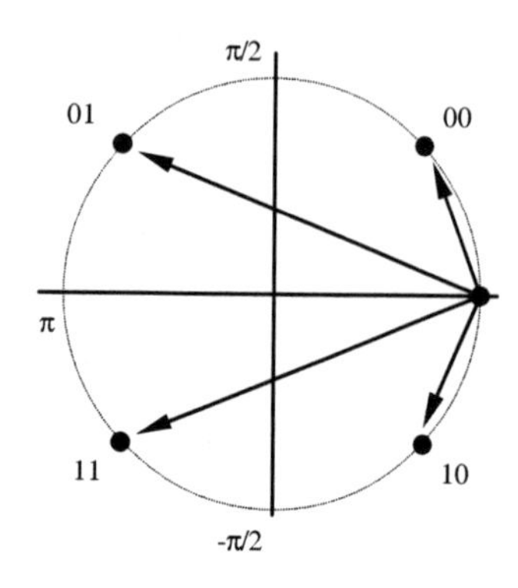

Figure 6.11 Mapping of transmission bits to symbols

Consider the transmission of the sequence 01100100. Splitting this sequence into pairs yields 01, 10, 01 and 00. Reference to Figure 6.11 shows that converting these bits pairs into phase changes gives $+3\pi/4$, $-\pi/4$, $+3\pi/4$ and $+\pi/4$. The first symbol is a reference and transmits no information. It has a phase of 0, so the transmitted symbols are 0, $3\pi/4$, $\pi/2$, $-3\pi/4$, and $-\pi/2$. The transmitted symbol points and transitions are shown in Figure 6.12.

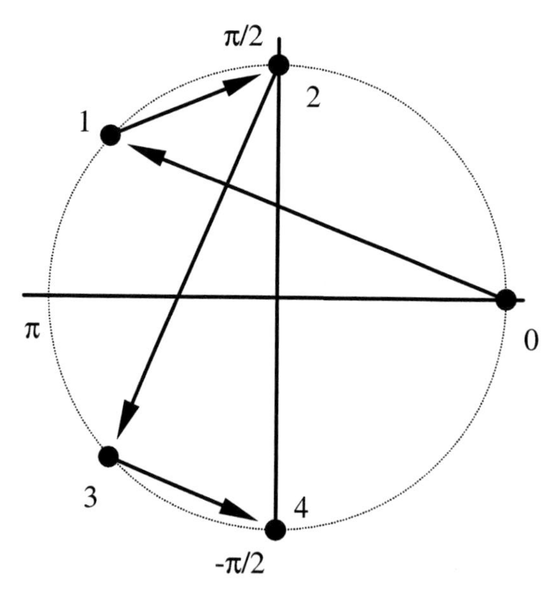

Figure 6.12 Sequence of symbols for the example bit stream

The disadvantage of this modulation scheme is that it has a very wide bandwidth unless it is filtered. This filtering takes the form of pulse shaping (see Section 1.6). TETRA uses a raised cosine filter, split between a root raised cosine filter at the transmitter and another root raised cosine filter at the receiver. Filtering reduces the bandwidth very effectively, but results in a non-constant signal envelope (i.e. the amplitude of the signal changes so that it is reduced at the end of the symbols where the phase transitions occur). The disadvantage of a varying signal envelope is that linear amplification is required in order to avoid spreading the signal. The effect of raised cosine filtering on a TETRA π/4–DQPSK is shown in Figure 6.13.

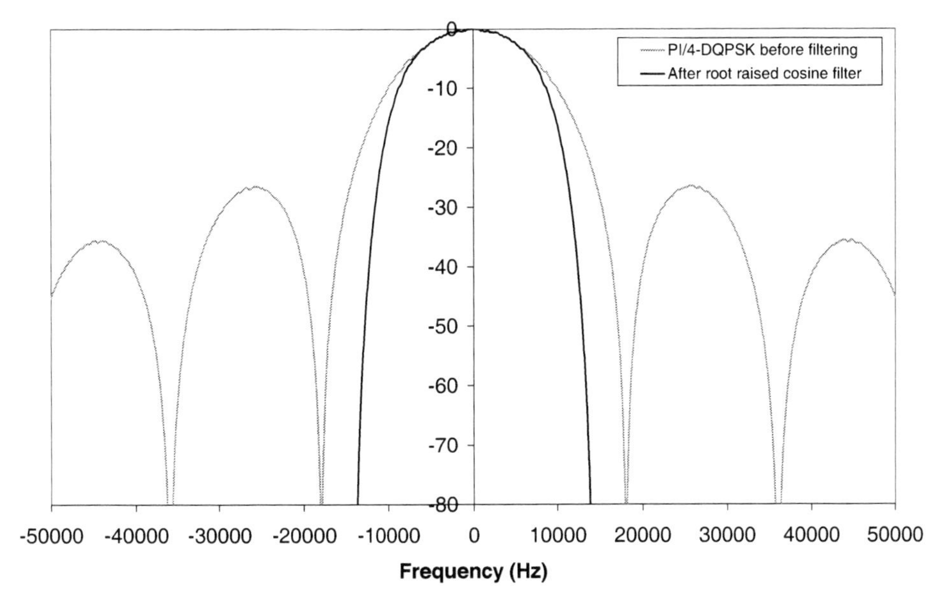

Figure 6.13 π/4–DQPSK spectrum with and without filtering

The sequence of operation of the modulator is therefore:

- Split the input bit stream into pairs of bits.
- Calculate the change in phase required to encode each bit pair.
- Starting with an initial phase of 0, calculate the required modulation symbols.
- Modulate the carrier with these symbols.
- Filter with a (root) raised cosine filter to reduce the bandwidth.
- Convert to the transmission frequency and transmit.

Table 6.2 Comparison of PMR modulation methods

How TETRA compares...

	TETRA	TETRAPOL[2]	APCO25 (ph 1)	APCO25 (ph 2)
Modulation	π/4–DQPSK	GMSK	C4FM	π/4–DQPSK
Carrier rate	36 kbit/s (18 ksymbols/s)	8 kbit/s	9.6 kbit/s	9.6 kbit/s
Carrier spacing	25 kHz	12.5 kHz	12.5 kHz	6.25 kHz
Efficiency (Bit/s/Hz)	1.6	0.64	0.77	1.54

[2] Assuming 12.5 kHz carrier spacing

6.6 FRAME, SLOT AND BURST STRUCTURES

6.6.1 TETRA Frame Structure

Each of the four slots in a TETRA frame consists of 255 symbols or 510 bits. On the uplink only, these can be arranged as two subslots of 255 bit durations each. 18 frames to form a multiframe. In addition, TETRA V+D groups 60 multiframes into a hyperframe of 61.2 seconds which is used for encryption control amongst other things.

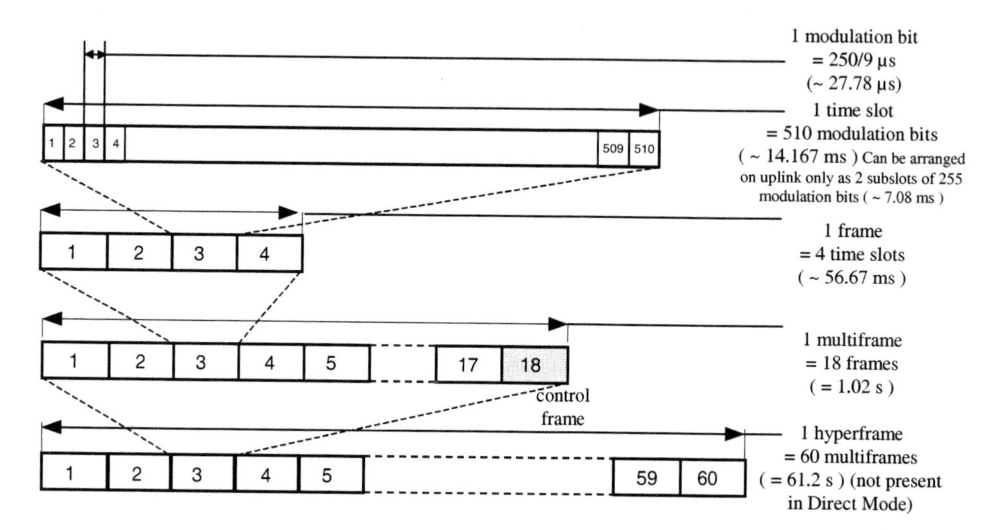

Figure 6.14 TETRA frame structure

In order to allow for signalling information, the speech data produced over the 1.02 second duration of 18 transmission frames is actually sent in 17 bursts over the first 17 frames in the multiframe. This leaves the last frame in the multiframe, frame 18, free for control signalling, so that some signalling capacity is always present even when data is being transmitted. Each speech frame lasts 30 milliseconds, and the information from two speech frames is formed into a single transmission burst (see Section 6.8 for a discussion of the transmission sequence).

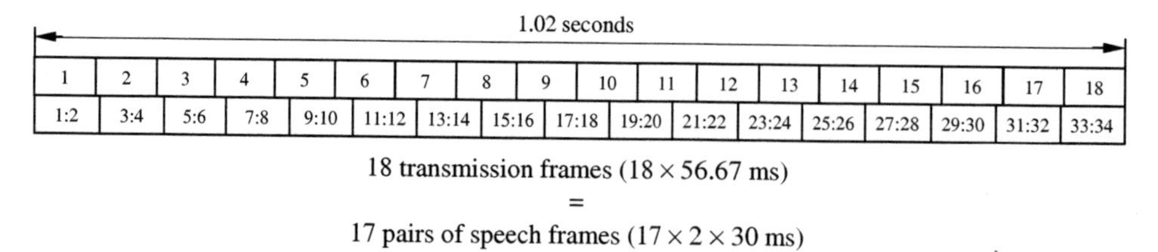

Figure 6.15 Temporal relationship between speech and transmission frames

Uplink and downlink frame timings are staggered so that uplink transmissions follow downlink transmissions by two slots. This serves two purposes. Firstly, full duplex transmission can be supported in corresponding uplink and downlink slots, since the mobile can receive in a slot, spend the next slot time switching to transmit mode, and then transmit in the same-numbered uplink slot. There is a remaining slot time for the mobile to switch back to reception mode before the downlink slot occurs. Secondly, transmission of the downlink in advance of the uplink allows the mobile to decode and act on the AACH before the uplink slot.

An example is shown in Figure 6.16. A duplex call is in progress on slot 1. The transmission in frame 2 slot 1 is received by the mobile station in its frame 1 slot 3. The mobile station then has time to decide what to do in its frame 2 slot 1 which follows two slots later. Its transmission in this slot is received by the base station in its slot 3.

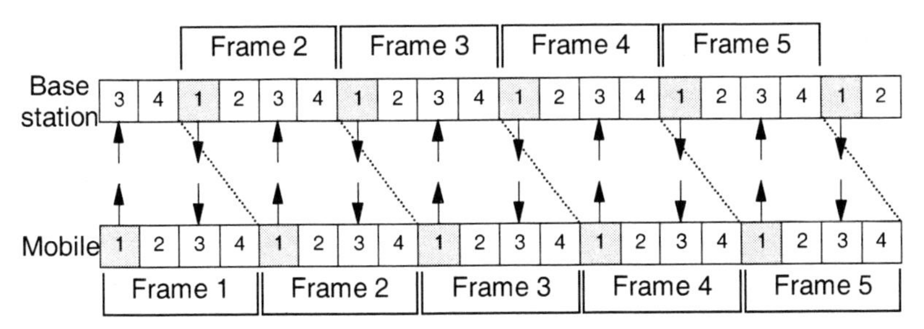

Figure 6.16 Timing relationship between uplink and downlink

6.6.2 TETRA Burst Structure

TETRA has a number of different burst types, but all have a similar structure. The different parts of a burst are as follows:

- **Ramping and linearisation**. In a TDMA system, the transmitter is switched on and off at the beginning and end of each slot. The burst structure must therefore include a period at the beginning and end of the slot for the transmitter to ramp-up to power and stabilise the transmitter.
- **Data bits**. This is the payload of the burst. In TETRA, the data bits are split to allow the training sequence to fit in the centre of the slot. It is possible for the two halves of the data to carry different logical channels.
- **Tail bits**. A couple of known bits are required at the start of the burst so that transmission of actual data can start with a known state. These bits help with symbol synchronisation and, if an equaliser is used, with channel equalisation. In TETRA, the tail bits also act to reduce transients in the filters, and are also included at the end of the burst.
- **Training sequence**. This is a known sequence of bits transmitted to allow the receiver to estimate the channel characteristics and for frame synchronisation (see Section 6.7). It is transmitted in the middle of the burst so that if it is used for equalisation, any estimate obtained will be close to the average characteristics experienced over the slot. Since the radio channel is changing, an estimate at one end of the slot may not characterise the other end very well

- **Guard period.** Time is required at the end of the burst for the transmitter to switch off before another slot begins. Also, in the case of uplink transmissions, time has to be left for different propagation delays between the mobile and the base station.

The uplink burst consists of these components as shown in Figure 6.17. There are two forms of uplink burst, a normal burst which occupies the complete time slot, and a shorter control burst, which occupies half a slot (a subslot). The overheads for a control burst are relatively high, since it needs the same ramping, tail and guard bits, and in fact has a longer training sequence since it is an access burst transmitted on its own, so synchronisation is more difficult. However, the control information to be carried is small, and having a half-slot burst doubles the opportunity for sending control bursts.

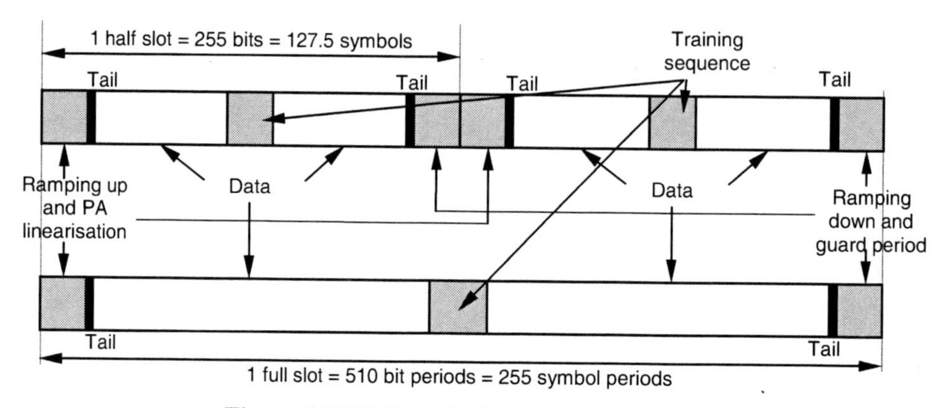

Figure 6.17 Uplink physical layer slot structure

Unlike the mobile, which has to switch its transmitter on and off at the end of each burst, the base station normally transmits continuously, with the intending recipient mobile changing for each slot. Downlink bursts therefore have an inter-burst training sequence to help with synchronisation of monitoring mobiles. They also have phase adjustment bits between the data and the inter-burst training sequence to ensure a known phase exists at that point.

If the base station does transmit discontinuously, as it would for time shared transmission where the carrier is shared between cells (see Section 6.14.4), it can ramp up and down its transmitter more quickly and accurately than the mobile. Furthermore, the downlink guard requirements are reduced when compared to the uplink, because all downlink bursts are transmitted from the same source. Uplink bursts are transmitted by different mobiles in different locations, and have to fit into the slot time at the receiving base station. Considering the example in Figure 6.18, a mobile at extreme range (75 km) transmits its slot exactly on time, but it is received by the base station 250 μs, or 9 bit times later. The mobile using the following slot is very near the base station and its burst is received only 1/3 bit time later.[3] This means that there are fewer ramp-up and guard bits on the

[3] Some systems, such as GSM, use a system of "timing advance", whereby the mobile is instructed to transmit its burst early so that it is received at the base station within the slot. TETRA, with its relatively low symbol rate, does not use this system.

downlink (10 and 8 bits instead of 34 and 14 bits), and so more bits of information can be transmitted in the downlink burst than on an uplink burst. This extra 30 bits of capacity is used for the "broadcast block", which is used to carry the Access Assignment CHannel (AACH), which is used for MAC information.

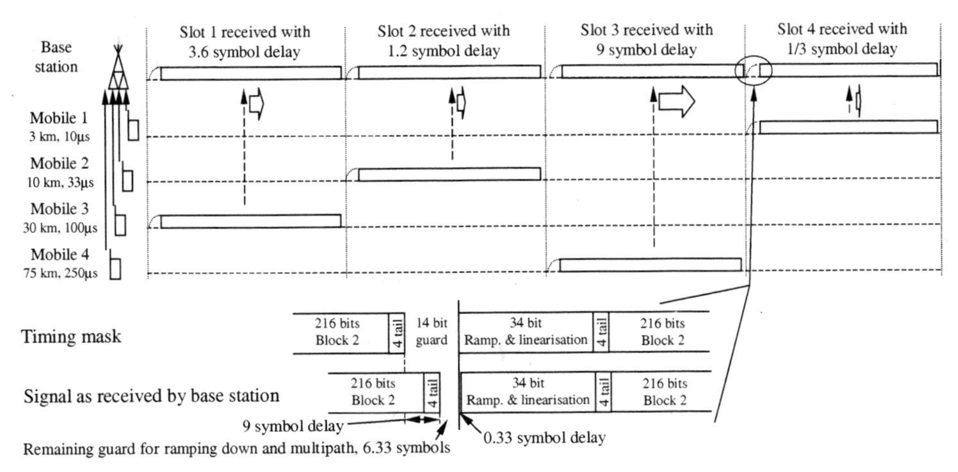

Figure 6.18 Use of the guard period to counteract propagation delay problems

Use of the Access Assignment CHannel differs between traffic and control channels. When the slot is carrying a traffic channel, the AACH carries the number of the mobile which is using the slot.[4] This notification prevents crossed calls which might occur if the signal was lost for a short time due to shadowing and the slot used by the mobile undergoing shadowing was allocated to another mobile. This occurs even if the mobile is in only transmitting on the uplink, in which case it would decode the corresponding downlink slot for the sole purpose of checking the AACH. Mobiles must monitor the AACH to ensure that they are still permitted to use the slot. The slot may have been re-allocated due to pre-emption by a higher priority call, or if contact was lost for a time due to fading. If the slot is re-allocated, the mobile cannot continue to use it and goes back to the Common Control CHannel to request a traffic channel again. On signalling channels, the AACH sends access control information (see Section 7.7.2.2).

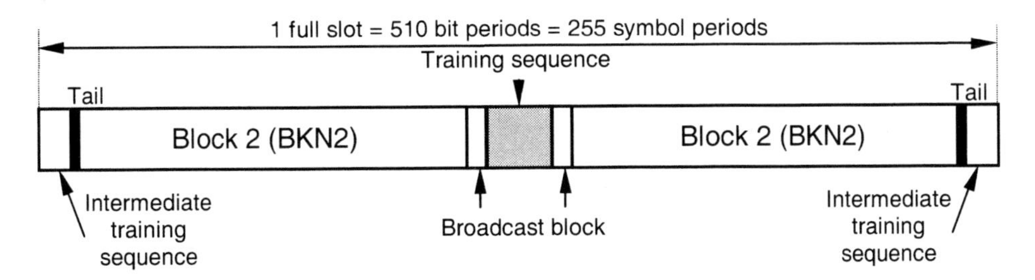

Figure 6.19 Basic downlink physical layer slot structure

[4] If the slot is used independently on the up- and downlinks by different mobiles, the AACH will indicate both mobiles.

6.6.3 Types of Burst of the Physical Layer

TETRA V+D has seven different physical bursts, three of which are used on the uplink and four on the downlink, while Direct Mode has three types of burst. These are as follows:

V+D uplink
- Normal uplink burst[5]
- Control half slot
- Linearisation half slot

V+D downlink
- Normal downlink burst[6]
- Discontinuous normal downlink burst (for time-shared operation)
- Synchronisation burst
- Discontinuous synchronisation burst (for time-shared operation)

Direct mode
- Normal burst
- Synchronisation burst
- Linearisation burst

Normal bursts contain two blocks of 216 bits. The error control coding uses these bits to transmit different signalling and data messages with different rates. Normally, both blocks are assigned to the same logical channel, but for signalling, or when stealing is in operation (see Section 7.5), a separate logical channel can be conveyed in each block. Downlink bursts also contain the 30 bits broadcast block which is used for the AACH. Control uplink half slots each convey 168 bits.

The downlink and direct mode have a special burst type to convey synchronisation and timing information. This is called the synchronisation burst, and carries the synchronisation channel in its first block. The second block can be used for other logical signalling channels. The synchronisation burst has an extended training sequence, which allows mobiles to align themselves to the transmitted bit stream more reliably.

The uplink and direct mode linearisation bursts are not bursts in the normal sense but simply times when mobiles can transmit in order to linearise their transmitters. Mobiles may need to do this when they change to a new frequency or change their power level after a power control command. Mobiles are not transmitting any information when they do this, and several mobiles can make use of the same linearisation burst opportunity. Linearisation is not needed so often on the downlink, since base stations are normally transmitting continuously. However, the base station can use the second block of a normal downlink burst when required for linearisation. No information is transmitted in this block.

---

[5] May be modified for multi-slot transmission by the inclusion of an inter-slot training sequence.
[6] Downlink linearisation takes place using a slightly modified downlink normal burst.

The bursts used in V+D mode are shown in Figure 6.20 for the uplink and in Figure 6.21 for the downlink.

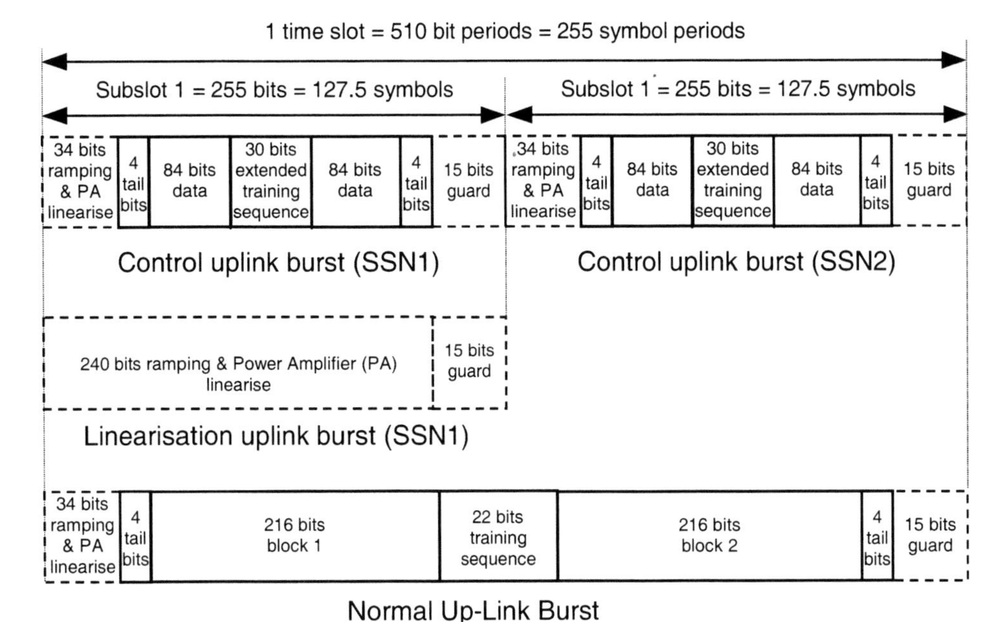

Figure 6.20 TETRA V+D Uplink Burst Types

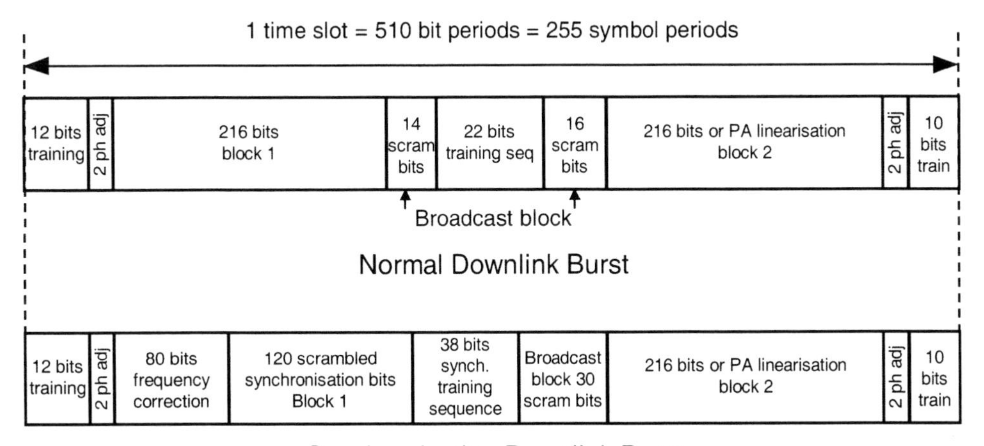

Figure 6.21 TETRA V+D Downlink Burst Types

In the case of discontinuous transmission on the downlink, which is used in timeshared transmission (see Section 7.11.4), the downlink normal and downlink synchronisation bursts are modified. The central 18 bits (9 symbols) of the inter-burst training sequence are replaced by an 8 bit guard followed by a 10 bit ramp and power amplifier linearisation period to allow one base station to ramp down and the other to ramp up. This is shown in Figure 6.22. The phase adjustment bits on either side of the training sequence are also changed. If neighbouring bursts are transmitted by the same base station in discontinuous mode, the base station need not ramp down and up, but simply transmits the full training sequence. However, the phase adjustment bits will match the discontinuous burst type, so this unramped form does not exactly match the continuous burst type.

Mobiles may also transmit continuously on the uplink if they are using multiple neighbouring slots. The ramp down and up sequence is replaced by the first part of the downlink inter-burst training sequence (training sequence 3), and the extended training sequence used in control bursts (see Figure 6.22).

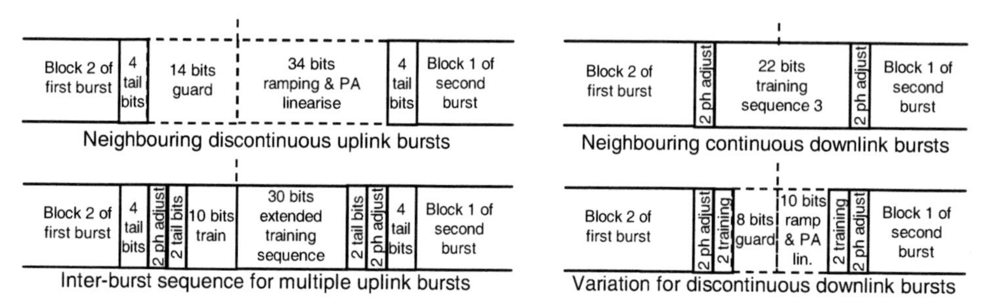

Figure 6.22 Burst variation for multi-slot uplink and discontinuous downlink bursts

Direct mode bursts are shown in Figure 6.23. These are normally discontinuous, but as in the case of neighbouring V+D uplink bursts, the transmitter does not need to ramp up and down if it is transmitting neighbouring bursts. In this case, a 40 bit frequency correction burst made up of a run of 0s, 1s and 0s is transmitted between bursts.

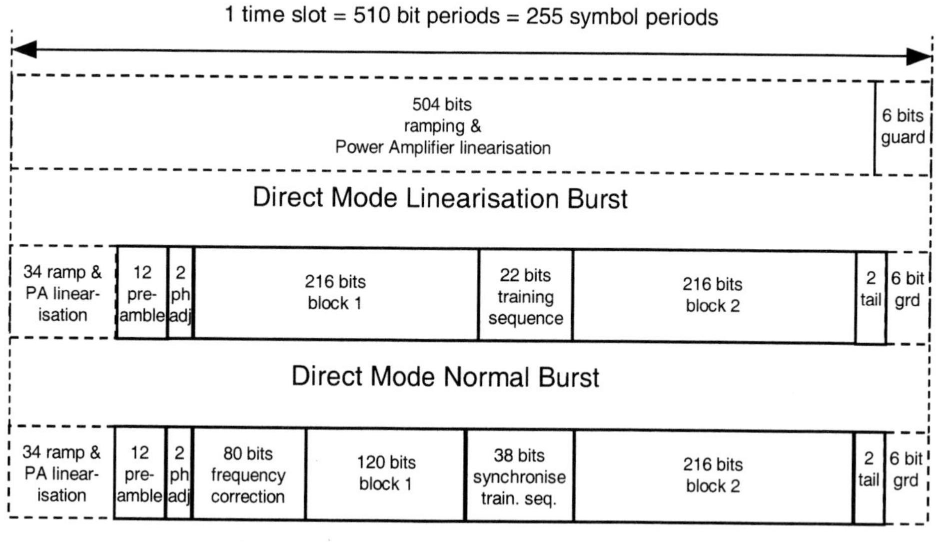

Figure 6.23 Direct Mode burst types

6.6.4 Logical Bursts Passed to the Physical Layer

The physical layer communicates with layer 2 via the TETRA virtual service access point. At this interface, several logical burst types are defined, onto which the MAC in layer 2 has mapped data. There is a one-to-one correspondence between physical layer bursts, or parts of burst, and the logical burst types, so the physical layer does not itself perform any routing or switching. Table 6.3 shows the different burst types available at the physical layer.

V+D		**Direct Mode**
Uplink	**Downlink**	
Normal uplink burst (NUB)	Normal downlink burst (NDB)	Normal burst (DNB)
Linearisation burst (LB)	*Block 2 of a NDB*	Linearisation burst (DLB)
Control burst (CB)	-	-
-	Broadcast block	-
-	Synchronisation burst (SB)	Synchronisation burst (DSB)
Slot flag (SF)	Slot flag (SF)	Slot flag (SF)

Table 6.3 Logical burst types passed to the physical layer

The slot flag is a one bit message conveyed by the use of one of two possible training sequences in normal bursts. It is used to indicate stealing and whether blocks 1 and 2 of the burst contain different logical channels.

6.7 SYNCHRONISATION

With digital modulation, synchronisation is necessary to ensure that the receiver is aligned on the transmitted symbols. In addition, *frame synchronisation* is required in a TDMA system to ensure that the receiver is aligned to the transmitted frames.

Symbol level synchronisation is achieved through the use of a training sequence. A training sequence is a known sequence of symbols, which the receiver attempts to detect. A circuit such as the one shown in Figure 6.24 will do this by multiplying a sequence of received symbols with tap gains that are based on the training sequence. When the received symbols match the tap gains, the sum will be large, and a threshold detector can be used to trigger an indication that the training sequence is found and the receiver can use this to decide on the frame timing. Setting the threshold slightly lower allows for some errors in the received training sequence at an increased risk of mistriggering.

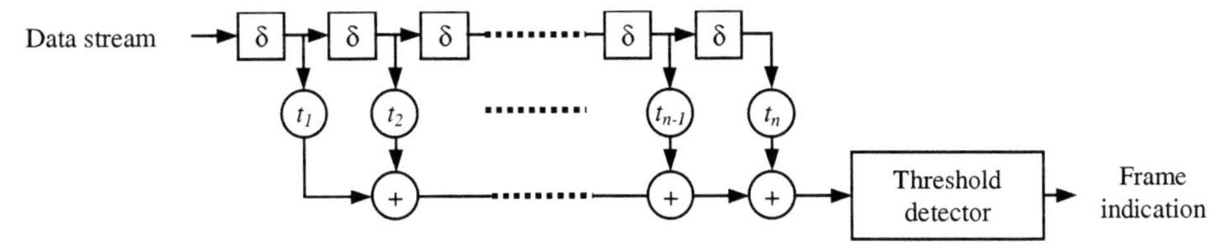

Figure 6.24 Basic training sequence detector and frame indicator

What this circuit is actually doing is cross-correlating the training sequence with the received data, and triggering when the result exceeds a threshold. The training sequences are designed with the property so that when the sequence is correlated with shifted versions of itself it yields very low values except when the shift is 0.

A difficulty with the design of training sequence is that the correlation value with other parts of the transmitted data stream should be low to avoid false triggering. In general, two independent random sequences will have a cross correlation of 0, but the transmitted data is not random. However, if through frame synchronisation, the receiver has a reasonably good idea about the location of the training sequence, the sequence can be designed so that for small shifts the correlation value is low, although this may result in higher correlation values for higher shift values.

TETRA has a number of different training sequences. There are two training sequences used on normal bursts, and the one used signifies the slot flag (see Section 7.8.3). Other training sequences are used between bursts when transmission is continuous, and for the uplink control burst. A specially extended training sequence is used for the synchronisation burst.

Note that in class E mobiles which have an equaliser (see Section 6.12.3), the training sequence also provides a known sequence for channel estimation.

TETRA terminals maintain a clock running at four times the symbol rate, and can therefore adjust in increments of a quarter symbol. This occurs if the mobile detects that there is a timing difference exceeding a quarter symbol. Base station clocks must be accurate to 0.2 parts per million (ppm), whereas mobiles, which normally receive synchronisation from the base station, must have clocks accurate to 2 ppm.

When mobiles first switch on, or come within range of a new base station, they have no information on the timing of the system. They therefore tune to a carrier and listen for a special synchronisation burst, which contains

- a frequency reference
- an extended training sequence to allow exact timing within the burst
- information on the burst and frame number

The extended training sequence gives better performance than the normal burst training sequence, and allows the frame structure to be recognised and the bit timing found. The V+D synchronisation burst is shown below (the continuous downlink burst is shown)

12 bits training	2 ph adj	80 bits frequency correction	120 scrambled synchronisation bits Block 1	38 bits synch. training sequence	Broadcast block 30 scram bits	216 bits or Power Amplifier linearisation Block 2	2 ph adj	10 bits train

Figure 6.25 Synchronisation continuous downlink burst

The frequency reference is transmitted using the frequency correction field. Note that if a sequence of 1s is transmitted, the modulation scheme will introduce a $-3\pi/4$ phase shift with each symbol, and the result after filtering will be a tone 6.75kHz below the carrier frequency. Similarly a series of 0s will advance the phase by $\pi/4$ with each symbol, producing a tone 2.25 kHz above the carrier. The frequency correction field transmits a series of 1s, 0s and then 1s which allows the receiver to synchronise on the required frequency.

The same system operates in Direct Mode, although in this case there is, of course, no base station to provide a reference. One of the mobiles, the one which initiates the call, is defined to be a master for the duration of the call transaction, and it provides the synchronisation reference. Any mobile that synchronises on the master is called a slave. The master sends synchronisation bursts (see Figure 6.23) in frame 18 in order to allow slaves to remain synchronised, and further synchronisation bursts will be transmitted at the start of a call transaction.

6.8 TRANSMISSION DELAYS

As was noted in Section 6.3.3, TDMA systems such as TETRA introduce delays. There are several different types of delay

- **Access delay**. This is the time it takes to access the system. This may be much longer for the first piece of information than subsequently. For this reason it is often defined either at a high level, where it would be the time to set up the channel, or at a much lower level, where it is the time between when the data is available and when the assigned burst begins.
- **Network transit delay**. This is the time it takes for a bit of information to get from the source to the destination. In a network with many switches which may repackage data for transmission this may be significant. Network transit delay is often called 'propagation delay' in fixed networks. It should not be confused with the propagation delay in a radio channel which is the time taken for a symbol to travel from the transmitter to the receiver. Radio propagation delay is very small in comparison with other delays and is already taken into account in the design of the burst.
- **Bit transmission delay (or packetisation delay)**. This is the time taken for a block or frame of data to be transmitted, and represents the difference in time between when all the information is available at the transmitter to when it is all available at the receiver. It also applies to circuit switched transmissions in systems like TETRA which organise data into blocks for coding and transmission.
- **Processing delay**. The time taken to process (usually encode and decode) a block of information.

Access delay is usually considered separately. This is because the time taken to set up the channel occurs only at the start, and by careful scheduling of the data in relation to the burst times (arranging the sampling time so that processed data is available just before the start of the transmission burst), the access delay at a frame level can be reduced to zero. The other delays are usually taken together as "end-to-end" delay, the time the data remains in the system from the transmitter to the receiver.

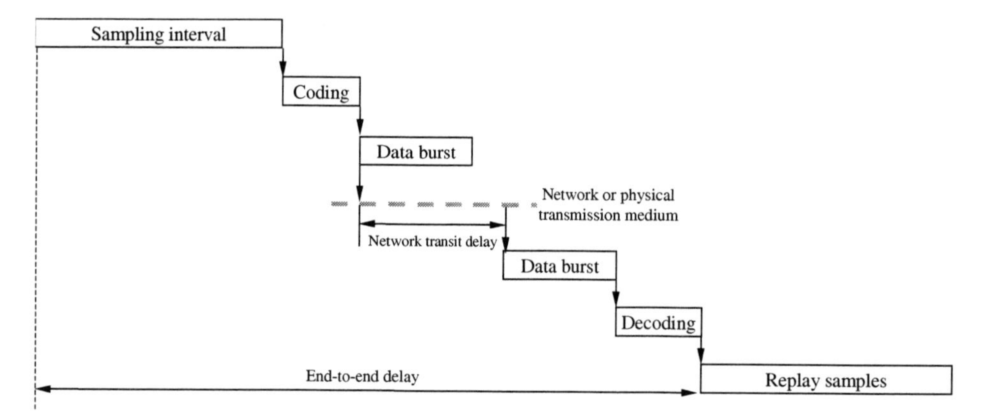

Figure 6.26 End-to-end delay

Note that these delays are not necessarily confined to a TDMA system. TETRAPOL, which has digital speech transmission, still has packetisation delay even though it is an FDMA system. In a TDMA system with digital speech transmission like TETRA, there are two instances of packetisation delay, one where a speech packet is formed, and one where a transmission burst is formed.

End-to-end delay in a TETRA system is made up as follows:

- Speech sampling interval (packetisation delay). In a TETRA speech channel, each speech packet is 30 ms long, and two are transmitted in one burst. The speech packetisation time is therefore 60 ms.
- Coding (processing delay). This is the time taken to encode the speech block, both the vector quantisation used for source coding, and the channel coding used for error protection. Vector quantisation is computationally intensive and therefore time consuming. The Designer's Guide gives typical values for these processes as a total of 35ms [1].
- Burst transmission (packetisation delay). This is the time taken to transmit the burst. In a TDMA system, the effect of packetisation along with coding and interleaving is that data is not available until the entire burst has been received.
- Transmission through the medium (network transit delay). This depends on the configuration and is discussed in greater detail below.
- Decoding (processing delay). This is the time taken to demodulate and decode the burst and reconstruct the speech waveform. The Designer's Guide gives a figure of about 13 ms for this task.

After this, the digital speech data is ready to be reproduced.

The network transit delay depends on the configuration of the TETRA system. In Direct Mode, there is no intervening infrastructure, and the network transit delay is effectively zero. In V+D mode, a single site communication will use one slot in the uplink and one in the downlink for each link. The network transit delay is then two slot lengths, since the data transmitted by the transmitting mobile will be sent on to the receiving mobile in the corresponding slot in the following frame. This network transit delay can be considered to be made up of packetisation delay (for the transmission to the base station, processing delay in the base station, and then access delay waiting for the downlink slot.) It would not be possible to have designed the TETRA system with the downlink slot immediately following the corresponding uplink slot as there would have been no allowance for processing delay. The transmission process on a single site is shown in Figure 6.27.

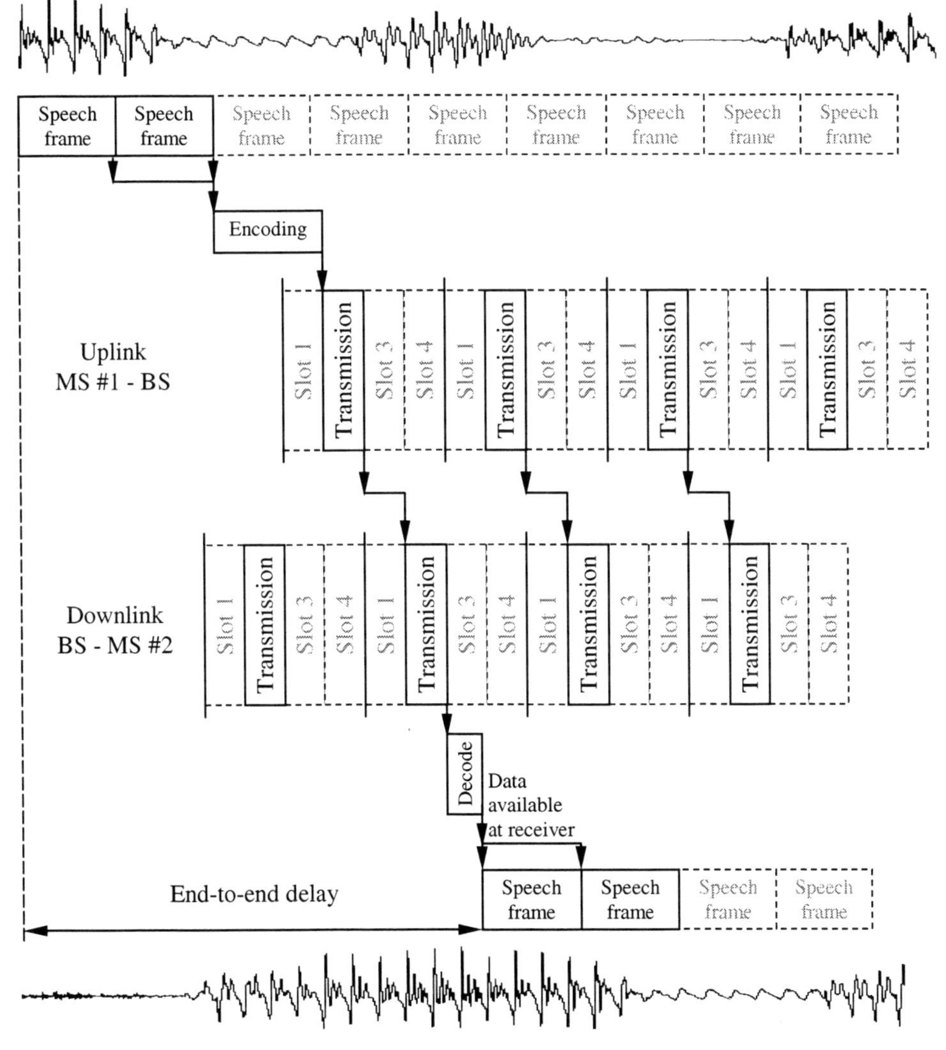

Figure 6.27 Speech coding delays

If the communicating mobiles are on different sites, the network transit delay will be increased by the time taken to send the data through the fixed network between the sites. If there is communication with other networks, there will be additional transit delays to these networks, and increased packetisation and processing delays if the data has to be transcoded.

Interleaving on the circuit switched data services (see Section 7.4.3.3) increases delay, since the packetisation delay will increase from a single transmission burst to the number of bursts the interleaving is over, including the intermediate burst intervals. For interleaving over 4 slots, this is 13 slot lengths, and for interleaving over 8 slots this is 29 slots, or over 410 milliseconds. To this have to be added the other delays to form the end-to-end delay.

The picture is complicated slightly by the use of frame 18 for signalling traffic. Seventeen pairs of speech packets take the same time (1.02 seconds) as 18 transmission frames, but data sampled just before frame 18 will need to wait until frame 1 to be transmitted. This is actually a case of a varying access delay due to the fact that the speech frame length and transmission frame length are different. The system is designed so that the speech data is sent slightly faster than it is produced by the speech coder in transmission frames 1–17, so that the transmit data buffer is empty after the transmission for the data in frame 17. It fills up again during transmission frame 18, and the cycle repeats.

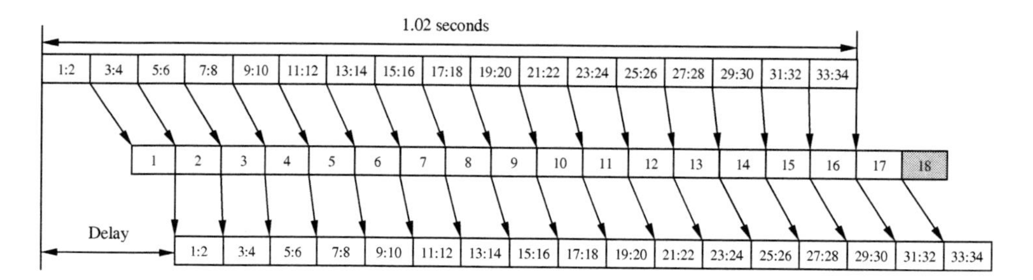

Figure 6.28 Scheduling of speech frames for transmission over a frame

The maximum delay the speech is subjected to is therefore one frame more than the simple case calculated above. Since the speech must be replayed at a constant rate, the network transit delay for a single site TETRA is therefore one frame and two slots, or six slots in total. For Direct Mode, the two slot uplink/downlink transit delay does not apply, so the network transit delay is 4 slots.

A typical end-to-end delay for a single site operation is therefore

Speech packetisation delay (60 ms)
+ encoding
+ transmission (14.667 ms)
+ network transit delay (6 × 14.667 ms)
+ decoding.

Using the figures in the Designer's Guide [1] gives 60 + 35 + (7 × 14.667) + 13 milliseconds, which gives a total of 207.2 ms.

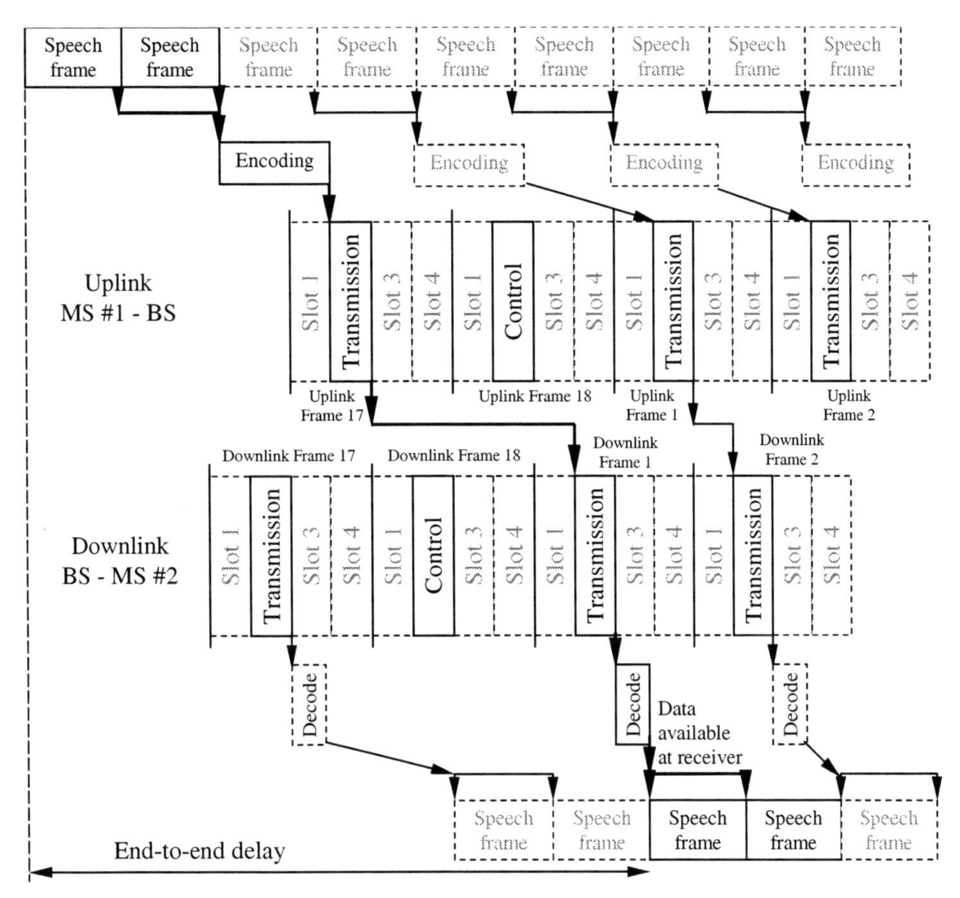

Figure 6.29 Speech delay from frame 17 to frame 1

6.9 SCRAMBLING

When transmitting data in a digital transmission system, it is often desirable that there are no regular patterns in the transmitted bit stream, since this ensures that such bit patterns do not cause unwanted frequency components in the transmitted signal. Also, in some modulation schemes regular runs of bits would cause synchronisation problems or introduce a dc bias. Although this does not apply in the case of $\pi/4$–DQPSK as it is a differential modulation scheme, continuous runs of 0s or 1s would cause tones in the output. Indeed, such runs are deliberately incorporated in the synchronisation burst.

Another issue is that in a TETRA system, frequency planning usually not possible and neighbouring areas may use the same carrier frequency. This could cause a problem if a mobile transmission intended for one base station was picked up by another, or vice versa. Consider the following case. While mobile 1 is communicating with base station A and mobile 2 is communicating with base station B, if mobile 2 is in a deep fade its signal may be lost and base station B may capture the signal from mobile 1 instead.

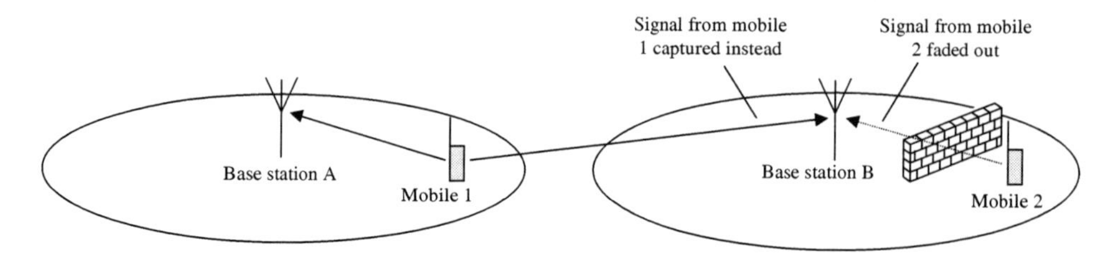

Figure 6.30 Example of signal capture from an adjacent cell

To prevent such a situation resulting in a false message, scrambling is undertaken to ensure that the transmitted sequence is different depending on the receiver the signal is intended for. Scrambling is undertaken by exclusive-ORing the transmitted sequence with a scrambling sequence.

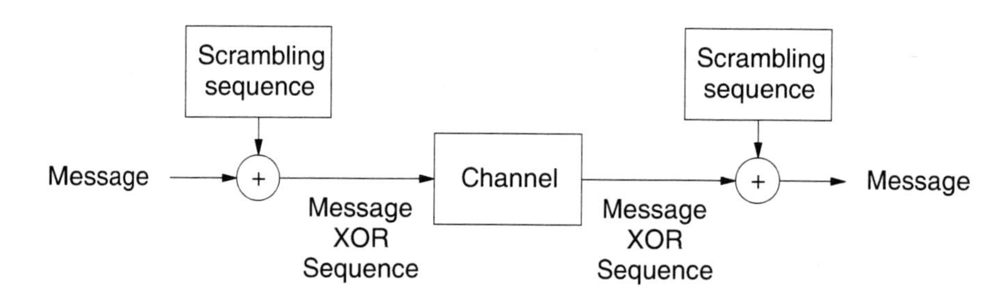

Figure 6.31 Scrambling process

The scrambling sequence is generated by the output of a linear feedback register. A small linear feedback register is shown in Figure 6.32.

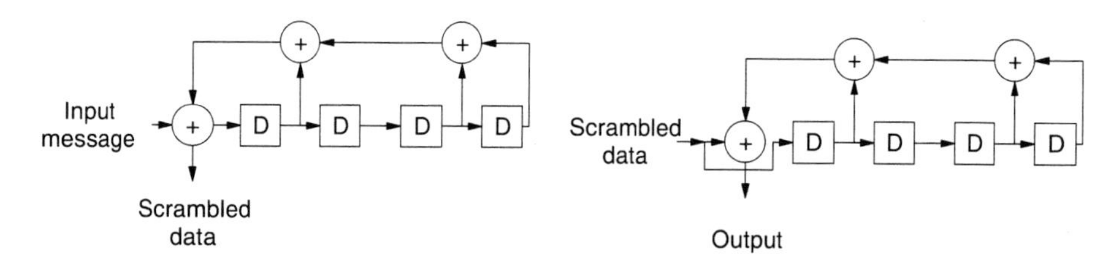

Figure 6.32 Example of scrambling using linear feedback registers

In TETRA, the registers are 32 bits long, and are initialised by the extended colour code (for V+D) or DM colour code (for Direct Mode) of the mobile (plus 2 zeros). The extended colour code is a 30 bit string made up of the mobile country code, the mobile network code and a 6 bit colour code for the individual mobile (see Section 5.8). The DM colour code is made up from the short subscriber identity and part of the Mobile

Network Identity [5]. These initialisers are used for all bursts except the broadcast channel in V+D, and the SCH/S and SCH/H carried on the synchronisation burst in Direct Mode, where the shift register is initialised to zero. This is so that the mobile can decode the messages required to set up the scrambling scheme.

While descrambling with the same scrambling sequence results in the original message, using a different scrambling sequence for encoding and decoding will result in a random sequence of bits. This is what would occur if a receiver picks up a message intended for a different receiver. Since TETRA uses channel coding (see Section 7.4), when this sequence of bits is passed up to the MAC, the channel decoding function will detect this as an error in the same way that it would if the transmitted bits were corrupted by a bad channel.

Note that scrambling is undertaken only to protect against false reception and not to prevent deliberate eavesdropping. Any eavesdropper can listen in to the initial control messages and discover the shift register sequence, and so gain sufficient information to decode the scrambling. In TETRA, protection against eavesdropping is the function of air interface encryption or cryptographic protocols at the application level.

6.10 TRANSMISSION POWER AND POWER CONTROL

6.10.1 Power Classes

The TETRA standard defines a number of different power classes for both mobiles and base stations. The different power levels for the different classes are given in Table 6.4.

Table 6.4 Mobile and base station power classes

Base stations		Mobile stations			
Power class	Nominal power	Power class	Nominal power	Applies to V+D Mode	Applies to Direct Mode
1 (40W)	46 dBm	1 (30W)	45 dBm	✓	
2 (25W)	44 dBm	2 (10W)	40 dBm	✓	✓
3 (15W)	42 dBm	3 (3W)	35 dBm	✓	✓
4 (10W)	40 dBm	4 (1W)	30 dBm	✓	✓
5 (6.3W)	38 dBm	5 (0.3W)	25 dBm		✓
6 (4W)	36 dBm				
7 (2.5W)	34 dBm				
8 (1.6W)	32 dBm				
9 (1W)	30 dBm				
10 (0.6W)	28 dBm				

Note that mobile power class 5 only applies to Direct Mode, while mobile power class 1 does not apply to Direct Mode Operation. Mobile power classes 1 and 2 would normally be vehicle-mounted mobile units, and mobile power classes 3, 4 and 5 would be hand-held portable units.

6.10.2 Power Control

Given the variable nature of the radio channel with the variation due to distance and shadow fading, the system has to be dimensioned so that the transmission link can be maintained in a poor radio channel. This means that when the radio channel is good – when the mobile is near the base station, for example – more power is being used to transmit the signal than is necessary to maintain quality. This excess power is a waste in radio resource as it causes interference to other mobiles. To counteract this problem, adaptive power control can be used by the mobile. Power control allows the system to minimise the transmit power required by the mobile whilst maintaining the quality of the radio uplink. By minimising the transmit power levels, interference to co-channel and adjacent channel users is reduced and MS power consumption can also be reduced. This second point is important as it minimises battery drain and extends operation time.

Two methods of adaptive RF power control are defined in TETRA. The first method, known as open loop power control, is implemented in the mobile. Using this method, the MS adjusts its transmit power based on the power level or equivalent signal quality being received by the mobile on the downlink from the BS. The term *open loop* refers to the fact that there is no feedback from receiving base station on the quality of the transmission from the mobile (see Figure 6.33). This lack of feedback allows open loop feedback to be used for initial random accesses, but has the disadvantage that should the propagation conditions on the uplink not be exactly the same as on the downlink, the downlink estimate may not be valid, due to the different fading the uplink and downlink will experience resulting from the frequency separation on uplink and downlink. This would cause the mobile to adjust its transmit power to the wrong value.

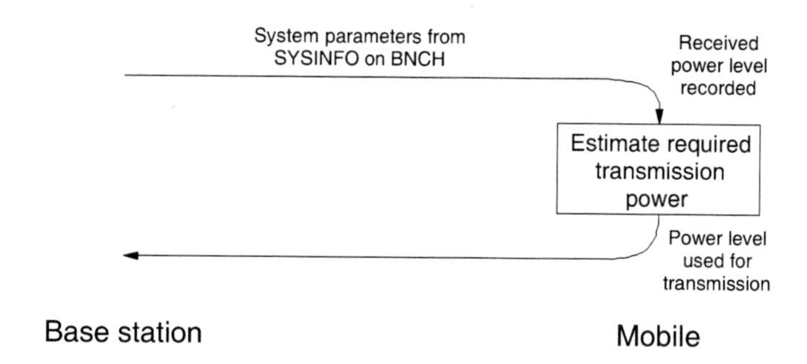

Figure 6.33 Open loop power control

A second method, known as closed loop power control, is supported by the mobile and implemented in the base station. This method avoids the difficulty of estimating the uplink channel from the downlink by including a feedback path from the base station. The mobile adjusts its transmit power according to instructions from the base station. The base station calculates the optimal mobile transmit power, for example based upon the power level being received on the uplink from that mobile. This has the advantage that the power level is based on the actual performance of the uplink.

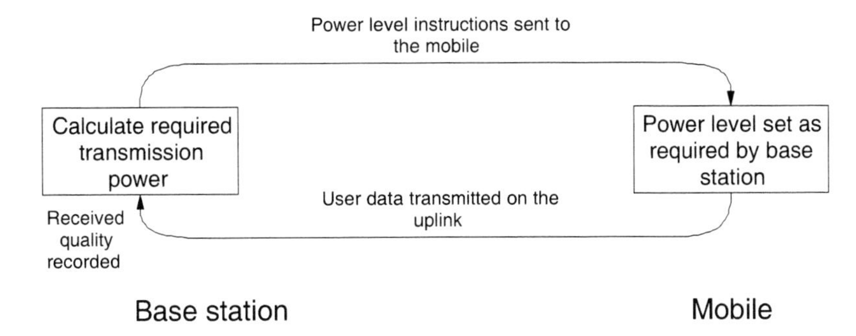

Figure 6.34 Closed loop power control

The actual power level used ranges from 45 dBm (step 1) to 15 dBm (step 7) in equal steps of 5 dB. However, mobiles are still limited by their power class as defined above, so only a class 1 mobile will have access to step 1, class 2, 3 and 4 mobiles being limited to steps 2 (40 dB), step 3 (35 dB) and step 4 (30 dB) respectively.

Note that adaptive power control is not used on the downlink in V+D mode, nor in Direct Mode, although the possibility exists for a Direct Mode slave to reduce its transmit power level.

6.11 CHANNEL QUALITY MEASUREMENT

The lower layers of the TETRA system provide a number of radio link measurements for use by the higher layers in system control. These measurements consist of signal strength, signal quality and round-trip MS-BS path delay measurements. Only the first two are used in Direct Mode.

6.11.1 Received Signal Strength Measurement

Received signal strength is measured on a 64 point scale over the range from –110 dBm to –48 dBm. The TETRA standards specify that an absolute accuracy of ± 4 dB should be obtained, with a relative accuracy between two measurements on the same carrier or on different carriers being ± 3 dB. The range is truncated, with signal strengths up to –110 dB are recorded as 0, between –110 dB and –109 dB as 1, and so on up to values exceeding –48 dB being recorded as 63. The sample duration of the signal strength measurement is either 1 ms or 4 ms.

6.11.2 Signal Quality Measurement

Signal quality measurement for V+D is undertaken by monitoring the rate of successful decoding of the AACH. The mobile maintains a running counter, called the Radio Downlink Counter (RDC), which is incremented each time an AACH message is received correctly by the number of slots since the last monitored AACH (i.e. four if the mobile is monitoring one slot per frame). The RDC is decremented by the number of slots multiplied by a factor, f_q, if the AACH is not decoded correctly. This factor defines the quality threshold for the mobile. There is an associated Radio Downlink Threshold, which defines the maximum value the RDC can take, as well as its initial value. If the message error rate (MER) on the AACH exceeds $1/(f_q+1)$, the RDC will fall on average. For example, if the factor is 4, the RDC will tend to decrease if the MER > 20%. If the RDC falls below 0, the MAC will consider that the radio link has failed, and inform layer 3 appropriately.

In Direct Mode, a similar system operates, only in this case there is no AACH, and the measurements are based on the success rate in decoding the SCH/S.

6.11.3 Round-trip MS-BS Path Delay

The round-trip path delay between the base station and the mobile may be measured by the base station as a criterion to relinquish a radio uplink. The path delay of the mobile is a representation of the distance of the mobile to the serving base station. This distance may be used to prevent the mobile exceeding the planned cell boundaries by a large amount. This information may be sent to the mobile by the base station when required.

6.12 PHYSICAL LAYER PERFORMANCE

6.12.1 TETRA Receiver Sensitivity

Operation of a mobile radio receiver demands that the received signal satisfies two constraints. The first requirement on the received signal is that it is powerful enough in absolute terms for the receiver to be able to discern it. The second requirement is that the signal power exceeds that of any interfering signals by a sufficient margin for the receiver to be able to decode the wanted signal successfully.

Which of these requirements will actually constrain the operation of the system depends on the way the system is operated. In low traffic areas with large cells and few users and without a cellular re-use pattern, there will be little if any interference and the minimum received signal constraint will apply. In high density areas, with a cellular re-use pattern, the interference constraint will apply as the interfering signals will swamp the desired signal long before the signal falls to the minimum power level.

The minimum received signal power that a mobile receiver can operate with is called the *receiver sensitivity*. In simple terms, the power at the receiver is the transmit power less the pathloss. Improving the receiver sensitivity – i.e. designing the receiver to work with smaller signal powers – allows either a lower transmit power for the same pathloss, or a larger pathloss for the same transmit power. The former is desirable from the point of view of safety, battery and regulatory constraints, while the latter provides increased coverage. However, increasing receiver sensitivity increases the cost and complexity of the receiver, and a compromise must be reached. In the TETRA system, the standard specifies that receivers must have a receiver sensitivity of at least the values given in Table 6.5.

Table 6.5 Receiver sensitivities

RX sensitivity	Mobile	Base station	Unit
static:	−112	−115	dBm
dynamic:	−103	−106	dBm

Note that there are different receiver sensitivities specified for the dynamic case, where mobiles are moving, and the static case. When mobiles are moving, fast fading will tend to average out, and the mobile will not remain in a deep fade for any length of time. However, the movement causes Doppler effects, which makes reception more difficult. Static mobiles will still experience fast fading due to the movement of surrounding objects, but the averaging effect will be significantly less.

The standard also specifies the minimum level of the desired signal relative to interfering signals. TETRA terminals should be able to operate with a signal to interference ratio down to 19 dB.

Table 6.6 Comparison of signal thresholds

How TETRA compares...

	TETRA	TETRAPOL	GSM
Dynamic sensitivity			
Base station	−106 dBm	−113 dBm	−104 dBm
Mobile	−103 dBm	−111 dBm	−103 dBm
Minimum SIR	19 dB	15 dB	9 dB

6.12.2 TETRA Reference Channels

The performance of a radio receiver is very dependent on the propagation environment. While mean signal strength levels can be defined relatively easily, and this can deal with pathloss and slow fading, there is still a dependency on fast fading. It is therefore necessary for the TETRA standards to define a number of propagation models for the purpose of setting standard performance levels. The propagation models are designed for a number of reference cases, and were developed for radio measurements and similar models used in the GSM specifications. The actual signal from the transmitter to the receiver would travel by a large number of different paths. This can be modelled by a much smaller number of signals, each of which has an average power and delay, and which undergoes a Rayleigh fading process. The overall received signal consists of the sum of these signals. The channels used to define operation in V+D mode are given in Table 6.7, while Direct Mode reference channels are given in Table 6.8. Each of these basic channels is defined at a number of speeds, which affect the rate of the Rayleigh fading processes. The faster the mobile is moving, the quicker the fading will effectively occur. Examples are TU50 (Typical Urban, 50 kph) or HT200 (Hilly Terrain, 200 kph).

Table 6.7 Channel models used for the V+D TETRA specification

Model	Tap 1	Tap 2	Tap 3	Tap 4	Tap fading
Static	0 dB at 0μs				Static
Rural area (RAx)	0 dB at 0μs				Rice
Typical urban (TUx)	0 dB at 0μs	−22.3 dB delayed 5μs			Class
Bad urban (BUx)	0 dB at 0μs	−3 dB delayed 5μs			Class
Hilly terrain (HTx)	0 dB at 0μs	−8.6 dB delayed 15μs			Class
Equaliser test (EQx)	0 dB at 0μs	0 dB delayed 11.6 μs	−10.2 dB delayed 73.2μs	−16 dB delayed 99.3μs	Class

Table 6.8 Channel models used for the Direct Mode TETRA specification

Model	Tap 1	Tap 2	Tap 3	Tap 4	Tap fading
Static	0 dB at 0μs				Static
Direct Urban (DUx)	0 dB at 0μs				Rice
Direct Rural (DRx)	0 dB at 0μs				Class

6.12.3 TETRA Receiver Classes

The trade off between complexity and performance has led the TETRA standard to specify different classes of mobiles with different performances. These are as follows

- **Class A**. This is designed for relatively harsh propagation environments, such as urban environments with large multipath components, or in mountainous terrain. This equipment is defined for operation using the static, TU50 and HT200 propagation models (see Section 6.12.2).
- **Class B**. This is designed for operation in less harsh radio environments, such as suburban or normal urban areas. However, this equipment would not be able to operate in areas with large multipath components. It is defined for operation using the static and TU50 propagation models.
- **Class E**. This equipment would contain an equaliser and is intended for very harsh propagation environments (with multipath delays in the order of two symbol durations). Such long "multipath" delays would also be found in quasi-synchronous operation (see Section 6.14.3). There is no corresponding base station classification. Operation is defined for the static, TU50 and EQ200 propagation models.

Class A equipment will be able to work in all areas except very harsh environments with severe multipath, whereas class B mobiles are limited to urban and suburban areas. However, class B mobiles are less complex, and therefore cheaper. Class E mobiles are the most complex, incorporating an equaliser, but do have the advantage of being able to operate over the widest range of environments.

6.12.4 Link Budget

The components of the link budget are slightly more complex than the transmit power – pathloss form given above. The transmit power is usually defined at the transmitter's power amplifier. Some of the signal is subsequently lost in the cable to the antenna. Also, pathloss is defined in terms of an isotropic antenna (see Section 1.10), and so transmit and receive antenna gains must be taken into account. Finally, the received signal strength is measured at the receiver circuitry, and so cabling losses have to be included. Figure 6.35 shows the relationship between these various values.

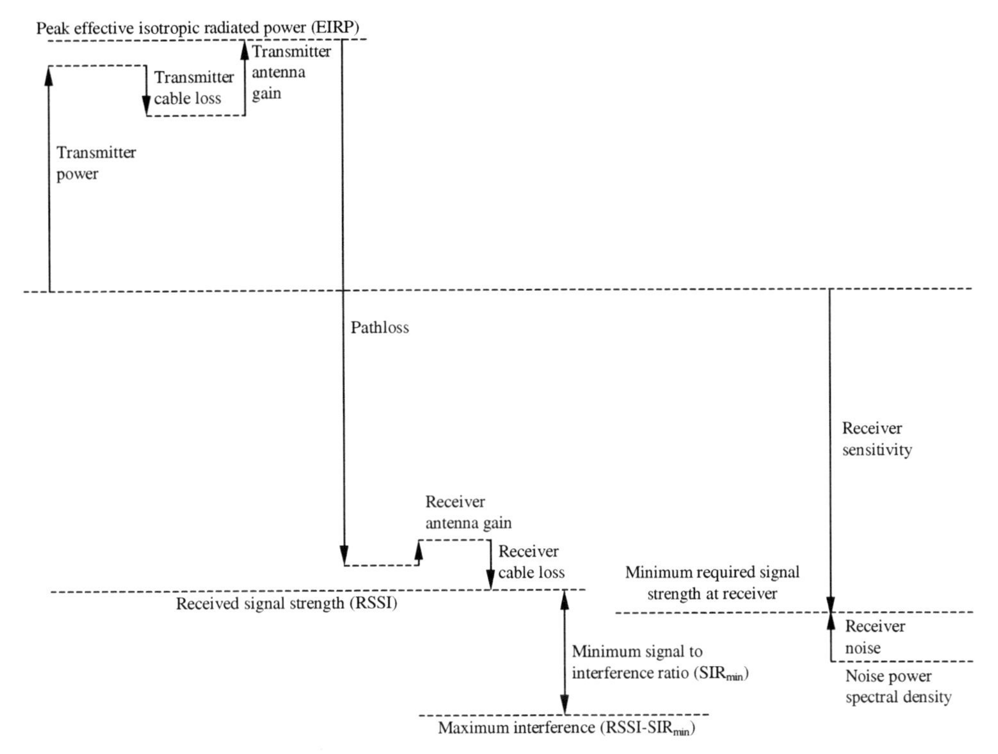

Figure 6.35 Make up of the link budget. Note that the antenna gains are assumed to be positive; however, the small size of hand-portable antennas means that their gains are usually negative

Receiver sensitivity depends on the background noise (the noise spectral density multiplied by the bandwidth of the TETRA channel, which forms a constant) and the receiver noise, which depends on the construction of the receiver. The standards [3][4] define maximum values of receiver sensitivity. The amount by which the received signal strength exceeds the receiver sensitivity defines the fade margin, the "slack" in the link budget to allow for fading. If fading exceeds this value, the signal strength will fall below the receiver sensitivity and performance will be poor or non-existent.

Another aspect which must be taken into account is the minimum value of signal to interference ratio. If the average value of interference is greater than the receiver sensitivity less the minimum signal to interference value, the system will be interference limited rather than noise limited. The fade margin is then the averaged received signal strength less the average interference and minimum SIR value (see Figure 6.36).

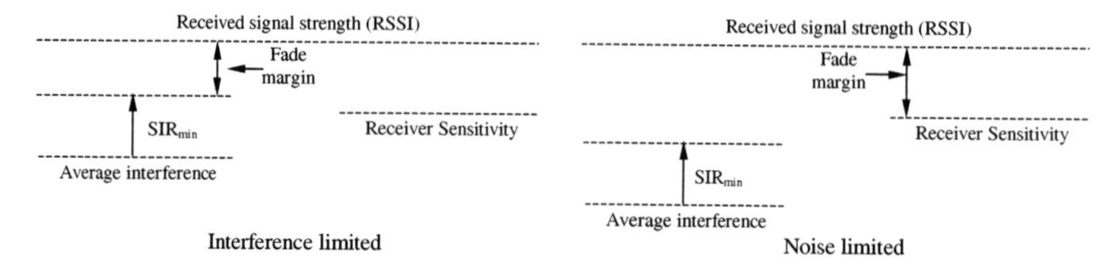

Figure 6.36 Fade margin in noise limited and interference limited systems.

The TETRA standards define that receivers must operate with a minimum SIR of at least 19 dB. Low density systems with large, widely spaced cells are likely to be noise limited, while high capacity systems with cellular re-use will be interference limited.

The following figures in Table 6.9, based on the Designer's Guide [1], are intended to be typical of real equipment. Five cases are shown: a mobile unit (MS) with mobile power class 2, communicating to and from a base station with BS power class 2; a hand-held portable unit (HH), with mobile power class 4, communicating with the same base station; and two power class 4 mobiles communicating in Direct Mode.

Table 6.9 Typical TETRA link budgets

	BS → MS	MS → BS	BS → HH	HH → BS	HH→ HH (Direct)	Unit
TX power	44	40	44	30	30	dBm
TX cable and filter loss	6	2	6	0	0	dB
TX antenna gain	8	2	8	–4	–4	dB
Peak effective isotropic radiated power	46	40	46	26	26	dBm
Propagation loss	L	L	L	L	L	dB
Signal level at RX antenna	$46 - L$	$40 - L$	$46 - L$	$26 - L$	$26 - L$	dBm
RX antenna gain	2	8	–4	8	–4	dB
RX cable loss	2	4	0	4	0	dB
RX input power	$46 - L$	$44 - L$	$42 - L$	$30 - L$	$26 - L$	dBm
RX sensitivity						
static:	–112	–115	–112	–115	–112	dBm
dynamic:	–103	–106	–103	–106	–103	dBm
Maximum acceptable median pathloss						
static:	158	159	154	145	138	dB
dynamic:	149	150	145	136	129	dB

From the above figures it can be seen that for operation to and from typical mobiles, a maximum median pathloss that can be tolerated is 158 dB for static operation, or 149 dB for dynamic operation. For operation from typical hand-held portable units, there is a maximum median pathloss of 145 dB for static operation, and 136 dB for dynamic operation. These figures can be used to estimate maximum cell sizes.

6.12.5 Area Coverage

A major issue is the area which can be covered by a base station under specific propagation conditions (for a given transmission power), or, perhaps, what transmission power would be required in order to cover a particular area. Both of these questions can be addressed from the point of view of the link budget, and the maximum pathloss that can be tolerated for the RSSI to be at least the value of the receiver sensitivity. If the pathloss from the transmitter to the receiver exceeds this maximum, the area will not be covered.

Since pathloss over an area is a statistical function, it is not possible to predict precisely whether coverage will exist at a specific location but rather give a probability that coverage will exist. Cells are planned so that the received signal will be satisfactory, either in terms of being above the receiver sensitivity (noise limited) or SIR threshold (interference limited) over a given proportion of their area. Existing systems usually use 90%, but some applications require a higher level. Future systems are expected to approach 99%, but achieving these levels requires much more infrastructure. Since pathloss consists of three components – distance based pathloss, slow fading, and fast fading, all three must be taken into account. In particular, fast fading will cause a change in performance depending on whether the mobile is moving or stationary. For stationary mobiles, the static receiver sensitivity is used, and fast fading must be considered. Fading therefore follows the Suzuki distribution, which is formed from the addition of lognormal and Rayleigh distributions.

For moving mobiles the dynamic sensitivity of the receiver must be used, but fast fading will be averaged out over the transmission burst and can be discounted, so the fading is simply the log normally distributed slow fading. If x is normally distributed, the probability density of x is

$$P(x \geq x_0) = \int_{x_0}^{\infty} \frac{1}{\sigma\sqrt{2\pi}} e^{-(x-m)/(2\sigma^2)} dx \tag{6.1}$$

where m is the mean and σ the standard deviation. The probability that x exceeds a threshold δ above the mean m is [12]

$$\begin{aligned} P(x \geq (m+\delta)) = \int_{m+\delta}^{\infty} p(x) \ dx &= \int_{m+\delta}^{\infty} \frac{1}{\sigma\sqrt{2\pi}} e^{-(x-m)/(2\sigma^2)} dx \\ &= \frac{1}{2}\operatorname{erfc}\left(\frac{m+\delta-m}{\sigma\sqrt{2}}\right) \end{aligned} \tag{6.2}$$

$$\therefore P(shadowing \geq \delta) = \frac{1}{2}\operatorname{erfc}\left(\frac{\delta}{\sigma\sqrt{2}}\right) \tag{6.3}$$

where erfc() is the complementary error function.

The probability that the shadowing will exceed the shadowing margin, and therefore that the received signal is not sufficient for communications, is therefore dependent on the standard deviation of the shadowing, and the shadowing margin. The shadowing margin is the difference between the mean received SIR at that point, and the minimum signal threshold, defined either by the noise or the interference. Since the system designer cannot control the shadowing, the only way to increase the probability of coverage is to increase the shadowing margin, which for a fixed link budget implies limiting the size of the cell. The shadowing margin for the system is defined as the shadowing margin at the cell border.

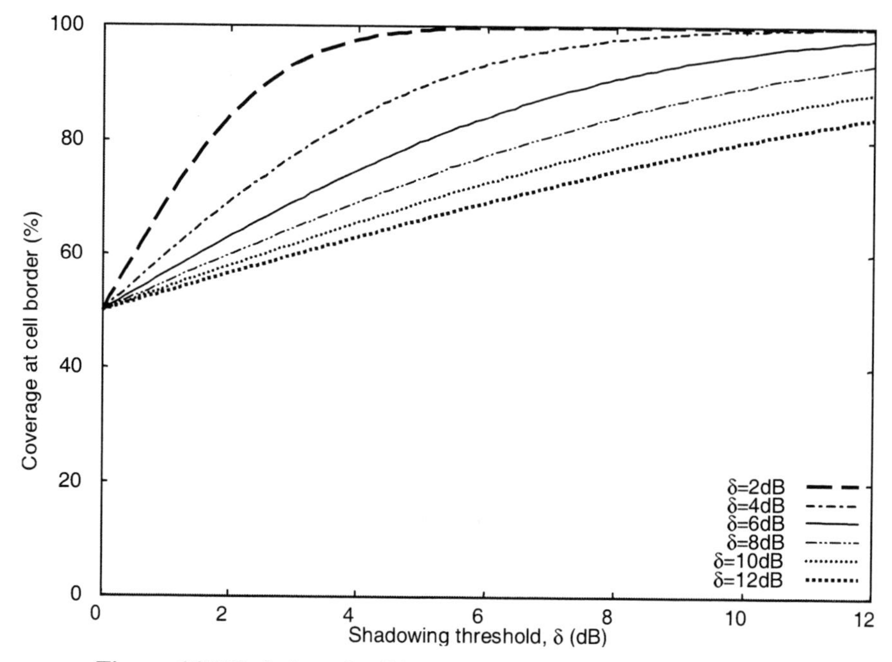

Figure 6.37 Variation of cell border coverage with shadowing margin

This gives the probability of coverage for a given shadowing margin at the edge of the cell. For coverage over the whole cell, this function must be integrated over the cell area, which depends on the pathloss function. For a complex pathloss function such as the Hata model, analysis is very complex, and is best performed by numerical integration on a computer. However, using the reasonable simplification that the mean pathloss follows a d^{γ} law (see Chapter 1), the coverage over the cell can be shown to be [6],[12].

$$P_{cell} = \frac{1}{2}\left[1 - \operatorname{erf}(a) + \frac{1}{2}e^{\left(\frac{1-ab}{b^2}\right)}\left(1 - \operatorname{erf}\left(\frac{1-ab}{b}\right)\right)\right], \quad \text{where} \quad a = \frac{-\delta}{\sigma\sqrt{2}} \tag{6.4}$$

$$\text{and} \quad b = \frac{10\log e}{\sqrt{2}} \cdot \frac{\gamma}{\sigma}$$

Assuming $\gamma = 4$, different values of shadowing margin yield the coverages shown in Figure 6.38. Small shadowing margins are sufficient to give 100% cell coverage when the shadow fading distribution has a small standard deviation, but for large standard deviations, very large margins are required to give high cell coverage. For this reason, increasing coverage beyond 95% is very expensive since the shadowing margin is effectively unused radio capacity. Equation (6.4) can be used to construct curves for other values of γ, or to relate coverage over the cell to coverage near the cell border.

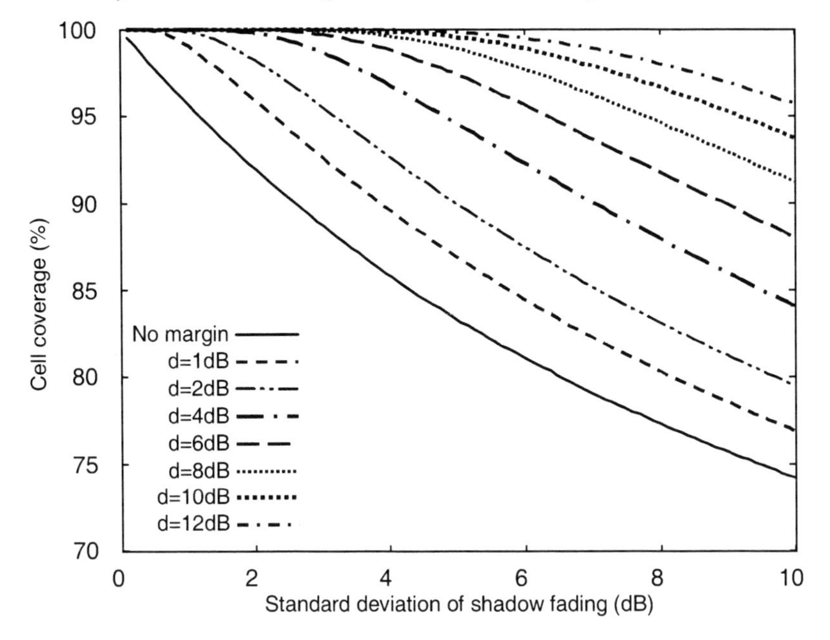

Figure 6.38 Cell coverage variation with shadowing margin

If mobiles are stationary, fast fading must be included, and shadowing follows a Suzuki distribution. However, since the static receiver sensitivity is greater than the dynamic one, noise limited cell size is constrained by moving mobiles rather than static mobiles. Cells dimensioned using the static case would not cover moving mobiles near the cell border. It may be possible to make use of this property if it can be ensured that lower powered hand-portable terminals are stationary when in operation. Maximum cell radii for static hand-portables are only slightly lower than those for moving mobile terminals. The tighter constraint of moving mobiles will be considered here. A discussion of coverage for stationary terminals can be found in [6]

6.12.6 Area Coverage in the Noise Limited Case

In the noise limited case, interference from other transmitters can be neglected and the constraint is the receiver sensitivity.

The cell size depends on the underlying radio environment. The equations given in Section 6.12.5 can be used if a simple d^{γ} pathloss approximation is used. Alternatively, the Hata model described in Section 1.14 is a good model for the large cells found in

PMR applications [11]. If this model is used, results are obtained as shown in Table 6.10. These figures show the coverage which can be expected for different cell sizes for mobiles and hand-portables. The difference caused by the different maximum acceptable pathlosses (as calculated in Section 6.12.4) can clearly be seen. The figures show the maximum cell size for mobiles and hand-portables for 75%, 90% and 95% coverage for different values of shadowing standard deviation. In all cases, moving terminals are considered (dynamic sensitivity and log-normal shadowing), since this results in smaller cell sizes than the static case.

Table 6.10 Maximum cell sizes for different propagation environments; noise limited case

	Range for mobiles (km)			Range for hand-portables (km)		
	75%	90%	95%	75%	90%	95%
Rural area						
$\sigma = 6$ dB	33.0	24.7	20.5	13.6	10.2	8.5
$\sigma = 8$ dB	31.3	22.3	17.0	12.9	8.8	7.1
$\sigma = 12$ dB	26.7	15.6	11.5	11.1	6.4	4.6
Suburban area						
$\sigma = 6$ dB	19.9	14.7	12.5	8.2	6.0	5.1
$\sigma = 8$ dB	18.7	12.8	10.4	7.8	5.2	4.2
$\sigma = 12$ dB	17.2	10.9	8.6	6.8	3.9	2.8

Figure 6.39 and Figure 6.40 show these results graphically.

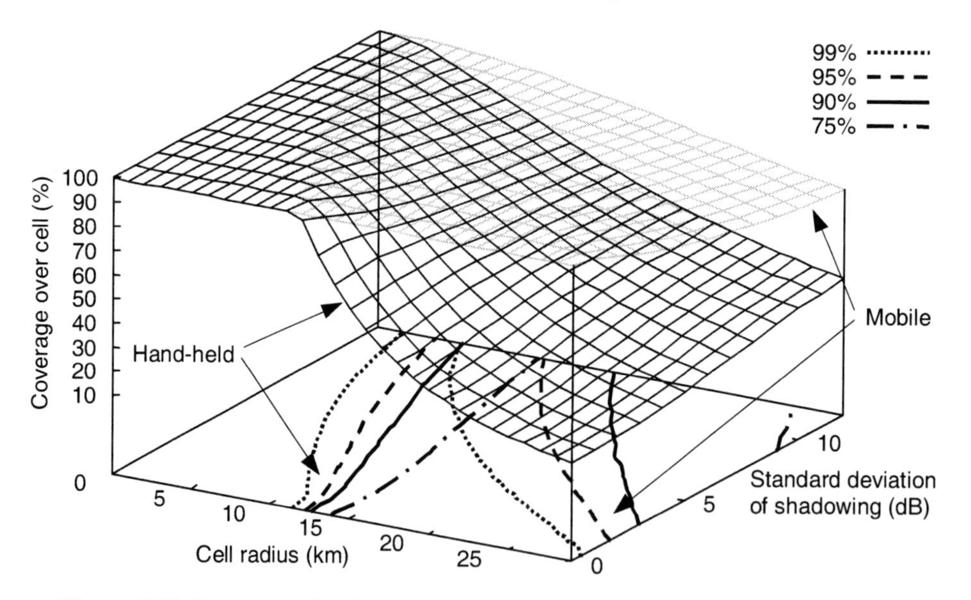

Figure 6.39 Coverage of mobiles and hand-portable devices in a rural environment

The base of the graphs shows the relationship between maximum cell radius and the standard deviation of the shadowing for 75%, 90%, 95% and 99% coverage. It should be noted in particular the reduction in cell radius required when the coverage requirement increases from 95% to 99%. This reduction in cell radius is considerably larger than the reduction for coverage increase from 90% to 95%. In system design terms, these reduced cell sizes equate to the requirement for more base stations and infrastructure. It can be seen that high coverage requirements are expensive.

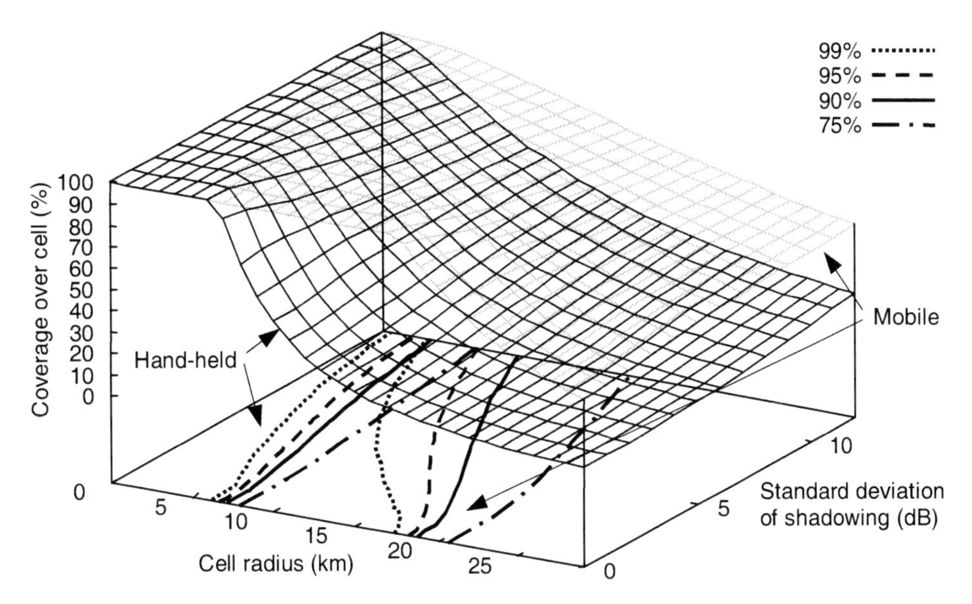

Figure 6.40 Coverage of mobiles and hand-portable devices in a suburban environment

6.12.7 Area Coverage in the Interference Limited Case

Referring to the discussion is Section 6.12.4, if the interference constraint is tighter than the received signal power constraint, the system is said to be interference limited. This will occur in particular if a cellular reuse strategy is adopted.

The interference limited case is considered for a general cellular system in Section 1.21. As was shown, for an average SIR at the cell border of $\mathrm{SIR_{thres}}$, the co-channel interference reduction factor is given by $q = (6 \times \mathrm{SIR_{thres}})^{0.25}$. This is for a pathloss exponent of 4, which is reasonable for the cell sizes under consideration. The minimum value of SIR required for TETRA is 19 dB. If this is the average value at the cell border, q is 4.67, and the cluster size, K, must exceed 7.27.

Using these assumptions, 50% of mobiles on the cell border will receive an acceptable signal. The coverage over the entire cell will depend on the standard deviation of shadow fading. A shadowing margin, δ, can be defined to increase mean SIR at the cell border. The average SIR at the cell border is now $\mathrm{SIR_{thres}} + \delta$, and

$$q = (6 \times (\mathrm{SIR_{thres}} + \delta))^{0.25}, \quad K = q^2/3 \tag{6.5}$$

If a mobile is on the cell border, the signal it receives will depend on the distribution of the shadowing of the desired signal and the interfering signals. Assuming moving mobiles, these shadowing distributions are log normal. Unfortunately, there is no closed form expression for the probability distribution of the resulting signal. However, if the shadowing is assumed to be uncorrelated with the same standard deviation, the distribution can be approximated by a log normal distribution with zero mean and a variance $\sqrt{2}$ times that of the log normal shadowing distributions [10]. This assumption is reasonable since the strongest interference will tend to come from the opposite direction to the desired signal, and at such angles the correlation between the shadowing is low [7].

Therefore,

$$P(interference \geq \delta) = \frac{1}{2}\text{erfc}\left(\frac{\delta}{\sigma'\sqrt{2}}\right) \quad \text{with} \quad \sigma' = \sqrt{2}\sigma \tag{6.6}$$

Equations (6.5) and (6.6) allow the calculation of co-channel interference reduction factors for different probabilities of interference at the edge of the cell. Examples are given in Table 6.11.

Table 6.11 Co-channel interference reduction factors for different cell border interference probabilities

Shadowing, σ		6 dB	8 dB	10 dB
Cell border interference probability				
5%	δ	9.9 dB	13.2 dB	16.5 dB
	q	7.7	9.7	12.4
10%	δ	7.7 dB	10.2 dB	12.8 dB
	q	6.6	7.9	9.5

6.13 INTERFERENCE TO OTHER SYSTEMS

It is very important that radio systems obey strict rules for the transmission of their signals to ensure that spurious signals do not interfere with the operation of other systems. There are three aspects to this:

- interference adjacent TETRA carriers;
- interference to other carriers within the radio band allocated to TETRA;
- interference outside the TETRA radio band to other radio systems and other equipment.

Interference in this last case is limited by general regulation (e.g. EMC regulations). The TETRA standards give a maximum value of interference outwith the band of −100 dBc (i.e. dB with respect to the carrier power). This limit applies to both mobiles and base stations.

6.13.1 Limits on Adjacent Carrier Interference

Adjacent channel interference is of particular concern in a PMR system like TETRA as it is not possible to use frequency planning to ensure that adjacent carriers are not used in the same geographical area. The TETRA standard limits emissions in the channels adjacent to the active transmit channel to the values given in Table 6.12.

Table 6.12 Limits on signal power for adjacent carriers

Frequency offset	Maximum level
25 kHz	–60 dBc
50 kHz	–70 dBc
75 kHz	–70 dBc

One of the important issues of TETRA is the fact that existing analogue FDMA systems may be converted over a period of time towards a TETRA system. This is one of the reasons why TETRA has been designed to use a 25 kHz frequency band, as it allows TETRA to share sections of the frequency spectrum with FDMA systems using 25 kHz channels (or sets of 12.5 kHz channels).

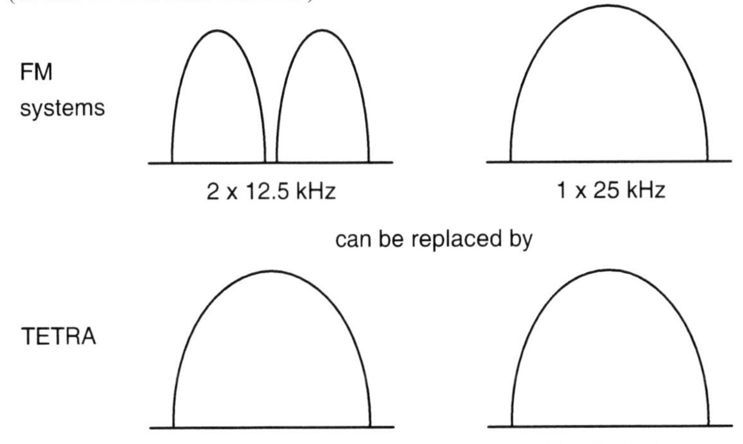

Figure 6.41 Replacement of FM systems with TETRA

ETSI specifies that analogue PMR systems have a maximum power of –60 dBc in the adjacent channel for a channel spacing of 12.5 kHz and –70 dBc for 25 kHz [2]. The –60 dBc limit is met by TETRA in all adjacent channels, thereby ensuring co-existence with analogue PMR systems with a channel spacing of 12.5 kHz. The limit for 25 kHz spacing is met by TETRA in the second and third adjacent channel. One 25 kHz channel must therefore be left unused between channels of a TETRA system and those of an analogue PMR system with 25 kHz spacing.

Concerning TETRA to TETRA interference in the adjacent channel, the allowed level of unwanted emissions close to the carrier as well as the other system parameters led to a reference adjacent channel interference ratio between the power of the signal on the active carrier and the power of the interferer on the directly adjacent channel of –45 dB for dynamic conditions, and –54 dB for static conditions. The adjacent channel interference ratio has to be respected when adjacent channels are used in the same area.

The co-existence of TETRA and other digital systems would depend on the requirements of the particular system, since adjacent carrier interference specifications vary between systems. For example, the TETRAPOL system, with its 12.5 kHz channel spacing, defines a maximum level of –60 dBc at 12.5 kHz from the carrier and –70 dBc at 25 kHz from the carrier, the same as for analogue systems.

6.13.2 Limits on Emissions Far From The Carrier Within the TETRA Band

As noted earlier, the filtered π/4–DQPSK modulation scheme has a non-constant envelope in contrast to analogue FM and some other less bandwidth efficient digital modulations like GSMK. This means that a highly linear transmitter is required in order to avoid non-linear distortions, as these would cause an intolerable broadening of the frequency spectrum.

If the transmitter power amplifier is not perfectly linear, there may be low-level wide-band noise-like emissions outside the RF carrier bandwidth. The TETRA specifications [3][4] define limits for this interference. These limits have been arrived at by probabilistic rather than by worst case analysis of several potential interference scenarios, and are given in Table 6.13.

Table 6.13 Limits on signal power far from the carrier

Frequency offset	Maximum level		
	MS power class 4 MS power class 5	MS power class 3	MS power class 2 MS power class 1 BS (all classes)
100kHz – 250kHz	–75 dBc	–78 dBc	–80 dBc
250kHz – 500kHz	–80 dBc	–83 dBc	–85 dBc
500kHz – nearest edge of TETRA band	–80 dBc	–85 dBc	–90 dBc

6.13.3 Limits on Emissions Outside the TETRA Band

The maximum value of interference outside the TETRA band is –100 dBc, or –70 dBm, whichever is greater.

6.14 AREA COVERAGE TECHNIQUES

The TETRA system is very flexible and is therefore able to incorporate the various options for system coverage provision which have been employed for PMR. The standard PMR system is a single cell. TETRA can cover large areas with a single cell, although it has a higher carrier bit rate and signal strength requirements than some other digital PMR systems. This restricts the range possible with a single cell. Options for covering larger areas include quasi-synchronous operation, which can be employed but is more complex on digital systems than on analogue ones. In high capacity scenarios, cellular re-use schemes, as used in public cellular systems, are possible. The TDMA structure of TETRA also allows carriers to be shared between cells in time-shared transmission, which can be used in low traffic areas. Antenna diversity and site diversity can be used to improve link performance, while repeaters, either fixed or mobile, can be employed to cover blackspots. In TETRA, a gateway between Direct Mode and V+D operation can be employed to extend operation to areas which cannot support portable operation. These various options are discussed in greater detail in the following sections.

6.14.1 Single Site Wide Area Coverage

If the area over which service is required is not too large, the simplest possible approach is to use a single large cell served by a base station. The limit on the cell size depends on the base station location, the terrain, the radio environment and the frequency band used, but using the bands around 400 MHz, cells sizes of more than 50 km are possible in ideal cases. However, the use of a single cell means that it is not possible to re-use carriers within the area so enough channels have to be provided to cope with the expected use over the entire area of the cell, which is not very efficient.

6.14.2 Cellular Channel Re-use

Cellular frequency re-use schemes are a standard method of planning mobile radio systems, and are discussed in detail in Chapter 1. Cell size is limited by the link budget as discussed in Section 6.12.4, but operating large cells reduces capacity, as a channel can only be used for a single mobile within a cell, and that channel cannot be used in any neighbouring cells which would suffer interference. In order to increase capacity, smaller cells can be used, so that radio channels can be reused over the area more frequently. The disadvantage of this approach is that smaller cells imply more base station sites (although sectored cells can be used as well), and an increase in handovers, which makes network switching more complex. However, unless the area to be covered is small, anything other than fairly light traffic will require a cellular approach to be used to give efficient use of the radio resource.

6.14.3 Quasi-synchronous Transmission

Quasi-synchronous transmission can be used to emulate wide area coverage from a single base station over a larger area than would be possible from a single transmitter alone. In quasi-synchronous transmission, transmitters on different sites are used to simultaneously transmit the signal at the same time.

While several base sites are used as in the cellular approach, different frequencies are not required since the same carrier is transmitted on all of them. This is an advantage when the number of available frequencies is low, since a large area can be covered. However, it shares the disadvantage of the single site approach by preventing the re-use of frequencies within the coverage area. Quasi-synchronous operation is therefore useful in areas where traffic density is low such as rural areas, but is wasteful of resources in areas of high user density.

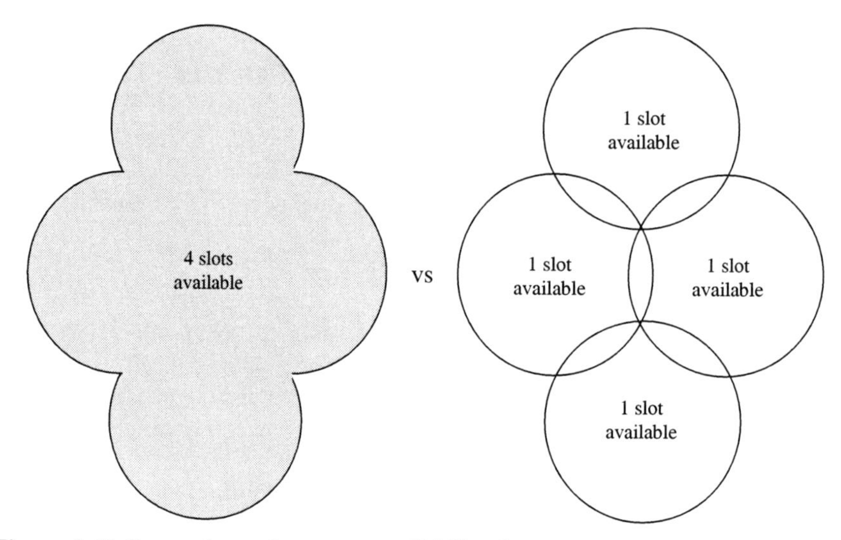

Figure 6.42 Comparison of resource availability for cellular and quasi-synchronous approaches

As an example, an area is covered with a regular arrangement of cells. In the first case, up to 12 users may have to use the system simultaneously. Using quasi-synchronous operation, 3 carriers would be required (using minimum mode). Using a cellular arrangement, the number of carriers would depend on the frequency re-use pattern, but would be 21 carriers with a re-use of 7, or 36 carriers with a re-use of 12, since a minimum of three carriers would be required in every cell. Quasi-synchronous operation therefore requires fewer carriers.

In a second case, 12 users have to be supported in each cell at the same time. Quasi-synchronous operation on its own does not employ re-use. This means that 3 carriers would be required for each cell, or a total of (3 × no of cells) carriers. The cellular system, since it employs re-use, would only require 21 carriers for a re-use pattern of 7 (or 3 × no of cells, if less), or 36 carriers (or 3 × no of cells, if less) for a re-use of 12. The cellular system therefore requires fewer carriers than the quasi-synchronous system if the number of cells in the system exceeds the number of cells in the cellular re-use pattern.

Quasi-synchronous operation has the advantage of simplifying system control, as the area can be considered to be one large cell, and so no handover is required. Mobiles can communicate as long as they can transmit to or receive from any of the base stations. This also avoids location management since the infrastructure does not need to know where the mobile is in order to communicate with it.

However, while control of the system is simplified, control of the physical layer transmission is complicated considerably. The problem occurs in areas where coverage from several base sites overlap and the mobile can receive several signals. These signals need to be transmitted at exactly the same time to ensure that they do not interfere with each other, which implies complex synchronisation between base stations. Also, even if they are transmitted at precisely the same time, differences in the propagation path between the base sites and the mobile will be such that they will be received at slightly different times. This time difference appears to the mobile like severe multipath interference. Unless the timing difference is less than about 20% of the symbol duration (or about 11 microseconds), an equaliser will be required in the mobile, which means that more complex class E mobiles, which have equalisers, may have to be used.

Analogue FM systems can implement quasi-synchronous operation more easily than digital systems. Since each base station is transmitting a constant tone, symbol interference does not occur, and the only issue is to ensure that interference between the signals does not cause a null at any specific location. To prevent this, a small frequency offset of a few hertz is used which results in a spatially moving interference pattern so that the signal will not suffer continuous destructive interference at any location. Digital systems require more complex control and setting up [8].

Just as transmitting at a lower symbol rate improves performance in the case of multipath interference, it also makes quasi-synchronous operation easier. TETRA has a symbol rate of 18 ksymbols/s, so signals can be received with a time offset of up to 10 μs while still maintaining acceptable performance without an equaliser.

Quasi-synchronous operation and cellular operation are not mutually exclusive, in that it is possible to set up the system using quasi-synchronous operation in low density areas as part of a cellular re-use scheme. The quasi-synchronous cells act in effect as single large cells as far as the overall system is concerned, which are then treated as single logical cells for the cellular re-use scheme. Such combinations allow a good trade off between providing capacity through re-use in high density areas and reducing resource requirements and simplifying network control in low density areas.

6.14.4 Time-shared Transmission

Rather than using the same radio resource at the same time in several cells as in quasi-synchronous transmission, an alternative is to take advantage of the TDMA structure of TETRA in order to split the transmission between cells in time sharing transmission. Different cells use different slots, so a carrier can be shared amongst up to four cells. Discontinuous downlink bursts are used in this case, which means mobiles loose the ability to listen to the inter-burst training sequence, and this may reduce performance slightly. Time-shared operation allows areas with low traffic to be served with very few carriers, and avoids the physical layer complexity of quasi-synchronous operation. While synchronisation is required between sites to ensure that transmission takes place at the correct time, it need not be as accurate as for quasi-synchronous operation, and there is no additional requirement for equalisers. There are three different modes of time shared operation: carrier sharing, main carrier sharing and MCCH sharing. These are detailed in Section 7.11.4

An interesting issue is how time-shared transmission compares with quasi-synchronous transmission. The quasi-synchronous solution makes slots available at the same time at all base sites to provide four channels per carrier. This provides better coverage than the time shared system, since all four time slots are available in any part of the coverage area, whereas for the time-shared transmission system, the slots are separated and used on four geographically separated base stations. This is not as efficient since if two mobiles are in the same area the time-shared system will not be able to serve both.

Quasi-synchronous operation moves complexity from the network to the mobile terminal. Since in quasi-synchronous operation, several base stations combine to act as a single large cell, routing is less complex, and handover between the base stations is not required. However, in contrast to this simplified infrastructure and control, the mobiles themselves will require equalisers to allow operation in overlap areas receiving signals from more than one base station.

6.14.5 Antenna Diversity

Another method of increasing the performance on mobile radio systems is to use antenna diversity. This is where several antennas are used on the same site and the best signal is chosen. Since all antennas share a common site, antenna diversity does not combat slow fading, but antennas mounted more than half a wavelength apart can help counter fast fading, and improve reliability. This means that coverage can be improved. The use of antenna diversity does not require any intervention by the system – in effect a more sensitive receiver is being used.

Antennas located at the same site can share the same infrastructure and physical mounting. This is important as most of the cost of a base station is it the mast and network connection, not in the transmission electronics.

6.14.6 Site Diversity

This is similar to antenna diversity, but rather than using several antennas at a single base station to combat multipath fading, antennas are sited in different locations with receive only base stations throughout the coverage area of a normal (transmit/receive) base station. This has the advantage of countering shadow fading as well as fast fading - something not possible with antenna diversity alone - but increases cost and complexity. It also requires selection across base stations, which is not defined in the standard.

Since the majority of the cost of base stations is the provision of the infrastructure, it is interesting to compare site diversity with quasi-synchronous operation, time-shared transmission and cellular re-use, which also require multiple base sites. Site diversity is similar from a system point of view to quasi-synchronous operation, since the system operates as a single large cell. Site diversity is simpler, and therefore less costly, since symbol level synchronisation is not required between sites. However, since site diversity makes use of the imbalance between the link budgets to and from portable units, the cell size is limited. Site diversity is also less complicated than time-shared or cellular operation, as handovers and related control and signalling are not required. Against this, cellular operation, with transmit and receive base stations on each site, would give a much greater capacity, but requires that enough frequencies are available to allow a re-use scheme. Time-shared transmission would not require additional frequencies, although like quasi-synchronous operation, site diversity gives flexibility advantages.

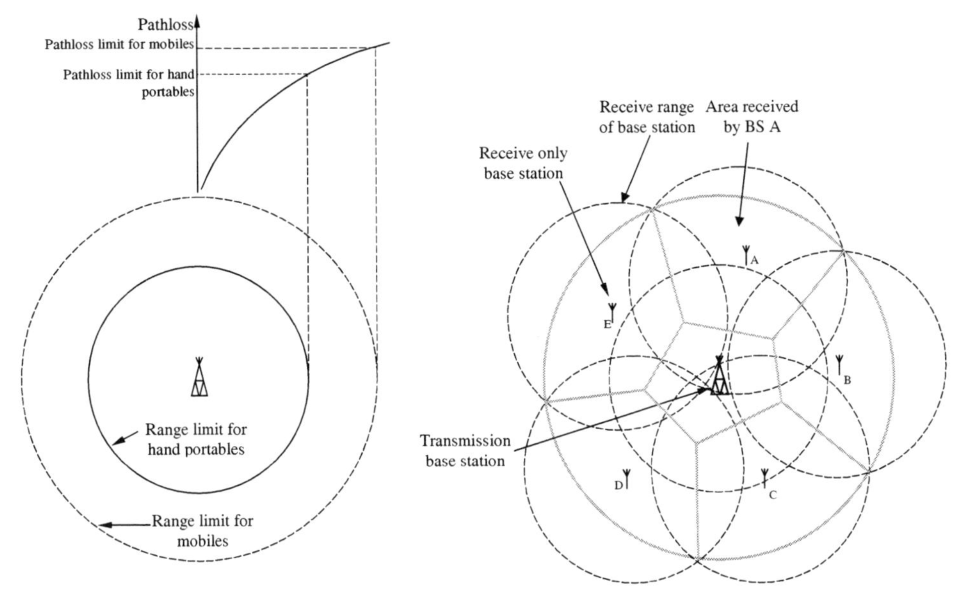

Figure 6.43 Site diversity to extend hand-held portable range

6.14.7 On-frequency Repeaters

Buildings and other structures cause shadowing, and if the shadowing is large enough then it will not be possible to provide a large enough signal to interference and noise ratio for the system to operate. If this occurs far from the base station, and there are many areas that are shadowed, then it becomes necessary to provide an additional base station to increase coverage. However, in many cases a large building or bridge can cause significant local shadowing in an area with good coverage otherwise, and on-frequency repeaters can be used as local "gap fillers" to avoid the cost of a new base station. The repeater picks up the signal from the base station using a directional antenna to reduce multi-path effects and repeats the signal over the shadowed area.

One issue with on-frequency repeaters is in the boundary zone of the gap where mobiles can receive signals both from the repeater and from the base station. Since the repeater is an on-frequency repeater, it is using the same frequency but will exhibit a delay due to its circuitry and the different path lengths via the repeater and direct from the base station. If the gap being covered is small, the signal spread is small and is unlikely to be a problem. However, in large gaps it becomes an issue, and is analogous to the problem with quasi-synchronous operation. Take the example shown in Figure 6.44. Here a base station is mounted on a hilltop with a smaller neighbouring hill. This neighbouring hill shadows the area near it, but signals further from the hill can receive coverage from the base station due to diffraction over and around the hill. An on-frequency repeater can be used to illuminate the shadowed area, but there is an overlap area that receives both signals. Operation in this area may require class E mobiles with equalisers.

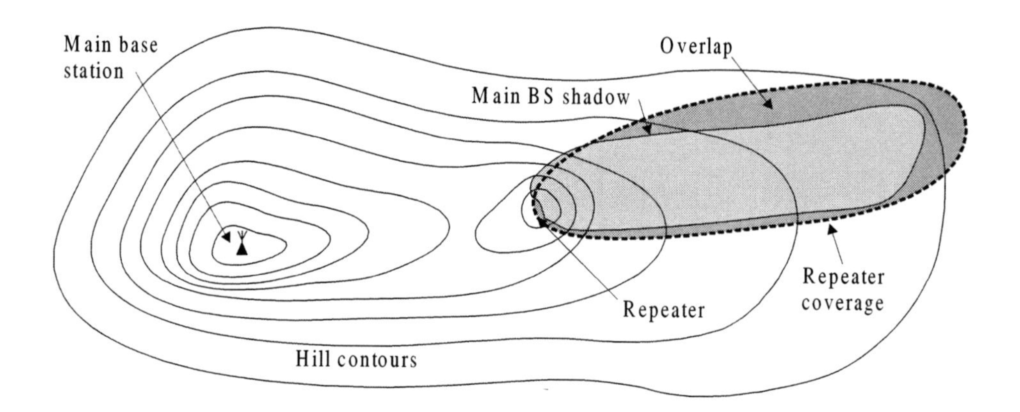

Figure 6.44 Use of a repeater to cover a gap caused by a hill shadow

6.14.8 Direct Mode / Trunked Gateway

Interworking between Direct Mode terminals and the TETRA infrastructure is possible through a Direct Mode / Trunked gateway. This makes it possible for the infrastructure to contact a Direct Mode mobile, as long as the mobile is in the coverage area of the gateway.

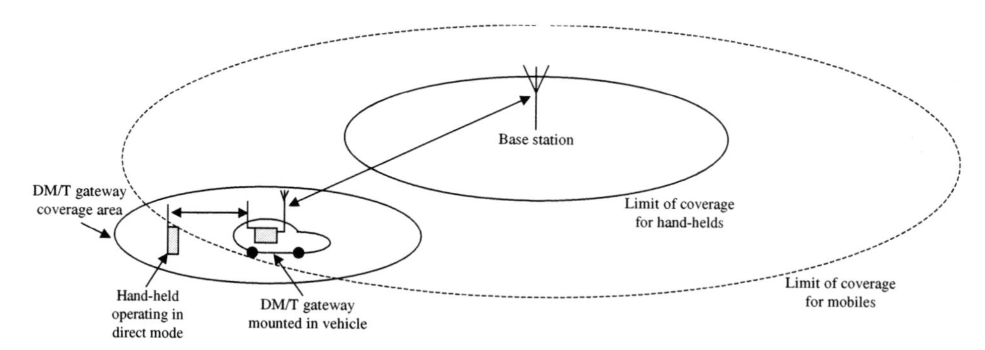

Figure 6.45 Extending coverage for hand portables using a gateway

Direct Mode can be programmed in the mobiles as a fall-back mode, so if a mobile can not find the infrastructure it goes to a predefined Direct Mode channel.

Networks that are projected for mobile use only, could give a virtual hand-portable coverage if a Direct Mode/Trunked gateway is mounted in the vehicle and the user always operates in the vicinity of this vehicle. This is similar to the use of a vehicle mounted repeater in an analogue PMR system.

A Direct Mode/Trunked gateway can also be used as fixed gap-filler instead of an on-frequency repeater in if the gap to be covered is small, such as an underpass. When mobiles cannot communicate with the infrastructure directly, they will go into fall-back mode, and will find the DM/T gateway. They can then use this to communicate with the infrastructure. Compared with the on-frequency repeater solution, the Direct Mode/ Trunked gateway avoids problems with signal spread in the overlap area since the channels used are different. However, Direct Mode will not support the full range of V+D services, so the services available through the Direct Mode/Trunked gateway will be limited.

6.15 CONCLUSIONS

This chapter has described the TETRA physical layer, and the choices made for modulation and TDMA transmission. Due to the nature of PMR requirements, the choices made in TETRA are significantly different from those made for public cellular systems such as GSM. In particular, TETRA has a narrower carrier spacing, and longer transmission frame, than GSM.

Other functions and features of the physical layer have been described, such as the delays introduced to the transmitted bit stream, synchronisation, power control and coverage techniques.

The function of the physical layer is to transmit the bit stream passed to it from layer 2 from the transmitter to the receiver. The physical layer itself runs continuously, irrespective of the information being transmitted, and does not take part in any routing or retransmission of bursts corrupted by errors. These routing functions are a key function of layer 2, which is described in the following chapter, Chapter 7.

REFERENCES

[1] ETR 300-1: 'Radio Equipment and Systems (RES); Trans-European Trunked Radio (TETRA); Voice plus Data (V+D); Designer's guide; Part 1: Overview, technical description and radio aspects', ETSI, 1997.

[2] ETS 300 086: 'Radio Equipment and Systems (RES); Land mobile group; Technical characteristics and test conditions for radio equipment with an internal or external RF connector intended primarily for analogue speech', ETSI, 1991.

[3] ETS 300 392-2: 'Radio Equipment and Systems (RES); Trans-European Trunked Radio (TETRA); Voice plus Data (V+D); Part 2: Radio Aspects', ETSI, 1996.

[4] ETS 300 396-2: 'Terrestrial Trunked Radio (TETRA); Technical requirements for Direct Mode Operation (DMO); Part 2: Radio Aspects', ETSI, 1998.

[5] ETS 300 396-3: 'Terrestrial Trunked Radio (TETRA); Technical requirements for Direct Mode Operation (DMO); Part 3: Mobile Station to Mobile Station (MS-MS) Air Interface (AI) protocol', ETSI, 1998.

[6] Gibson, J. D., (ed), *The Mobile Communications Handbook*, CRC Press, Boca Raton and IEEE Press, 1996.

[7] Graziano, V., 'Propagation Correlations at 900 MHz', *IEEE Trans. Vehicular Technology*, vol. 27, November 1978, pp. 182–189.

[8] Grier, R., 'Trunking Advances: Fine Tuning a Digital Simulcast Trunking System', *Communications*, October 1994, pp.64–70.

[9] Hess, G. C., *Land-Mobile Radio System Engineering*, Artech House, Boston, 1993

[10] Jakes, W. C., (Ed), Microwave Mobile Communications, Wiley, 1974 (IEEE Press Reissue, 1994)

[11] Lee, W. C. Y., *Mobile Communications Engineering*: Theory and Applications, 2nd Ed., McGraw-Hill, 1997.

[12] Rappaport, T. S., *Wireless Communications: Principles and Practice*, Prentice-Hall, 1996

[13] Sampei, S., *Applications of Digital Wireless Technologies to Global Wireless Communications*, Feher/Prentice Hall, 1997.

[14] Stüber, G. L., *Principles of Mobile Communication*, Kluwer Academic Publishers, Boston, 1996.

7

The Data Link Layer

7.1 INTRODUCTION

The second layer in the OSI model is the data link layer. The function of this layer is to provide error free communication between the transmitter and the receiver so that information can be passed by the higher layers. In order to do this, the data link layer uses layer 1, the physical layer, to provide a raw bit pipe for information and adds the following functionality:

- Error detection and error control techniques to provide reliable communication to the network layer, layer 3, instead of the unreliable communication provided by layer 1.
- Scheduling transmission of data provided by layer 3 over the physical layer.

Given that TETRA is a mobile system, the physical channel between the transmitter and the receiver is shared between a number of users. The data link layer therefore also has to deal with scheduling between users. Finally, the data link layer has a number of control functions to support its operation and the operation of the physical layer.

7.2 ORGANISATION OF THE DATA LINK LAYER

For systems with a shared transmission medium, such as mobile systems, it is usual to divide the functionality of the data link layer into two sub-layers; medium access control (MAC), which handles resource scheduling and logical link control (LLC), which communicates with the upper layers and ensures reliable transmission. In TETRA, the MAC sub-layer is further divided between an upper MAC and a lower MAC. The upper MAC handles the majority of the resource scheduling and logical channel multiplexing, while the lower MAC has some channel error protection functionality and also maps the different types of data burst provided by the upper MAC on to the physical layer.

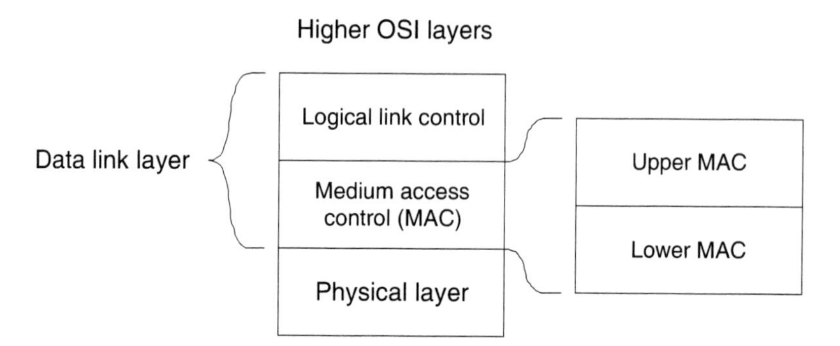

Figure 7.1 Subdivision of the data link layer in shared transmission medium systems

The functionality of the various parts of the data link layer is as follows:

- **Logical Link Control (LLC)**
  This layer is responsible for data transmission and retransmissions, segmentation and re-assembly, and organisation of the logical links. It provides error free data to layer 3.

- **Upper Medium Access Control (MAC)**
  The upper MAC layer performs TDMA random access procedures, fragmentation/ re-association and frame synchronisation. In addition, it measures the quality of the link by measuring signal strength and signal quality (which is defined in terms of the number of AACH messages which can be correctly received). It provides data to the logical link control layer with errors detected (so that the LLC can retransmit if required).

- **Lower Medium Access Control (MAC)**
  The lower MAC layer performs TDMA channel coding, interleaving, and mapping between upper MAC channels and the slots provided by the physical layer. This includes arranging for slot stealing if necessary.

Layers communicate via service access points (SAP). The SAP between the upper and lower MAC is called the TETRA MAC Virtual SAP, TMV-SAP (or DMV-SAP in Direct Mode). There are four SAPs between the upper MAC and the LLC sub-layer, one for user data (TMD-SAP), and one each for broadcast signalling (TMB-SAP) and signalling for specific mobiles (TMA-SAP). The corresponding Direct Mode SAPs are DMD-SAP for Direct mode MAC data, and DMA-SAP, for Direct Mode signalling messages. There is no equivalent to the TMB-SAP in Direct Mode since there is no concept of broadcast messages.

In both V+D and Direct Mode there is an additional SAP which deals with inter-layer management functions rather than transmitted data. This is the TETRA MAC control SAP, TMC-SAP for V+D and DMC-SAP for Direct Mode. The TETRA data link layer has the functional entities shown in Figure 7.2.

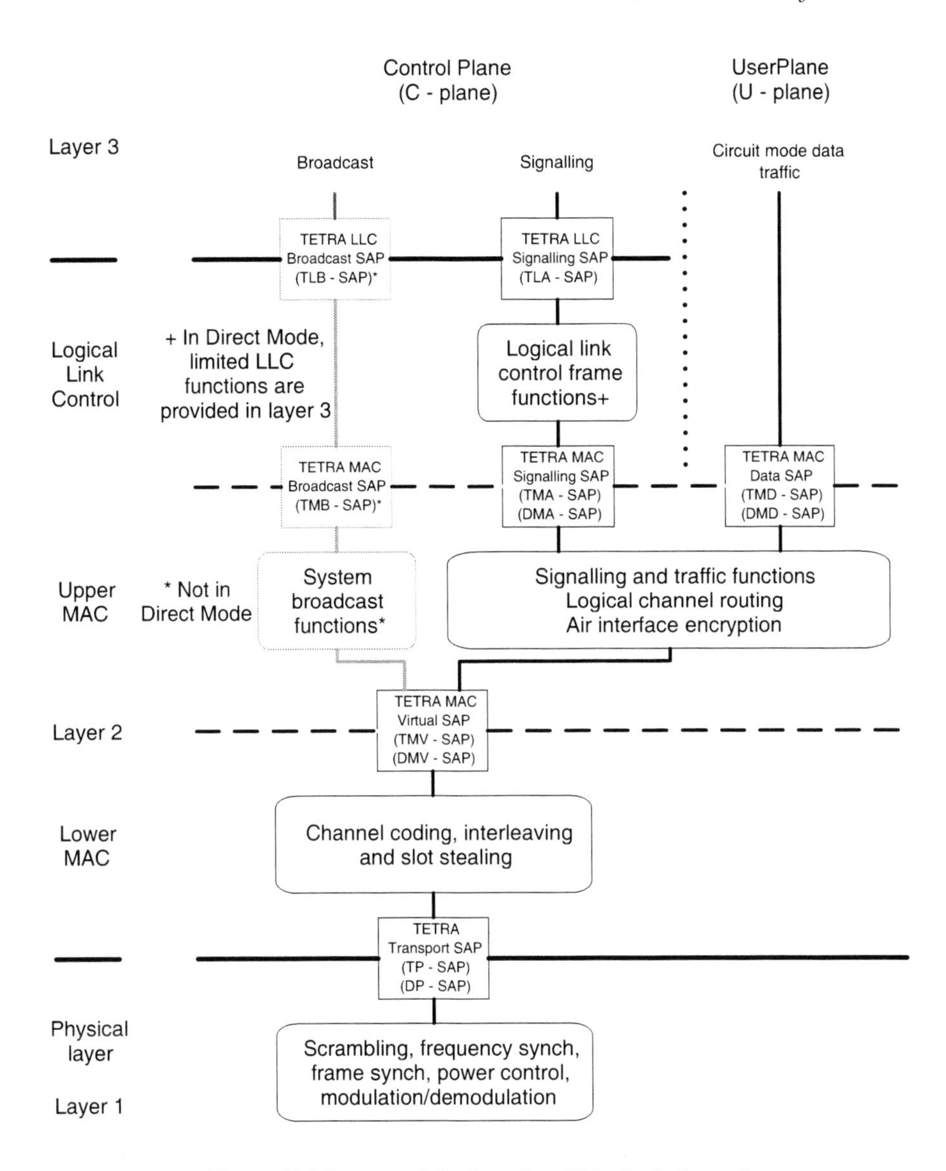

Figure 7.2 Layout of the functional blocks in layer 2

7.3 ORGANISATION OF THE MAC

As well as the information flow itself, the MAC has to carry out a number of control functions including routing, access control, control of the power control algorithm executed in layer 1, measurement of the signal quality and implementation of the air interface encryption. These last two functions are carried out in the MAC but are used by layer 3 processes such as encryption and handover. The division of control processes in the MAC is as shown in Figure 7.3. The V+D MAC is shown, but Direct Mode is similar.

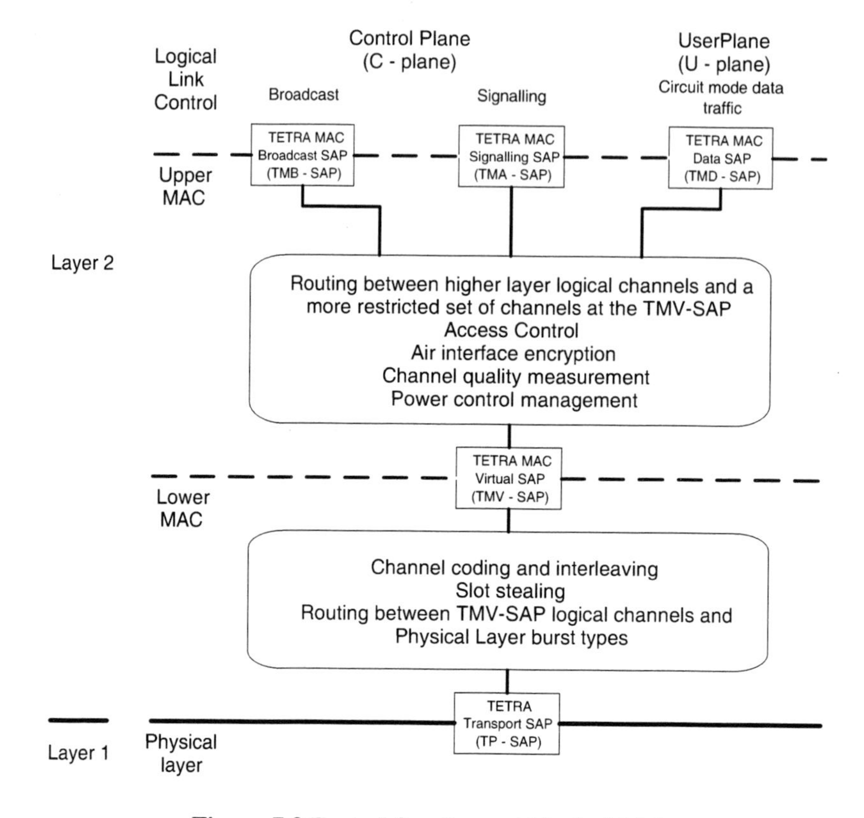

Figure 7.3 Control functions within the MAC

The lower MAC routes data which appears at the TMV-SAP for transmission, encoding it with the appropriate channel code and stealing capacity for signalling if necessary.

Upper MAC functionality can be grouped into two main categories

- radio channel access control
- radio resource management

The control functions handle control of radio channel. This includes controlling access to the channel, high level routing and co-ordination between channels. The management functions assist these processes, and also provide additional information that is available to high level management functions in layer 3, such as handover. Specifically, radio channel control consists of the following:

1. Random access procedures (contention control on a particular physical channel and flow control of uplink random access).
2. Multiplexing/demultiplexing of the logical channels onto bursts.
3. Fragmentation/re-association (if a message from the LLC is too long to be sent within a single burst).
4. Frame and multiframe synchronisation, keeping track of the frame number within a multiframe, and multiframe building and synchronisation.

Radio resource management functions include the following:

1. Bit Error Ratio (BER) and BLock Error Rate (BLER) measurements.
2. Pathloss calculation of the serving cell and monitoring and scanning of adjacent cells.
3. Power control management (execution is in the physical layer).
4. Address management for individual, group or broadcast calls.
5. Radio path establishment and radio resource allocation on instruction from the upper layers.
6. Buffering of control information and speech frames until transmission.

Transmitted and received data undergoes the processes shown in Figure 7.4.

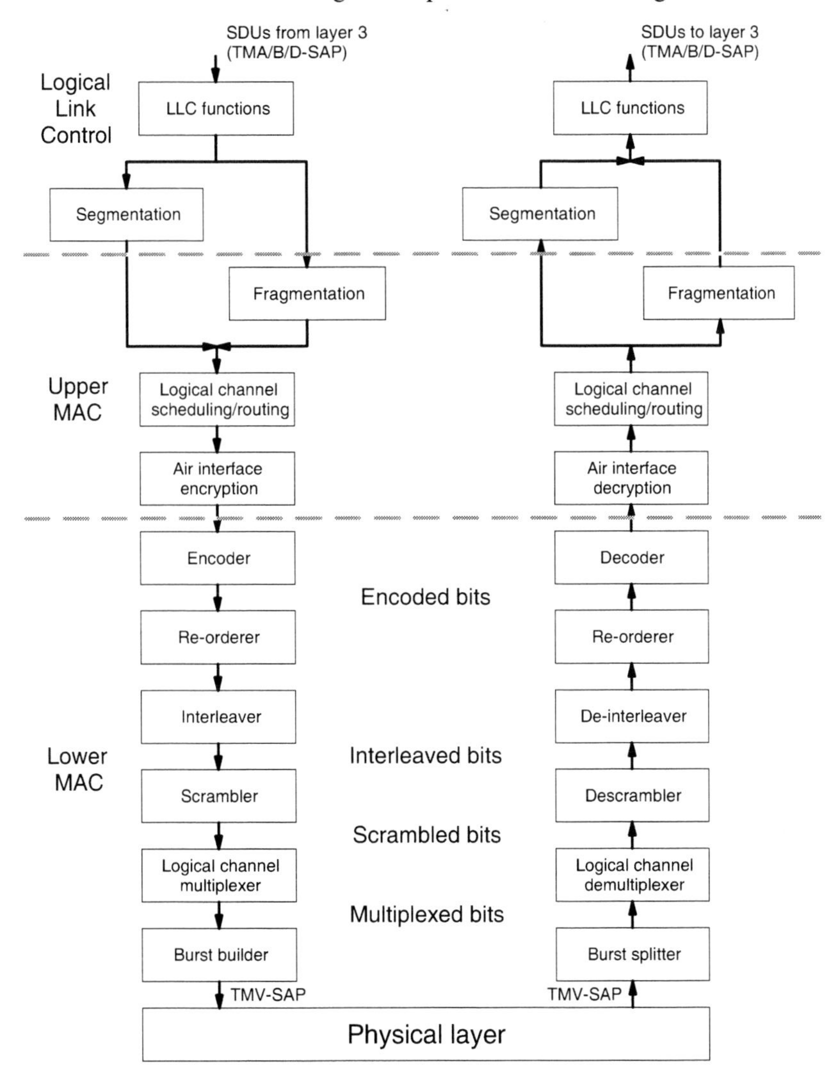

Figure 7.4 Data processing steps in layer 2

Information is passed to layer 2 via service access points in the form of service primitives. The overall responsibility of the MAC sub-layer is to transmit this information to the receiver and pass it back to the upper layers with information on its reliability.

Table 7.1 Service primitives at the upper and lower MAC

Upper MAC Service Primitive (passed at TMA/B/D – SAPs)	Lower MAC service primitive (passed at the TMV-SAP)
TMA-UNITDATA request or TMB-SYNC request (BS only) or TMB-SYSINFO request (BS only) or TMD-UNITDATA request	TMV-UNITDATA request
TMA-UNITDATA indication or TMB-SYNC indication (MS only) or TMB-SYSINFO indication (MS only) or TMD-UNITDATA indication	TMV-UNITDATA indication

7.4 CODING AND INTERLEAVING

TETRA uses a comprehensive range of error control techniques to make efficient use of the air interface. Four basic systems are used, which have different parameters depending on the logical channels being sent. Coding is undertaken by the lower MAC, and the upper MAC maps the its logical channels on to a reduced set of logical channels for the lower MAC to encode. Section 7.8.4 deals with this in detail, but the channels present at the lower MAC are:

On full slots

- Traffic CHannel (TCH)
- Full slot Signalling Channel (SCH/F)
- STealing CHannel (STCH)

On half slots

- Half slot Downlink Signalling CHannel (SCH/HD) (V+D only)
- Half slot Uplink Signalling CHannel (SCH/HU) (V+D only)
- Half slot Signalling CHannel (SCH/H) (Direct Mode only)

On synchronisation bursts

- Broadcast Synchronisation CHannel (BSCH) (V+D only)
- Synchronisation CHannel (SCH/S) (Direct Mode only)

On the broadcast block

- Access Assignment CHannel (AACH) (V+D only)

Linearisation bursts are just time masks and have no transmitted data or coding.

The lower MAC does not choose what is sent but applies the correct coding to the logical channel passed to it via the TETRA MAC Virtual SAP. Figure 7.5 gives a summary of the different coding schemes available to the lower MAC, and the corresponding logical channel.

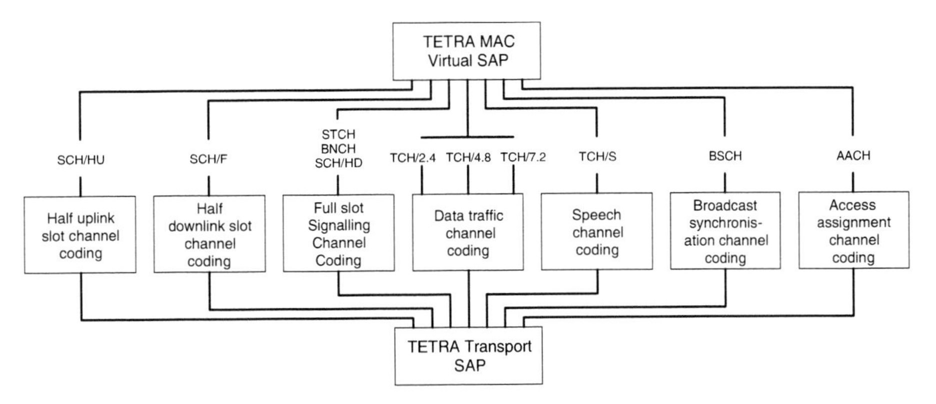

Figure 7.5 Different codes available in the lower MAC (V+D mode)

A short introduction to different types of coding will be given before a description of the specific schemes used in TETRA.

7.4.1 Error Control Strategies

Error correcting coding is included in a transmission system to protect against errors introduced by the transmission medium. A radio system is more error-prone than most. Three different strategies can be identified in relation to error control.

- **Error concealment.** Here errors are detected and the corrupted information identified so that it can be discarded. The remaining information is available and in some cases can be used to mask the missing corrupt data. This can be done by repeating a previous sample, muting the corrupt sample, or by trying to interpolate from the surrounding values. Error concealment works well in speech and audio systems as long as the error rate is low. General errors of the type shown in Figure 7.6 cause clicks which can be annoying to the listener, but muting or interpolation means that the errors cannot be heard. Error concealment can also be used for images in some cases, but not for general data, as it makes use of the properties of the sound or image and the senses and understanding of the human listener.

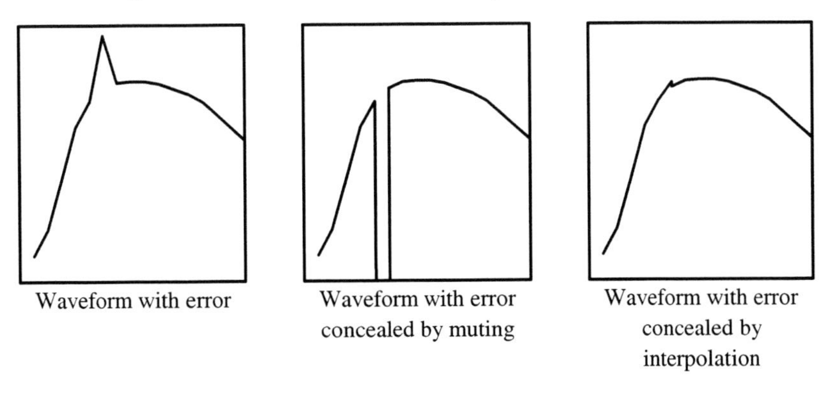

Figure 7.6 Effect of different error concealment procedures on a waveform

- **Automatic Repeat reQuest (ARQ).** As in the case of error concealment, errors are detected, but in this case the transmitter is informed and requested to send the data again. Various strategies are possible for this repetition, and are discussed in greater detail in Section 7.10.2. ARQ has a number of disadvantages, in particular that it requires a feedback channel to the transmitter, and that in the case of an error a delay is introduced to the transmission of the data. The transmitter also has to store the data until it has been informed that it has been received correctly, and depending on the type of ARQ the receiver may have to store data as well. This buffering requirement may be large. However, ARQ schemes can achieve very low error rates with relatively low overheads and complexity.

- **Forward Error Correction (FEC).** In this case additional message symbols are added to the message being sent so that it is possible for the receiver to reconstruct the message in the event that part of it was corrupted. FEC is computationally complex and requires an additional transmission overhead but has the significant advantage of not requiring a feedback path to the transmitter. Also, little additional delay is introduced over that involved in transmitting the additional symbols.

It is possible to use combinations of these strategies, with FEC being attempted first, and ARQ or error concealment being employed if this fails. In TETRA, the lower MAC undertakes FEC for some channels, and the logical link control layer undertakes ARQ for some control messages and packet mode data (see Section 7.10.2).

7.4.2 Forward Error Correction and Detection

If the transmission channel introduces errors, the messages received after transmission will be equal to the transmitted messages with some errors added. For error detection, it is necessary to detect that these errors have been added, and for FEC, it is a further requirement to detect their location.

The code designer cannot control the errors. However, it is still possible to define a code in such a way that the message can be recovered from the received codeword for most errors which will occur. In simple terms, if the messages are sufficiently "different", they will still be recognisable as long as few errors occur. If every possible combination of symbols forms a codeword, it is not possible to detect the presence of errors because one valid codeword is changed into another valid codeword. For example, it is easy to spot the error in 'Thss phrase', because not all combinations of letters form words. If the same were to occur with a telephone number it would be much harder to recognise.

This concept of difference can be quantified as follows. Take as an example a code made up of pairs of numbers between 0 and 5 (this could represent, for example, the amplitude and phase of a sinusoidal signal). When the numbers are sent from the transmitter to the receiver they can be corrupted by the channel. Let the codewords be A(0,0), B(0,5), C(5,0) and D(5,5). These codewords can be plotted as shown in Figure 7.7. If (5,0) is received, this is recognised as a codeword; no error has occurred and the message was C. A received message P(0.1,4.6) shows that an error has occurred, but since small errors are more likely than large ones, it is still possible to estimate that the message was B. Q(2.5,2.5) could have resulted from either A, B, C or D.

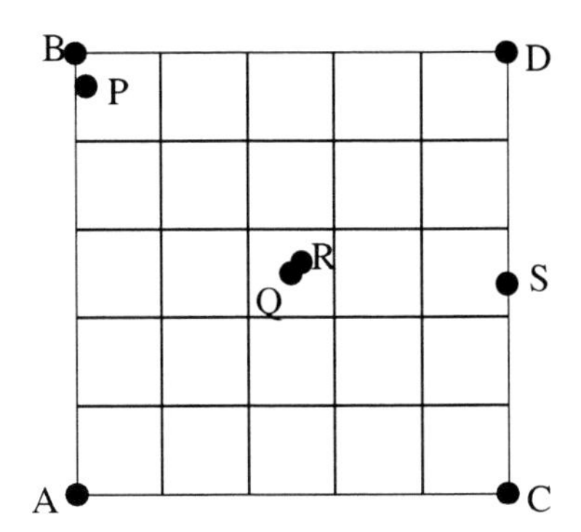

Figure 7.7 Distance between messages

This intuitive process can be quantified by looking at the distance from the received codeword to all possible valid codewords, and choosing the nearest. Using such a coding, errors can always be corrected if they corrupt codewords by less than half the distance between the nearest two codewords. Alternatively, errors are always detected if they do not corrupt codewords by as much as the smallest distance between two codewords.

Errors further from a codeword can sometimes be corrected. Consider point R(2.6,2.6): this can be decoded to (5,5) even though its distance from (5,5) is greater than half the minimum distance of 2.5. However, if an error were to occur to message D to take it to S(5,2.4), it would be incorrectly decoded as C, even although the distance of the error is less than in the case of R. It is therefore usual to constrain the code as shown in Figure 7.8 by only assuming a particular letter was intended if the received message is within a distance d_c of a codeword, i.e. it is within one of the shaded circles in the figure.

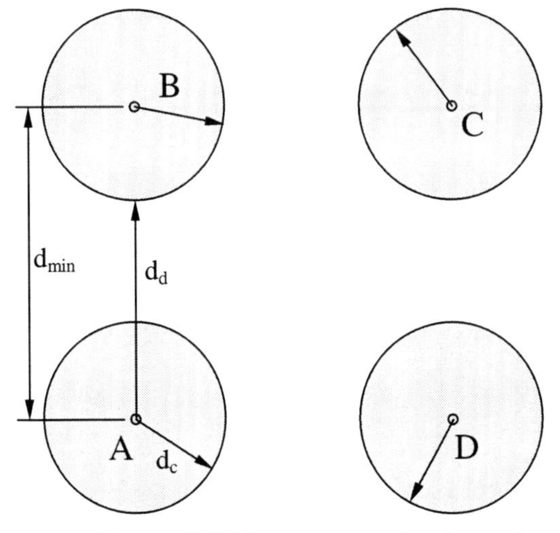

Figure 7.8 Distance around codewords

If the minimum distance between any pair of codewords is d_{min}, it is possible to correct errors up to d_c and detect errors up to d_d, as long as $d_{min} \geq d_d + d_c$. This is because if a received message lies outside the circles, it is possible to recognise that there is an error, although it cannot be corrected because it is not within d_c of any circle. By varying the size of the circles, error correction capability can be traded off against error detection. Note that for a fixed d_{min}, increasing d_c reduces d_d, and one issue to be recognised is the possibility of incorrectly "correcting" a received message. If a message was so badly corrupted that it was moved more than d_d, and so fell within d_c of another codeword, it would be decoded incorrectly.

Mathematically, distance corresponds to a metric. TETRA is a digital system and the codes it uses are binary. The simplest and most common metric for binary signals is Hamming distance, which is the number of coefficients where the binary bit streams differ. In other words, the Hamming distance between two bit streams *a* and *b* equals the weight of *a-b*, where the weight of a bit stream is its number of non-zero components.

Consider three codes using symbols made up of three binary digits. The error detection and correction possibilities are as shown in Figure 7.9.

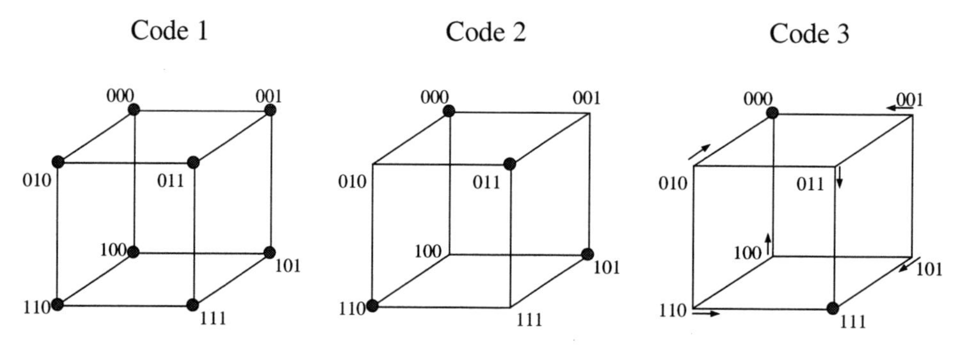

Figure 7.9 Three 3-bit codes

The first code uses all possible combinations of the three bits and so has a minimum Hamming distance of 1. It cannot correct or detect any errors because any error would form a new codeword. Code 2 has a minimum Hamming distance of two, and was constructed by deleting all codewords with a Hamming distance of one from another codeword. This code can detect a single error, although not correct it (010 could be 000, 011 or even 110). Two errors would not be detected. Code 2 is actually an even parity check code, and will detect three errors as well. Code 3 has distance 3. It will correct one error or detect 2 (but not both). If the code were used for correction, 010 would be corrected to 000, as would 100 and 001 as shown by the small arrows.

This code shows that by not using some combinations of symbols, distance can be introduced between the symbol combinations that remain. The symbols added to introduce this distance are known as *parity* symbols; the other symbols are *information* or *message* symbols. Table 7.2 lists the parity and resulting codewords for the three codes shown in Figure 7.9.

Table 7.2 Parity and codewords for example 3 bit codes

Code 1			Code 2			Code 3		
Message	Parity	Code-word	Message	Parity	Code-word	Message	Partity	Code-word
000		000	00	0	000	0	00	000
001		001	01	1	011	1	11	111
010		010	10	1	101			
011		011	11	0	110			
100		100						
101		101						
110		110						
111		111						
Minimum distance 1			Minimum distance 2			Minimum distance 3		

This is an example of a *block* code. This term is used when the code considers symbols in groups or blocks (3 bits in this case). It is also possible to consider symbols sequentially. This is the case for *convolutional* codes. The example codes are also all in a form where the codeword is formed from message symbols followed by parity symbols. This is known as the *systematic* form.

Error correcting codes are normally described in the form (n, k), where number of information symbols is denoted by k, and the total number of symbols is denoted by n, although the form (n, k, d), where d is the minimum distance, is also used. The *rate* of the code, R, is the proportion of output bits to message bits, is n/k.

The existence of a distance between codewords does not necessarily imply that the message can easily be found from the received codeword, even if fewer errors occur than the code can theoretically correct. In the worst case, it may be necessary to compare the received bit stream with all possible codewords in order to see to which it is nearest. If the block length is large, there are many possible codewords, and so this can be a very time-consuming process. A choice of coding function must therefore be made so that there is a simple method of decoding.

For example, consider a relatively small code with 120 information bits, 23 check bits, giving a total of 143 bits [9]. The code has 2^{120} codewords, so an exhaustive search, where the received bit sequence was compared to all possible codewords, would require in the order of 10^{36} comparisons. At a rate of one billion comparisons per second, this would take in the order of 10^{19} years. The universe is only of the order of 10^9 years old. Fortunately, this type of code can be decoded using its coding rules in 264 simple steps.

7.4.2.1 Linear codes

If coding is a linear function with respect to the information message, i.e., *codeword* = *f(message)*, there is a simple method of encoding and testing if received words are codewords. A linear code function does not necessarily make the code simple to decode, however, but it does simplify construction.

Consider a two-dimensional grid of integers taken modulo 5.

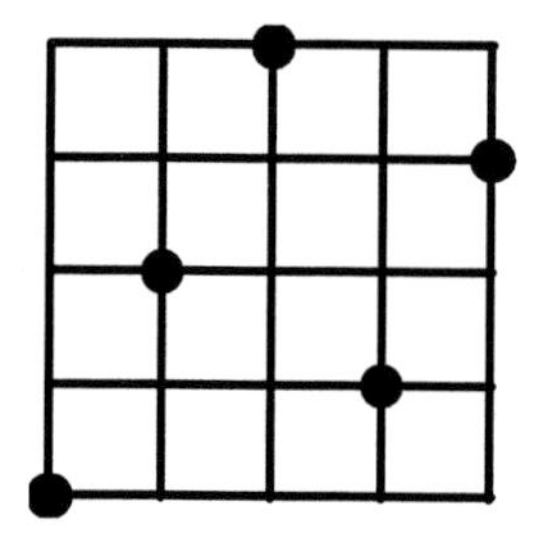

Figure 7.10 Linear code

A two-dimensional code can be defined such that a message i is encoded into a two-dimensional codeword $(i, 2i)$. This is a linear code, since every linear combination of codewords forms another codeword. The codewords are A(0,0) (i=0), B(1,2) (i=1), C(2,4) (i=2), D(3,1)=(3, 6 mod 5) (i=3), E(4,3) = (4, 8 mod 5) (i=4). The best codes are formed by those functions which cause a wide spread of codewords with a large distance between them, since the error correcting ability is dependant on the minimum distance.

The above example forms a one error correcting code, if errors in this case can change one dimension by one unit. Each one of the grid points which is not a codeword is at most one horizontal or vertical position away from one that is. This is shown in Figure 7.11 by labelling each non-codeword with the small letter of the codeword which it is nearest. If any of the small letters is received it can be assumed that the corresponding capital letter was sent.

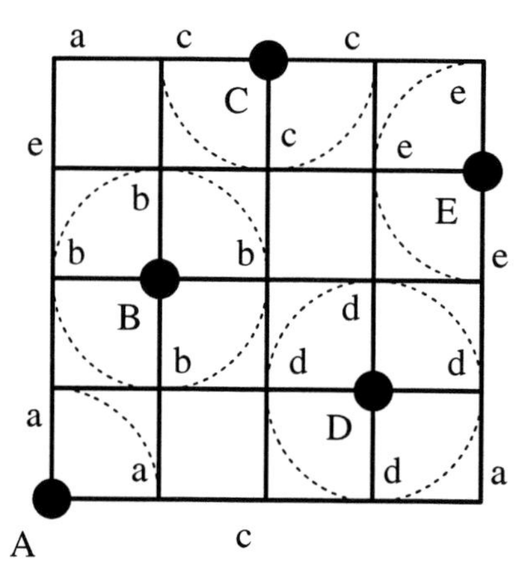

Figure 7.11 Error correcting mapping for code

Linear codes are much easier to implement because instead of having to record details of every codeword, it is simply necessary to record details of the linear function, and multiply the message by that. However, error correction is not in general simple. Some linear codes do have defined methods to detect the nearest codeword in the presence of errors, but in general the better the code (in terms of the number of errors it corrects and the additional bits added), the more complicated are the decoding algorithms required.

Linear codes can be defined by a *generator matrix* made up of code words. k of these codewords are independent, so the generator matrix G has k rows and n columns. All the codewords can be formed by adding rows, so the codeword, c, is formed from the message, m, by $c = m$ G. The structure is G is such that when it is put in standard echelon form it has the structure [I|P] (this results in a systematic code). There is a corresponding matrix $H = [P^T|I]$ called the parity check matrix. $GH^T = 0$. H is used to decode the code. $cH^T = 0$ if there are no errors, since $cH^T = mGH^T = m.0 = 0$. If c had been corrupted by errors in transmission, e, what will be received is $c' = c + e$. Decoding c' gives $(c + e)H^T = mGH^T + eH^T = 0 + eH^T = eH^T$. This result is called the *syndrome*, and is only equal to zero if no errors occur. The syndrome is used to correct any errors. If an error does occur, the syndrome is only dependant on the errors, and not on the message. This makes error correction significantly easier.

Special types of linear codes, called cyclic codes, have a well defined structure and are simple to implement. A cyclic code is a code where in addition to the property that every linear combination of codewords forms another codeword, every cyclic shift of the codeword forms another codeword. For example, in the binary case, if 01001011 is a code word, so is 10010110, 00101101, and so on. Such codes can be therefore be constructed from shifts and linear combinations of a single codeword, called a *generator*. If properly designed, the generator ensures that the codewords are evenly distributed throughout the code space and the code has good distance properties. It is also easy to check if a received codeword has any errors, as all codewords are divisible by the generator. For this reason, special cyclic codes called *cyclic redundancy checks* (CRC) are used for error detection and not error correction.

Well designed cyclic block codes can detect a very large proportion of bursts of errors. Error bursts are groups of errors such that the first and last bits of the group are errors, and the intermediate bits may or may not be errors. Burst errors are very common in a radio environment, being caused by fades, etc.

7.4.2.2 Linear code example

A (7,4) single error correcting Hamming code has the following generator matrix.

$$G = \begin{pmatrix} 1 & 0 & 0 & 0 & 1 & 1 & 0 \\ 0 & 1 & 0 & 0 & 1 & 0 & 1 \\ 0 & 0 & 1 & 0 & 0 & 1 & 1 \\ 0 & 0 & 0 & 1 & 1 & 1 & 1 \end{pmatrix} = (I_4 \mid P)$$

The Hamming code output is $c = m$G, where G is the generator matrix of the Hamming code. The code has 16 (2^4) possible messages and code words are as given in Table 7.3.

$$c = mG \Rightarrow \begin{pmatrix} c_1 & c_2 & c_3 & c_4 & c_5 & c_6 & c_7 \end{pmatrix} = \begin{pmatrix} m_1 & m_2 & m_3 & m_4 \end{pmatrix} \begin{pmatrix} 1 & 0 & 0 & 0 & 1 & 1 & 0 \\ 0 & 1 & 0 & 0 & 1 & 0 & 1 \\ 0 & 0 & 1 & 0 & 0 & 1 & 1 \\ 0 & 0 & 0 & 1 & 1 & 1 & 1 \end{pmatrix}$$

The parity check matrix is therefore

$$H = \begin{pmatrix} 1 & 1 & 0 & 1 & 1 & 0 & 0 \\ 1 & 0 & 1 & 1 & 0 & 1 & 0 \\ 0 & 1 & 1 & 1 & 0 & 0 & 1 \end{pmatrix} = \left(P^{\mathrm{T}} \mid I_3 \right)$$

Table 7.3 Possible messages and code words for (7,4) Hamming code

Message	Codeword	Message	Codeword	Message	Codeword	Message	Codeword
0000	0000000	0100	0100101	1000	1000110	1100	1100011
0001	0001111	0101	0101010	1001	1001001	1101	1101100
0010	0010011	0110	0110110	1010	1010101	1110	1110000
0011	0011100	0111	0111001	1011	1011010	1111	1111111

If there is no error, the syndrome is 000, and the message is simply the first four bits of the codeword. The syndromes for all single errors are shown in Table 7.4. Since they are all distinct, all single errors can be corrected. If the syndrome is non-zero, the appropriate bit is corrected, and then the message can be found from the first four bits.

Table 7.4 Syndromes for single bit errors

Single error in position	1	2	3	4	5	6	7
Syndrome	110	101	011	111	100	010	001

7.4.2.3 Convolutional codes

The codes discussed so far are block codes, where information is processed block at a time. Block codes are efficient but can be hard to decode. An alternative is to encode and decode sequentially.

It would be possible to generate a number of output bits each time an input bit is received. A very simple code, called a repetition code, would repeat on the output the input bit several times. If the rate of repetition is 2, the result would be

Input 1 1 0 1
Output 11 11 00 11

It should be noted that the output has twice as many bits, so the rate of the code, $R = ½$.

This is actually a very poor code. It can detect one error (though not two, if they corrupt the same pair of output bits), but cannot correct even one error, for if 11 is corrupted to 10, it is just as likely to have been 00 as 11. The performance can be improved greatly by adding some memory to the coder and making the current output bits dependent not only on the current input but also on previous bits. Such codes are called *convolutional codes.* The number of bits taken into account – the memory of the code plus one – is called its constraint length.

A simple convolutional code is a (2,1,3) code (2 output bits for each 1 input bit with a constraint length of 3) such that the first bit is the binary sum of the current input, the input on the last bit, and the input two bits previously, and the second bit is the sum of the current bit and the bit two bits ago (see Figure 7.12). This code has a distance of 5 (no two output sequences differ by less than 5 bits), and so it can correct two errors. It can correct more errors than this if the errors are sufficiently widely spaced (i.e. further apart than the code's memory). Note that this is the code which is used in the TETRAPOL system.

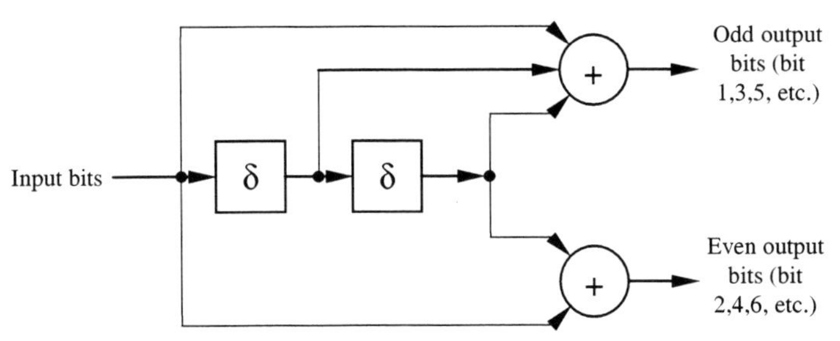

Figure 7.12 (2,1,3) convolutional code encoder

Convolutional codes are simple to encode and decode, but if they do make a mistake, then a large number of output bits will be corrupted (since the decoder will attempt to correct the bit stream and add additional errors in the process). For this reason, some form of additional error detection is desirable.

The memory in the shift registers of a convolutional coder gives the system a state. For the (2,1,3) convolutional code, the constraint length K is 3, and so there are $2^{K-1} = 4$ states. It is possible to draw a state transition diagram for the code. Such a diagram has a node for each possible state of the system, connected by lines depicting the possible changes of state the system can undergo. These lines are labels with the inputs which would cause this change of state, and the output that is produced. For example, if the coder is in state 00, it can remain in state 00 by receiving a 0, whereupon 00 will be output, or it can move to state 10 if a 1 is received, outputting 11 as it does so.

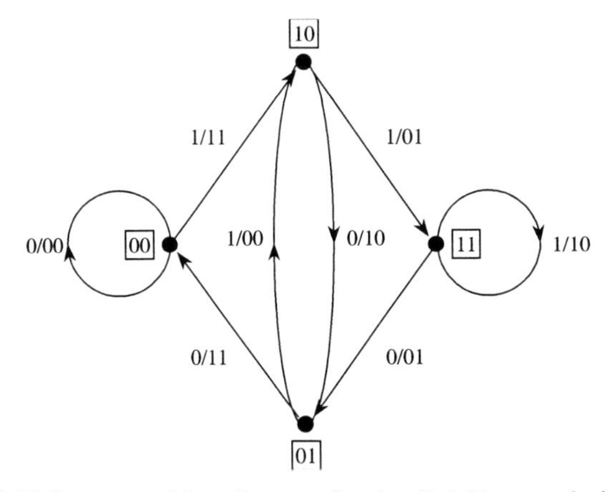

Figure 7.13 State transition diagram for the (2,1,3) convolutional code

While a state transition diagram is very useful for showing possible movements between states of a system, for a convolution code it is usually more important to record the sequence of state changes and corresponding outputs in response to different inputs, and for this purpose a code trellis is useful. This is similar to a state transition diagram but with the existing states listed on the left and the next states on the right, with state transitions drawn in between.

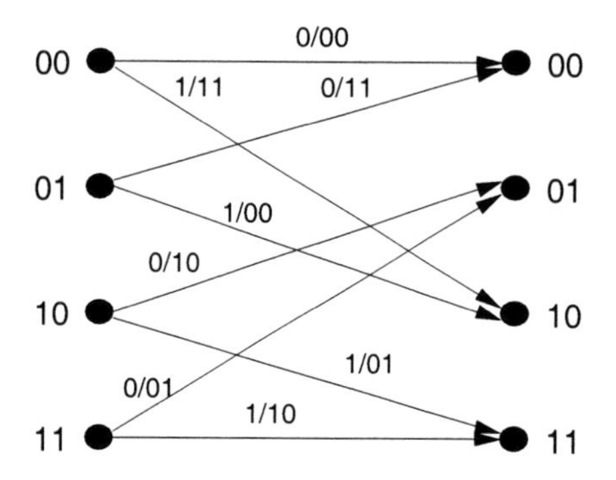

Figure 7.14 Trellis diagram for the (2,1,3) convolutional code

Example of Encoding Convolutional Codes

The following example shows the states and output of the (2,1,3) code when the binary stream to be encoded is 101001011100. The state transitions are denoted by the thick line. Note that the two shift registers require two tail bits at the end of each sequence to be encoded to read out the bits in the code's memory and return the encoder to the known 00 state.

	1	2	3	4	5	6	7	8	9	10	11	12
Input	1	0	1	0	0	1	0	1	1	1	0	0
Current state	00	10	01	10	01	00	10	01	10	11	11	01
Next state	10	01	10	01	00	10	01	10	11	11	01	10
Output	11	10	00	10	11	11	10	00	01	10	01	11

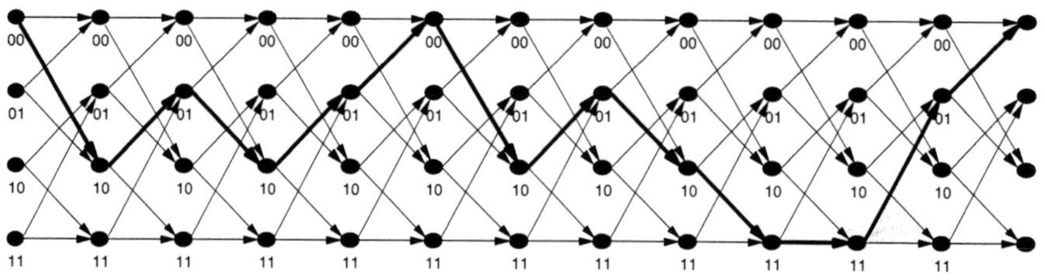

Figure 7.15 Encoding example for a (2,1,3) code

Decoding Convolutional Codes

There are several different methods of decoding convolutional codes, and as would be expected there is a trade off between decoding complexity and performance. A relatively simple method, called sequential decoding, is to trace through the encoding trellis based on the received bits, taking the path at each node with results in the lowest number of discrepancies between the received bits and the bits that would have been generated. This number is termed the branch metric. If two paths have the same metric, a random choice is made between them and decoding continues. If the metric exceeds a threshold based on the expected number of errors the decoder decides that it must have made a mistake on one of its random choices, deletes that path, and tries again.

The threshold is usually taken to be the expected number of bit errors at that point (the bit error probability P_e times n), plus a margin of a few bits. This margin can be increased if it is found that no path survives.

This approach can be extended to give maximum likelihood decoding (i.e. choosing the transmitted sequence that would produce the smallest number of differences between its codeword and the received sequence). If every possible path from each node is followed, then all possibilities are taken into account. However, if two or more paths reach the same node they will all continue in exactly the same way, so it is possible to delete all arriving paths except the one with the lowest metric. This means that the number of surviving paths is kept to, at most, the number of states, and so the complexity is reasonable. This algorithm is the Viterbi algorithm, and while it is more complex than the sequential algorithm, it can be implemented quite easily.

Example of Decoding Convolutional Codes

The operation of the Viterbi algorithm can be shown using the example of the (2,1,3) code. If a bit stream of 11 10 10 10 11 10 10 10 01 10 01 11 is received, the problem is to find the most likely message sequence. This is done by tracing all possible messages, and comparing the output they would have produced with the received sequence. The most likely message sequence is then chosen as the one which has the fewest differences between the output it would have produced and the received sequence.

For the first message bit, the system is starting, and is therefore known to be in the initial state of 00. The trellis in Figure 7.14 gives two possibilities for what the encoder could have produced: either a 0 was encoded, producing a state change to state 00, outputting 00, or a 1 was encoded, producing a change to state 10, and outputting 11. If 0 was encoded, and 00 outputted, then there are two differences to the sequence received. Therefore, the path from state 00 to state 00 has weight 2. If the encoder actually produced 11, then no errors have occurred, and so the path weight from 00 to 10 is 0. If the decoding process was to stop at this point, it would be possible to conclude that the path with weight 0 was the more likely, and that a 1 had been sent, since this would correspond to the change from state 00 to state 10. However, as more bits are received, the path weights may change, so it is normal not to decode the output as being 1 until more bits are received.

For the second message bit, the encoder could either have been in state 00 or state 10. Starting in state 00 would result in transitions to states 00 (with output 00) or 10 (with output 11). Both these possible outputs differ by one bit from the received bit pattern 10. This gives total path metrics of 3 in each case. Starting in state 10 would allow transitions to states 01 or 11 with total path metrics of 0 and 2 respectively. On the third step, all the states are possible as starting points. The total path metrics are as shown in Figure 7.16.

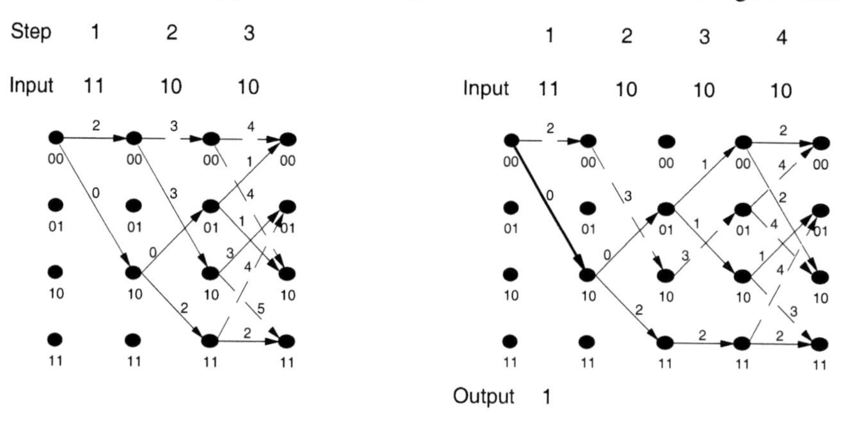

Figure 7.16 Decoding example after three steps

Figure 7.17 Decoding example after four steps

At this point, there is more than one path to a state, and those paths which reach a state with a higher weight than another path reaching the same state can be deleted. This is because the paths will continue from that state in the same way, so the higher weight path has no possibility of having a lower overall weight than the lower weight one. When this is done (the dotted paths), one of the possible paths at step 2 is also deleted, as no valid path continues through state 00 in step 2.

After a fourth step, there is only one remaining path at step 1. This is called the survivor, and since this transition results from an input of 1, it is possible to decode the input up to that point as 1. This process can be repeated, with further bits being decoded as the paths become unique up to particular points.

Practical decoder implementations cannot allow for the storage of infinite path lengths. The maximum length of the stored path is usually limited, both to limit complexity and delay (the output bit cannot be guaranteed to be available until after a number of input bits equal to the maximum path length times $1/R$ have been input after it). Very little performance is lost if this limit is about 4 or 5 times the constraint length.

The tail bits inserted at the end of the transmitted bit stream to read out the final bits in the shift registers mean that the coder will end in state 00. This allows the decoder to use this knowledge to delete any paths that do not end in state 00. When this final state is reached, a process called traceback is initiated, following the lowest weight path and outputting the relevant input bits. All the possible paths for the example are shown in Figure 7.18, with the maximum likelihood path highlighted.

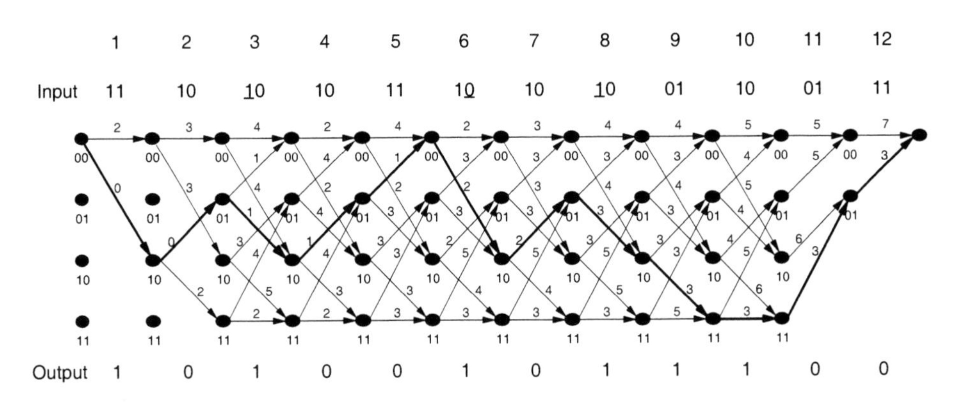

Figure 7.18 Viterbi (maximum likelihood) decoding example for the (2,1,3) code

7.4.2.4 Puncturing

An issue with convolutional codes is that they have a rather poor rate. The rate can be improved by "puncturing" the code to change the proportion of information bits to parity bits added to allow errors to be corrected. Puncturing is a process of removing some of the additional parity bits. Some of the ability of the code to correct errors will be lost by doing this, but the process allows a tradeoff between the number of bits added, and the resulting ability of the code to correct errors, and the transmission requirements in terms of number of bits. In TETRA, three different TCH rates are defined – 2.4 kbit/s with high protection, 4.8 kbit/s with low protection, and 7.2 kbit/s with no protection.

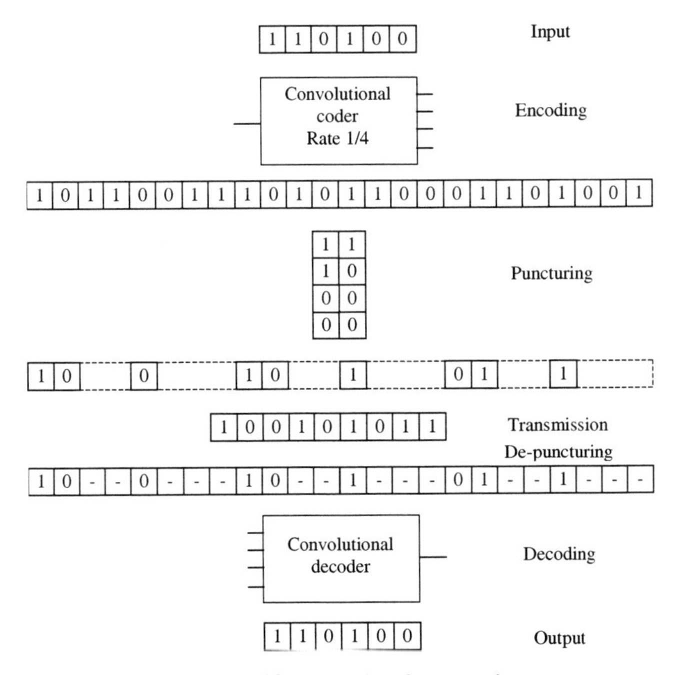

Figure 7.19 Example of puncturing

In the example in Figure 7.19, a 1/4 rate convolutional code encodes six message bits into 24 bits. Puncturing is used with a puncturing matrix which deletes 2 bits from the first four produced by the coder, and three from the second four, and so on in a repeating cycle. This is the puncturing scheme used for the TETRA TCH/4.8 channel. The overall result is three bits out for every two bits in to the convolutional coder, or an overall rate of 2/3. This leaves only 9 bits to be transmitted. At the receive, blanks are reinserted into the data stream to replace the punctured bits. These blanks are marked as unknowns, and so when the Viterbi convolutional decoder is calculating path metrics, it will ignore these blanks. As an example, a path of ?0 would have a distance of 0 from 00, 1 from 01, 0 from 10 and 1 from 11. This reduces the decoders ability to detect and correct errors, but does not otherwise affect the decoding algorithm.

The code used in TETRA is a Rate Compatible Punctured Convolutional (RCPC) Code. This means that a single mother code is punctured in a number of different ways to produce codes with different rates. All these codes share the same basic encoder and, more importantly, decoder, which reduces complexity.

7.4.2.5 Interleaving

It may be recalled that multipath fading can cause short term denigrations of the received signal. The convolutional coder used to protect the transmitted bits is most effective against errors which occur randomly. This is because if the errors occur in groups, it is not possible for the coder to calculate what the correct sequence should have been. The errors produced by multipath fading are likely to be grouped together, so it is wise to spread out, over several bursts, bits which are adjacent in the sequence being encoded. It is then far less likely that errors will occur together in groups. This re-ordering is called *interleaving*. An example of interleaving over 8 bursts is shown in Figure 7.20.

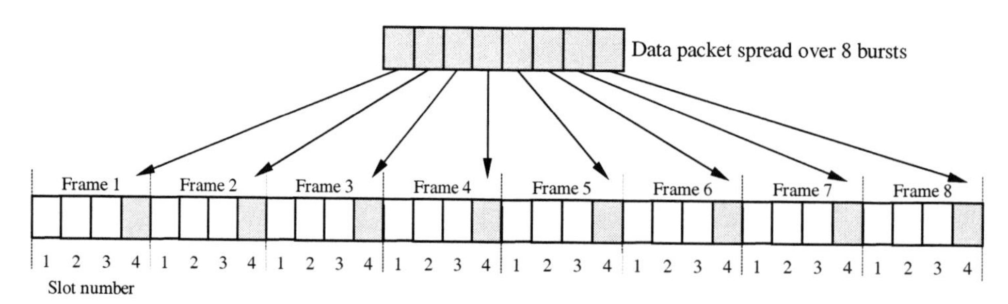

Figure 7.20 Interleaving over 8 bursts

A burst of errors that corrupted the whole for frame 5, say, would only corrupt $^1/_8$th of the data packet. In contrast, had the packet been transmitted continuously, it would either not have been affected at all, or half the packet would have been corrupted. Interleaving can therefore be seem to average out the errors over the transmitted data packets. If the average error rate is below the error correcting capability of the code, the code will correct the errors and interleaving will improve the overall error rate. If, however, the error correcting code cannot correct this number of errors, interleaving will actually increase the error rate. In particular, in the special case of a code with no error correcting capability being used, interleaving is unwise. It would simply spread the errors to affect a larger number of packets.

One method of interleaving, called block interleaving, is to read the transmitted bit stream into an array in one direction (along the rows, for example), and then reading them out in the other (along the columns). At the receiver, the process is reversed. Neighbouring bits in the transmitted bit stream are separated by the length of the column when the bit stream is de-interleaved. The number of codewords interleaving takes place over gives the separation between previously neighbouring bits in the transmitted bit stream is called the *depth* of interleaving.

The following example uses a small 6 × 4 array to carry out interleaving. 24 bits are read in to the array row by row, and then read out column by column. The interleaving depth is 6 bits. The bits are transmitted, and a burst of four errors occurs. When the interleaving process is reversed, the errors are spread out, which will make decoding them easier.

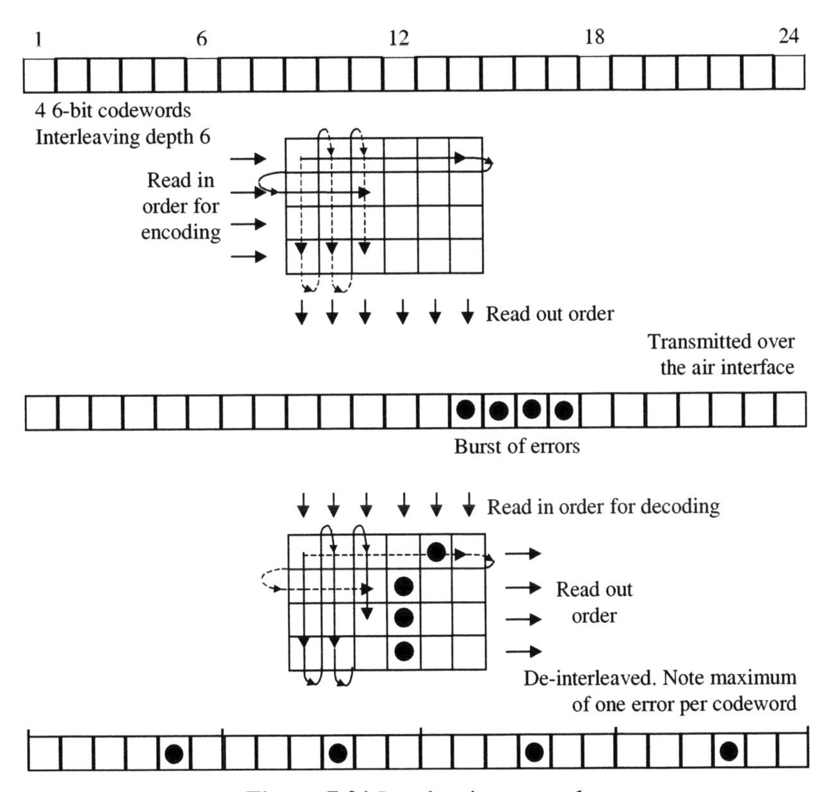

Figure 7.21 Interleaving example

It takes 24 bits to fill the array, which means that bits must be processed in blocks of 24. This introduces a delay. This delay will increase if the size of the array in increased. However, the amount of spreading, and the effectiveness of the interleaving against bursts of errors, depends on the number of rows. There is a trade off between interleaving depth and delay.

7.4.3 TETRA Data and Signalling Channel Coding Schemes

TETRA signalling and data channels use a four stage coding scheme for forward error correction and error detection which makes the best use of different coding strategies as described in the previous section. An overview of the stages is shown in Figure 7.22.

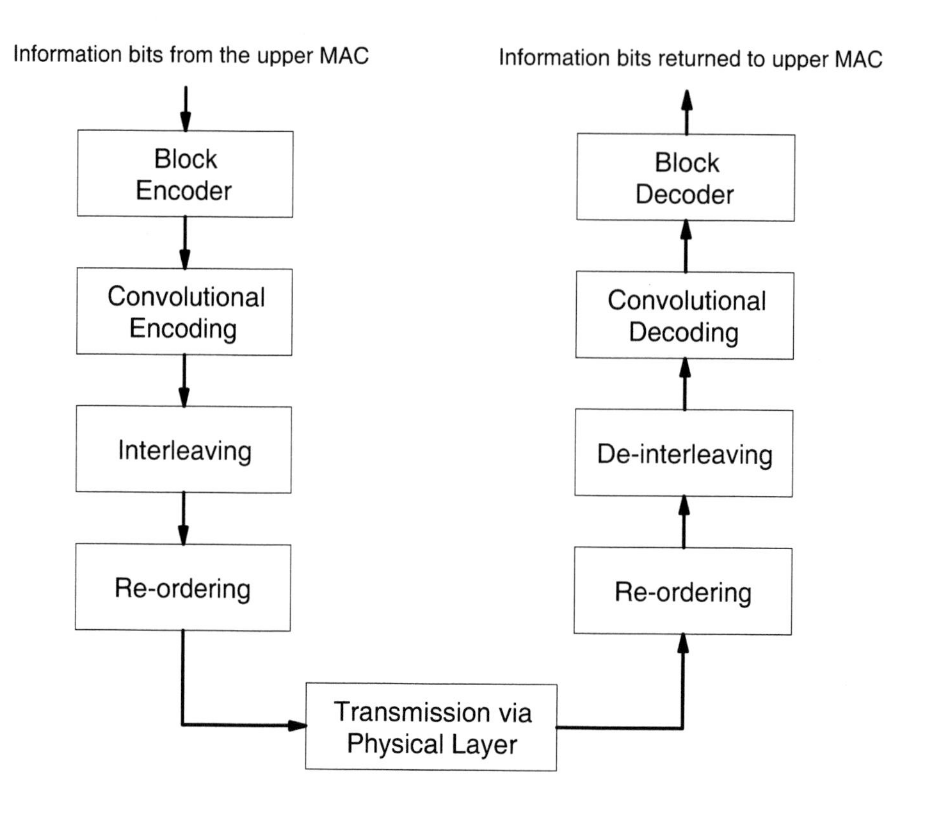

Figure 7.22 Outline of the TETRA data coding scheme

The first stage in coding, and therefore the last stage in decoding, is a block code. This provides error detection, and is performed as the last step to ensure that if any errors remain after FEC they can be detected.

FEC itself is provided by RCPC convolutional coding. Convolutional codes have the advantage of simple implementation, and TETRA uses puncturing to offer a variety of transmission rates and levels of protection.

Convolutional codes perform best with random errors, and interleaving and re-ordering is undertaken to try to make the errors as uncorrelated as possible. Interleaving across bursts adds considerable delay, and so although it is most effective at spreading errors, it is not always employed. Re-ordering changes bit positions within the burst, but involves much less delay since only one burst is involved.

Table 7.5 Coding steps used for different signalling and data channels

	SCH/ F,HU,HD	AACH	STCH	BSCH	TCH/ 2.4	TCH/ 2.4	TCH/ 7.2
Block coding	✓	✓	✓	✓	✗	✗	✗
Convolutional coding	✓	✗	✓	✓	✓	✓	✗
Interleaving	✗	✗	✗	✗	✓	✓	✗
Reordering	✓	✗	✓	✓	✓	✓	✗

Different logical channels have different coding strategies. These are summarised in Table 7.5. Details of the coding steps are given in the following sections.

7.4.3.1 Block coding

TETRA signalling channels use a block code for error detection to ensure that received blocks are error free. Two types are used, a (K+16, K) cyclic redundancy check (CRC) code, and for the ACCH, a shortened Reed Muller code.

The CRC code used in TETRA is a CCITT code [7], which adds 16 bits to each block of data. This code can detect the following error patterns:

- all error bursts of length up to 16 bits
- 99.997% of bursts of length 17 bits
- 99.998% of bursts longer than this
- all possible combinations of 3 random errors or fewer
- all possible combinations of odd numbers of errors

For the ACCH, a shortened (30,14) Reed Muller code is used [7]. The Reed Muller code corrects as well as detects errors.

It should be noted that block coding is only used on the signalling channels. It is not performed on either data or speech channels, because since retransmission is not used for these services due to delay constraints (see Section 7.10.3), the knowledge that an error had occurred can not be acted upon. Application layers can add error detection if they require it.

7.4.3.2 Convolutional coding

Error correction for data and signalling channels is performed by a rate compatible punctured convolutional (RCPC) code. The base code is a 1/4 rate convolutional code with a constraint length of 5; a (4,1,5) code. The circuit to generate this code is shown in Figure 7.23, and the output depends on the current input and the previous four input bits.

The output bits produced on each input are as follows, where b_i is the current input, b_{i-1} is the previous input bit, b_{i-2} is the input bit prior to that, and so on.

$$\text{Bit } 1 = b_i \oplus b_{i-1} \oplus b_{i-4}$$
$$\text{Bit } 2 = b_i \oplus b_{i-2} \oplus b_{i-3} \oplus b_{i-4}$$
$$\text{Bit } 3 = b_i \oplus b_{i-1} \oplus b_{i-2} \oplus b_{i-4}$$
$$\text{Bit } 4 = b_i \oplus b_{i-1} \oplus b_{i-3} \oplus b_{i-4}$$

At the start of the coding process, the encoder's registers are initially set to 0, so b_{-1}, b_{-2}, b_{-3}, and b_{-4} are 0. In order to read out the final bits from the convolutional coder's memory, four additional "tail" bits (with value 0) have to be added to the bits to be encoded.

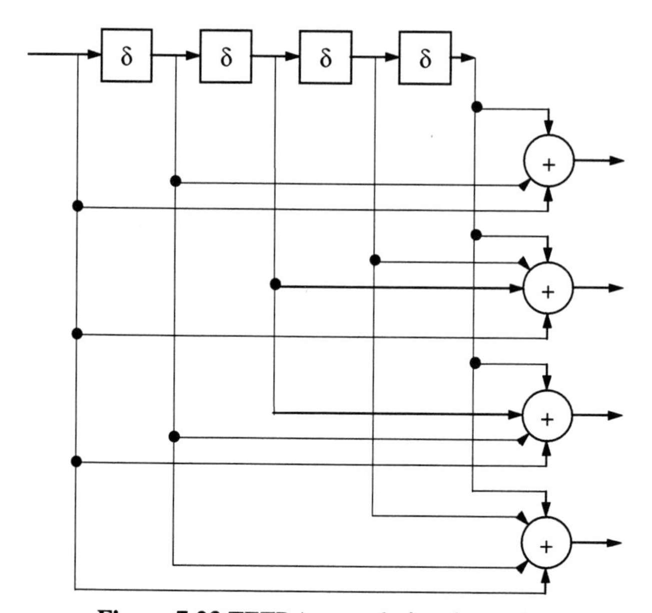

Figure 7.23 TETRA convolutional encoder

Four different puncturing schemes are defined, resulting in four different codes of rate 2/3, 1/3, 292/432 and 148/432. The 292/432 and 148/432 schemes are basically 2/3 and 1/3 rate schemes with some additional puncturing to make way for the four tail bits. The puncturing scheme for the 2/3 rate and 1/3 rate codes is as follows.

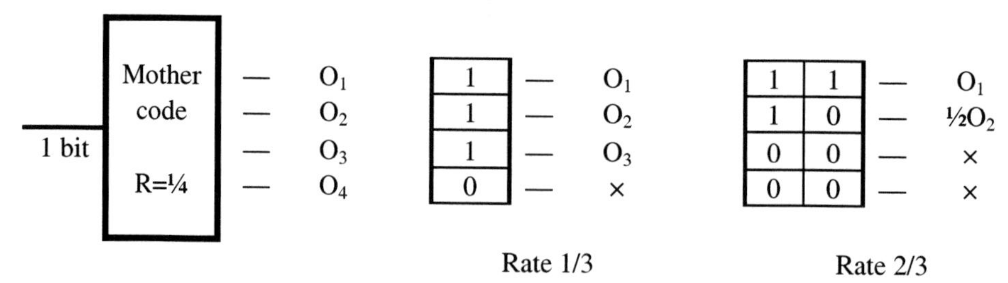

Figure 7.24 Basic puncturing schemes on the data and control channels

Puncturing the code reduces its ability to correct errors, but if done carefully, the reduction in error correcting ability is small compared to the change in rate. Figure 7.25

shows the error correcting ability of the TETRAPOL 1/2 rate code and the TETRA codes in the presence of random errors (i.e. assuming perfect interleaving). Note that puncturing the basic TETRA 1/4 rate code to 1/3 rate has very little effect, and also that the TETRA 2/3 rate code works better that the TETRAPOL 1/2 rate code most of the time even though it has fewer error correcting bits. This is because it is based on a more powerful code with a longer constraint length.

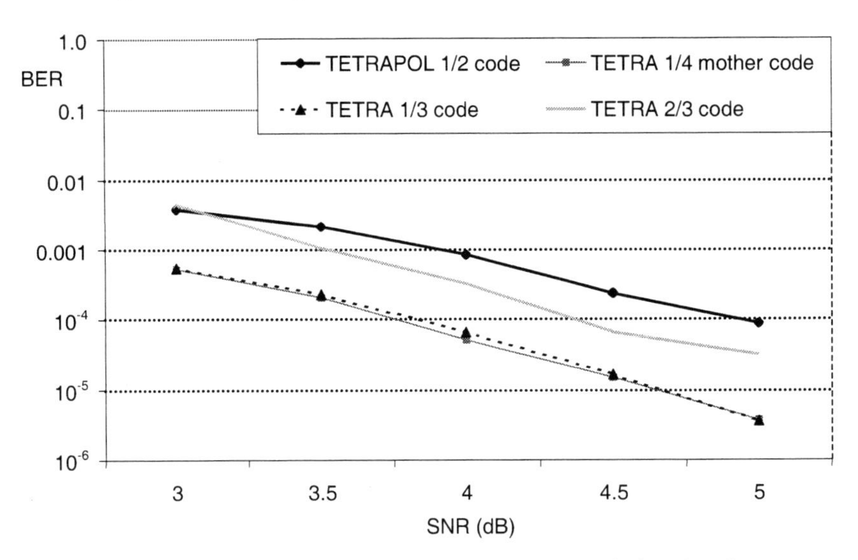

Figure 7.25 Performance of the punctured convolutional codes

7.4.3.3 Interleaving

The protected data traffic channels (i.e. TCH/4.8 and TCH/2.4) have three different possible interleaving schemes; interleaving over 8 bursts, interleaving over 4 bursts, or not using interleaving at all. TCH/7.2, which passes circuit mode data through the system without coding, does not use interleaving.

Block i
Block i, part A (Bits 1-108)
Block i, part B (Bits 109-216)
Block i, part C (Bits 217-324)
Block i, part D (Bits 325-432)

. . . .

6:A	5:A	4:A	3:A	2:A	1:A
6:B	5:B	4:B	3:B	2:B	1:B
6:C	5:C	4:C	3:C	2:C	1:C
6:D	5:D	4:D	3:D	2:D	1:D

Transmitted block i
Block i, part A (bits 1-108)
Block (i-1), part B (Bits 109-216)
Block (i-2), part C (Bits 217-324)
Block (i-3), part D (Bits 325-432)

. . . .

Transmitted block 6	Transmitted block 5	Transmitted block 4	Transmitted block 3	Transmitted block 2	Transmitted block 1
6:A	5:A	4:A	3:A	2:A	1:A
5:B	4:B	3:B	2:B	1:B	108 zero bits
4:C	3:C	2:C	1:C	108 zero bits	108 zero bits
3:D	2:D	1:D	108 zero bits	108 zero bits	108 zero bits

Figure 7.26 Interleaving for $N = 4$

Part
Part A (Bits 1-54)
Part B (Bits 55-108)
Part C (Bits 109-162)
Part D (Bits 163-216)
Part E (Bits 217-270)
Part F (Bits 271-324)
Part G (Bits 325-378)
Part H (Bits 379-432)

8:A	7:A	6:A	5:A	4:A	3:A	2:A	1:A
8:B	7:B	6:B	5:B	4:B	3:B	2:B	1:B
8:C	7:C	6:C	5:C	4:C	3:C	2:C	1:C
8:D	7:D	6:D	5:D	4:D	3:D	2:D	1:D
8:E	7:E	6:E	5:E	4:E	3:E	2:E	1:E
8:F	7:F	6:F	5:F	4:F	3:F	2:F	1:F
8:G	7:G	6:G	5:G	4:G	3:G	2:G	1:G
8:H	7:H	6:H	5:H	4:H	3:H	2:H	1:H

Transmitted block 8	Transmitted block 7	Transmitted block 6	Transmitted block 5	Transmitted block 4	Transmitted block 3	Transmitted block 2	Transmitted block 1
8:A	7:A	6:A	5:A	4:A	3:A	2:A	1:A
7:B	6:B	5:B	4:B	3:B	2:B	1:B	54 0 bits
6:C	5:C	4:C	3:C	2:C	1:C	54 0 bits	54 0 bits
5:D	4:D	3:D	2:D	1:D	54 0 bits	54 0 bits	54 0 bits
4:E	3:E	2:E	1:E	54 0 bits	54 0 bits	54 0 bits	54 0 bits
3:F	2:F	1:F	54 0 bits	54 0 bits	54 0 bits	54 0 bits	54 0 bits
2:G	1:G	54 0 bits	54 0 bits	54 0 bits	54 0 bits	54 0 bits	54 0 bits
1:H	54 0 bits	54 0 bits	54 0 bits	54 0 bits	54 0 bits	54 0 bits	54 0 bits

Figure 7.27 Interleaving for $N = 8$

Note that interleaving over N blocks requires N-1 additional bursts to transfer the information. It also adds to the transmission delay by N-1 frame lengths. This is why interleaving is not always used for traffic channels. The minimum number of transmitted bursts, even for the shortest messages, is N. For this reason, and the delay, interleaving is not used for signalling channels at all.

It is possible to group together 2, 3 or 4 traffic channels to provide higher net rate data communications. If this is done, the interleaving takes place over the constituent traffic channels individually. For example, a 9.6 kbit/s net rate traffic using TCH/4.8 on slots 3 and 4 would interleave on slots 3 and 4 independently – data on slot 3 would not be interleaved with that on slot 4. This ensures that the spreading effect is equally effective.

7.4.3.4 Re-ordering

After this interleaving takes place, the bits are re-ordered within a burst. This process is sometimes referred to as "block interleaving". A bit in position i moved to position j according to the following rule

$$j = 1 + (103 \times i) \bmod 432$$

This re-ordering is done in all case for TCH/2.4 and TCH/4.8 even when there is no interleaving (i.e. N = 1). It is not used for TCH/7.2.

Interleaving of control channels, which occupy a single burst, is not an option. However, their bits are re-ordered according to various different schemes based round a (K, a) block interleaver. This moves a bit in position i moved to position j according to the following rule

$$j = 1 + (a \times i) \bmod K$$

Note that under this definition the traffic channels undergo a (432, 103) block interleave.

Taking the example of the (216, 101) block interleaver used for the stealing channel, amongst others, the first bit (bit 1) would be moved to bit 102, bit 2 to bit 203, bit 3 to bit 88 (304 mod 216 = 88), and so on.

7.4.4 Coding Parameters for Data and Control Channels

By changing the parameters of each of these blocks the characteristics of the channel code can be changed as required for each different type of logical channel. Each logical channel has its own code format, as shown in Figure 7.28 and Figure 7.29.

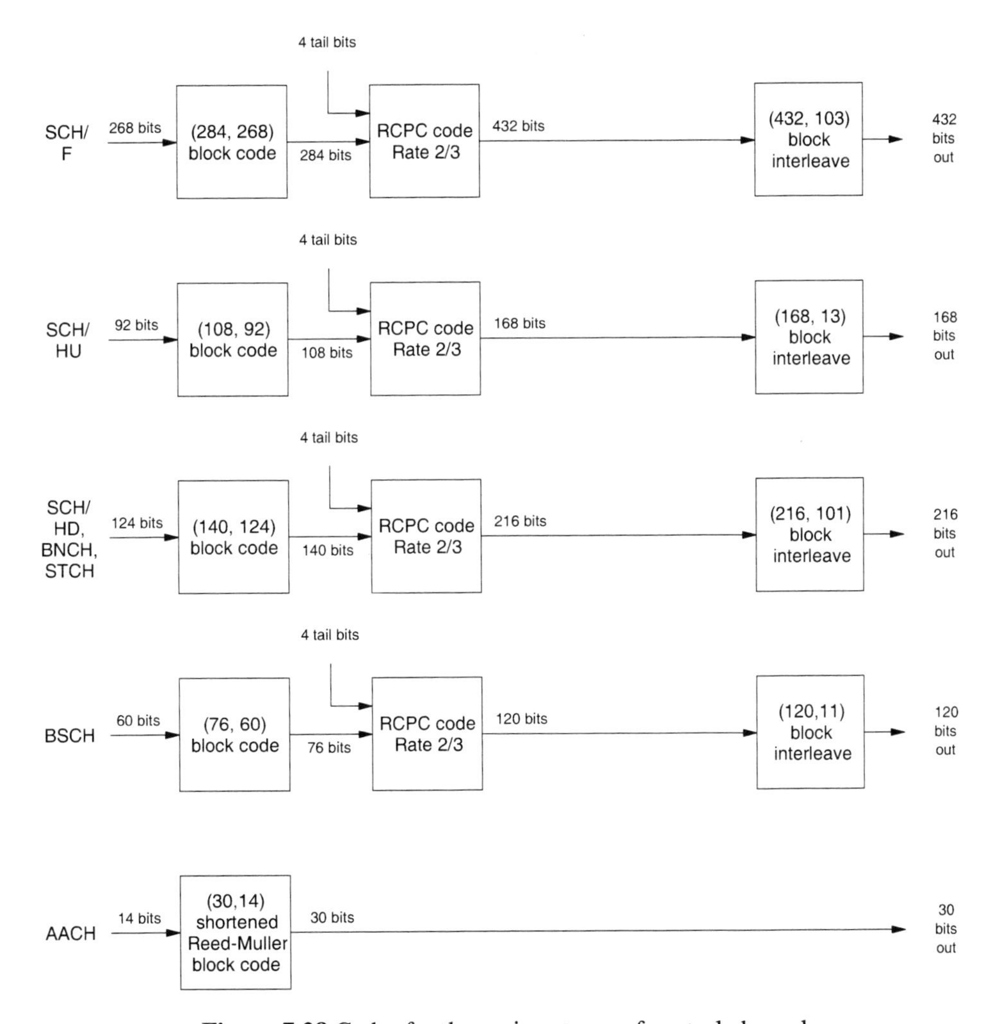

Figure 7.28 Codes for the various types of control channel

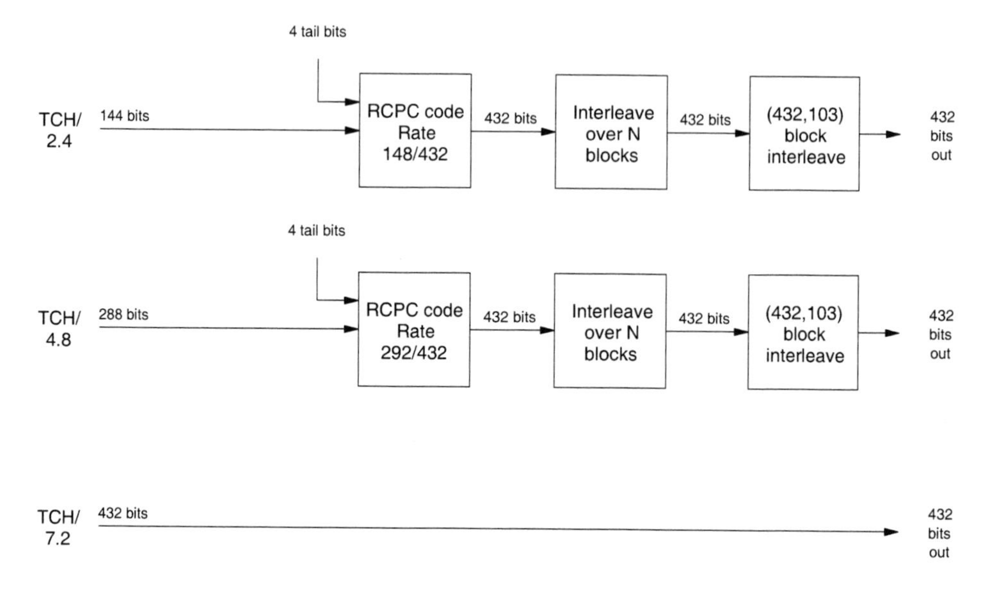

Figure 7.29 Codes for the various types of data traffic channel

In Direct Mode, the coding is very similar. The traffic channels TCH/7.2, TCH4.8 and TCH/2.4 have the same coding, as does the Stealing Channel (STCH) and the Full slot Signalling CHannel (SCH/F). The Half slot Signalling CHannel (SCH/H) has the same coding as the downlink half slot signalling channel in the Voice+Data mode. The synchronisation channel in direct mode, the SCH/S, has the same coding as the V+D Broadcast Synchronisation CHannel, the BSCH.

7.4.5 TETRA Speech Coding Scheme

A standard TETRA speech frame consists of 137 bits. Normally, data from two speech frames are transmitted in a transmission burst, which means that 216 bits are available for the speech frame and channel coding.

The speech bits passed to the MAC from the speech coder are classed in three different categories depending on their importance to speech quality (see Section 4.9.1). Class 2 bits are those bits which have the greatest effect on speech quality, class 1 having less importance, and class 0 having the least effect on the quality of the speech. In order to make the most efficient use of the transmission capacity, the different classes of bits are treated differently, with the class 2 bits receiving the most error protection, and the class 0 bits receiving none.

The speech coding scheme is similar to the coding for the data and signalling channels described in the previous section, although there is no interleaving (as the speech blocks would be delayed too much), and only class 2 bits have all the remaining coding components (see Table 7.6).

Table 7.6 Coding steps used for different classes of speech bits

	Class 2	Class 1	Class 0
Block coding	✓	✗	✗
Convolutional coding	✓	✓	✗
Interleaving	✗	✗	✗
Re-ordering	✓	✓	✓

The coding stages are as follows. The two 30 bit blocks of class 2 bits and the two 56 bit blocks of class 1 bits are combined together by taking one bit from each speech frame in turn. A (68, 60) block code is used for the 60 class 2 bits. This code is an extended cyclic code with generator $g(x) = (1 + x + x^7)(1 + x)$. Four tail bits are added and then a RCPC code is used. The mother code of the RCPC code is a (3,1,5) convolutional code as follows (see Figure 7.30)

$$\text{Bit 1} = b_i \oplus b_{i-1} \oplus b_{i-2} \oplus b_{i-3} \oplus b_{i-4}$$
$$\text{Bit 2} = b_i \oplus b_{i-1} \oplus b_{i-3} \oplus b_{i-4}$$
$$\text{Bit 3} = b_i \oplus b_{i-2} \oplus b_{i-4}$$

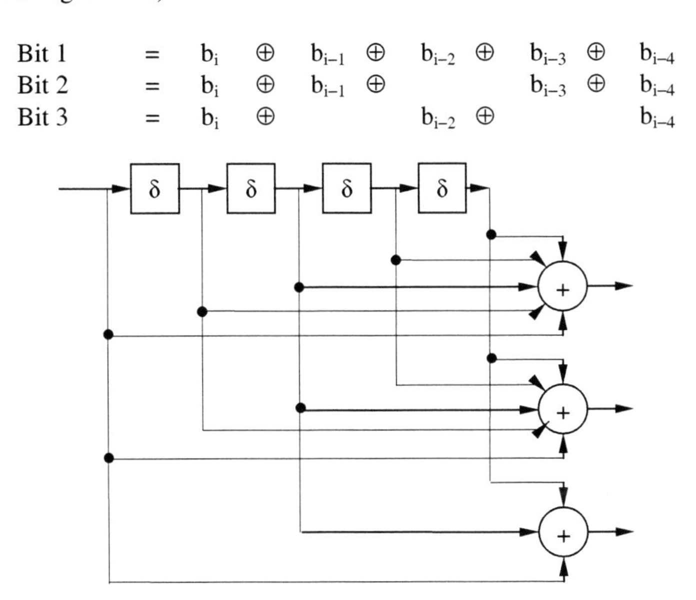

Figure 7.30 Convolutional encoder for the TETRA speech traffic channel

Two different puncturing schemes are used: 4/9 for the class 2, parity check and tail bits, and 2/3 for the class 1 bits. The puncturing matrices for these are as follows.

1 bit — Mother code R=1/3 — O_1, O_2, O_3

Rate 2/3:

	1	2		
	1	1	—	O_1
	1	0	—	½O_2
	0	0	—	×

Rate 4/9:

	1	2	3	4		
	1	1	1	1	—	O_1
	1	1	1	1	—	O_2
	1_a	0	0	0	—	¼O_3

Figure 7.31 Speech traffic channel puncturing schemes

Figure 7.33 shows a summary of the complete coding scheme for a speech burst.

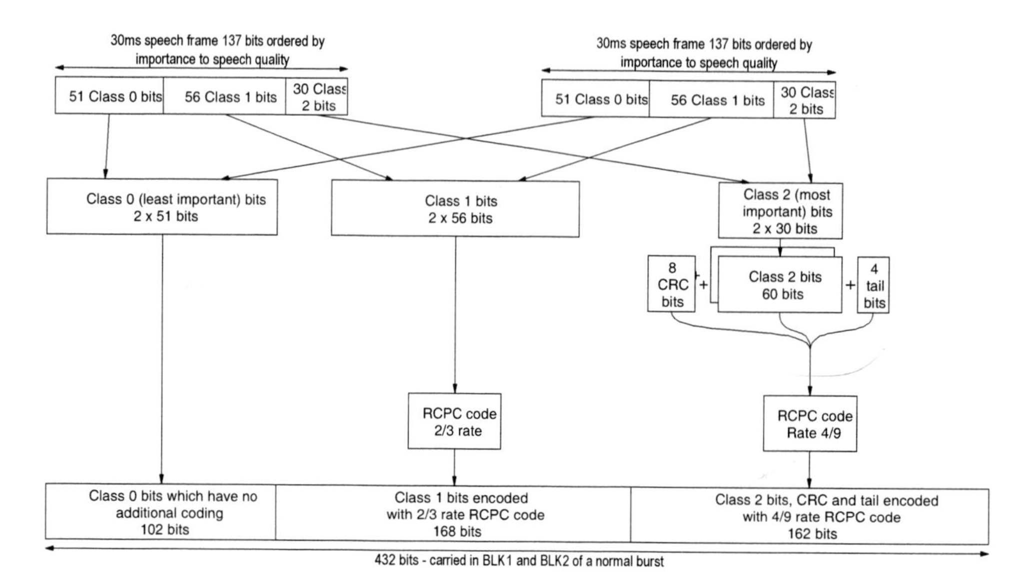

Figure 7.32 Overall speech traffic channel coding scheme

The re-ordering of speech frames uses a standard rectangular bit interleaving scheme. The block size is 18 rows by 24 columns, with bits being read in row by row and read out for transmission column by column.

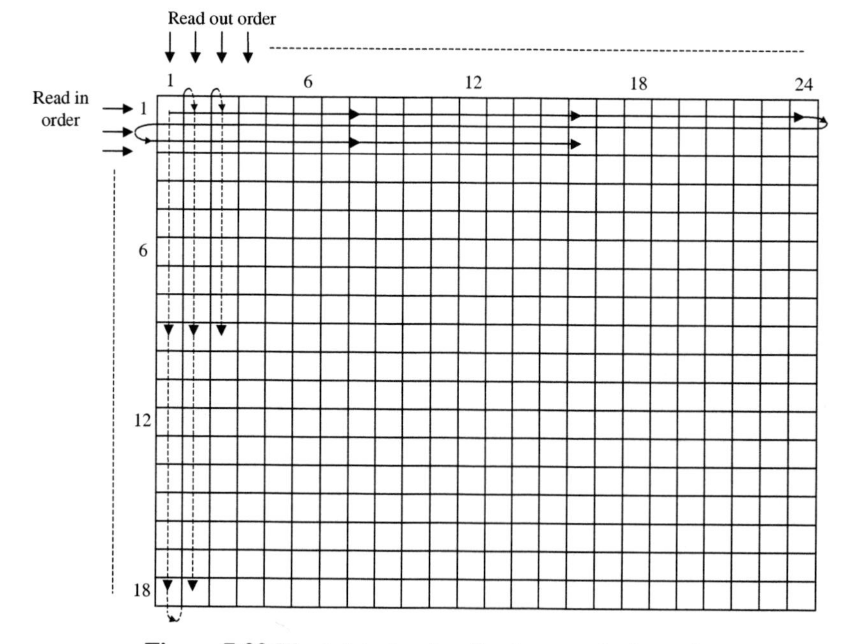

Figure 7.33 Block interleaving for the speech channel

When stealing is used, one or both of the speech frames may be replaced by the STCH. One STCH takes 216 bits, so if one STCH block has to be transmitted, the first of the two spee[illegible] scheme is used to transmit the othe[illegible]

Rat[illegible]RC block code generated by $g(x)$ = 1[illegible]3 rate puncturing on the class 1 bits[illegible] is increased slightly by deleting eve[illegible] an 8/17 puncturing scheme.

To[illegible]oded speech bits is the (216,101) blo[illegible] 7.4.3.4).

7.4[illegible]ls

Ta[illegible]al channels and the physical burst typ[illegible]

Tab[illegible]ls and physical layer bursts (V+D)

Logical channel		Bits	Physical layer burst*	Bits
Traffic CHannel				
7.2 kbps net rate	TCH/7.2	432	NB	432
4.8 kbps net rate	TCH/4.8	288	NB	432
2.4 kbps net rate	TCH/2.4	144	NB	432
Speech	TCH/S	432	NB	432
STealing CHannel	STCH	124	½ NB	216
Signalling CHannel (SCH)	SCH/F	268	NB	432
	SCH/HD	124	½ NB	216
	SCH/HU	92	CB	168
Broadcast Synchronisation CHannel	BSCH	60	SB	120
Access Assignment CHannel	AACH	14	BBK	30
Broadcast Linearisation CHannel	BLCH	–	BLK2 of NB or SB	–
Common Linearisation CHannel	CLCH	–	LB	–

* The available physical layer bursts for V+D are
Uplink – Normal burst (NB), Control burst (CB), Linearisation burst (LB)
Downlink – Normal burst (NB), Synchronisation burst (SB), Broadcast block (BBK)

7.4.7 Example of Coding Steps

As an example of the coding system, the full rate signalling channel (SCH/F) is considered. A block of information on this channel consists of 268 bits.

The first coding step is to encode these bits with a (284, 268) block code to form a 284 bit block. Four tail bits, all set equal to zero, are appended to give 288 bits.

The Rate Compatible Punctured Convolutional code with rate 2/3 encodes these 288 bits into 432 bits.

A (432,103) block interleaver re-orders the 432 bits into another 432 bits block, and then these bits are passed to the physical layer where they are scrambled with the scrambling sequence. The first 216 bits of the 432 bit block form the bits of block 1 (i.e. the first half of the burst) while the next 216 for the second half, block 2.

7.4.8 Coding Performance

The performance of the coding scheme depends on the radio channel conditions, and on not only the number of errors it introduces but also on their positioning. However, it is possible to define the performance of the coding system in some reference cases.

Table 7.8 gives the maximum Message Error Rates (MER) and Bit Error Rates (BER) for a class A mobile with a signal to interference ratio of 19dB (the lowest the system is rated to operate with). Two types of reference channels are considered (see Section 6.12.2), TU50 (Typical Urban channel at 50 kph) and HT200 (Hilly Terrain channel at 200kph).

Table 7.8 Maximum message error rates (MER) and bit error rates (BER) for a class A mobile with a signal to interference ratio of 19dB

Logical channel		TU50	HT200
AACH	MER	9%	16%
BSCH	MER	6%	10%
SCH/HD	MER	7%	9.2%
BNCH	MER	7%	9.2%
SCH/F	MER	6.5%	7.5%
TCH/7.2	BER	2%	3.8%
TCH/4.8 N=1	BER	4%	4%
TCH/4.8 N=4	BER	1.2%	4%
TCH/4.8 N=8	BER	0.4%	4%
TCH/2.4 N=1	BER	1.2%	1.3%
TCH/2.4 N=4	BER	0.02%	0.4%
TCH/2.4 N=8	BER	0.01%	0.2%
STCH	MER	7%	9.2%

N gives the interleaving depth in number of blocks. Note that increasing the interleaving depth increases the performance, as would be expected.

Table 7.9 gives a comparison of the TETRA coding scheme with the error control coding schemes of TETRAPOL and APCO25. It can be seen that the coding scheme used by TETRA is by far the most flexible in terms of the availability of different error correcting parameters. This means that TETRA can adjust to different data transfer requirements and levels of protection, although it does result in a more complex system. APCO25 includes some options, while TETRAPOL has relatively limited options. The FDMA structure of TETRAPOL limits data communication capabilities (see Section 6.3.2), and TETRAPOL data services have a low rate. However, a high level of protection is used, which is in line with TETRAPOL's low minimum signal to noise and signal to interference requirements. This means that TETRAPOL can operate with larger cells and reduced infrastructure requirements.

Table 7.9 Comparison for different PMR coding systems

How TETRA compares...

TETRA	TETRAPOL	APCO25
Options for different coding rates and error protection Options for different interleaving depths	Rate 1/2 convolutional code ⇒ low data rate of 3.2 kbit/s Better error protection No interleaving	Rate 1/2 (unconfirmed) Rate 3/4 (confirmed) Interleaving

7.5 STEALING

It may be the case that when user data is being transmitted, urgent signalling information needs to be transmitted. If this signalling cannot wait until frame 18, or if the capacity available in frame 18 is insufficient, another solution must be found, and this is send the data on the traffic channel instead of the user data. This process is called 'stealing', and is indicated using the Slot Flag. The logical channel of the signalling data sent in place of the traffic data is the STealing CHannel (STCH). The channel coding on the STCH is different from the TCH it replaces (see Section 7.4.3), but the burst format is the same (i.e. normal uplink, downlink or direct mode burst).

Stealing operates on a half slot basis. The Slot Flag indicates whether stealing has occurred, in which case the first half slot will contain signalling information rather than user data. Stealing can take both half slots; this is indicated by the message in the first half slot. Note that speech coding normally interleaves the two speech blocks over a burst, but if the first half slot is stolen, the first speech block is discarded and the second speech block is transmitted in the second half slot using the special scheme described in Section 7.4.5.

Note that stealing only applies to traffic channels, and not to the control channels. This is because if priority signalling is to be accommodated on a signalling channel, the transmitter can schedule it ahead of other signalling messages.

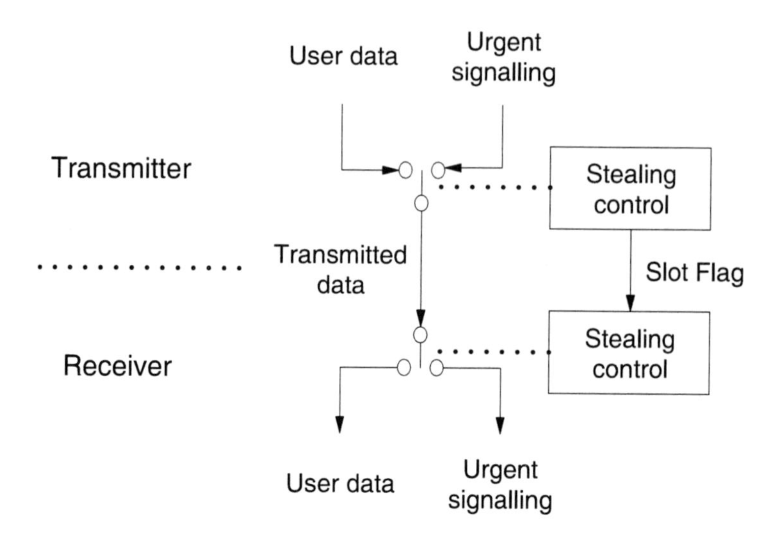

Figure 7.34 Stealing system overview

Stealing replaces the information that would have been transmitted. This information is discarded rather than queued, and if the MAC receives the STCH rather than the TCH, it informs the receiving layer 3 that the information has been discarded. The application can then take necessary action to re-send the lost information if it needs to. This is done so that the constant delay characteristics of the speech or circuit switched data are maintained, but means that stealing degrades the quality of the traffic channel. Frequent stealing should therefore be avoided. Packet switched data is sent on the signalling channels, and as such is subject to rescheduling rather than stealing.

Control messages have a priority which allows messages requiring stealing to be identified. These priorities can define that a message should be sent within a number of frames rather than immediately to allow the MAC to preserve TCH capacity.

Stealing can take place in V+D mode and in Direct Mode. In V+D mode, there are subtle differences in the operation of stealing between the uplink and the downlink, because on the downlink messages can be directed to mobiles other than the mobiles receiving the original traffic channel.

7.5.1 Use of Stealing on the Uplink

In the uplink, stealing can only take place by the mobile authorised to transmit on the corresponding traffic channel. Stealing may take place for a number of different reasons, in particular for passing urgent layer 3 signalling messages (terminating a call, for example), or to send user end-to-end signalling.

Stealing is indicated by the Slot Flag in the normal way. When the MAC in the base station decodes the STCH, it is able to direct the message accordingly.

7.5.2 Use of Stealing on the Downlink

Unlike the uplink, the downlink can be received by many mobiles. The base station may make use of this fact when stealing. Stealing could be used to send signalling relating to the call of the TCH that is being stolen (as in the uplink), or it could be used to send other information (such as an urgent call waiting indication) to a member of the group receiving the call the TCH relates to. In certain cases stealing may be used to send messages to other mobiles if it is known they are listening to the relevant slots. In this case, stealing would be used to poll a mobile if the MCCH was replaced by a TCH in minimum mode (see Section 7.11.3). This means mobiles that are receiving the common or associated signalling channel need to check whether the messages are addressed to them and, if they are, process the messages as appropriate.

7.6 TRUNKING METHODS

In a non-trunked system, channel allocation is relatively simple, since a specific channel is used by a terminal and if it is not free, the transmission must be delayed or aborted. In a trunked system, a pool of channels are shared by many terminals, and any terminal can access any spare radio channel, which is returned to the common pool when the transmission is completed. Due to the fact that channels are shared in this common pool, it is likely that other users may be able to make use of released resources, even for short periods. This raises the issue of when the channel is released – whether at the end of the complete call, or at the end of individual transactions (activations of the pressel). The TETRA standard supports three different methods – message trunking, transmission trunking and quasi-transmission trunking. It is up to the network operator or manufacturer to decide which trunking method to use, as different methods will provide solutions depending on whether it is desired to optimise resources use, minimise access time or minimise call dropping, or some other characteristic. Note that these different trunking methods only affect the allocation of traffic channels by the base station. The mobile does not perceive any difference between the different strategies, but simply carries out channel change instructions as directed by the infrastructure.

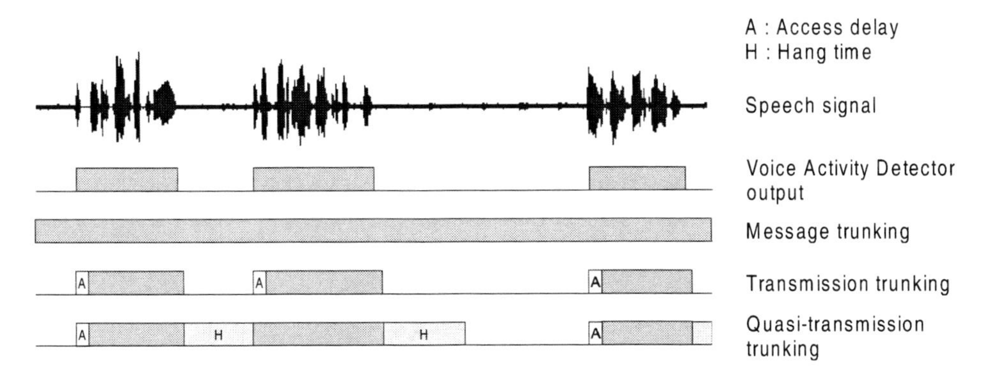

Figure 7.35 Different trunking methods

7.6.1 Message Trunking

Message trunking is a traffic channel allocation strategy in which the same traffic channel is continuously allocated for the duration of a call. This may include several separate call transactions i.e. pressel activations by separate terminals. The traffic channel is only de-allocated when the call is explicitly cleared by the call owner in the case of a group call, either party hanging up during an individual call or if an activity timer expires.

The main advantage of message trunking is that once a traffic channel has been allocated the users will not experience any delay at each "over" (each new call transaction) since there is no queuing for the allocation of channel resources. This absence of delay ensures that a conversation can proceed without interruption. Message trunking also minimises processing and signalling overheads in the infrastructure, and fits well with telephone calls to external networks, which will almost always use message trunking.

However, message trunking makes inefficient use of the radio resource, as the channel is still allocated even between pressel activations when there is no speech to transmit. This means that there are fewer traffic channels available for other conversations, and initial call set-up times will be longer, particularly when the system is heavily loaded.

7.6.2 Transmission Trunking

Rather than set up a channel for the duration of a call consisting of several transactions, with transmission trunking a traffic channel is allocated only for the duration of each individual transaction (i.e. for each activation of the pressel). The traffic channel is de-allocated at the end of each transaction and control signalling for the next transaction takes place on the control channel.

Transmission trunking makes efficient use of the available traffic channels since a traffic channel is only allocated when the users are actually speaking. Studies of conversations indicate that on average people only speak for about 45% of the time – the remaining time is made up of the silence while they listen to the other party [4]. Transmission trunking is possible on a "talkspurt" level, which de-allocates the channel even for short silence periods (orders of tenths of seconds), but to operate effectively such systems need to be able to allocate channels very quickly (tens of milliseconds) and trunk over a large number of voice calls. This is more appropriate in a cellular environment, and such systems are proposed for future public cellular systems.

PMR transmission trunking is usually based on pressel activation. Studies have shown that on average such activations follow a negative exponential distribution with a mean of about five seconds [7].

The problem is that if the channel needs to be allocated on each pressel activation, there is a delay while the channel is set up. If the available traffic channels are heavily loaded, there may be a noticeable delay before the channel can be re-established, and this may prove disconcerting to users and result in a loss of the flow of the conversion. It might even extend the conversation. Also, on systems with few traffic channels, the averaging effect between users will be less, contributing to the difficulty in obtaining a channel.

The only way of avoiding these problems is to restrict the allocation of channels so that there is a very high probability that a traffic channel will be available for an ongoing conversation. On systems with few traffic channels, there may be little, if any, gain over transmission trunking.

Transmission trunking is not restricted to point-to-point radio or dispatch mode calls. Calls to the PSTN can use transmission trunking on the air interface even though the network to which the call is connected is using message trunking.

7.6.3 Quasi-transmission Trunking

A compromise between the two approaches of message trunking and transmission trunking is *quasi-transmission trunking.* Here a traffic channel is allocated for each call transaction but the channel de-allocation is delayed for a short period, called the channel "hang" time, at the end of each transaction (i.e. after each pressel release). During the hang time, the TCH reverts to a FACCH, and may be re-allocated as a TCH for a new call transaction that is part of the same call. If the "hang" time expires the traffic channel is released and the mobile returns to the common control channels for signalling. Quasi-transmission trunking therefore offers a compromise between message trunking and transmission trunking, offering some of the improved traffic throughput of the latter without as great a risk of breaking up the flow of conversations.

However, satisfactory operation relies on careful selection of the channel "hang" time. Some existing systems use times of around two seconds, but this is about 40% of the mean transaction time for pressel based transmission trunking, which means that there is still a significant overhead. Selecting a suitable "hang" time may be particularly difficult if the system carries a mix of traffic, for example dispatcher traffic and point-to-point conversations. Users with short query-response transactions may find that they are inhibited from access to the system by longer two-way conversations which are effectively message trunked.

7.7 RANDOM ACCESS

When a mobile wants to transmit information, some method of allocating resources for the transmission must be used. Unless each mobile in the system is assigned a unique radio resource for its transmissions, the mobile has to transmit on a common radio resource in order to request resources for its transmission. This process is called random access.

There are a number of possible strategies for this process, and before looking at the possibilities it is useful to review the relevant requirements for random access. The main requirements are:

- That the system should be reliable.
- The protocol should be stable.
- Ongoing calls should not be interrupted (except by planned pre-emption).
- That delays should be minimised – quick call set up is an important requirement in safety related PMR applications.
- Resources should be used efficiently.
- It should be possible to change the access priorities of different mobiles, and therefore their grades of service.

In an FDMA system it is not possible to assign a different radio resource for the exclusive use of a single mobile (thus avoiding the access problem) without a very inefficient use of the radio resource and imposing a considerable restriction on the number of mobiles in the system. In a TDMA system, such a system is possible and is called polling. Each mobile would be assigned a different slot in a very large TDMA frame (like a hyperframe) and when it wishes to start a call it transmits in this slot. A slot is required for each mobile which may become active. The access slot allocation to a mobile, and therefore the priority can be broadcast by the base station and can therefore be changed dynamically, and since slots are assigned to only one mobile it avoids the problem of more than one mobile trying to access the system on the same radio resource and meets the stability requirement. However, with anything more than a very small number of mobiles the delays involved as the mobile waits for its unique slot can be very long, and resource use is poor, since access slots are assigned to mobiles whether they are required or not. Since access attempts are relatively rare, the access slots will normally not be used.

To avoid this waste in resource, random access may be used. In this case there exists a common access resource (i.e. radio channel or set of slots) which is shared by all the mobiles. The fact that it is shared means that it is possible that more than one mobile will wish to transmit a message to the base station at the same time and the messages will overlap. Such messages are said to "collide".

A simple random access protocol is ALOHA (see Section 4.7.1). Here transmitters send out information on a shared transmission medium with no concern as to whether or not a transmission is taking place. If no collision takes place, then the transmission is successful, but if any other transmission takes place during the transmission then there will be a collision.

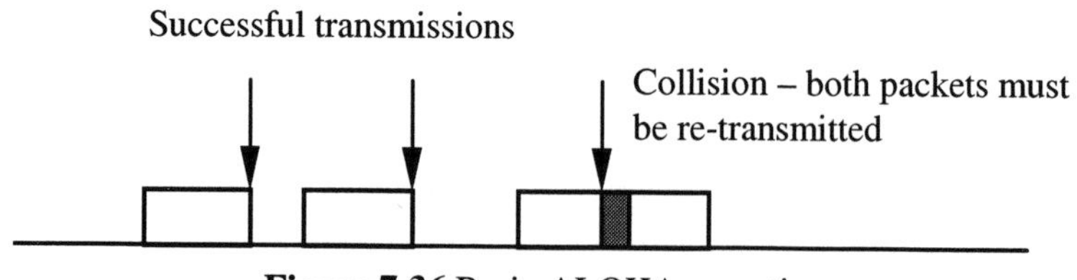

Figure 7.36 Basic ALOHA operation

ALOHA is very inefficient. In order to get through without error, no other transmission must have been ongoing, nor must one start for the duration of the packet. This will only work if transmission is not in fact very likely.

An alternative approach is to listen to the channel before transmitting, and only transmit if it is clear. This is carrier sense multiple access, or CSMA. Collisions can still occur if two parties decide to start at almost the same time, but this problem can be reduced by getting the transmitters to listen to the signal to see if a collision occurs. If it does, the transmission stops and the transmitter starts again after a random period. This is CSMA with collision detect – CSMA/CD.

7.7.1 Random Access for Direct Mode Operation

Direct Mode random access uses CSMA. Resources are not assigned by the infrastructure, and any Direct Mode mobile is permitted to use any radio carrier it can tune to. Channels can be free, occupied or reserved. A free channel is available for use. An occupied channel is in active use by a group or individual or group call. After a call transaction the channel is placed in a reserved state where the Direct Mode master – the mobile initiating the transaction – continues to provide synchronisation until a time-out or another mobile takes over the call.

Mobiles detect if a channel is free by attempting to decode synchronisation bursts on the channel. If the mobile detects that a signal is present, it will mark the channel as occupied, even if the reception of the signal is too poor to successfully decode it.

7.7.2 Random Access for V+D Mode Operation

In V+D, the situation is more complicated. On the downlink, resources are controlled by the infrastructure, so when the base station has to transmit to the mobile it can assign resources as necessary. Random access is not required.

Random access is required on the uplink, but the CSMA scheme used in Direct Mode is unsuitable. The problem is that CSMA depends on a truly common transmission medium, because each party has to be able to detect is someone is transmitting and if a collision occurs. A radio channel does not meet these criteria because shadowing may mean that a mobile will not detect another mobile's use of the channel. If it starts to transmit, however, it would interfere with the signal of the other mobile. This is known as the "hidden terminal" problem.

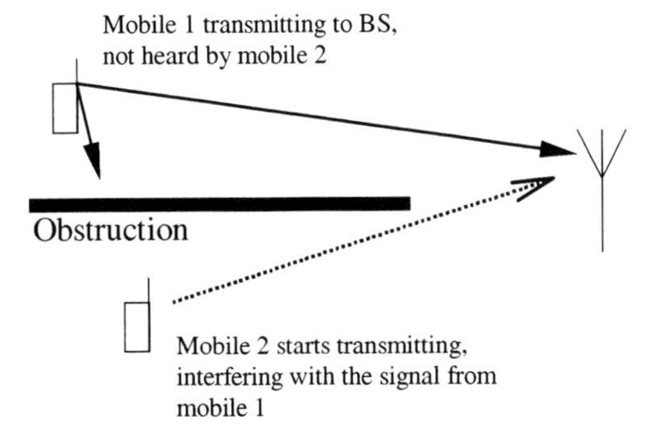

Figure 7.37 Problem of CSMA in mobile environments

The hidden terminal problem may occur in Direct Mode, but over the short ranges involved it would be unlikely that a signal would be hidden to the point that a mobile would not detect it, even if the signal was of too low a strength to decode. However, over the area of a cell in V+D operation, the hidden terminal problem cannot be avoided. This means that ALOHA must be used for random access in V+D. A refinement of the basic ALOHA scheme is to use slotted ALOHA to reduce the period vulnerable to collision as the TETRA access messages are all the same size. The improvement this gives is described in Section 4.7.1.

Random access takes place on separate control channels, so that any collisions do not affect on-going calls. The bursts used are control uplink bursts, and so there are two subslots per full slot. This increases the number of access opportunities.

The TETRA random access protocol is very flexible, and by a suitable choice of access parameters, it is possible for the base station to achieve all the requirements listed above. In particular, the slots which mobiles can use for access are controlled. This means that different groups of mobiles can be prioritised, and this priority can be changed.

Random access is only used for unsolicited messages from the mobile, whereas messages solicited by the base station are generally sent in a slot reserved by the base station for the response. This is called a reserved access, and is also used for other services which have been prearranged with the base station, such as packet data transfer.

In order to access the system, the mobile waits for an access slot which it is allowed to use. A mobile will continuously monitor the downlink to receive information of the access slots which are available to it so that it will be ready when it does need to access the system. Unlike some other mobile systems which use slotted ALOHA with access slots, the access slots are grouped into four different groups A, B, C and D using "access codes" on the Access Assignment CHannel (AACH). Mobiles may only use the access slots with the access code they have been assigned. This system allows TETRA to offer different grades of service and priorities to different groups of mobiles.

Mobiles are allocated to groups in a process called binding, grouping together group subscriber identity, subscriber class and priority (see Section 4.7.2). This is an operator option and the definition of the binding parameters is broadcast periodically in an ACCESS DEFINE message (see Section 7.7.4.1).

The fact that the binding of mobile stations to access codes is dynamic makes the system very flexible. The binding defines the minimum valid priority for an access code. It may also restrict use of the access code to a set of subscriber classes, or to a group of mobile stations. A mobile may use a subslot designated for a particular access code only if the message priority and the subscriber class or mobile identity conform to the current binding.

Take the following example (Figure 7.38). Mobiles may access only in subslots they are allowed to use, so mobiles from group A have 6 opportunities, B have 3, C have 2 and D 1.

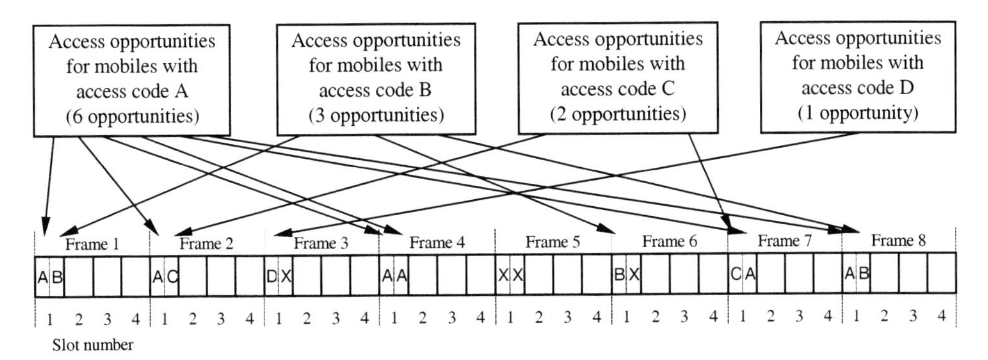

Figure 7.38 Uplink access subslots example

Group A would have the highest priority and grade of service (assuming that there were not very many more mobiles in that group) as it has the most opportunities. Group D has a low priority, and on average mobiles in this class would have to wait longer to access the system. Note that X indicates that there is no access opportunity (which would be the case if the slot was reserved or if it is used for the CLCH (Common Linearisation CHannel)).

The system must be designed so that there are not too few access slots in a particular group, as if mobiles collide in their access attempts they will have to retry. If this happens, the retries will increase the number of attempts and cause instability. However, since the access opportunities can be changed dynamically it is possible to adapt the system to counteract this problem if it occurs.

Obviously, should a collision occur, mobiles cannot simply retransmit in the next permitted slot because the colliding mobile would do the same and the collision would re-occur. To counter this, and decrease access times (the time taken to gain access to the system), mobiles access at a random point within "access frames", consisting of a number of access slots.

7.7.3 Access Frames

Mobiles can only belong to one access group at a time and can only use subslots relevant to their particular access code. However, mobiles may not necessarily make a request on the first subslot corresponding to their access code after they wish to access resources. Rather, requests are made at a random from one of the subslots of that class within an "access frame". The access field in the ACCESS-ASSIGN message indicates the number of following uplink subslots which make up an access frame for this access group. A special value ("ongoing frame") is used when the field does not mark the start of a new access frame.

When a user access request is initiated, for example a valid request for access code A, the mobile MAC is permitted to send a first random access request in the next available code A subslot provided that this occurs within a designated time (given by the ACCESS-DEFINE parameter, IMM). It should be noted that with slotted ALOHA, a request must be made at the start of a subslot – it cannot be made at some point during the slot, so if a subslot has already started when the request was initiated, the mobile must wait until the next subslot of that class.

If an immediate first access attempt is not made then the mobile MAC must wait for an ACCESS-ASSIGN message containing a frame marker for its access code, and then choose a subslot randomly from that access frame for its access request. If a mobile is unsuccessful in its access attempt, it must again wait until the next frame marker and choose another subslot randomly from that frame if it wants to make another attempt. This is to try to ensure that colliding mobiles do not collide again in their next attempt.

There are two types of access frame – discrete and rolling. In the first case access frames do not overlap. The length of each frame is transmitted at the start of the frame and subsequent subslots are marked with an ongoing frames mark to indicate that the frame is continuing. A mobile will wait for a time, designated WT, after which is decides that the access attempt was unsuccessful. It will then wait until the start of the next access frame, note it size, and access at random on one of the access slots in the frame.

The second option is to have a rolling access frame. Here the length of the access frame is transmitted each access subslot, and the mobile then randomly attempts access at one of the next subslots making up that frame.

The following example illustrates the sequence of events for a mobile in group A.

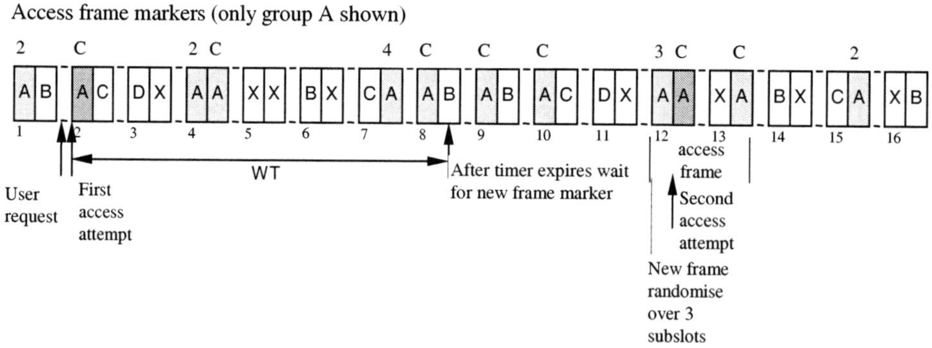

Access frame markers (only group A shown)

User request
First access attempt
WT
After timer expires wait for new frame marker
access frame
Second access attempt
New frame randomise over 3 subslots

Figure 7.39 Random access procedure using distinct access frames

The access subslots are the same as those given in Figure 7.38, but the other slots are omitted for clarity. Each access subslot contains a marker giving the length of the access frame from that point. "C" indicates that the previous frame is continuing. The mobile finds that it requires access during frame 1. The first subslot of frame 2 belongs to group A, and so the mobile can attempt access there. The mobile then waits WT, but in this example does not receive an acknowledgement. The mobile wishes to try again, but has to wait for the next access frame. The next A access subslot shows it to be the middle of an access frame, and so the mobile waits for the next access frame. This starts at the beginning of transmission frame 12. The access frame has length 3, so the mobile has to pick one of these three to make an access. It randomly chooses number 2, and accesses again then.

The access frame markers are transmitted as part of the access assign message in the AACH (see 7.7.4.2). It is therefore received in the corresponding downlink slot, which is two slots ahead of the uplink slot. Mobiles therefore have time to contend on the first slot of the access frame if they choose this slot in the random choice procedure.

The second example, shown in Figure 7.40 is similar, but in this case rolling access frames are used. Access frames overlap, so all (or most) access subslots start an access frame. In this case, the mobile does not have to wait for the access frame ongoing when WT times out to finish before the start of an access frame and the randomising process.

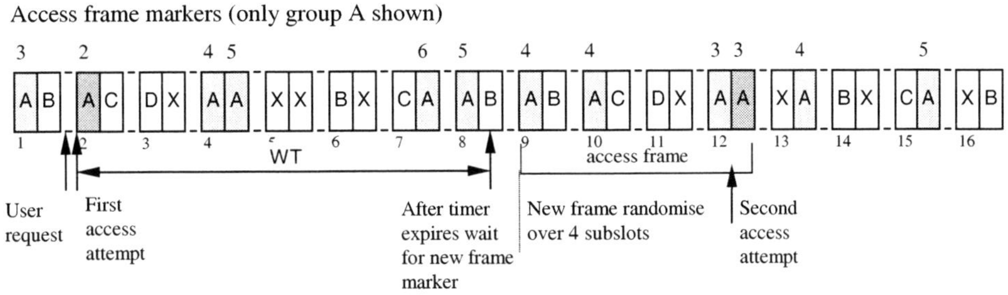

Figure 7.40 Random access procedure using rolling access frames

The option of using discrete or rolling access frames is not defined in the standard; the choice is up to the operator. Rolling access frames offer the possibility of faster access for a given size of access frame, but are more difficult to control, since estimation of the number of mobiles requiring access at any point in time is more complicated than in the discrete case. Irrespective of the type of access frame, the optimal size of the access frame will depend on the number of mobiles contending for access and the traffic mix [10].

7.7.4 Access Control Channels

Access opportunities are defined by ACCESS-ASSIGN and ACCESS-DEFINE messages. ACCESS-DEFINE messages contain the slowly varying information on the binding, etc., whereas ACCESS-DEFINE messages specify the access opportunities themselves.

7.7.4.1 The ACCESS-DEFINE message

The ACCESS-DEFINE message is transmitted at intervals on the BNCH (see Section 7.8.2.3). It contains slowly changing information about the random access parameters for an access code. The message includes:

- The priority and mobile binding to the access code.
- A parameter (IMM) defining when immediate access is permitted for the first transmission. There are three option: mobiles are always permitted to access

immediately when first accessing; or they never are (i.e. they must always randomise over an access frame); or they may transmit immediately if an access opportunity comes up within a set number of frames (between 1 and 14).

- The waiting time (WT) before the mobile decides to retry. This is counted in terms of access opportunities for the mobile and can be between 1 and 15.
- The permitted number of retries before the mobiles abandons the random access attempt. Valid values are between 0 and 15. If the value is 0 the mobile never retries – it is only successful it its access attempt if it works on the first attempt. While this avoids any additional load on the access channel from retransmissions it results in poor performance due to a high rate of blocked calls.
- A framelength multiplier flag. This value is used to scale the access frame length values transmitted on the AACH. The values given in the access field on the AACH (see Section 7.7.4.2) are either used as they are, or multiplied by a factor of 4, which makes the access frames much longer for use when the system is congested.
- The uplink channel configuration. This is used to inform the mobile on which frames on the downlink it should monitor the AACH.

7.7.4.2 The ACCESS-ASSIGN message

The ACCESS-ASSIGN message is transmitted on the AACH in the broadcast block of every downlink slot. It has two purposes: to specify the use or destination of that downlink slot, and to give access rights for the corresponding uplink slot.

The uplink slot structure is the same as that on the downlink but follows it in time by two slots. This means that when slot 1 of frame 2 is being transmitted on the downlink, slot 3 of frame 1 is being transmitted on the uplink. The access rights of a particular slot are sent in the corresponding (i.e. same numbered) slot on the downlink. This means that there is a delay of two slots between the slot when the access rights are conferred and the slot that they correspond to, so that the mobile has time to prepare and does not have to transmit and receive at the same time.

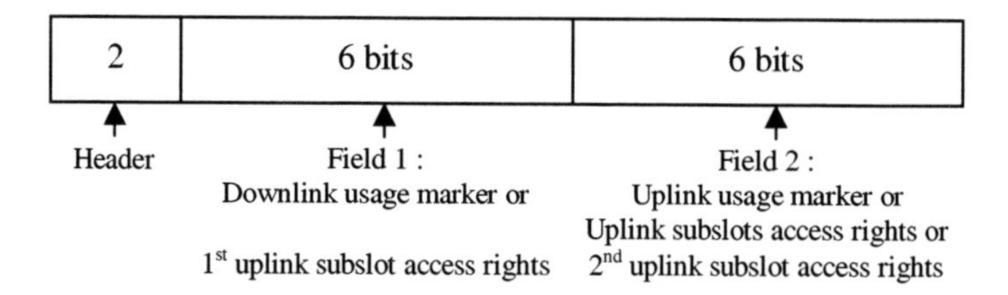

Figure 7.41 ACCESS-ASSIGN message structure

The ACCESS-ASSIGN message consists of three sections: a two-bit header followed by two six-bit fields (see Figure 7.41). These fields either contain usage markers or access fields. An access field defines the use of uplink access subslots, giving the access group to which the subslot is assigned and the access frame length. It can also indicate a CLCH subslot (see Section 6.6.3) or reserved subslot. The usage marker identifies the traffic in the slot, which for the uplink implies permission for mobiles to use the slot. It also identifies reserved, common and assigned control slots. Normally, the first field contains the downlink usage marker and the second field contains the uplink usage marker. If the downlink is not assigned, both fields can be used for access fields and so different access rights can be given for each uplink subslot. However, if the downlink is assigned but the uplink is used for access, the first field contains the usage marker for the downlink slot and the access field applies to both subslots.

7.7.5 Reserved Access Procedures

The major problem with access is the initial access, when the base station is unaware of the fact that the mobile requires a channel. However, once this initial random access to the system has been performed, all subsequent accesses can use reserved accesses. These exchanges can be planned, which avoids the problem of contention, and makes more efficient use of radio resources. When a mobile is required to respond to the base station, or when it has further signalling to send after the initial access, the base station may reserve slots for that particular mobile, possibly on request from the mobile.

The AACH indicates which subslots are reserved and therefore not available for random access by other mobile stations. The mobile for which a subslot or slot(s) are reserved is informed separately on the downlink signalling channel.

7.7.6 Independent Allocation of Slots on the Uplink and Downlink

Unlike traditional systems where uplink and downlink radio channels are paired to form a single logical channel, TETRA allows uplink and downlink channels on the same slot of the same frequency pair to be allocated for different purposes. If, for example, a group call originates in one cell and has receiving mobiles in another cell, only a downlink is required in that second cell (see Figure 7.42). The corresponding uplink channel may be allocated to call which only requires an uplink channel in that cell, or for signalling.

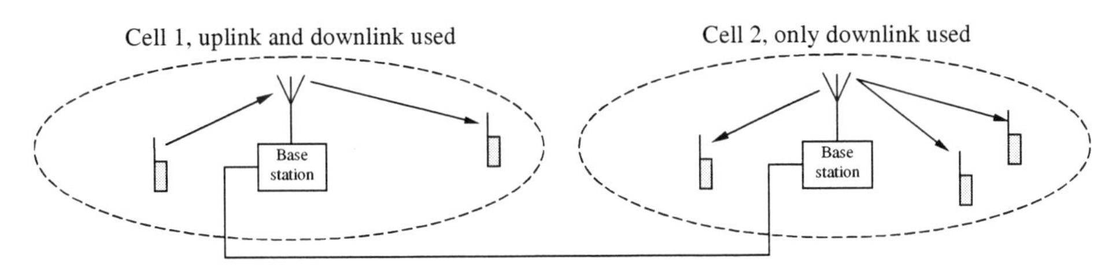

Figure 7.42 Group call spanning two cells

The ACCESS-ASSIGN message in the AACH gives access rights to uplink and downlink slots. Permitted combinations on normal channels are as follows:

1. Uplink and downlink assigned to the same circuit mode call.
2. Uplink and downlink assigned to different circuit mode calls.
3. Uplink assigned to a circuit mode call, with the downlink assigned to a Secondary common Control CHannel (SCCH).
4. Uplink assigned to a SCCH, with downlink assigned to a circuit mode call.

In addition, in minimum mode (see Section 7.11.3), it is possible to allocate only one direction of the MCCH to a circuit mode call, rather than both directions. The unused direction is available for normal MCCH use.

7.8 LOGICAL CHANNEL ROUTING

7.8.1 Introduction

As discussed in Section 4.5, TETRA has a complex hierarchy of logical channels. At the higher layers it is more convenient to treat different signalling channels, for example, as different logical channels. At lower layers of the system, the number of logical channels reduces to the point where at the physical layer there is only a Traffic Physical, Control Physical and Unallocated Physical channel. The mapping between high-level logical channels and the lower levels is carried out by the MAC.

In addition to the information passed to the MAC from the upper layers, the MAC also generates its own messages to transmit for peer-to-peer communication, and also provides some resources (messages) to support to operation of layer 1. The upper MAC multiplexes control and user information from all these sources before passing it in a restricted number of forms to the lower MAC for coding, from where it is passed to layer 1 for transmission.

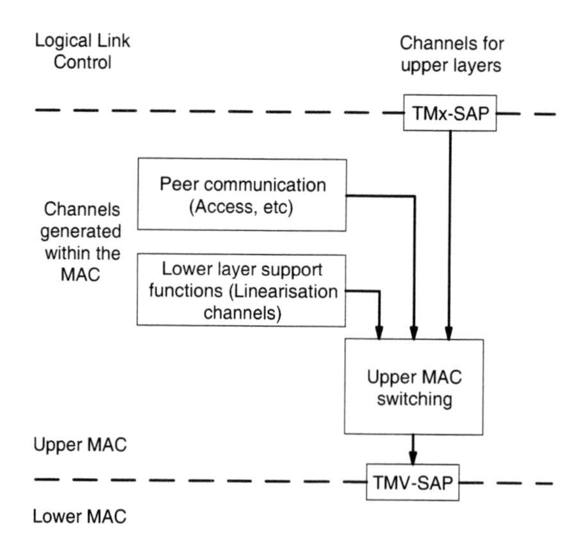

Figure 7.43 Upper MAC switching functions

7.8.2 Logical Channels Passed to the Upper MAC

The information that the MAC has three different types. Each type of information is passed from the LLC via a different SAP.

- User traffic, for example speech or circuit switched data, which arrives at the MAC via the TMD-SAP.
- Control messages intended for a specific mobile, and packet data, which arrives at the TMA-SAP
- Broadcast control messages intended for a number of mobiles (and which are therefore unacknowledged and occur only on the downlink) which are routed to the TMB-SAP.

7.8.2.1 Traffic channels

TETRA has several different types of traffic channels for circuit mode voice and data traffic (packet mode data uses the Common Control CHannel). The traffic channels are

- Speech Traffic CHannel (TCH/S)
- 7.2 kbps net rate data Traffic CHannel (TCH/7.2)
- 4.8 kbps net rate data Traffic CHannel (TCH/4.8)
- 2.4 kbps net rate data Traffic CHannel (TCH/2.4)

It is possible to group together traffic channels to increase the net rate. Data rates of up to 28.8 kb/s, 19.2 kb/s or 14.4 kb/s can be achieved by combining 4, 3 and 2 consecutive slots on the same carrier.

7.8.2.2 Mobile specific signalling channels

There are two basic types of signalling channel – common control channels which can be assessed by any mobile as required and associated control channels which are paired with traffic channels.

Common control channels are used by mobiles which are not active, or which are active but have no traffic channel. These channels are also be used to transmit packet mode data, which allows data to be transmitted without setting up a circuit switched traffic channel.

There are two different types of Common Control CHannel (CCCH) as follows:

- Main Control CHannel (MCCH)
- Extended Control CHannel (ECCH)

The common control channels are always present. One carrier at each base station is designated the main carrier, and slot 1 on this carrier is used for the Main Control CHannel (MCCH). In systems with only a few carriers, this could be unacceptably inefficient, so TETRA defines a “minimum mode”, where slot 1 on the main carrier is used for mobile traffic if no other spare slots are available (see Section 7.11.3). In such cases, the system can steal from this traffic channel to provide main control channel signalling as and when required. The MCCH is used, among other things, to provide opportunities for mobiles to access the system, and for a system with a large number of carriers a single slot on a single carrier may not provide sufficient capacity. If this is the case, TETRA can use an extended mode where additional common control channels called Extended Control CHannels (ECCH) are provided on other carriers, again in slot 1. Different system modes of operation are discussed in Section 7.11.

The other type of signalling channel is the Associated Control CHannel (ACCH), which comprises:

- Fast Associated Control CHannel (FACCH)
- Stealing CHannel (STCH)
- Slow Associated Control CHannel (SACCH)

The associated control channels are paired with a traffic channel. The normal signalling channel is the Slow Associated Control Channel (SACCH), which is the name for the signalling sent in the 18 frame of the multiframe. The SACCH is therefore always present when the corresponding traffic channel is in use.

At the start and end of an activity period on a traffic channel there is a requirement for additional signalling (to set up and clear down the call, for example). To allow for this, the traffic channel itself (i.e. the channel in slots 1 to 17) is used for signalling as the Fast Associated Control CHannel (FACCH). Note that the FACCH replaces the TCH, whereas the SACCH exists in combination with it.

When the traffic channel is in use, there may still be a requirement to send additional signalling, either because the SACCH does not have a large enough capacity or because the message cannot wait until the next occurrence of frame 18. In this case bursts can be stolen from the TCH to send signalling information, and this forms the STealing CHannel (STCH).

The STealing CHannel (STCH), also, therefore exists in combination with the TCH. On the uplink, only the transmitting mobile can steal, and would do so either for signalling to the infrastructure or to the end user. However, on the downlink, the system can steal capacity for sending signalling messages either to the transmitting mobile or to any member of the ongoing call (this may for instance be used to signal an urgent call waiting message to an individual member of the group). In minimum mode (see Section 7.11.3), the system may steal capacity from the traffic channel using the MCCH resources to update the waiting mobile stations or to contact specific mobiles.

The specific control channel used depends on the mobile's activity. If a mobile does not have a traffic channel, signalling is conducted via the common control channels. It is through this that the mobile requests permission to transmit, or is signalled by the base station to receive.

When the base station assigns a traffic channel, it is set up as a Fast Associated Control Channel (FACCH). After initial signalling is carried out, the channel changes to a TCH (frames 1 to 17) and SACCH (frame 18). While the traffic channel is in use, it may be stolen for additional signalling. When the data has been transmitted or the speech activity is completed, the channel is stolen again to allow the TCH to be replaced with the FACCH for the signalling required to clear down the call or activity period.

An example of this process is should in Figure 7.44. The activity starts at ❶, when a channel is allocated, which is set up in signalling mode as a FACCH. The channel is in slot 2. Having conducted any necessary signalling, the channel is transferred into a TCH (❷). Each TCH has an associated SACCH in frame 18, which occurs at ❸, ❹, ❻, ❼ and ❾. In addition to associated signalling, frame 18 can also carry broadcast signalling (see Section 7.8.2.3). At ❺, stealing is undertaken to transmit urgent signalling. At ❽, stealing is undertaken to transfer to a FACCH to clear down the transaction, which ends at ❿.

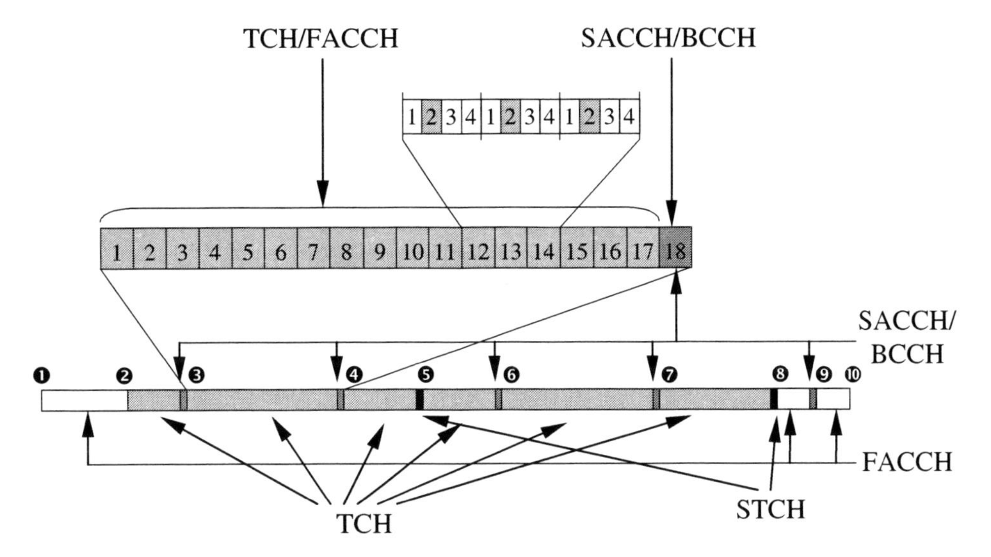

Figure 7.44 Use of control channels during an activity period

7.8.2.3 Broadcast signalling channels

The Broadcast Common Control Channel (BCCH) is used for two types of broadcast control channel:

- Broadcast Synchronisation CHannel (BSCH)
- Broadcast Network CHannel (BNCH)

These channels, which are present only on the downlink, transmit information about the serving base station. The Broadcast Synchronisation CHannel (BSCH) transmits the SYNC message. This provides physical layer synchronisation information, including an extended training sequence for synchronisation and slot alignment, frequency correction bits to set the carrier frequency, along with MAC layer information on frame timing (i.e. the number of the current slot, frame and multiframe) as well as network layer information. Due to the physical layer synchronisation, the BSCH is transmitted on a synchronisation burst rather than a normal downlink burst.

The Broadcast Network CHannel (BNCH) either carries the ACCESS-DEFINE message (see Section 7.7.4.1) or the SYSINFO message. The SYSINFO message gives the main carrier frequency so that mobiles can tune to it after picking up any SYSINFO message, information on any secondary control channels, power information and network layer information.

The broadcast control channels are normally transmitted in frame 18. Each takes up one block of 216 bits, leaving the other block within the burst for the SACCH on the corresponding assigned channel or for the base station to send signalling to other mobiles. On each slot, the sequence shown in Figure 7.45 is used, and the sequence is offset by one for each slot in the frame so that every frame 18 has one BSCH and one BNCH. However, transmitting the network information takes several half slot BNCH instances and it will take some time therefore for a mobile to acquire all the information it needs. This is not normally a concern since mobiles will be switched on before they are used and an initial delay can be tolerated. The standards allow the option of transmitting the BNCH on other physical control channels (indicated by the AACH) as necessary. Mobiles continue to monitor the BCCH after they are switched on and also during transmissions so as to ensure that they receive network information and remain synchronised.

Multiframe number	1	2	3	4
DL Frame 18, First half slot		BSCH *		
DL Frame 18, Second half slot		*		BNCH

* Transmitted on a Synchronisation Burst

Figure 7.45 BSCH and BNCH transmission sequence

7.8.3 Logical Channels Generated Within the Upper MAC

There are a number of logical channels present in the MAC either for the purpose of peer-to-peer communication or to support correct action of the physical layer. These are:

- **Access Assignment Channel** (AACH). This gives access rights on control channels and usage markers on traffic channels (see 7.7.4.2). This is carried on every downlink slot on the broadcast block. There is no corresponding channel in Direct Mode.
- **Linearisation channels**. These are not channels in the normal sense but simply a time interval allowed by the system for mobiles to linearise the power amplifiers in their transmitter (see Section 6.6.2). They may need to do this after switching to a new frequency or changing the transmission power level. In V+D, the CLCH (Common Linearisation Control CHannel) is used for linearisation. The CLCH can be mapped on to the Linearisation Burst (LB), a half slot burst which can occupy the first half slot of an uplink burst; the remaining half block of the burst can be used for a half slot signalling channel. There is also a corresponding channel, the Base station Linearisation Channel (BLCH), which can be used by base stations to linearise. Base stations may need to linearise if they are operating in discontinuous mode. The BLCH

is transmitted as the second block of a normal downlink burst or a synchronisation burst. Direct Mode has a Linearisation Channel (LCH) which is mapped to a DM Linearisation Burst (DLB). Unlike the LB, the DLB takes a full slot length.

- **Slot Flag** (SF). This is a one bit channel which is sent by using one of two possible training sequences in a normal up- or downlink burst (see Section 6.6.2). It is used to indicate whether a normal uplink burst or normal downlink burst contains a single logical channel or two logical channels, one in each block of the burst. In Direct Mode, the burst preamble changes as well as the training sequence. Possible combinations which can be used on normal bursts are shown in Table 7.10.

	Block 1	Block 2	Slot Flag
V+D uplink and downlink	TCH		No
Direct Mode	STCH	TCH	Yes
	STCH	STCH	Yes
	SCH/F		No
V+D Downlink only	SCH/HD	SCH/HD	Yes
	SCH/HD	BNCH	Yes

Table 7.10 Possible combinations of logical channels in a slot

Other block combinations are possible but they do not use normal bursts. These include BSCH+SCH/HD (or SCH/S + SCH/H in Direct Mode) on a synchronisation burst, SCH/HU+SCH/HU (two control uplink bursts) and CLCH+SCH/HU (linearisation burst and control uplink burst).

7.8.4 Logical Channel Routing to the Lower MAC

From the preceding sections it can be seen that there are a number of logical channels in the upper MAC for V+D mode.

- Main Control CHannel (MCCH)
- Fast Associated Control CHannel (FACCH)
- Slow Associated Control CHannel (SACCH)
- STealing CHannel (STCH)
- Traffic CHannel (TCH)
- Broadcast Synchronisation CHannel (BSCH)
- Broadcast Network CHannel (BNCH)
- Access Assignment CHannel (AACH)
- Common Linearisation CHannel (CLCH)
- Broadcast Linearisation CHannel (BLCH)

Many of these channels have the same requirements with regard to transmission and coding, and by grouping logical channels together it is possible to have a more restricted set of logical channels for the lower MAC to consider. The channels present in the lower MAC, and their corresponding upper MAC channels, are shown in Table 7.11.

Table 7.11 V+D Logical channels between the upper and lower MAC sub-layers

Channel		Direction	Corresponding upper MAC channel
Signalling CHannel (SCH)		Up and downlinks	MCCH FACCH SACCH BNCH
Access Assignment CHannel (AACH)		Downlink only	AACH*
Broadcast Synchronisation CHannel (BSCH)		Downlink only	BSCH
STealing CHannel (STCH)		Up and downlinks	STCH
Common Linearisation CHannel (CLCH)		Uplink	CLCH*
Base station Linearisation CHannel (BLCH)		Downlink	BLCH*
Traffic Channel	(TCH)	Up and downlinks	TCH
7.2 kb/s net rate 4.8 kb/s net rate 2.4 kb/s net rate Speech	(TCH/7.2) (TCH/4.8) (TCH/2.4) (TCH/S)		

* Generated within the upper MAC and not passed to higher layers

The range of logical channels in Direct Mode is far more restricted, lacking broadcast and access channels, and having a more restricted set of signalling channels. The mapping of Direct Mode channels between the upper and lower MAC is shown in Table 7.12.

Table 7.12 Mapping upper MAC SAP and the lower MAC (Direct Mode)

SAP	Definition		DMV-SAP Logical channel	Definition
DMA	Signalling CHannel		SCH/F	Signalling CHannel (Full slot)
			SCH/H	Signalling CHannel (Half slot)
	Synchronisation CHannel		SCH/S	Signalling CHannel (Sync.)
	STealing CHannel (signalling)		STCH	STealing CHannel (signalling)
DMD	Traffic CHannel	(circuit mode)	TCH	Traffic CHannel (circuit mode)
	7.2 kb/s net rate 4.8 kb/s net rate 2.4 kb/s net rate Speech	(TCH/7.2) (TCH/4.8) (TCH/2.4) (TCH/S)		
	STealing CHannel (user signalling)		STCH	STealing CHannel (user signalling)
MAC generated	Linearisation CHannel		LCH	Linearisation CHannel

A mapping between the different SAPs for the different logical channels at the upper MAC and the lower MAC is given in Table 7.13.

Table 7.13 Mapping upper MAC SAP and the lower MAC (V+D)

SAP	Definition	TMV-SAP Logical channel	Definition
TMA	Main or Secondary Control CHannel	SCH/F	Signalling CHannel (Full slot)
		SCH/HD	Signalling CHannel (Half slot Downlink)
		SCH/HU	Signalling CHannel (Half slot Uplink)
	Fast Associated Control CHannel	SCH/F	Signalling CHannel (Full slot)
	Slow Associated Control CHannel	SCH/HD	Signalling CHannel (Half slot Downlink)
		SCH/HU	Signalling CHannel (Half slot Uplink)
	STealing CHannel (signalling)	STCH	Stealing CHannel (Signalling)
TMD	Traffic CHannel (circuit mode)	TCH	Traffic CHannel (circuit mode)
	Stealing CHannel (user signalling)	STCH	Stealing CHannel (user signalling)
TMB	Broadcast Synchronisation CHannel	BSCH	Broadcast Synchronisation CHannel
	Broadcast Network CHannel	BNCH on SCH/HD	Signalling CHannel (1/2 slot downlink)
MAC generated	Access Assignment CHannel	AACH	Access Assignment CHannel
MAC generated	Common Linearisation CHannel	CLCH	Common Linearisation CHannel

7.8.5 Channel mapping in the lower MAC

The lower MAC does not perform any routing, as there is a mapping between logical channels at the TETRA MAC Virtual SAP and the Physical Layer burst. These mappings are shown in Table 7.14 and Table 7.15.

Table 7.14 TETRA Direct Mode mapping from the TMD-SAP to the DP-SAP and physical layer bursts

TMD-SAP Channel	Definition	Physical burst	Definition
SCH/S	Synchronisation CHannel	BKN1 of DSB	1st half of DM Synchronisation Burst
SCH/H	Half slot Signalling CHannel	BKN2 of DSB	2nd half of DM Synchronisation Burst
SCH/F	Full slot Signalling CHannel	DNB	DM Normal Burst
STCH	STealing CHannel	DNB + SF	DM Normal Burst and Slot Flag
TCH	Traffic CHannel	DNB	DM Normal Burst
LCH	Linearisation CHannel	DLB	DM Linearisation Burst

Table 7.15 TETRA V+D mapping from the TMV-SAP to the TP-SAP and physical layer bursts

TMV-SAP Channel	Definition	Physical burst	Definition
SCH/F	Full slot Signalling CHannel	NDB, NUB	Normal Downlink Burst, Normal Uplink Burst
SCH/HD	Half slot Downlink Signalling CHannel	NDB+SF BKN2 of SB	Normal Downlink Burst and Slot Flag 2nd half of Synchronisation Burst
SCH/HU	Half slot Uplink Signalling CHannel	CB	Control uplink Burst
STCH	Stealing CHannel	NDB+SF, NUB+SF	Normal Downlink Burst and Slot Flag, Normal Uplink Burst and Slot Flag
TCH	Traffic CHannel	NDB, NUB	Normal Downlink Burst, Normal Uplink Burst
BSCH	Broadcast Synchronisation CHannel	SB	Synchronisation Burst
AACH	Access Assignment CHannel	BBK	Broadcast BlocK
CLCH	Common Linearisation CHannel	LB	Linearisation Burst

The lower MAC has to perform the relevant coding and interleaving functions for each channel type, and then pass each to the physical layer at the correct time. It also has to ensure that the Slot Flag is sent by using the correct training sequence. The set of logical channels and physical layer bursts varies between the V+D downlink, V+D uplink and Direct Mode, and this is shown in Figure 7.46, Figure 7.47 and Figure 7.48 respectively.

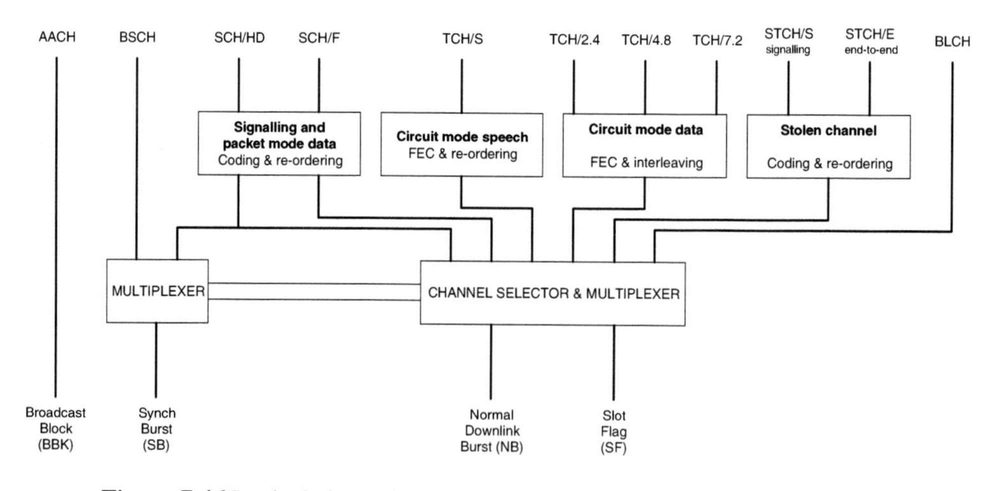

Figure 7.46 Logical channels supported in the lower MAC: V+D downlink

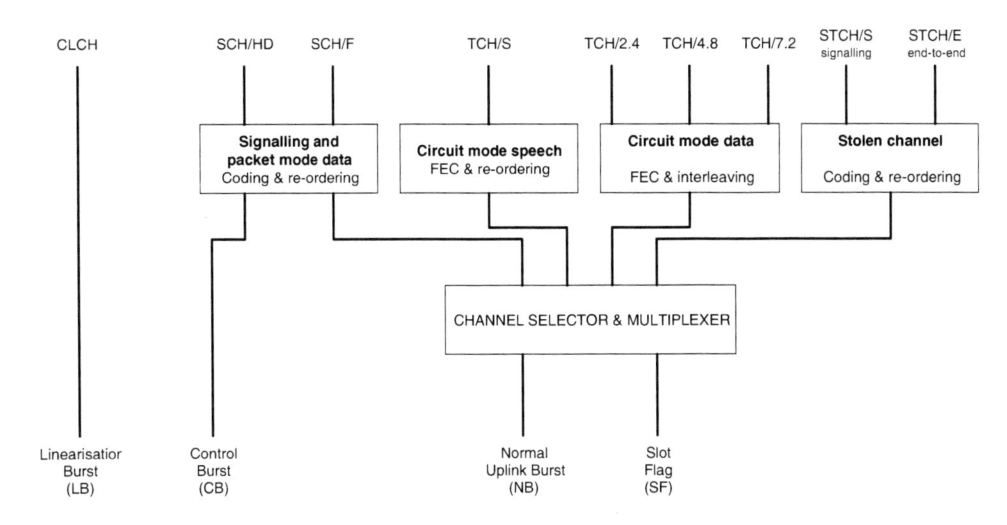

Figure 7.47 Logical channels supported in the lower MAC: V+D uplink

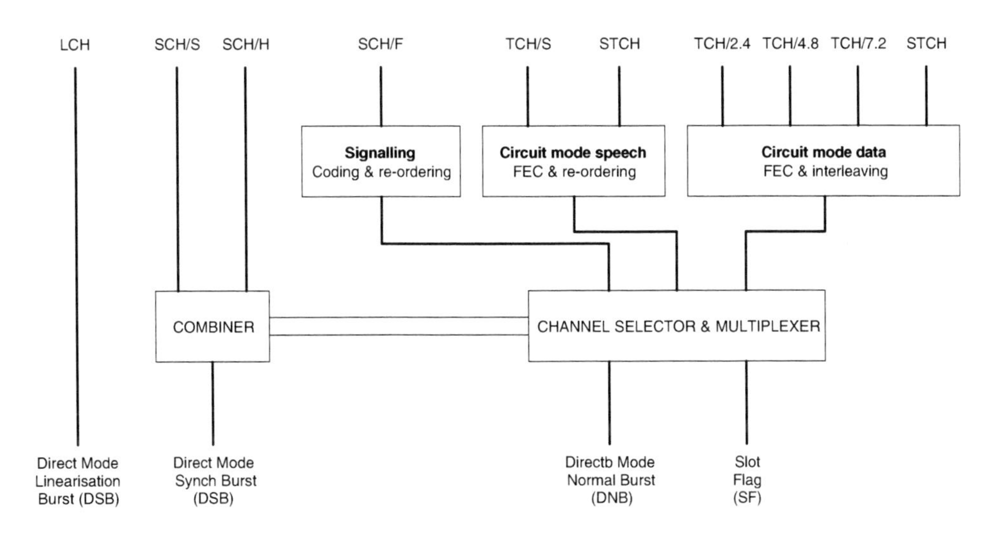

Figure 7.48 Logical channels supported in the lower MAC: Direct Mode

7.8.6 Inter-working Between MAC Layers

The various control channels inter-working within the MAC are shown in Figure 7.49, Figure 7.50 and Figure 7.51. It should be noted that the transformation caused by the upper MAC in order to prepare signalling for the lower MAC. The upper MAC receives messages to be transmitted from layer 3, and it has to decide how best to send them. It is therefore in control of this message passing. Also shown is TETRA MAC Control SAP (TMC-SAP) and TETRA Physical Control SAP (TPC-SAP), used for communicating layer management information for control purposes.

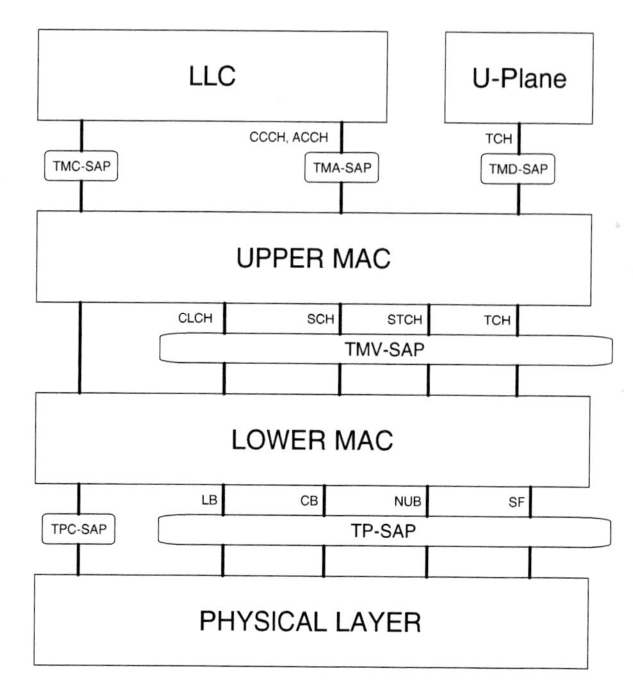

Figure 7.49 Inter-working between MAC sub-layers and logical channels for V+D uplink transmission

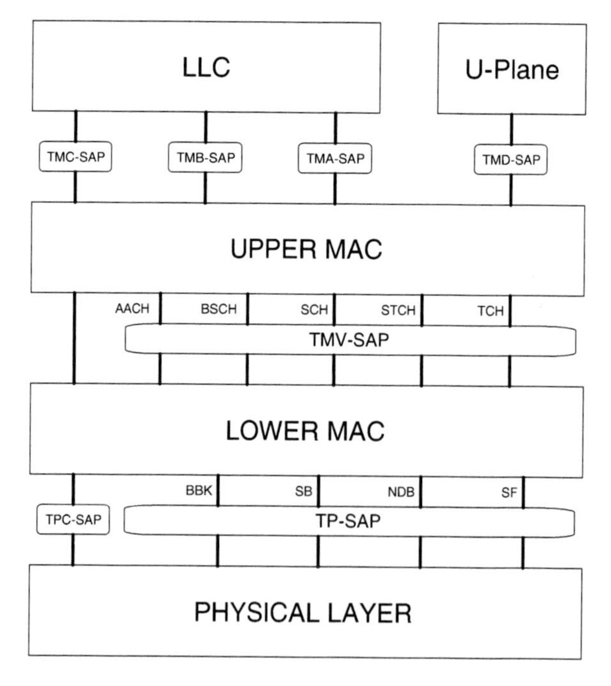

Figure 7.50 Inter-working between MAC sub-layers and logical channels for V+D downlink transmissions (i.e. MS receive)

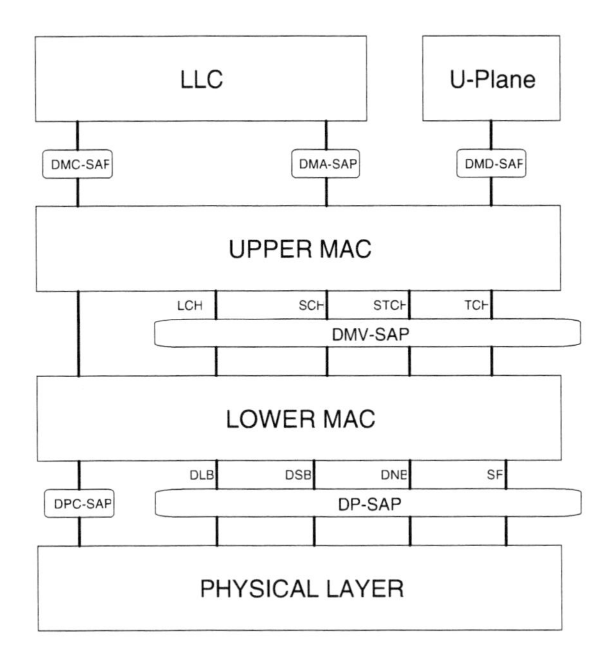

Figure 7.51 Inter-working between MAC sub-layers and logical channels for Direct Mode operation

7.9 AIR INTERFACE ENCRYPTION

Air interface encryption is one of the comprehensive security features of the TETRA standard, which also includes end-to-end encryption and authentication (see Section 9.2).

The air interface encryption scheme encrypts the TETRA MAC SDUs by exclusive-ORing each of them with a Key Stream Segment (KSS) generated from a cipher key and an offset. Only the SDU itself is encrypted, the MAC header remains in the clear. The KSS is restarted for each SDU, even if two are sent in the same burst. An overview of the process is shown in Figure 7.52.

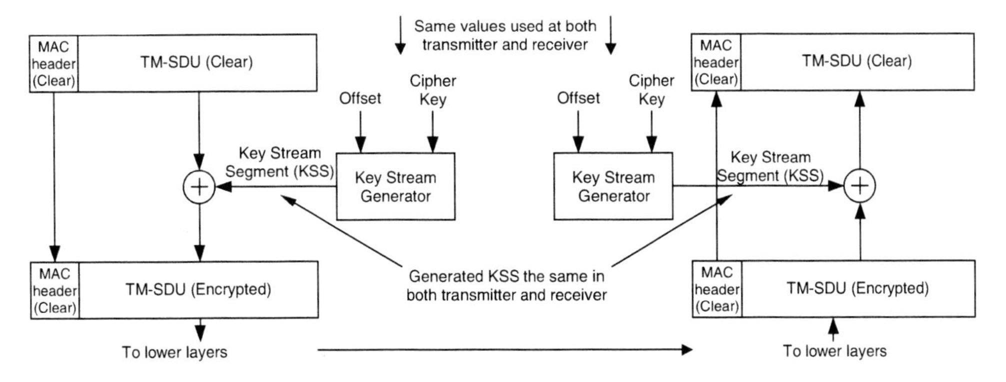

Figure 7.52 Air interface encryption overview

The Key Stream Generator takes two parameters in order to generate a KSS, a cipher key and an offset parameter. In V+D mode, the offset is called the "Initial Value". This is constructed from the slot, frame, multi and hyperframe number, as well as a flag for uplink or downlink. This means that the Initial Value, and the resulting KSS, will not repeat for more than 540 hours for a particular key value, and that the uplink and downlink will use different KSS. This long repetition value makes analysis of the encrypted data very difficult, and prevents replay message attacks on the system whereby an attacker simply retransmits a previously recorded encrypted message in the hope of fooling the recipient. The composition of the Initial Value is shown in Figure 7.53.

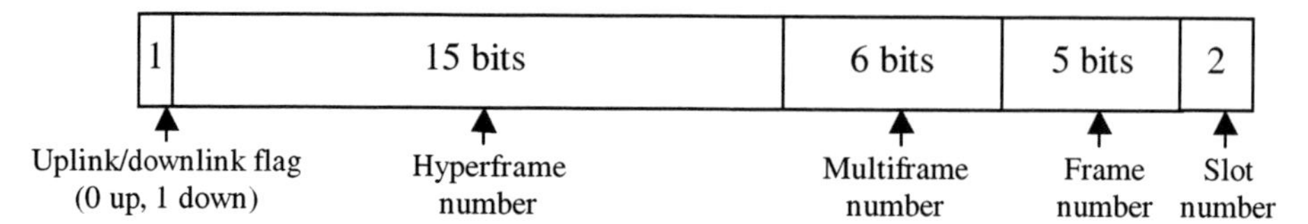

Figure 7.53 Initial Value for the air interface encryption scheme

Direct Mode does not have hyperframes, nor the concept of an uplink and downlink. The offset in Direct Mode is therefore different, and is termed the "Time Variant Parameter" (TVP). This parameter is transmitted on the SCH/S. Note that even though Direct Mode does have frame and multiframe numbering, these are not used to form the TVP. This is because the frame number and multiframe number are chosen arbitrarily by the master at the start of a call, and the values could repeat often and this would compromise security. The Direct Mode master randomly chooses a new TVP at the start of each call, which is then incremented on each frame.

The other parameter to the Key Stream Generator is the Cipher Key. TETRA has four different types of key (see Section 9.2):

Static Cipher Key (SCK). This is a fixed key used throughout the system when no session authentication is in operation. The are 32 possible keys, only one of which may be used at any time. The key in use is indicated on the SYSINFO message. The key used by the Key Stream Generator is derived from the SCK, a version number that is incremented on each use of the key, and a number derived from the user's authentication key (such as a PIN number, or number stored on a SIM card).

Derived Cipher Key (DCK). This is a key that is produced as a result of an authentication process between the mobile and the infrastructure. Either the mobile or infrastructure alone can be authenticated, or a mutual authentication of both mobile and base station can be undertaken [5]. The DCK is different for each mobile and communication session. The DCK is derived in part from the user's authentication key.

Common Cipher Key (CCK). The use of the user's authentication key for the DCK makes it unsuitable for use as a key for group calls involving a number of terminals, as the users' authentication keys will be different. For this purpose, a CCK is for each location area. Only one CCK may be used in any location area at any time. The CCK is distributed to mobiles using the DCK after authentication.

Group Cipher Key (GCK). Specific groups may have a group cipher key. The GCK is never used on its own but may be used to modify the CCK for use in calls to that group for additional security.

Air interface encryption is optional. Even if it is used, downlink broadcast signalling in V+D mode, and SCH/S in Direct Mode, is transmitted in the clear, as these are used to send information necessary to initialise the encryption scheme.

7.10 LOGICAL LINK CONTROL

7.10.1 Introduction

The Logical Link Control (LLC) sub-layer uses the error detected data stream provided by the MAC layer in order to provide error free data to layer 3. In order to do this it adds a frame check sequence (FCS) to the data packets being transmitted and then checks on reception to see if the data has been received correctly. If the data has not been received correctly it requests that the data be sent again. The LLC is therefore responsible for data transmission and retransmissions, segmentation and re-assembly, and organisation of the logical links.

Unlike the MAC, which uses forward error correction to protect against channel errors, the LLC requests that data be sent again if the information is in error using a system known as ARQ (Automatic Repeat reQuest). The retransmission of information to replace messages with errors leads to a variable delay. For this reason, it is not used for circuit mode data or speech traffic, since these services are sequential with fixed delay requirements and retransmitting information would not be appropriate.[1] Also, broadcast messages go to several mobiles and there is no guarantee of reception so these are not processed by the LLC. The only messages which are processed by the LLC are signalling and packet mode data messages. Therefore, the LLC does not perform any function on messages passed to the TMD-SAP or TMB-SAP, and only operates on messages via the TMA-SAP. This explains why the LLC only appears in the path from the TLA-SAP in the logical model of the TETRA lower layers.

The LLC provides two types of link – basic and advanced. In the basic link, messages are passed to the MAC to be split into fragments small enough for transmission, and then re-assembled at the other end of the link. The resulting message is checked for errors, and if an error is detected, the complete message must be retransmitted. In this case, the MAC undertakes most of the work. In the advanced link, the LLC splits the message, and checks each message part on receipt. If an error occurs, only the section in error need be retransmitted.

Direct Mode does not have an advanced link, so these functions are absent. The remaining functions for the basic link undertaken in the LLC of V+D are undertaken by

---

[1] Applications may add further error protection to circuit mode data at higher layers and this may include retransmissions.

Layer 3 in Direct Mode, and the Direct Mode standard does not specify a logical link control sub-layer.

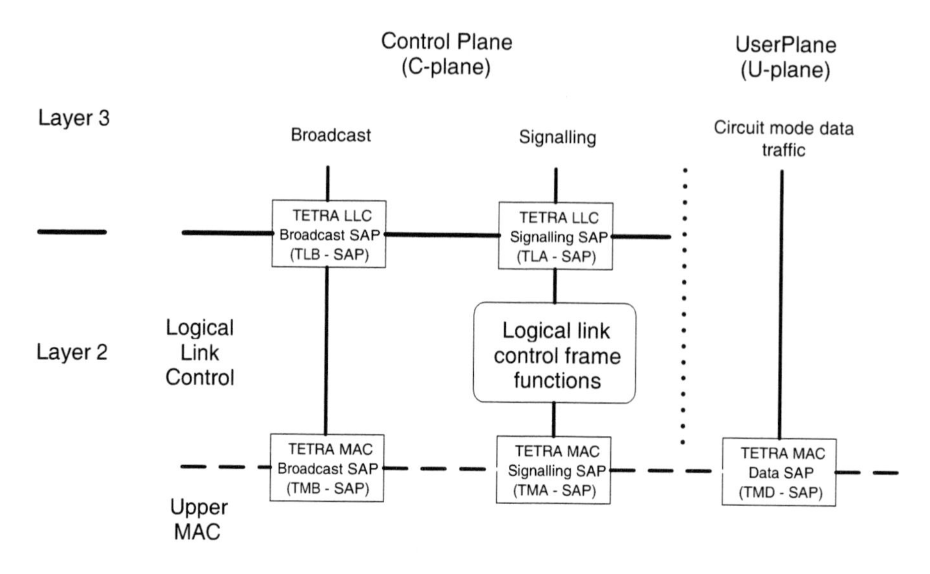

Figure 7.54 Location of TETRA link control functionality

7.10.2 ARQ

As mentioned in Section 7.4.2, ARQ is an error control technique. The error control strategies examined so far have been error detection and forward error correction strategies. A disadvantage of FEC is that the additional redundancy must be dimensioned so that the worst case of expected errors can be corrected, and if fewer errors occur, this redundancy is wasted. The result is a lower channel capacity than would otherwise be obtainable.

When ARQ is used, if an error occurs, the receiver requests that the transmitter resends the information. ARQ schemes will work only if there is a reasonable chance that a block will be transmitted without error. If a block is likely to be in error, this will also be the case when it is retransmitted. ARQ schemes therefore work best where there are occasional severe errors which occur infrequently but would require too much redundancy to be added if FEC was used.

The amount of redundancy used for FEC can be compared with that required for ARQ. In the FEC case, the redundancy is fixed at the number of error correcting bits in the code. In the ARQ scheme, there is also a fixed redundant element – the error detection bits – but these will be fewer than in the FEC case. However, in the case of ARQ, errored blocks are also effectively redundant, so if blocks have to be retransmitted too frequently the overall throughput will be lower than the FEC case. This is shown graphically in Figure 7.55, with the redundant information shaded.

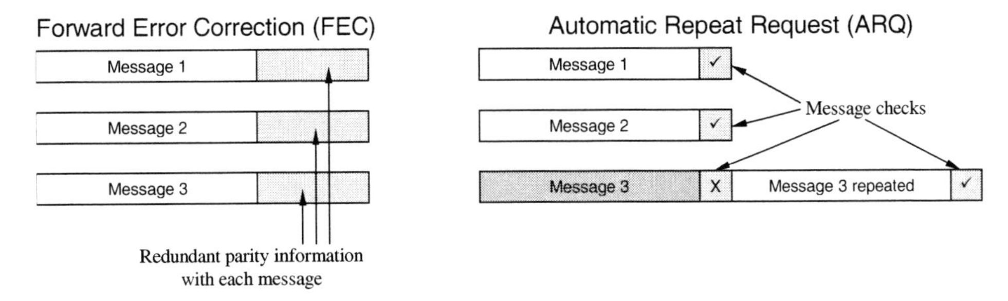

Figure 7.55 Redundant information for FEC and ARQ

ARQ systems tend to be less complex than FEC schemes, for as was noted in the Section 7.4.2, error detection is simpler than error correction. However, ARQ schemes require a return path to the transmitter, which must have a low error rate. This implies that the return path must have a large amount of FEC to keep its decoded error rate low, which may require significant transmission resources. Also, while FEC schemes have a fixed delay, ARQ introduces a variable delay depending on the number of retransmissions that are required. This would not be suitable for constant delay services like speech.

There are three types of ARQ – stop and wait, go back *n*, and selective repeat. The stop and wait system is the simplest form. Each message is transmitted from the source to the destination, and the destination acknowledges the message back to the source. The source does not transmit another message until it receives the acknowledgement for the previous message. To guard against lost messages, a time out is usually used, so if the source hears nothing from the destination, it transmits the message again.

Stop and wait is wasteful of channel capacity, since the source has to stop transmitting to wait for acknowledgements. "Go back *n*" is an alternative where the source assumes the message will reach the destination intact and continues to transmit. If this assumption proves incorrect, the source will have to retransmit all the messages from the one which had the error, even though subsequent messages may have been received correctly, as the receiver has no way to store the subsequent messages. The source therefore has to store the messages until the acknowledgement is received. Up to *n* messages can be stored (buffered), and the source will have to stop transmitting if positive acknowledgements have not been received before the buffer is full.

Go back *n* is wasteful in transmitted resources in that while the source is transmitting most of the time, some good messages may be discarded (and have to be retransmitted) if they occur after a message with an error. A more efficient system is "selective repeat", where only blocks with errors are retransmitted. The disadvantage of this technique is that blocks have to be buffered at the destination because error-free blocks will be received out of order. However, while it is more complex, selective repeat makes the most efficient use possible of resources.

The following diagram illustrates the different types. In each case, the channel suffers from errors at the same point. For stop and wait, the 1st, 2nd and 3rd blocks are transmitted and received correctly (the first channel error burst occurs during a wait period). The 4th

message is corrupted by errors, and has a negative acknowledgement. It is therefore retransmitted. In the 'go back *n*" case, the first error burst corrupts the 2nd message block, but the source is aware of this until it receives an acknowledgement, by which time it has transmitted the 3rd and 4th message blocks. The source has to go back to the 2nd block and transmit it again. In this case, *n* must be 3 or more. Had it been only 2, the source would have had to stop after transmitting message 3 until it receives the acknowledgement of block 2. The selective repeat system also assumes a minimum buffer of at least 3, and only retransmits blocks that had errors. Note that this system transmits 11 blocks in the time the other systems managed only 7 or 5.

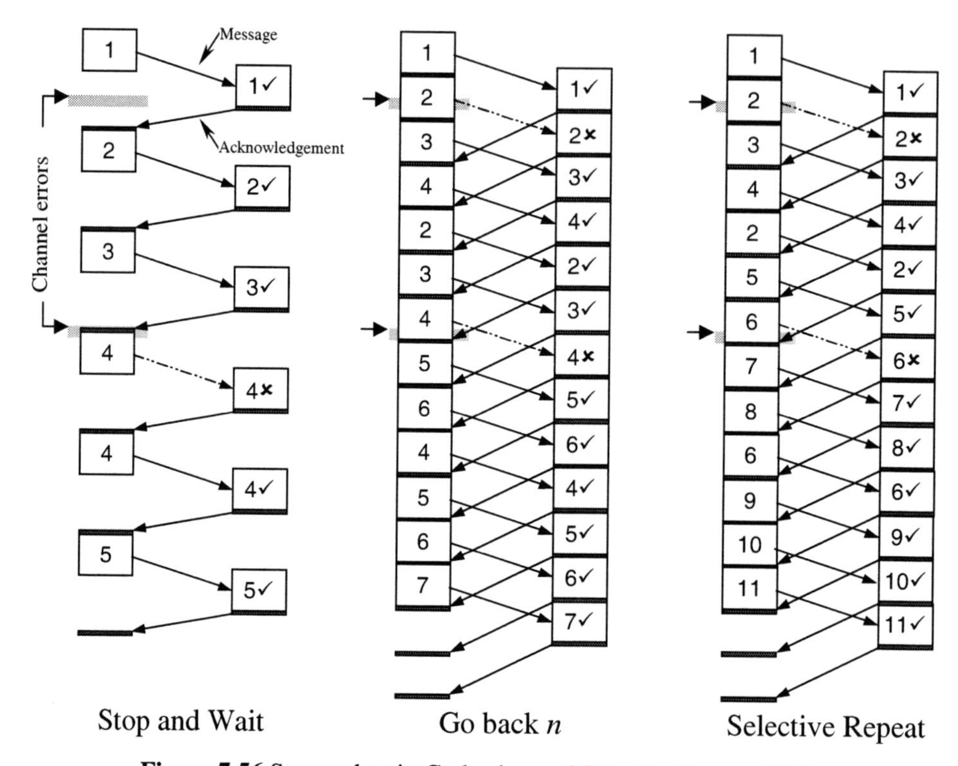

Figure 7.56 Stop and wait, Go back *n* and Selective Repeat ARQ

The size of the buffer at the transmitter and receiver is very important to the efficient operation of the system. Stop and wait requires no buffering, but leaves the channel idle most of the time. Go back *n* is better, and requires a transmitter buffer. Selective repeat is most efficient, but requires a transmitter and a receiver buffer. The transmitter knows the size of the receiver buffer, which defines a "window size", the maximum number of messages the receiver can store. There is no point in the transmitter sending more messages than the window size beyond an acknowledged message because if an error had occurred, the receiver would not have been able to store the additional messages and they would have to be resent. In the example shown in Figure 7.57, there is a window size of 8. An error occurs in the transmission of block 3. The receiver can only store blocks 4 to 10, and block 11 cannot be transmitted until a retransmission of block 3 has been received correctly.

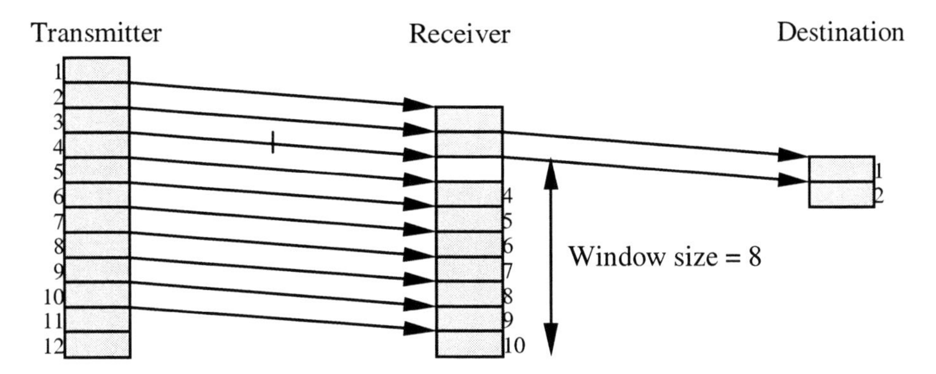

Figure 7.57 Effect of window size on transmissions

If a block is in error, it must be retransmitted. It is therefore sensible to keep the blocks short, to reduce the amount of information retransmitted. However, short blocks mean more messages, and higher signalling overheads (in terms of acknowledgements, etc). As usual, a compromise must be reached.

It is possible to combine FEC and ARQ. FEC is the most efficient strategy when the error rate is reasonably constant, while ARQ is good for variable error rates. If the error characteristics of the channel are such that there is a constant background error probability, along with some bursts of higher numbers of errors, an FEC scheme can be used within an ARQ scheme with the FEC designed to clean up the background errors conforming to the constant error rate. Should a burst of errors occur, the FEC will fail, and the ARQ scheme will then operate to resend the block.

A development of this system is a so-called "hybrid" ARQ scheme. Again FEC is employed, but if it fails, instead of resending the block, additional error correcting information is sent which is combined with the data already sent in order to try to correct the errors. Hybrid ARQ schemes require complex error correcting coding schemes, but are very efficient for services which can tolerate a variable delay.

7.10.3 LLC Operation

TETRA V+D defines two types of communication links – a basic link and an advanced link. The basic link is typically used for signalling messages whereas the advanced link may be used for packet mode data. Direct Mode only uses the basic link.

In the basic link, the LLC adds a header and some checking before passing the message to the MAC for transmission. The MAC layer carries out most of the work of the protocol and the LLC checks for correct reception and requests the information again should an error occur. The LLC provides logical link handling, along with scheduling data transmission and retransmission where required, and acknowledgement of received data.

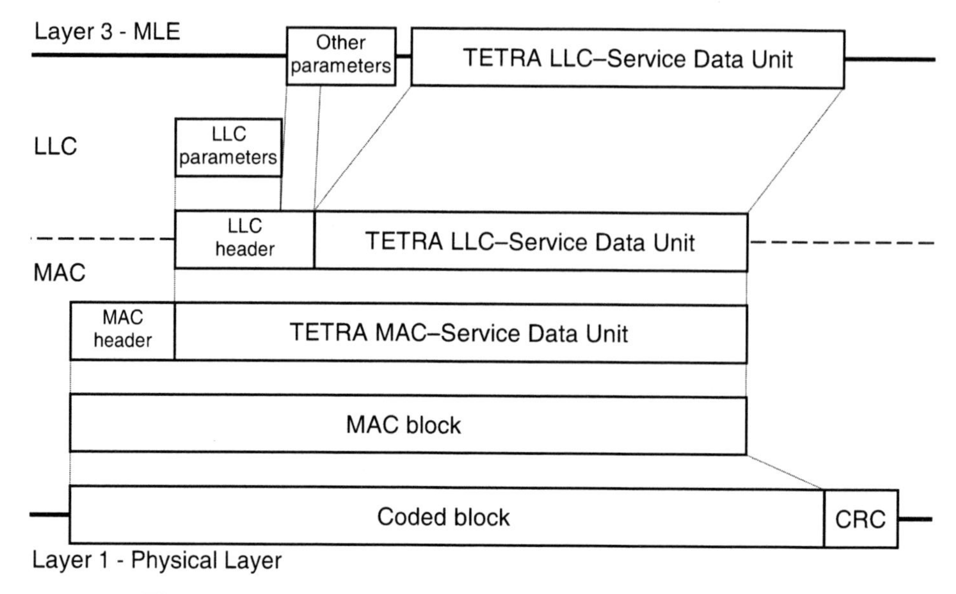

Figure 7.58 Stages of transmission of an SDU passed to the LLC

If the Service Data Unit (i.e. data packet) is too large to be sent in a single MAC block, it is "fragmented" by the MAC, which splits the LLC messages into smaller sections so that it can be passed as shown below.

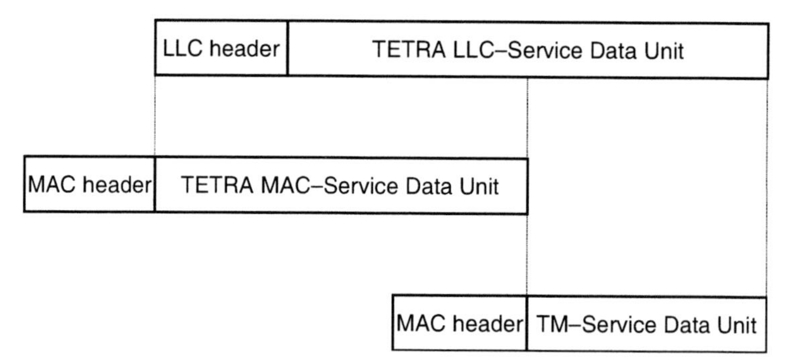

Figure 7.59 Fragmentation of an LLC-SDU

The basic link will reject a message if any part of it is in error, so long messages have to be retransmitted in their entirety. This is acceptable for short signalling messages, but for longer packet data messages this is inefficient. TETRA therefore offers the option of an advanced link where long messages are "segmented" in the LLC and passed in MAC Service Data Unit sized sections to the MAC for transmission. Each segment has an LLC header and can therefore be individually checked, so if a failure does occur, only the individual segment needs to be retransmitted. The sending LLC can continue to send data with the retransmissions up to a set maximum beyond the oldest outstanding incorrectly received block, i.e. up to the window size.

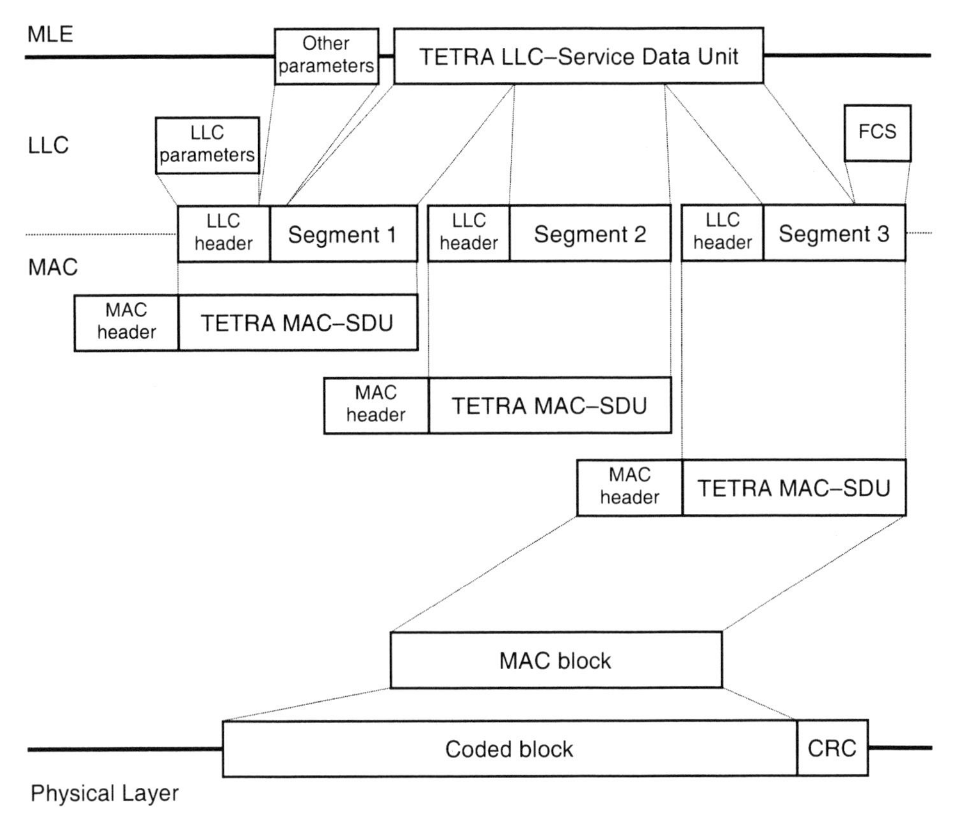

Figure 7.60 Segmentation of an LLC-SDU

The selective re-transmission is based on the segments into which the transmitting LLC divides a TL-SDU for sending. The receiving LLC informs the transmitting LLC which segments are not received correctly, and then the transmitting LLC sends the missing segments in the later transmissions until the whole TL-SDU is received correctly as recognised by the MAC layer error detection. The whole TL-SDU may still be erroneous and the receiving entity shall ask a re-transmission, when an error in the TL-SDU is detected by the frame check sequence added by the LLC itself. The example in Figure 7.61 from [3] shows the operation of the advanced link.

Three TL-SDUs are to be transmitted. Each SDU is segmented. The first SDU is the longest, and is broken into six segments. The second and third SDUs are split into four segments each. The LLC calculates a frame check sequence for each SDU and buffers them for transmission.

In this example the window size is two – i.e. the sending LLC can send on information from the second block before the first block is received in its entirety, but it may not send on any information on the third block until the first block is complete, and so on. A complete SDU is indicated by the receiving LLC returning an acknowledgement showing that it has received all segments and the FCS is verified.

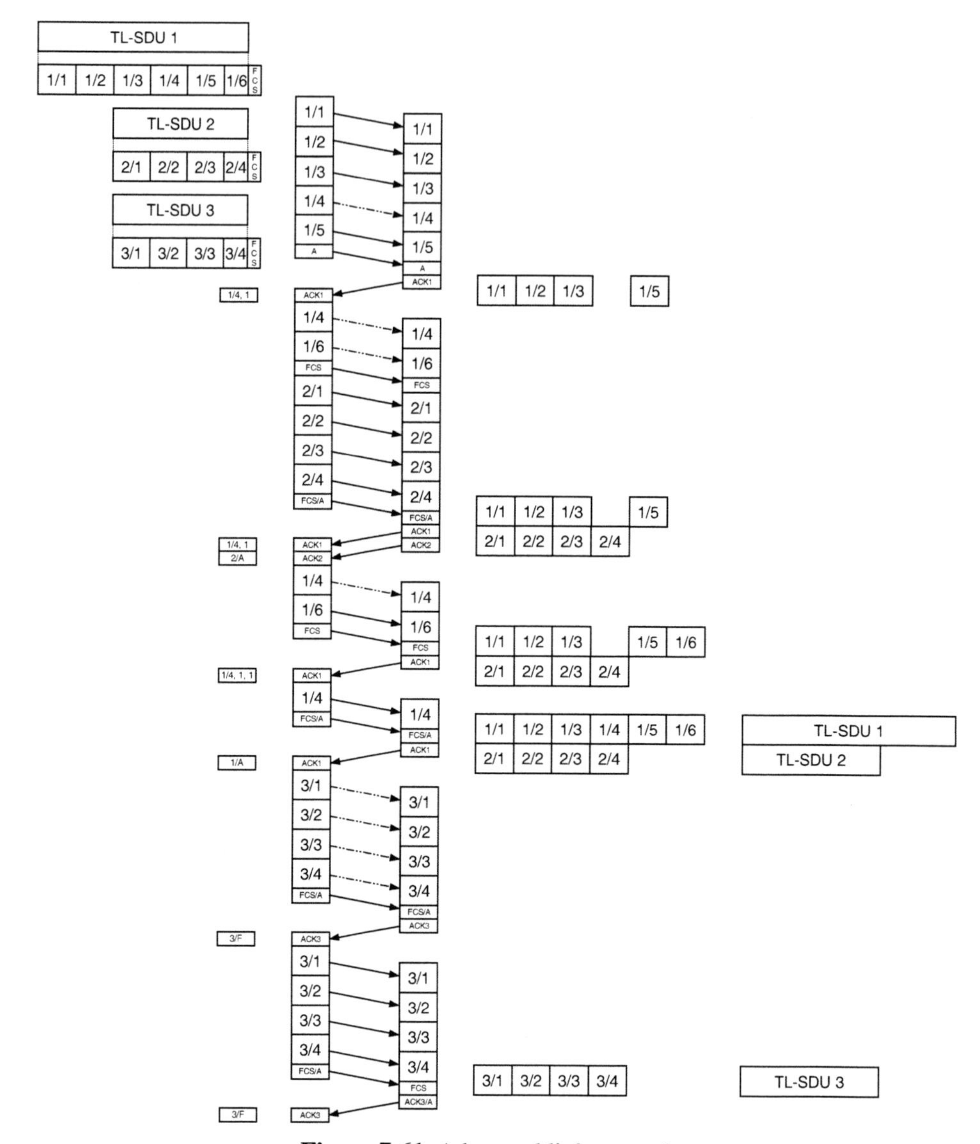

Figure 7.61 Advanced link example

The transmitting LLC sends the first group of five segments, followed by an acknowledgement. The receiving LLC responds to this acknowledgement with an acknowledgement of its own (ACK1), indicating that segments 1, 2, 3 and 5 have been correctly received, but that segment 4 was not received correctly. The correctly received blocks are stored in a buffer at the receiver, but are not deleted from the transmit buffer until an acknowledgement of the complete SDU is received, since a failure in the overall FCS would require the retransmission of the entire SDU.

The acknowledgement contains bit maps as shown, the first part of the ACK (1/4) indicates the first segment which is not received correctly, i.e., segment number 4 in the first SDU and a bit map from that segment onwards, 1 in this case as only one further segment was transmitted (segment 5) and this received correctly.

The transmitting LLC then modifies its transmission and adds the segments which were not acknowledged or sent i.e. 4 and 6 from the first SDU (1/4 and 1/6), before continuing transmission of new segments in this case from the second SDU. The last segment of the second SDU is sent marked "FCS/A". The second and third acknowledgements indicate that again segments 4 and 6 from the first SDU are not yet received correctly, but the second SDU is received totally and correctly (2/A). The transmitting LLC then resends segments 4 and 6 of the first SDU, but cannot continue to the third SDU as the window size of two in this example.

The receiving LLC again misses the segment number 4 of the first SDU and the receiving LLC sends acknowledgement after receiving the sixth and final segment of the first SDU. The first acknowledgement indicates that only the fourth segment of the first SDU is not yet received correctly. The transmitting LLC then resends the missing segment of the first SDU and after receiving the second acknowledgement can send the third SDU, which fits into the new SDU window. That SDU is received correctly as indicated by each segment CRC, but the total frame check sequence does not match and the receiving LLC sends an acknowledgement indicating a failure (3/F) of the third SDU. After resending, the third SDU is this time received correctly and acknowledged (3/A).

The receiving LLC delivers the first TL-SDU to layer 3 only after all segments are received correctly. The second SDU is already correctly received and acknowledged by the second acknowledgement, but the receiving LLC cannot deliver it before the first SDU is received to keep SDUs in the correct sequence. When the first SDU is received and acknowledged, the receiving LLC passes this, and the waiting second SDU, to layer 3, while the transmitting LLC can deletes SDUs from its buffer as soon as an acknowledge for the complete SDU is received.

7.11 SYSTEM MODES OF OPERATION

Business mobile radio requirements are very variable, and as a result, the TETRA standard has to be very flexible to allow for the construction of systems including very lightly loaded rural systems, heavily loaded systems with many speech calls, systems with predominantly data traffic, and so on. The TETRA standard allows for this flexibility by defining a number of operating modes described in this section.

Normal mode, extended mode, and minimum mode are distinguished by their use of the common control channels. In normal mode, the MCCH forms the common control channel, and this is satisfactory in most cases. However, if there are a large number of users and many RF carriers, additional common signalling capacity may be required, and this is provided by extended mode where additional common signalling channels are defined. At the other end of the usage spectrum, in low load conditions with only one or two carriers, a permanent common control channel may be inefficient, and minimum mode allows the common control channel to be temporarily replaced by a traffic channel to support an additional call.

The other operating modes address the problem of efficient use of the radio spectrum over large, lightly loaded areas, and are main carrier sharing mode, and carrier sharing mode. These are where the main or an additional carrier is shared between different sites in a time multiplexed manner, either with separate main control channels, or with the main control channel itself being shared to improve efficiency at the expense of signalling capacity.

7.11.1 Normal Mode

The normal mode of operation is designated as a standard TETRA implementation. A typical installation is expected to have about 4 or 5 radio frequency pairs present on a site (i.e. 16 to 20 voice channels). In this mode, the common control channel on the main carrier is the main control channel (MCCH) and is present in timeslot 1 of all frames 1 to 18. This common control channel is used for all common control signalling. All mobiles not involved in a call listen to the downlink transmissions of the MCCH. The base station transmits on all the downlink slots of the main carrier during normal mode.

In normal mode, the main carrier will be continuous (i.e. the base station will transmit on all the slots in the frame). Other carriers may use a time shared mode so that slots are shared between base stations, but this will not be apparent to mobiles.

7.11.2 Extended Mode

A single MCCH will be satisfactory for most installations. However, with installations with large numbers of carriers, or with high levels of packet data, or a large number of short transactions (as may occur with short messages and transmission trunking), the access capacity available from a single slot MCCH may be insufficient. In such cases, additional common control channels, Secondary Control Channels (SCCH), can be provided, whereupon the system is said to be in extended mode.

There are two ways the SCCH can be used, either as Common SCCH or Assigned SCCH.

7.11.2.1 Common SCCH

Common SCCH have the same functionality as the MCCH, but apply to a subset of the mobile population, effectively reducing the loading on each channel. Like the MCCH, a Common SCCH is a single slot on the main carrier. One, two or three Common SCCHs can be defined, which occupy slots 2, 3 and 4 of the main carrier respectively. The maximum configuration therefore has common signalling on all four slots of the main carrier.

Each mobile is assigned one of twelve numbers, from 0 to 11, as an SCCH allocation number either at registration or subscription. The common control channel they use is then given by that number modulo the number of Common SCCH channels plus 1. This has the effect of sharing the mobile population between the MCCH and the Common SCCHs, which then all act similarly. For example, if a mobile was assigned an SCCH allocation number of 7, it would use the MCCH if there are no Common SCCHs (7 mod 1) + 1 = slot 1, SCCH on slot 2 if there is one Common SCCH (7 mod 2) + 1 = slot 2, SCCH on slot 2 if there are two Common SCCH (7 mod 3) + 1 = slot 2, and SCCH on slot 4 if there are three Common SCCH (7 mod 4) + 1 = slot 4.

The number of Common SCCHs can change during system operation, and so the slot on the main carrier the mobile receives may change accordingly. Such a change is indicated on the BNCH.

7.11.2.2 Assigned SCCH

The base station may also operate Assigned SCCHs after the mobile has made an initial access on the MCCH (or appropriate Common SCCH). This increases signalling capacity. Assigned SCCHs may be used for particular purposes and may be multi-slot (up to four), although the same number of slots must be used for uplink and downlink. This increased signalling capacity could be used to support a general packet data channel of a mobile or group of mobiles.

7.11.3 Minimum Mode

Minimum mode is where the MCCH is replaced by a traffic or assigned control channel. It is intended for use in low traffic density areas. Such areas will normally have a single carrier, and to reserve a slot in every frame would mean that more than 25% of the available capacity would be used for signalling.

Minimum mode operation allows a base station to allocate all time slots on the main control carrier for traffic or dedicated control purpose. Therefore in minimum mode, only frame 18 would be available for common control, although stealing is still possible and can be used to contact waiting mobile stations. The lack of uplink capacity for access caused by minimum mode is not a major problem, as if minimum mode is being used there would be no spare slots to allocate to a mobile in any case. The only difficulty is in the case of high priority uplink calls which may pre-empt other calls, which will note be able to access the system. However, the system may go in and out of minimum mode several times within a multiframe as transactions begin and end, and so high priority calls could inform the base station of their presence at these points.

Mobiles know when minimum mode is in operation, and therefore when the MCCH is unavailable, as they are monitoring the AACH for the MCCH in order to receive access information.

Minimum mode is not restricted to single carrier base stations, but the relative capacity gain of its use reduces if there are more traffic channels. Minimum mode can be used with main carrier sharing, when the relative gains are even greater. However, it cannot be used in MCCH sharing mode, as the MCCH resources do not exist in the cell all the time to be allocated to a traffic channel.

7.11.4 Discontinuous Transmission – Time, Carrier and MCCH Sharing

TETRA defines a number of modes for discontinuous transmission on the downlink. Base stations normally transmit continuously on the downlink, but using discontinuous transmission allows resources to be shared between cells. In all cases, good synchronisation is required between cells that are sharing resources.

There are three types of time sharing transmission

- Traffic channel sharing
- Time-shared carrier
- MCCH sharing

In traffic channel sharing, each base station has its own main carrier, with an MCCH, but additional carriers may be shared between base sites on a slot by slot basis. In the following example, three carriers are available, with each base station having one as its main carrier and using two slots of the remaining carrier for traffic channels. Traffic channel sharing is completely transparent to the mobile.

In the case of time-shared carrier transmission, each base station may share its main carrier with another base station. The MCCH is still used, existing in slot 1 of the frame, with slot numbers being adjusted if necessary. Figure 7.62 shows an example where time slots 1 and 2 on a carrier are used for one base station, and slots 3 and 4 are used on another, where they are renamed 1 and 2 to suit the MCCH.

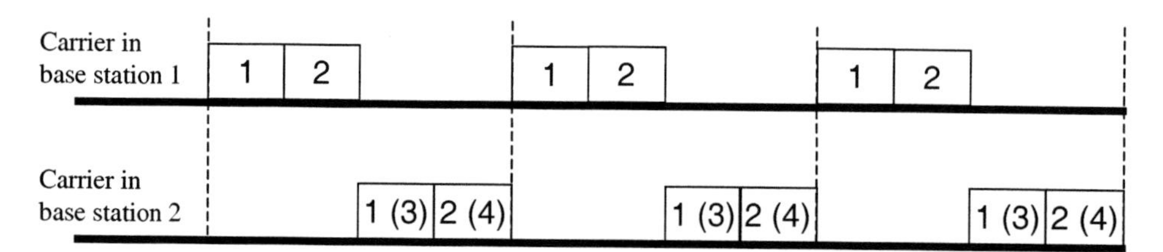

Figure 7.62 Time-shared carrier

Minimum mode is possible when the system is using a shared main carrier.

Allocating a complete MCCH for a base station with very low traffic is not very efficient, which leads to the use of MCCH sharing. Here, the MCCH is multiplexed between 2, 3, 4, 6, 9, 18 or 36 different cells. Slot 1 is left vacant when it is used in a different cell. MCCH capacity is divided between cells by this multiplexing so its capacity from the point of view of individual mobiles is reduced, lengthening signalling delays and operations like call set-up. However, it frees slots for use as traffic channels. Figure 7.63 shows the use of a shared MCCH between two cells.

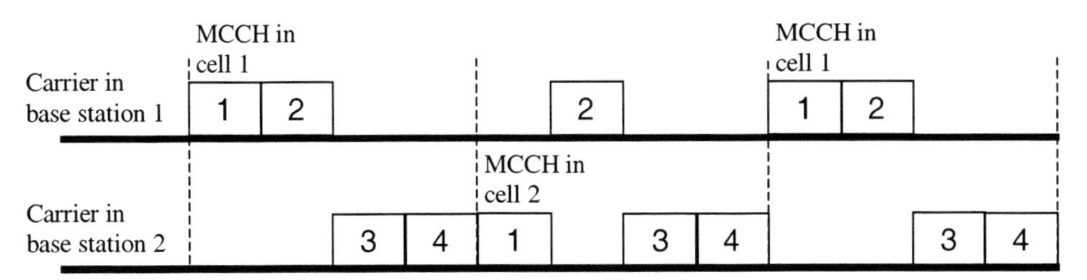

Figure 7.63 Shared MCCH

7.12 MOBILE STATION MODES OF OPERATION

TETRA defines four modes of operation for mobiles. This takes into account the fact that mobile terminals require at least a minimum level of activity even when they are not in a call in order to listen for network initiated calls or to inform the network of their current location. Other modes are common signalling and packet mode, when a mobile is involved in a transaction but does not have an assigned traffic channel, traffic mode, when the mobile does have an assigned traffic channel, and energy economy mode, which is a variation of the idle mode which allows the mobile to sleep for a number of frames to reduce power consumption. Energy economy mode exists only in V+D mode.

7.12.1 Idle Mode

The "idle" mode is the state of a registered mobile not actively in communication with a base station. It is continuously listening to the Main Control CHannel (MCCH) or to any of the other Common Control signalling CHannels. All mobiles must be capable of monitoring adjacent cell signal strength in this mode.

7.12.2 Signalling and Packet Mode Data

This mode is used for signalling, including signalling messages transferring packet mode data. There are three types of signalling mode

- Common Control CHannel
- Assigned Control CHannel
- Associated Control CHannel

Common Control CHannels, either the MCCH or common SCCH, are available for all common control signalling (including transport of user packet data) by the mobile. The mobile will also listen for paging messages from the base station on its assigned Common Control CHannel.

Assigned control channels are SCCH used for additional signalling which the base station may instruct the mobile to use. This could be used for packet mode data or for additional signalling in support of the set up of a circuit mode call.

Associated control channels are the FACCH and corresponding SACCH on an allocated slot and are used for signalling in conjunction with an established connection.

7.12.3 Traffic Mode

Traffic mode is the mobile state when it has been assigned a traffic channel and user speech or circuit mode data is being transferred. When in traffic mode capacity may be stolen by the transmitting party for signalling traffic by the stealing mechanism (see Section 7.5).

7.12.4 Energy Economy Mode

Even when a mobile is in idle mode, it still has to monitor the infrastructure in order to receive signalling, for a new call, for example. In practice, the messages directed at the mobile will only occur infrequently, and if the equipment is battery operated, this results in an inefficient use of battery power. In energy economy mode, the mobile temporarily switches off, or "sleeps", for a predetermined number of frames, before waking to monitor a frame.

The base station instructs a mobile to enter energy economy mode, but it does so in response to a request from the mobile. Since the base station is in control of the mode, it knows when the mobile is asleep and therefore when it would not be listening to signalling messages. There are seven different energy groups defined for energy economy mode. For each group, the mobile monitors its slot in a frame and then sleeps for the number of slots defined for that mode, as given in Table 7.16.

Table 7.16 Energy groups and slot monitoring frequencies

Energy Group	Frames to sleep	One frame monitored ...
"Stay alive"	–	Every frame (not in energy economy mode)
1	1	Every second frame
2	2	Every third frame
3	5	Three times per multiframe
4	8	Twice per multiframe (0.51 sec)
5	17	Once per multiframe (1.02 sec)
6	71	Every four multiframes (4.08 sec)
7	359	Every twenty multiframes (20.4 sec)

Energy economy mode saves mobile power at the expense of signalling capacity. For group 7 in particular, there may be a considerable delay in contacting a mobile in order to inform it of a call, for example. This reduction in functionality has to be set against the increased standby time possible using a particular energy group. When a base station instructs a mobile to enter energy economy mode, it tells the mobile which frame and multiframe number to start the sequence on. It is therefore able to adjust the sequencing of different mobiles to ensure CCCH capacity will be available if necessary for the monitored frames.

An energy economy mode allocation is usually valid in all cells within a location area, so if a mobile changes cell it will continue the same monitoring pattern after acquiring slot and frame synchronisation.

It is not expected that a mobile in energy saving mode would have to modify its monitoring behaviour if the system entered minimum mode. However, the infrastructure would need to take account of the monitoring pattern of energy economy groups and may need to page mobiles by using the stealing channel.

Although the sleeping mobile cannot be contacted from the infrastructure during its sleep period, it can be locally aroused by the application to initiate a call or data transaction.

The mobile terminates energy economy mode by requesting the base station to assign it to "stay alive" mode, i.e. idle mode. Alternatively, energy economy mode may be implicitly terminated by making a call.

7.13 CONCLUSIONS

This chapter has described layer 2 of the TETRA system. Layer 2, and in particular the MAC sub-layer, contains key functionality for the efficient and successful operation of the system. As well as undertaken coding, random access, logical channel routing and stealing, the MAC undertakes a number of support functions for layer 3 in respect of encryption and channel quality measurement. Chapter 8 will describe the remaining layer defined in the TETRA system, layer 3.

REFERENCES

[1] Brady, P. T., 'A Statistical Analysis of On-Off Patterns of Speech', The Bell Systems Technical Journal, 47, pp73–91.

[2] ETR 300-1: 'Radio Equipment and Systems (RES); Trans-European Trunked Radio (TETRA); Voice plus Data (V+D); Designers guide; Part 1: Overview, technical description and radio aspects', ETSI, 1997.

[3] ETS 300 392-2: 'Radio Equipment and Systems (RES); Trans-European Trunked Radio (TETRA); Voice plus Data (V+D); Part 2: Radio Aspects', ETSI, 1996.

[4] ETS 300 396-2: 'Terrestrial Trunked Radio (TETRA); Technical requirements for Direct Mode Operation (DMO); Part 2: Radio Aspects', ETSI, 1998.

[5] ETS 300 396-7: 'Terrestrial Trunked Radio (TETRA); Technical requirements for Direct Mode Operation (DMO); Part 2: Security', ETSI, 1996.

[6] ETS 300 396-3: 'Terrestrial Trunked Radio (TETRA); Technical requirements for Direct Mode Operation (DMO); Part 3: Mobile Station to Mobile Station (MS-MS) Air Interface (AI) protocol', ETSI, 1998.

[7] Haykin, S., *Digital Communications*, John Wiley & Sons, New York, 1988.

[8] Hess, G. C., *Land-Mobile Radio System Engineering*, Artech House, Boston, 1993

[9] Lin, S, and Costello, D, *Error Control Coding: Fundamentals and Applications*, Prentice Hall, 1983.

[10] Namislo, C, Analysis of Mobile Radio Slotted ALOHA Networks, *IEEE Transactions in Vehicular Technology*, VT-33, August 1984, pp 199–204.

8

TETRA Network Layer Protocols

8.1 INTRODUCTION

The focus in Chapters 6 and 7 has been on the transmission functions of TETRA close to the radio medium. Under the physical layer, the bit level signal representation, including the modulation and demodulation schemes and the characteristics of RF amplifiers have been described. The data link layer in Chapter 6 has detailed the various schemes of how the bit streams at the physical layer are formatted and transmitted with clearly defined control mechanisms for sharing the FDM/TDMA radio channels. It should therefore be clear that layers 1 and 2 (of the ISO protocol stack) are responsible for most of the radio technology that specifically characterises TETRA.

In this chapter the focus will be mainly on the network layer which corresponds to layer 3 of the ISO protocol architecture. The network layer can be characterised as networking procedures that rely on the layer 2 techniques for providing basic communication services, known as *bearer services*. Bearer services essentially hide the details of the underlying technology (which is radio for TETRA) and functionally appear identical to higher layer protocols, in effect providing the convergence for various communication technologies such as wireless and wireline communications. As we move up the ISO protocol hierarchy, the bearer services will be used to construct more sophisticated communication services and applications.

8.2 NETWORK LAYER CONCEPTS

8.2.1 Generic Functions of the Network Layer

The network layer provides the transfer of information between two communicating entities. In the structured OSI model representation, the object of the network layer is to relieve higher layers of the OSI functions of the need to know about the underlying data transmission and switching technologies used to connect systems. The upper four layers are *end-to-end protocols* between the communicating entities and the information used at these layers is transparent to the network protocols. Figure 8.1 makes a clear distinction between network protocols (layers 1 to 3) and end-to-end protocols (layers 4 to 7).

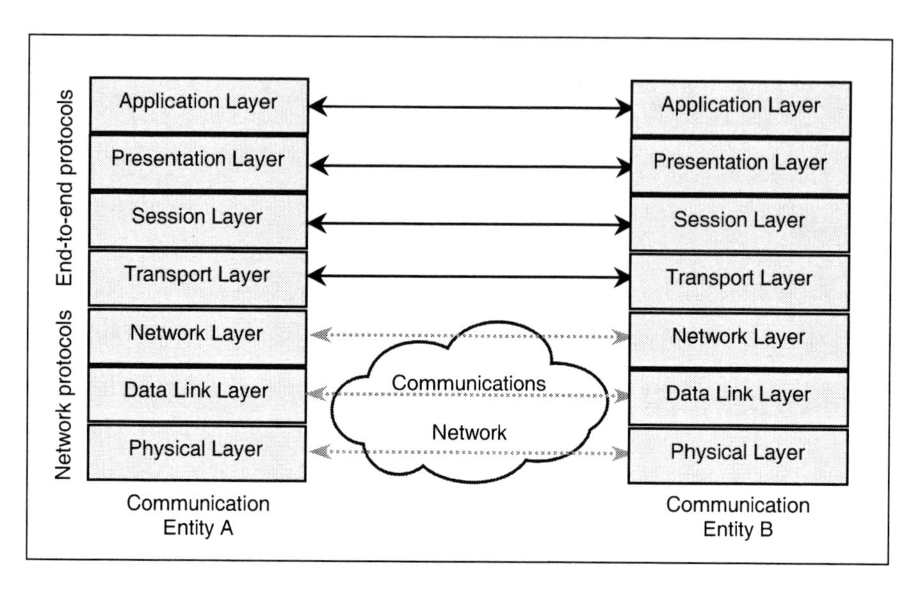

Figure 8.1 OSI network protocol architecture

The network layer provides additional communication functions that extend the capabilities of the underlying data links into more flexible inter-networking facilities. That is, the network layer is responsible for the flow of information across the network regardless of the physical complexity of the underlying connections. Various network functions are involved to realise an efficient and orderly flow of information. The main ones are described below:

- **Naming and addressing.** Examples of this include IP address for Internet networks, and *network service access poin*t (NSAP) for telecommunication networks implementing an OSI model. In mobile communications, in addition to generic network service access points, which are used for routing packets within the network, there are also various identities and addressing schemes for locating user terminals, subscribers (through the subscriber identity module or SIM), and networks.

- **Connection control.** This deals with connection establishment, data transfer, and termination. In the case of mobile communications, additional and specialised tasks such as radio link management functions will be required.

- **Message routing and flow control.** This involves various routing and flow or congestion control strategies depending on the type of communication service. Numerous techniques can be cited under this function of the network layer protocol including *hierarchical routing*, *sliding window* flow control schemes (used in X.25

networks), traffic shaping techniques (for congestion control), call handling policy and call admission control (e.g. telephone calls), and so on. With regard to mobile communications, the most notable tasks in this category of the network layer functions include mobility management and call admission control strategies.

8.2.2 Control and User Plane Separation

As described in the preceding section, the network layer builds on the data link layer with additional functionalities that enable the provision of network services to user applications. In situations where a direct link exists between two communicating entities, the data link layer alone could provide the necessary information transport. Indeed, circuit-switched speech and data services in TETRA rely only on the data link layer for user information transfer. If on the top of circuit-switched speech service the user wishes to use supplementary services such as a three-way conversation or a group call in TETRA, then, the user must somehow interact with the network for such additional services to be enabled. Among other things such additional service provision entails verification of user access permission to the requested services and routing of the user request to an appropriate server within the network. That is, some sort of network connection management would be required over and above the connection for user traffic which in this instance is provided by the data link layer.

The network connection management tasks are otherwise known as *network signalling* and the associated protocols are called (network) signalling protocols. There are, therefore, two types of network traffic which are distinct from each other:

- signalling data traffic
- user data traffic

There are significant differences between signalling and user traffic as summarised below:

- For the majority of the user traffic, e.g. speech, information flow is a continuous stream of data and usually with a constant delay requirement and this demands a regular physical link. Signalling traffic, on the other hand, tends to be bursty or discontinuous with more activity during connection establishment and with a much-reduced activity while the connection is active. Hence, signalling information needs only a discrete physical link for transmitting control information but also a continuous virtual link so that various control functions can be invoked whenever the user application or the network demands.
- Different channel structures tailored to the characteristics of signalling traffic and user traffic are used. This point should be evident from the detailed descriptions of the TETRA data link layer in Chapter 7. For instance, the various control signalling schemes based on *dedicated channels*, *slow associated channels*, *traffic-channel stealing*, etc. are all mechanisms for optimising the performance of system signalling by providing some balance between signalling and user traffic bandwidth.
- Signalling traffic requires a different type of error protection treatment from user data traffic. Signalling data packets with bit errors are usually rejected due to the impact of

such errors could have on system operation, whereas user data may tolerate bit error rates to a certain degree (e.g. speech codec).

Because of these basic differences, it is often convenient to separate network protocol stacks into signalling traffic and user traffic. The protocol stack corresponding to the signalling traffic is usually known as the *control plane* (C-plane) and that corresponding to the user traffic is known as the *user plane* (U-plane). The concept of C-plane and U-plane separation is illustrated by TETRA's network protocol stacks which will be described in Section 8.3.

Although the C-plane and U-plane separation are provided for a good reason, in practice the functions of the C-plane and U-plane are not rigidly adhered to. It is possible to transmit user data in the C-plane or signalling data in the U-place. For instance, the short data service (SDS) in TETRA transmits messages by taking advantage of the spare capacity within the C-plane. Likewise, for transmitting signalling information in the traffic channel, recall the time slot stealing mechanism (see Section 6.5) in which traffic slots are stolen for additional signalling provided by the SACCH logical channel.

8.2.3 Users' View of the Network Layer

From the user's perspective the network layer may be identified with respect to two key points:

- The network layer provides the bearer services which are basic to network service provision. For instance, the TETRA standard encompasses layers 1 to 3.
- The network layer is generally the level at which equipments of different capabilities and/or different manufacturers inter-operate. That is, the network software developed for layer 3 and above could, in principle, be used for other equipments with different physical and data link as long as conformance to layer 3 interface is achieved.

Specifically for TETRA:

- Inter-operability at layer 3 is assured by use of common layer 3 protocols for V+D, DMO and PDO systems.
- For the above reason, conformance testing for TETRA equipment is performed at the network layer.

Looking from a different angle, the above two factors should explain why the TETRA standardisation is devoted to the lower three layers of the OSI model.

8.3 OVERVIEW OF TETRA NETWORK LAYER

8.3.1 Types of Information Transported

The types of information transported at or below the network layer of a TETRA network are summarised below.

1. Signalling control messages (C-plane). These are the most important information at the network layer and they are primarily concerned with the task of network connection management, including control for call connection, supplementary services and mobility management. Signalling messages initially use the user traffic channels and then resort to SACCH by releasing the traffic channels to user traffic. That is, they are transmitted both in signalling and traffic modes.

2. Short data service (SDS). The SDS is offered by the network signalling entity within the C-plane. The SDS has been described in Section 5.8.4.

3. Packet mode user data services. Packet mode data services in TETRA are handled over the C-plane rather than on the U-plane. This approach was taken so that the air interface encryption provided for signalling messages could also be used for packet mode data.

4. Circuit mode speech services. These are transmitted in the traffic mode. A slow signalling will be in operation while circuit mode speech service is active.

5. Circuit mode user data services: These are transmitted in the traffic mode. As for circuit mode speech, a slow signalling will be in operation while circuit mode speech service is active.

6. User-to-user data. This is the signalling information between users and transmitted in traffic mode. User-to-user signalling is intended to support added features such as end-to-end encryption synchronisation during a call.

Having presented an overview of all the information types encountered at the lower three layers of the TETRA protocol stacks, the last three are handled at the data link layer and do not propagate to the network layer and will not be considered further. They will be referred to indirectly however, in connection with the signalling protocols that provide them with the necessary connections and associated supplementary services.

8.3.2 TETRA Protocol Stacks

The TETRA infrastructure involves various communication links as shown by its reference points (Section 5.5). Each of these reference points can be described by the signalling protocols applicable to that interface. Three main categories of communication protocols can be identified with TETRA:

- TETRA-specific air interface protocols – These are comprised of the MS-BS interfaces at the R2 reference point. These protocols exist in three forms depending of the type of a TETRA system: Voice plus Data (V+D) system, packet data optimised (PDO) system, or Direct Mode Operation (DMO).
- User interface protocols – There are communication protocols at the user interfaces designated by reference points R1, R3 and R4 (see Section 5.6).

- Inter-networking protocols – These are used for inter-system signalling and data communication mainly based on X.25 and IP connections.

Of main interest here are the TETRA-specific protocols over the air interface. The protocol stacks for the air interface are summarised below, with important features of these protocols first identified and then treated under separate sections. Subsequent to the air interface protocols, inter-networking protocols are briefly described.

8.3.3 V+D Protocol Stack

The MS to BS interface protocol stack for V+D is shown in Figure 8.2. The detailed functions of layers 1 and 2 have already been covered in the previous chapters and most of the discussions here will be confined to the layer 3 functions.

In the figure, the lines extending to the left of the MS (or to the right of the BS) represent the service access points. For instance, it can be seen that circuit mode data and speech services are tapped from the data link layer at the medium access control (MAC) sub-layer. Due to the user-to-user circuit-switched data transport, these services are established with the data link layer circuits of the communicating nodes without the network layer functions being required.

It should be noted that, in line with the principle of separating the C-plane and U-plane traffic (Section 8.2.2), the V+D protocol stack is divided into C-plane and U-plane. The network layer, which is part of the C-plane, is divided into two sub-layers as described below.

- **Mobile/Base Link Control Entity (MLE).** This sub-layer is common to all the other layer 3 functions and its main function is to shield the sub-layers above it from the communication disruptions caused when the MS changes a cell through handover procedures. The MLE functions will be discussed in more detail under Section 8.4.

- **The Sub-Network Access Functions.** These consist of four protocol entities with distinct functions, namely, the Mobility Management (MM), Circuit Mode Control Entity (CMCE), connection-oriented packet data service (CONP), and connectionless packet data service (CNLS).

  - The MM protocol entity provides mobility services at the layer 3 of the air interface and deals with mobility-related network signalling such as authentication and registration of subscribers as their locations need updating from time to time. MM will be treated in detail in Section 8.5.

  - The CMCE protocol entity provides signalling for controlling circuit mode calls between the MS and the network. It also provides the control for supplementary services and short data service. CMCE will be introduced in Section 8.6.

  - Connection-oriented and connectionless data services provide packet mode user data. These will be described in Section 8.7.

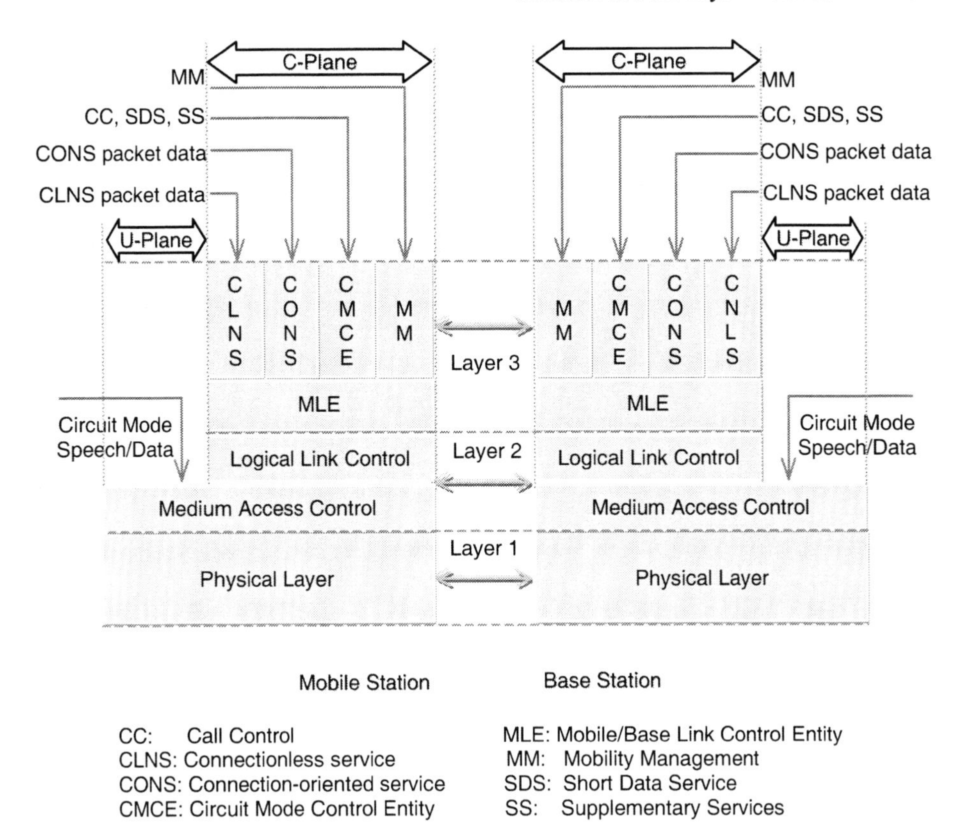

Figure 8.2 Mobile to base station protocol stack for V+D TETRA system

8.3.4 PDO Protocol Stack

The V+D and PDO specifications are based on the same physical radio platform but because of the difference in the way the time slots are used in framing the bits on the air interface, implementations do not inter-operate at the physical or data link layer. However, full inter-operability of V+D and PDO equipment is expected at layer 3. Equipment conforming to the V+D specification will, depending on the options supported, provide a wide range of bearer, teleservices, and supplementary services relevant to a joint voice/data capability. Equipment conforming to the PDO specification, on the other hand, will support only packet data services.

Figure 8.3 depicts the protocol stack for a PDO TETRA system. From a protocol architecture viewpoint, the basic difference between PDO and V+D systems is the CMCE protocol entity, which only exists for the latter. This is because a PDO system does not support voice and voice-related supplementary services. Otherwise, at the network layer both V+D and PDO provide packet data services that may be handled with a common protocol. In other words, the network layer provides a convergence point for V+D and PDO data services which are inherently different at the data link layer.

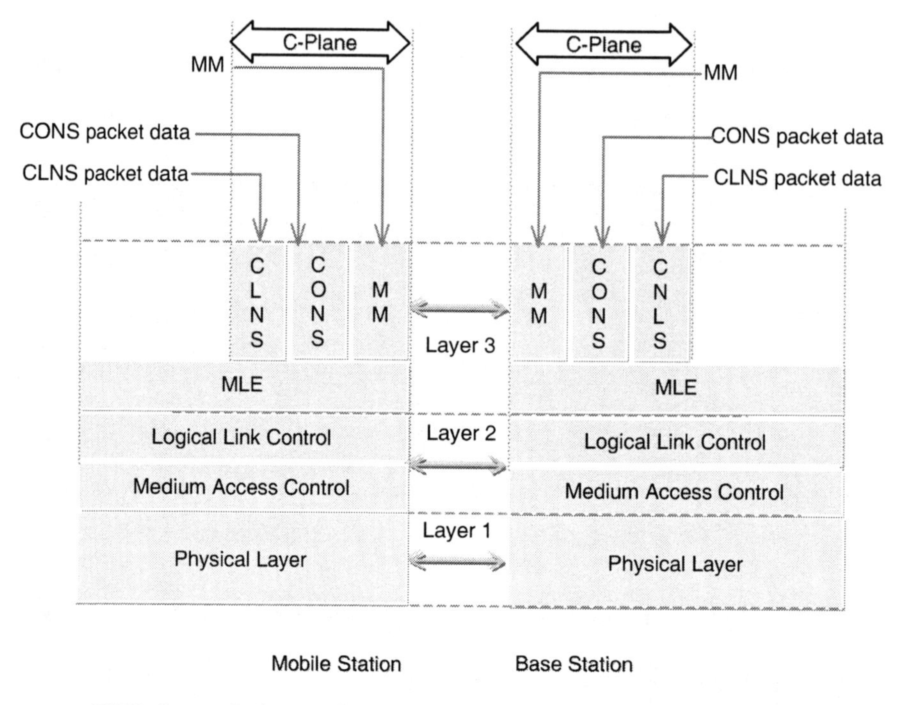

Figure 8.3: Mobile to base station protocol stack for PDO TETRA system

For full description of the PDO protocol stack shown in Figure 8.3, the reader is referred to Section 8.4 for the MLE, Section 8.5 for MM, and Section 8.7 for packet data services.

8.3.5 DMO Protocol Stack

Direct Mode Operation (DMO) provides mobile-to-mobile communications without using the trunked mode network infrastructure (Section 5.3.7 for details). Figure 8.4 depicts the protocol stack for DMO. It should be noted that data link layer speech and data are supported, with the short data service also provided at the network layer, under the management of the Direct Mode Call Control (DMCC). Protocol entities such as the Mobile/base Link control Entity (MLE) are also non-existent in DMO since mobility management and cell selection procedure are not relevant.

Since the DMO in comparison to the V+D trunked mode of operation has a limited functionality, its layer 3 protocols are also much reduced and simplified. From a conceptual point of view, therefore, the description of DMO network layer protocols could be regarded as a subset of the trunked V+D protocols with their functional differences simply attributed to the mode of operation (i.e. trunked mode versus direct mode). For instance, the DMO short data service is identical to that of V+D at the network layer.

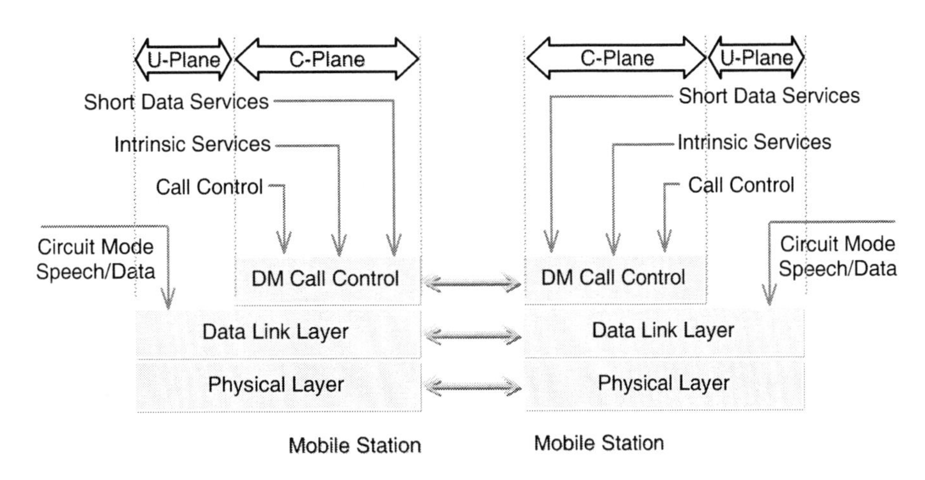

Figure 8.4 Protocol stack for mobile to mobile TETRA DMO

Similarly, the function of the call control (CC) protocol entity within DMO is conceptually similar to that of V+D except that the functionality is much reduced since call trunking is not in operation. With a detailed description of the V+D protocols in the following sections, the reader should find the DMO protocols relatively easy to grasp by noting the basic differences between the DMO and V+D protocols as summarised below.

1. The direct mode of operation dispenses with the radio link management and cellular mobility functions since only mobile-to-mobile communication is intended within the DMO coverage area. Therefore, the MLE and MM protocol entities that account for the major part of the V+D protocols are absent in DMO. For DMO mobiles that communicate with V+D gateways, only a much-reduced version of the MM functionality would be required for the purpose of authentication and registration.

2. DMO supports only circuit mode services, and therefore network layer protocols for packet mode services (CONP and CNLS) are non-existent for DMO. Furthermore, unlike the V+D system, DMO does not support a multi-slot operation, i.e. two or more time slots cannot be combined for increased capacity.

3. Layer 3 bearer services are provided with the DMCC which interacts with the data link layer air interface protocol. Again, this is a watered-down version of CMCE for the V+D system. The DMCC handles call control, SDS, and *intrinsic services*. Intrinsic services are supplementary services available to DMO.

8.3.6 Remarks on Air Interface Protocols

The brief description of TETRA air interface protocols outlined above has identified the following categories of protocols for further discussion. These will be presented in the sections to follow and in the order shown below.

- Mobile/base Link Entity (MLE) protocols;
- Mobile Management Protocols;
- Circuit mode connection entity;
- Packet mode data service protocols.

As remarked in the preceding subsections, it should be noted that some of the above protocols apply to more than one type of TETRA system, e.g. V+D and PDO systems.

8.4 MOBILE TO BASE LINK ENTITY

8.4.1 Overview

An important aspect of the TETRA air interface protocol is that it is the MS that makes the handover decisions, not the infrastructure. There is a significant advantage with this approach in that the mobiles are made to share the radio link management tasks for call re-establishment which otherwise would overload the network, particularly when dealing with group call management. Although the mobiles carry out handover decisions, in normal circumstances the infrastructure will be informed of the mobile's intention to select another cell. This is in sharp contrast to the operation of GSM where handover decisions are carried out by the base station on the network side.

The MLE sub-layer within the mobile is concerned with the radio link management. As its main task, it manages the connection between the mobile station and the base station while the call is in progress. The fact that the mobile is in charge of a cell selection and reselection task suggests that the function of the MS-MLE becomes vital to the overall system operation. The MS relies on the MLE to gain access to all communication resources according to requests received from the MM entity (Section 8.5), which is in control of activating and deactivating the MLE.

From the protocol layering point of view, the purpose of the MLE sub-layer is in fact to hide most of the radio aspects of the air interface. The resulting MLE services are therefore intended to be comparable to wire-line layer 2 protocols.

8.4.2 MLE Functions

As one of it main tasks, the MLE protocol entity measures the signal strengths in adjacent cells. These measurements, together with other related information broadcast by the cells concerned, will be used to maintain a reliable air interface, if necessary by using intercell handover. At the lower layer, the MLE protocol communicates with the LLC (refer to the network protocol stacks of Figure 8.2 or Figure 8.3) with data exchange at the LLC service access point (SAP). LLC SAPs have been described in Chapter 7 with an introduction in Section 7.2. The MLE also assists the Mobility Management protocol of the MS in keeping track of the movements during cell reselection. For instance, if the MLE encounters a new location area, which can be determined by decoding of system information from network broadcast burst, then, it will inform the MM to register and update its location with the network. Registration and location updates are handled by the Mobility Management protocol entity which is described in Section 8.5.

Figure 8.5 depicts the MLE functional model with four internal sub-entities. The detailed MLE functionalities can be described with reference to these sub-entities which comprise of the attachment management, data transfer, network broadcast and network management. These are briefly described below.

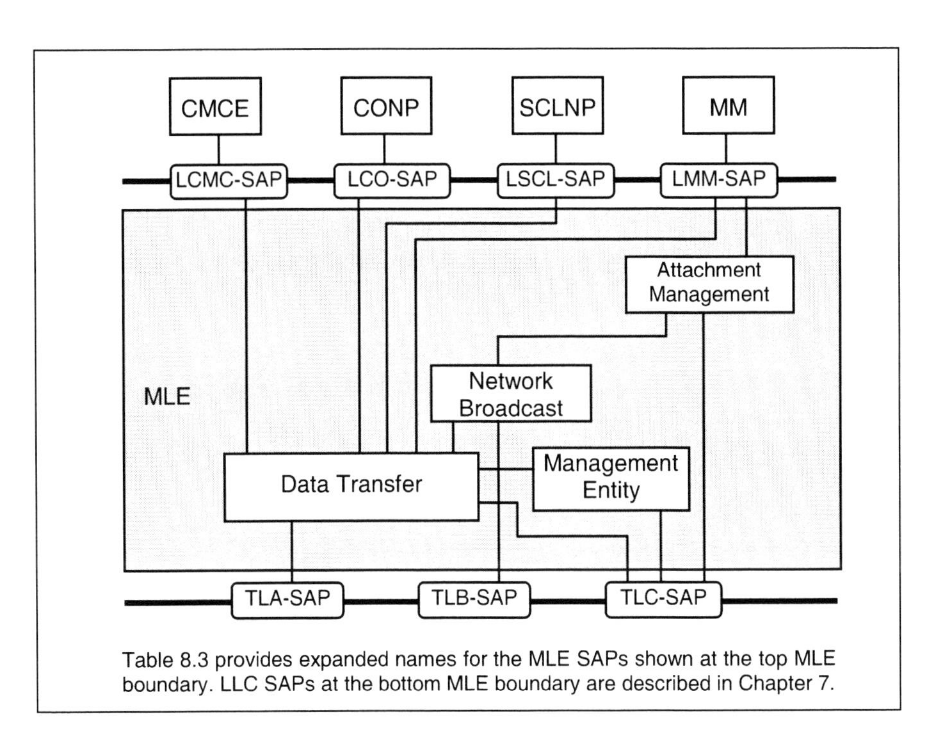

Table 8.3 provides expanded names for the MLE SAPs shown at the top MLE boundary. LLC SAPs at the bottom MLE boundary are described in Chapter 7.

Figure 8.5 MLE functional model

8.4.2.1 Attachment management

This sub-entity deals with the radio link management, the main functions being:

- management of monitoring and scanning procedures;
- surveillance of the serving cell for signal quality;
- management of the adjacent cell ranking procedure for selection;

- management of the cell relinquishable, improvable and usable radio criteria;
- management of the roaming announcements and declarations; and
- informing upper entities of CMCE and CONP of broken and restored MLE connections via the Data Transfer sub-entity.

The attachment management deals with the radio link establishment and maintenance aspects and deserves a detailed treatment as presented in Section 8.4.3 below. Subsequent to the description of the attachment management, an overview of MLE primitives and peer-to-peer protocol data exchange is presented. This overview is intended to provide some insight into how the TETRA radio link management tasks are performed based upon neighbour cell monitoring and mobile-initiated handover decisions.

8.4.2.2 Data transfer

This sub-entity deals with the transfer of data with the underlying LLC entity and upper entities. More of the functions provided with this sub-entity will be described in connection with the protocols which are the service recipient of this sub-entity (see Figure 8.5) in their respective sections. Here, it will suffice to say that the following processes represent the main functions of the data transfer sub-entity.

- selection of the underlying LLC service;
- address handling (ISSI, GSSI and TMI as defined in Section 5.8);
- informing the upper protocol entities (CMCE, CONP and SCLNP) of enabled and disabled access to the communication resources;
- routing and multiplexing to layer 2 service end points, including addition or removal of MLE protocol control information;
- routing and multiplexing to MLE SAPs and other MLE functional entities; and
- quality of service mapping, e.g. priority, throughput, transfer service.

8.4.2.3 Network broadcast

On the network side, this sub-entity broadcasts system information to all mobiles. The main functions include:

- formatting and broadcasting of the network information;
- reception and analysis of network information;
- configuring of layer 2 MAC with synchronisation and *System Information* broadcast described below.

The *System Information* is a series of messages that are broadcast at regular intervals from the infrastructure to the MS-MLEs, and exists in two formats:

- Immediate system information:
  - MNC;

  - MCC;
  - LA Code (LAC);
  - subscriber class, one of 16 user population subdivisions which is operator's option to implement;
  - cell service level, designating the traffic loading in the cell as high, medium, low, or unknown; and
  - late entry information availability for group calls.
- Network broadcast system information:
  - frequencies of adjacent cells for cell selection and reselection; and
  - parameters for cell selection and reselection (detailed in Section 8.4.3).

Network broadcast procedures are based on the unidirectional signalling of the infrastructure. If required, the MS can also request the infrastructure for system information through MLE signalling. More on this is described in connection with MLE PDUs in Section 8.4.5.

8.4.2.4 Network management

As the name suggests, this sub-entity is responsible for communication of management information between the MS and the infrastructure. The main functions are:

- handling network management procedures, e.g. those addressed to the TETRA Management Identity (TMI); and
- handling local management information from the management entity to the lower layers.

At present, network management PDUs are not defined in the TETRA standard. As a result, the signalling procedures and PDUs for the network management sub-entity could be network specific.

8.4.3 Attachment Management

At the heart of the radio link management is the measurement of the radio link strength between the mobile and neighbouring cells as well as the serving base station. Given that a TETRA mobile is in charge of choosing a serving cell, it is required to carry out different types of radio link measurement procedures and in different modes. To distinguish the various link measurement procedures and modes, the TETRA standard [1], [2] uses terminologies with important concepts attached. For a clear understanding of the radio link management functions, it is therefore important to be acquainted with some of these terminologies as summarised in the following paragraphs.

8.4.3.1 Definition of MLE terms

In order to perform cell selection and reselection, which is the basis of cellular operation, the mobile must carry out radio link signal strength measurements with regard to various cells. In TETRA, the radio link strength is represented by a pathloss parameter called *C1* or *C2*, which will be explained shortly. Both *C1* and *C2* are calculated from a received signal strength indicator (RSSI) measurement at the mobile and power control parameters, which the mobile acquires from the serving cell or directly from the neighbouring cells being monitored.

The mobile is said to be *scanning* when it is measuring the power of neighbouring cells and calculates the pathloss parameter based upon the power control information which is broadcast by the neighbouring cells themselves. To be scanning a cell, the mobile must have synchronised itself with that cell, and then decoded the system information of the cell. The mobile is said to be *camped on* a cell if it is able to decode the system information of that cell. The pathloss parameter calculated through scanning is known as *C1*.

The mobile is said to be *monitoring* when it is measuring the power of the neighbouring cells and calculates the pathloss parameter based upon information on the neighbouring cells but broadcast by the serving cell (and not directly by the neighbouring cells themselves!). Monitoring, as opposed to scanning, is used if the mobile is unable to synchronise itself with the neighbouring cells. The pathloss parameter calculated through monitoring is known as *C2*.

The process of monitoring the quality of the radio link to the serving cell through the measurement of the downlink power and calculation of *C1* is known as *surveillance*.

Three types of scanning are possible. *Foreground scanning* is used when the MS is in idle mode in the serving cell and wishes to scan an adjacent cell. *Background scanning* is applied when the MS wishes to scan an adjacent cell and maintains any current service on the serving cell. This procedure requires that the MS switches from the serving cell main carrier to the adjacent cell carrier to be scanned in between any transmissions or receptions on the serving cell. *Interrupting scanning* is similar to foreground scanning except that the MS is in an active call on the serving cell and temporarily suspends service in order to scan an adjacent cell.

8.4.3.2 Acquiring cell synchronisation and system information

Figure 8.6 summarises the procedures involved in cell synchronisation and subsequent acquisition of the system information. The mobile acquires synchronisation by locating the sync training sequence in the Broadcast Synchronisation CHannel (BSCH). The BSCH logical channel is described in Section 4.5.1. It is the MAC that decodes the SYNC protocol data unit (PDU), which contains various information types as summarised in Figure 8.6. Once the mobile is able to decode the SYNC PDU, then it can acquire the network information by decoding the Broadcast Network CHannel (BNCH) for the SYSINFO PDU. This decoding is performed by the MLE. The SYSINFO PDU contains, among other things, parameters for cell selection/reselection and power control, which are essential for the scanning and monitoring procedures. In fact, having decoded both the SYNC and the SYSINFO PDUs, the mobile has all the information it needs to communicate with the network on the uplink radio path.

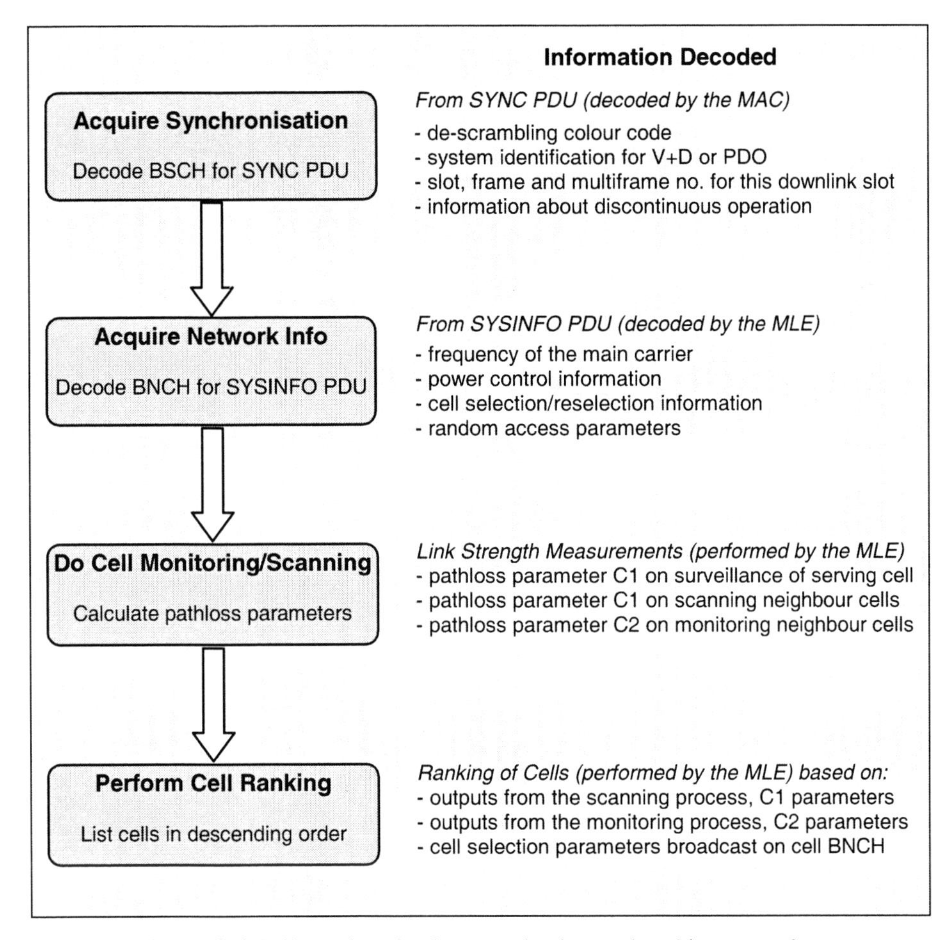

Figure 8.6 Cell synchronisation, monitoring and ranking procedure

8.4.3.3 Pathloss calculation and neighbouring cell ranking

The MLE protocol, through its MAC, performs signal strength measurements both on the serving cell and on selected neighbouring cells. The signal strength measurement is an approximation of the radio signal pathloss and represented by parameters, *C1* and *C2*, as defined below.

Pathloss Parameter C1

The pathloss parameter *C1* is calculated for the serving cell and for adjacent cells by *scanning*. *C1*, in dB, is calculated according to the formula:

$$C1 = RSSI - RXLEV\_ACCESS\_MIN - Max\,(0, MS\_TXPWR\_MAX\_CELL - P_{MS})$$

where:
RSSI is averaged received signal level at the MS or equivalent signal quality measurement;
RXLEV_ACCESS_MIN is a minimum permissible received level at MS in this cell;
Max (0, x) designates maximum of the two values, separated by a comma in the bracket;
MS_TXPWR_MAX_CELL is a maximum MS transmit power permissible in this cell; and
P_{MS} is a maximum transmit power of the MS.

All the parameters after the equal sign of the formula are in dBm.

RXLEV_ACCESS_MIN and *MS_TXPWR_MAX_CELL* are cell selection parameters, which are transmitted in all cells using the BNCH and decoded by the MS (see Figure 8.6) for *C1* calculation.

Path loss Parameter C2

The pathloss parameter *C2* is calculated for adjacent cells by *monitoring*. *C2*, in dB, is calculated according to the formula:

$$C2_n = RSSI_n - RXLEV\_ACCESS\_MIN\_MCELL_n - Max\,(0, MS\_TXPWR\_MAX\_MCELL_n - P_{MS})$$

where:
n indicates the nth adjacent cell carrier;
$RSSI_n$ is averaged received signal level at the MS or equivalent signal quality measurement;
$RXLEV\_ACCESS\_MIN\_MCELL_n$ is a minimum permissible received level at MS;
$MS\_TXPWR\_MAX\_MCELL_n$ is a maximum MS transmit power permissible in the cell; and
P_{MS} is a maximum transmit power of the MS.

As for *C1*, all the parameters after the equal sign of the above formula are in dBm. Unlike for the scanning procedure, though, the cell selection parameters for the adjacent cells, $RXLEV\_ACCESS\_MIN\_MCELL_n$ and $MS\_TXPWR\_MAX\_MCELL_n$, are transmitted in the serving cell by using an MLE broadcast message. Where these parameters are not known by the serving cell or where the MS has not received them on the serving cell, the cell selection parameters for the serving cell (which are broadcast on the serving cell BNCH) will be used as default values.

Ranking of Neighbouring Cells

Once the values of *C1* and/or *C2* are determined through scanning and/or monitoring, the neighbouring cells are ranked in descending order of their suitability for communication. Thus, a neighbouring cell with the best pathloss parameter will be a candidate for selection by the mobile. However, an understanding of the radio environment dynamics shows that pathloss measurements alone may not be adequate for reliable cell selection.

Additional techniques would be necessary for a more reliable cell selection as explained below.

8.4.3.4 Cell selection criterion

In the situation where a mobile is at a cell boundary, it is possible that the signal strength measurements from the serving cell and neighbouring cell(s) are within a close range of each other. With a rapid variation of the radio signal, it is possible that the pathloss parameter of the serving cell can go back and forth relative to the adjacent cell being monitored. This can set the cell selection procedure between the adjacent cells into a *ping-pong* situation – a phenomenon that must be avoided due its needless handover signalling and possible link interruptions. A combination of other factors, e.g. poor radio signal quality, shadowing effects of various types, and so on, within the radio environment can also lead to similar undesirable effects.

To tackle the problems associated with the radio environment dynamism, additional techniques are employed in addition to the signal strength measurements just described. As a cell selection criterion, the TETRA standard [1] employs a combination of three additional parameters:

- a threshold level – a level against which the pathloss parameters of the serving cell and neighbouring cells are compared for the cell selection procedure;
- a hysteresis control parameter – an offset value that makes a neighbouring cell pathloss parameter appear worse than it actually is, in order to minimise the ping-pong effect just explained; and
- a time out control – a time window for a given condition to persist before some action or decision with cell selection process is carried out. In effect, short-term conditions that satisfy the threshold and hysteresis conditions but fail to meet the timeout condition will be filtered out.

The threshold and hysteresis parameters are broadcast by the network, just like the cell selection parameters discussed earlier, while the time out values are fixed as described below. The standard specifies only the procedures and not the implementation issues, and therefore the threshold and hysteresis values are network operators' options. In fact, this approach is quite advantageous in that the threshold and hysteresis values could be varied in situations where the cell selection procedures are required to be adaptive to the radio environment.

Taking into account the above points, and depending on the relevant pathloss parameter (that of the serving cell, neighbouring cell, or both), the following criteria are defined [1] to assist with the cell selection process:

- link failure;
- radio relinquishable serving cell;
- radio improvable serving cell; and
- radio usable neighbour cell.

The *link failure* criterion is straightforward in that this condition is declared whenever the pathloss parameter $C1$ is less than zero.

A serving cell is declared *radio relinquishable* when the following conditions are met simultaneously:

- the serving cell pathloss parameter $C1$ falls below a specified threshold denoted by *FAST_RESELECT_THRESHOLD* for a period of 5s;
- the pathloss parameter, $C1$ or $C2$, of at least one of the neighbouring cells in the ranking list exceeds the pathloss parameter, $C1$, of the serving cell by a hysteresis offset value *FAST_RESELECT_HYSTERESIS* for a period of 5s; and
- no cell re-selection has taken place within the previous 15 seconds.

In the above, replacing the threshold and hysteresis parameters with the corresponding parameters shown in the right hand column of Table 8.1, a serving cell is instead declared to be *radio improvable*.

Table 8.1 Parameters for radio relinquishable and improvable criteria

	Radio Relinquishable Criterion	Radio Improvable Criterion
Threshold	*FAST_RESELECT_THRESHOLD*	*SLOW_RESELECT_THRESHOLD*
Hysteresis	*FAST_RESELECT_HYSTERESIS*	*SLOW_RESELECT_HYSTERESIS*

And finally, a neighbouring cell is declared *radio usable* when the following conditions are fulfilled:

- for a period of 5s, it has a pathloss parameter, C1 or C2, which is greater than:

  FAST_RESELECT_THRESHOLD + FAST_RESELECT_HYSTERESIS

- and no cell re-selection has taken place within the previous 15s.

It should now be clear that the above criteria could be easily used by the cell selection algorithm within the MS. For instance, in poor signal quality areas the MLE, based on the above procedures, would declare a serving cell as radio relinquishable, in which case the MS must change if it can find a better serving base station. On the other hand, radio improvable criterion could apply to a reasonable service area where the MS may change the cell only if it wishes to improve the signal quality. For such a middle ground decision, it could be the case that other factors such as quality of service (QoS) management functions, strategies for load balancing between cells, and various other operational factors are also considered as part of the decision process.

8.4.3.5 Cell selection and reselection procedures

Oncc a valid ranking list is available and the preferred cells are identified with the reselection criteria, the mobile can then initiate signalling (through its MLE protocol entity) in order to carry out the selection process. Although the selection process (which has commenced from Section 8.4.3) is now almost complete, some final but crucial tasks remain to be carried out. This is due to the fact that the selection process involves very important interactions between the mobile (which is in control of the handover) and the currently serving, as well as a new, base station. As such, this process is required to account for a number of eventualities as outlined below.

A number of factors influence how cell reselection is carried out, including:

- whether the MS is idle or in traffic mode, and if so whether it is a group call or individual call and whether the mobile is the transmit mobile or a receive mobile;
- whether the mobile is capable of background scanning while in traffic mode;
- the circumstances of the MS, whether it has had time to gain information on the new cell; and
- the grade of service of call re-establishment supported by the infrastructure, whether the infrastructure supports any planned call re-establishment, and if so, whether the channel allocation is negotiated by the mobile with the new cell or transferred to the mobile via serving cell.

If the MS is idle, it is not very important that it informs the current cell of its intention for selecting another cell. If however the mobile is engaged in a call, it should make every effort for minimising interruption to active connections, so communication with the serving cell and a new cell becomes essential. Smooth handover is not always possible, however, for instance, with mobiles that do not support background scanning. Such mobiles have to temporarily suspend the traffic connection and perform forward scanning for a new cell and attempt to restore the suspended connection in a new cell.

Since the mobile decides cell selection in TETRA, whether or not the serving and/or the new cell are told of the mobile's intention in advance can have a significant impact on the quality of service. Informing the serving cell of the intention to reselect a new cell can help release the resources immediately. It would be better still if the new cell is also informed of the mobile's intention to change cell so that a channel can be allocated in advance.

Based on combinations of the above factors, five reselection categories have been specified for TETRA. These options are intended to accommodate different mobile capabilities (e.g. mobiles without background scanning) as well as reselection difficulties that may arise from the radio environment or traffic conditions. Erratic pathloss measurements for instance can force a mobile to take a less optimal course of action in cell reselection. Likewise, a mobile overloaded by a group call can have less opportunity to interact with the serving cell or with the new cell for carrying out a smooth handover.

The five categories of cell reselection supported within the TETRA link management protocol are listed below.

- **Undeclared.** With this option, the MS does not inform the old cell or the new cell that cell selection is performed. This type of selection is appropriate when an MS is not engaged in a circuit mode call or connection-oriented data, and therefore no communication link is affected as a result.

- **Unannounced.** With this procedure only the new cell is informed about the cell selection. This can take place when the MS is engaged in a circuit mode connection (CMCE) or connection oriented data (CONP) but for a number of reasons the mobile does not have the opportunity to inform the serving cell of its intention to find service in another cell. The MS may attempt to recover the disrupted CMCE and CONP connections in the new cell.

- **Announced type 3.** With this procedure the serving cell is only informed of the intention to change the cell, but the MS does not know the new cell before changing to it. This option is provided for MSs which are unable to perform background scanning of a selected neighbouring cell. In this situation, the MS is forced to break the call(s) for a period and perform foreground scanning in order to acquire broadcast and synchronisation information for a new cell. Upon selecting the new cell, call restoration signalling can be used to regain the disrupted connections. This can occur when an MS is currently engaged in a circuit mode call but does have the opportunity to inform the infrastructure of its intention to find service in another cell. A typical case is when the MS is engaged in a group call.

- **Announced type 2.** With this reselection the MS knows the new cell before changing to it, an improvement over announced type 3, but it does not know its channel allocation on the new cell in advance. This can occur when the MS is currently engaged in CMCE and CONP connections and unable to make prior negotiation for channel allocation with a new cell, but informs the serving cell of its intention to find service elsewhere. The MS then negotiates directly with the new serving cell. This reselection can result in a temporary interruption of connection.

- **Announced type 1.** With this option, the MS knows a new cell and the traffic channel allocations on that cell before deciding to leave its serving cell. It also informs the serving cell of its intention to find service on another cell and the details of its preferred new serving cell. Negotiations for access to the new serving cell are performed via the serving cell and the infrastructure. Channel allocations on the new serving cell are issued via the present serving cell. This reselection procedure essentially provides a seamless handover and is referred to as *forward registration*.

Table 8.2 summarises the above cell selection types and their distinguishing features with respect to interaction with the serving and new cells, and mobile's prior knowledge of channel allocation in the new cell. The TETRA standard specifies that the MS-MLE protocol should support as a minimum the first three cell selection procedures, i.e. undeclared, unannounced and announced type 3. However, it is not stipulated for the network to know which type of cell reselection procedures the MS can support in order for the cell reselection procedures to work. The MS may be able to determine which types of reselection procedures are supported by the network from the neighbouring cell information transmitted by MLE signalling. For details on this, the reader is referred to Section 8.4.5 on the MLE PDU description for the D-NWRK-BROADCAST message. If the network does not support neighbouring cell information in the MLE signalling, then it will be assumed that only the first three procedures will be supported by the network.

Table 8.2 Cell selection types

Cell reselection types	Serving cell informed?	New cell informed?	Mobile knows its channel in advance?
1. Undeclared	No	No	No
2. Unannounced	No	Yes	No
3. Announced Type 3	Yes	Yes	No
4. Announced Type 2	Yes	Yes	No
5. Announced Type 1	Yes	Yes	Yes

8.4.4 MLE Services and Access Points

The MLE provides services[1] to layer 3 protocol entities above it with the MS-network peer protocol interaction as depicted in Figure 8.7. Each MLE message contains a protocol discriminator in the first octet that allows the routing of messages to the service access points as shown in Table 8.3 with the protocol discriminator encoding shown in Table 8.4. From the Table, all messages destined for the MLE are preceded by binary 101_2 as an identifier. Protocol messages destined for MLE sub-layer functions are called MLE PDUs while those destined for the upper protocol entities (MM, CMCE, CONP and SCLNP) are called MLE service user (MLE-SU) PDUs.

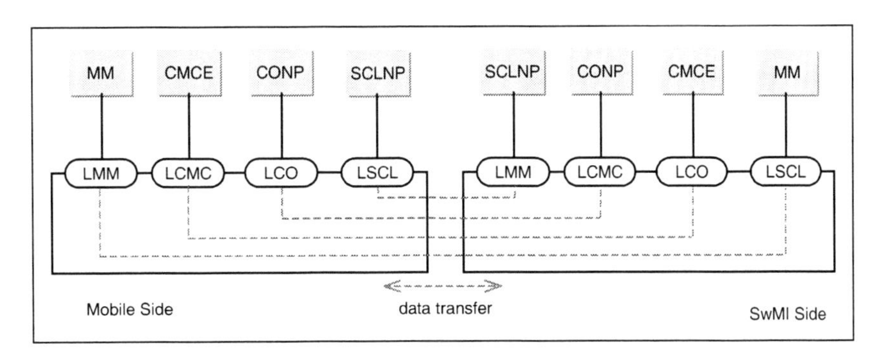

Figure 8.7 Services relationships offered by the MLE at the air interface

Table 8.3 MLE SAPs

SAP Name	Upper layer 3 potocol service user
LMM-SAP	Mobility Management (MM)
LCMC-SAP	Circuit Mode Control Entity (CMCE)
LCO-SAP	Connection Oriented Network Protocol (CONP)
LSCL-SAP	Specific Connectionless Network Protocol (SCLNP)

[1] The MLE relies on primitives from data link layer below it which will not be described here.

Table 8.4 Protocol discriminator encoding

Binary ID	Protocol
000	Reserved
001	MM protocol
010	CMCE protocol
011	CONP protocol
100	SCLNP protocol
101	MLE protocol
110	TETRA management entity protocol
111	Reserved for testing

Because of the importance of radio link management functions in mobile communications, in the following section details of primitive functions executed by MLE PDUs will be described. These will provide more insight into how the radio links are managed, including some of the unique features built into TETRA (for instance handover and power control procedures) at the radio link management. MLE-SU PDUs will be briefly mentioned in later sections under the protocols that utilise the MLE services as shown in Table 8.3.

8.4.5 MLE PDU Descriptions

The MLE functions described in the preceding section, also shown in Figure 8.7, are made possible with the execution of service primitives and consequent exchange of PDUs. MLE PDUs are concerned with the radio link management. These tasks are carried out with a set of commands, each of these identified by a protocol discriminator with the coding as shown in Table 8.5. MLE PDUs which have the protocol discriminator 101_2 (for MLE protocol as shown in Table 8.4) comprise both cell change PDUs and network broadcast PDUs. It should be noted that the same PDU type code, shown in Table 8.5, may be interpreted differently, depending on whether the PDU is from BS to MS, or MS to BS. Figure 8.8 depicts the flow of MLE PDUs between the TETRA SwMI and the MS-MLE. The MLE PDUs below are presented in a logical order of their invocation, rather than in the order of their binary encoding.

Table 8.5 MLE PDU types

Binary ID	MLE PDU type (downlink)	MLE PDU type (uplink)
000	D-NEW CELL	U-PREPARE
001	D-PREPARE FAIL	Reserved
010	D-NWRK-BROADCAST	Reserved
011	Reserved	Reserved
100	D-RESTORE-ACK	U-RESTORE
101	D-RESTORE-FAIL	Reserved
110	Reserved	Reserved
111	Reserved	Reserved

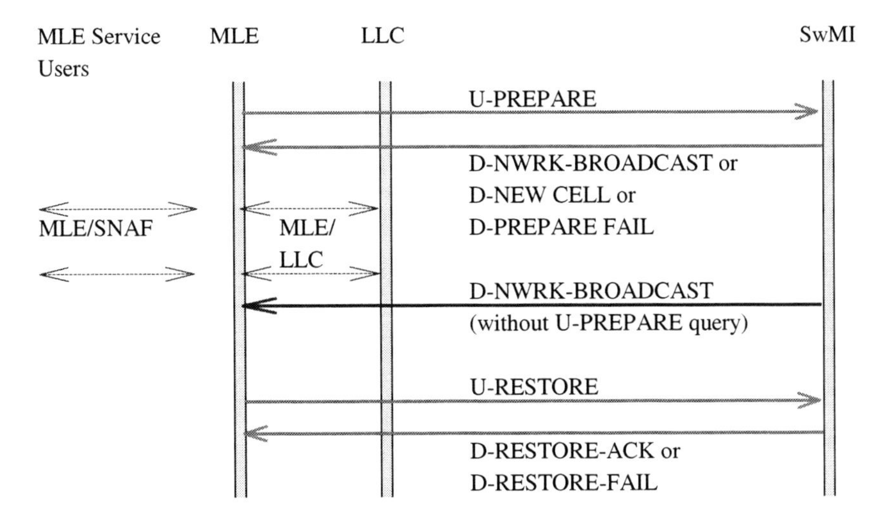

Figure 8.8 MLE PDU exchange procedures

8.4.5.1 D-NWRK-BROADCAST

This is a broadcast message from the SwMI that informs the MS-MLE about parameters for the serving cell and parameters for one or more neighbour cells. The MS may also enquire for the same message by sending the U-PREPARE PDU (see below) and without having to wait for D-NWRK-BROADCAST.

PD	PDU Type	Cell reselect parameters	Cell service level	TETRA Network Time	Number of neighbour cells	Neighbour cell info.
(3)	(3)	(16)	(2)	(48)	(3)	[...]

shaded fields are optional

PD (protocol discriminator): 101_2 for MLE

PDU type: D-NWRK-BROADCAST = 010_2

Cell reselect parameters: refer to Section 8.4.3 for definitions and Table 8.7 for coding.

Cell service level: indicates traffic loading in a cell. 00_2: load unknown; 01_2:low load; 10_2: medium load; 11_2: high load.

TETRA network time: the absolute network time used for time stamping in the SCLNP protocol. Formed from 24-bit UTC (universal time coordinate) time and 24-bit local time offset. All 1s reserved to flag invalid timestamp value on network malfunction etc.

Number of neighbour cells: indicates how many *neighbour cell information* elements follow. Up to 8 neighbour cells is possible.

Neighbour cell information: very extensive information for a neighbour cell, including cell identities (MCC, MNC, LAC), max transmit power, minimum receive access level, main carrier number, etc. Refer to Table 8.6.

Figure 8.9 D-NWRK-BROADCAST PDU message

Cell reselecting parameters definitions and their importance in pathloss parameter calculation are described in Section 8.4.3. Close inspection of the D-NWRK-BROADCAST PDU in Figure 8.9 shows that this protocol message can provide essential radio link management information about neigboring cells. This point is apparent from additional encoding schemes shown in Table 8.6 and Table 8.7.

Table 8.6 Neighbour cell information element

Information Element	Bit Length	Remarks
Cell identifier	5	Identifies a cell. See also under U-PREPARE PDU below.
Announced cell reselection types supported	2	00: only announced type 3 01: announced type 2 and 3 10: announced type 1 and 3 11: announced type 1 and 2
Neighbour cell synchronisation	1	0: neighbour cell not synchronised 1: neighbour cell is synchronised
Cell service level	2	Cell traffic loading – see above in Figure 8.9
Main carrier number	12	Indicates the carrier number relative to the base frequency defined by the frequency band*
Main carrier number extension	10	If present defines an offset of the above carrier number*
MCC	10	If not present assumed to be the same as that of the serving cell
MNC	14	If not present assumed to be the same as that of the serving cell
LA	14	If not present assumed to be the same as that of the serving cell
Maximum MS transmit power	3	If not present assumed to be the same as that of the serving cell
Minimum RX access level	4	If not present assumed to be the same as that if the serving cell
Subscriber class	16	If not present assumed to be the same as that if the serving cell
BS service details	12	If not present assumed to be the same as that if the serving cell
Timeshare cell information	5	If not present, it shall be assumed that the neighbour cell is not operating in a discontinuous mode of operation.
TDMA frame offset	6	If present, the neighbour cell shall be synchronised to the serving cell and this value indicates the frame offset for the neighbour cell. If the cells are synchronised and this element is not present, it shall be assumed by the MS that the frame offset is zero.

* downlink main carrier frequency = (base frequency + (main carrier number x 25) + offset) kHz
uplink main carrier frequency = (downlink main carrier frequency + duplex spacing) kHz

8.4.5.2 U-PREPARE PDU

The message is sent on the serving cell to the SwMI by the MS-MLE, when preparation for cell reselection to a neighbouring cell is in progress. In response, D-NEW CELL or U-PREPARE FAIL message is expected from the network. A timer of 5 seconds is used with this message to await for the response from the network; if no response is received within this time, then, a different course of action may be taken by the MS (e.g. reselecting a new cell directly).

Table 8.7 Cell reselect parameter coding

PDU information element	Length	Value	Remark
		0000_2	0 dB
SLOW_RESELECT_THRESHOLD	4-bit	0001_2	2 dB
		...	... etc
		1111_2	30 dB
FAST_RESELECT_THRESHOLD	4-bit	same as for row 1 element	same as for row 1 element
SLOW_RESELECT HYSTERESIS	4-bit	same as for row 1 element	same as for row 1 element
FAST_RESELECT_HYSTERESIS	4-bit	same as for row 1 element	same as for row 1 element

The MS can request transmission of the D-NWRK-BROADCAST PDU by sending a U-PREPARE PDU to the SwMI. This may be necessary when the mobile has not yet received the neighbouring cell information and needs this in order to initiate cell reselection procedures. This can occur if the current serving cell signal level is falling and the MS cannot wait for the normal D-NWRK-BROADCAST broadcast to be sent. The U-PREPARE PDU sent to the SwMI to request the broadcast PDU should not contain an SDU or other optional elements as this procedure is simply intended to prompt the SwMI.

PD	PDU Type	Cell Identifier	SDU
(3)	(3)	(5)	[...]

PD (protocol discriminator): 101_2 for MLE

PDU type: U-PREPARE PDU = 000_2

Cell Identifier: Identifies a neighbour cell. This cell identifier, if 00001_2 to 11111_2, may have been obtained from the **D-NWRK-BROADCAST** procedure, above or alternatively, the MS-MLE can set this value to 00000_2 to initiate a neighbour cell enquiry.

SDU: Service data unit is for forward MM registration.

Figure 8.10 U-PREPARE PDU message

8.4.5.3 U-PREPARE FAIL message

This is one of the two possible responses to the U-PREPARE message just described. Upon receipt from the SwMI (through the D-NWRK-BROADCAST procedure described above) this message is used by the MS-MLE as a preparation failure, while announcing cell reselection to the old cell. The message PDU is depicted in Figure 8.11.

PD	PDU Type	Fail Cause
(3)	(3)	(2)

PD (protocol discriminator): 101_2 for MLE
PDU type: U-PREPARE FAIL PDU = 001_2
Fail Cause: Indicates to the MS-MLE the failure cause as a result of requesting an MLE service in the SwMI. These are encoded as follows:
00_2: Neighbour cell enquiry not available 10_2: Subscriber class (note 1) not allowed
01_2: Cell reselection type not supported 11_2: Restoration cannot be done on cell

(note 1):
The subscriber class is intended to offer a population subdivision of some sort in up to 16 classes; the operator defines the values and meaning of each class. The MS can obtain this information with the D-NWRK-BROADCAST message described above.

Figure 8.11 U-PREPARE FAIL message

8.4.5.4 D-NEW CELL message

This message, from the SwMI in response to the U-PREPARE message, informs the MS-MLE that it can select a new cell. The new cell is as previously indicated in the U-PREPARE message as a cell identifier as shown in Figure 8.10.

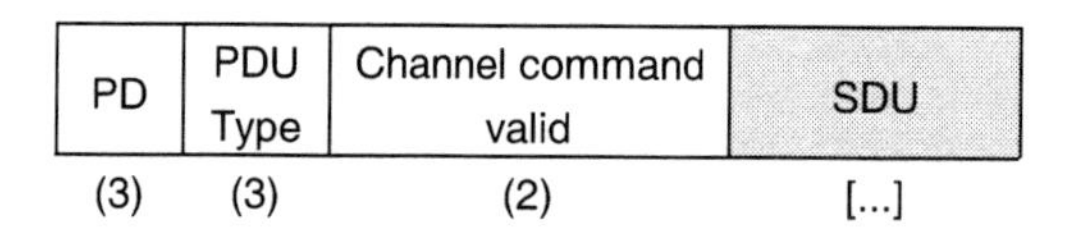

PD (protocol discriminator): 101_2 for MLE
PDU type: D-NEW CELL PDU = 000_2
Channel command valid: Indicates to the MS-MLE when to initiate a channel change as a result of cell reselection with the following binary coding:
00_2: Follow channel allocation in MAC header
01_2: Change channel immediately
10_2: No channel change; wait for next D-NEW CELL
11_2: Reserved
SDU: An optional field to carry an MM registration PDU which is used to forward register to a new cell during announced type 1 cell reselection.

Figure 8.12 D-NEW CELL message

8.4.5.5 U-RESTORE message

This message is sent by the MS-MLE, when restoration of the C-Plane towards a new cell is in progress. In response to this message, D-RESTORE-ACK or D-RESTORE-FAIL message is expected from the network.

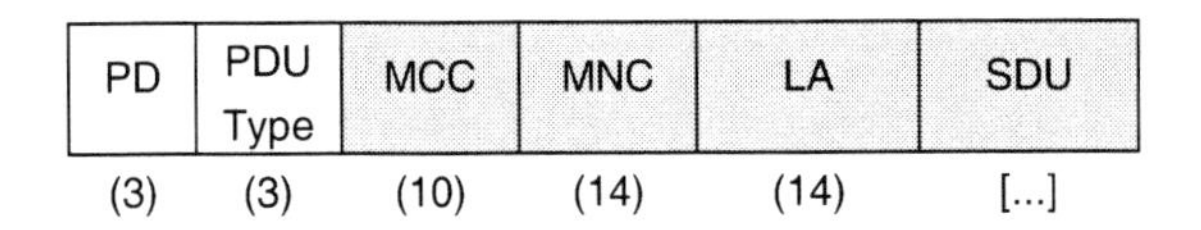

PD (protocol discriminator): 101_2 for MLE
PDU type: U-RESTORE PDU = 100_2
The remaining fields are optional and present if their values on the new cell are different from those on the old cell.
MCC: Mobile Country Code; **MNC**: Mobile Network Code; **LA**: Location Area
SDU: An optional field to carry CMCE protocol information used to restore a call after a cell reselection.

Figure 8.13 U-RESTORE message

8.4.5.6 D-RESTORE-ACK message

This message, from the network, acknowledges the C-Plane restoration on the new selected cell when requested by MS-MLE. It is in response to the D-RESTORE message.

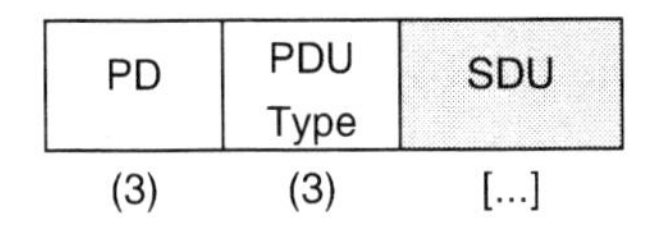

PD (protocol discriminator): 101_2 for MLE
PDU type: U-RESTORE-ACK PDU = 100_2
SDU: An optional field to carry CMCE protocol information for restoring a call after a cell reselection.

Figure 8.14 U-RESTORE-ACK message

8.4.5.7 D-RESTORE-FAIL message

This message is identical to the U-PREPARE FAIL message described under Section 8.4.5.3 above, except for the PDU type field, which is set to 101_2.

8.4.6 Concluding Remarks on MLE

The MLE represents one of the most extensive specifications of the TETRA document. This is not surprising considering that the radio link management represents a kernel of mobile cellular operation. The TETRA MLE attempts to account for conceivable pitfalls in the radio environment, for instance, as demonstrated by the pathloss calculation, cell ranking and cell reselection procedures and as a result this leads to a detailed MLE protocol. By way of an example, it is interesting to note that the attachment management sub-entity described in 8.4.3 alone runs over 20 pages of the standards document but only represents a small fraction of a rather hefty protocol. The description in Section 8.4 is therefore an attempt to distil important aspects of the MLE protocol.

The main tasks involved in the radio link management functions have been highlighted with an outline of how the MLE functions relate to the various protocol entities in control signalling. Assisted with the MLE, the mobile performs continual cell synchronisation, scanning, signal strength measurements, and neighbour cell ranking since it is in charge of cell handover. The MLE PDUs have also been described in order to reveal the radio link management functions through the interaction of the MS and the SwMI. Five types of cell selection schemes are specified for TETRA, and together with the cell selection criteria, defined in terms of signal threshold values, ample opportunities exist for implementing resource management strategies that are adaptive and flexible.

As important as the MLE in mobile communications protocol functions is the mobility management. This is the subject of the Section 8.5 below.

8.5 MOBILITY MANAGEMENT PROTOCOLS

Basic concepts in mobility management have been described in Section 5.10 and this background reading is assumed for the protocol descriptions presented in this section. It is therefore useful that the reader is familiar with the related coverage in Section 5.10.

8.5.1 Mobility Management Functions

Mobility management (MM) deals with network signalling aspects of authentication and registration procedures. On the MS side, the MM sub-layer controls which cells the MS is allowed to roam and decides whether the MS is successfully connected to the network. As part of mobility management functions, TETRA defines an additional task for energy economy management within the MM sub-layer. This function allows the MS to switch into modes which consume less power when not involved in a call.

In total, the protocol services provided by the MM sub-layer comprise five major tasks as summarised below.

- *Registration procedure (due to user request or MLE indication).* This procedure allows a user (through user application or radio link management indication) to register to the network and then be informed of the result of the registration. When a user roams or migrates he or she will also be informed that the MS is ready for use or that registration was not possible. With the same procedure, cancellation of the registration or *deregistration* for network detachment can be requested whereupon the user will be notified with an indication for success or failure.
- *Attachment/detachment of group identities (both due to user and network request).* This service informs the user application to either activate or deactivate already defined group identities in the MS.

- *Terminal disable/enable notification (network control).* This procedure notifies the infrastructure of the MS's status and this status is accordingly updated as attached/detached in the network database. This is in contrast to simply powering up or down during which the infrastructure is not explicitly informed. With this notification, the infrastructure avoids unnecessary paging since only attached MSs will be paged. If detached, all the traffic to the MS are rejected or possibly rerouted to an alternative destination (e.g. to voicemail) if such feature is supported.
- *Terminal attach/detach notification (terminal control).* This procedure informs the user application of the temporary or permanent disabling performed by the network.
- *Terminal energy saving mode change (due to user request).* With this procedure, the user requests for changing the energy saving mode.

Most of the above MM functions are illustrated with a peer PDU exchange between the MS and the network as summarised in Section 8.5.8.

8.5.2 MM Subsystem Elements

Mobility management subsystem is an important part of the TETRA infrastructure. It relies on the following elements to carry out the MM functions outlined in Section 8.5.1.

- **Mobile database.** This consists of the user authentication key (UAK) which is an important parameter for the authentication routines[2] during the various registration procedures; ITSI for terminal addressing for authentication and detach/attach MM functions; and *Search Area* which may have been defined by the user. (TETRA has a Supplementary Service for defining a call area, known as *Select Area*, as described in Section 5.9.6).
- **Timer within the MS.** This is used for periodic registration which may be invoked by the infrastructure.
- **Network database.** This is the most important element of the MM subsystem. An overview of the basic concepts for such databases has been presented in Section 5.10.2. The network side uses this database (which could be distributed and cached for fast access) for authenticating the user and subscribed services. In visited networks, an equivalent function will be carried out by a visitor database (VDB) by downloading essential user information from the home database (HDB).

8.5.3 Functional Organisation of the MS-MM

At the MS side there are two entities involved with MM procedures, namely the user application at the higher level, and radio link management, through the MLE protocol entity, at the lower layer. The functionalities of the MM protocol entity can be explained with reference to the interaction with these two entities as well as the peer interaction with the network side as depicted in Figure 8.15. The local interaction is effected by communication service primitives at the appropriate service access points – on the MLE

[2] Authentication procedures are discussed in Chapter 9.

side shown by the LMM-SAP and on the user application side by the TETRA Network MM SAP (TNMM-SAP).

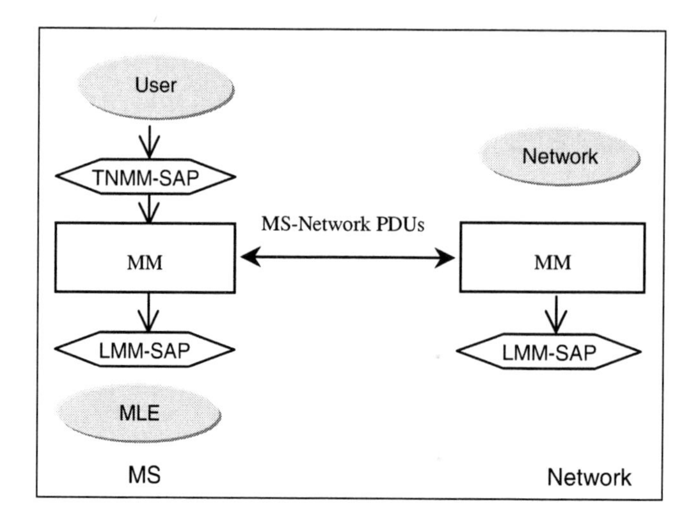

Figure 8.15 MS-Network MM peer protocol interaction

Three levels of interaction can be identified with this representation:

- **MM at LMM-SAP.** This interacts with the MLE sub-layer that manages the radio link. If, for instance, the MLE detects a new location area, then it will request the MM to initiate a registration procedure for location update. This update will be carried out with a peer protocol mentioned below.
- **MM at TNMM-SAP.** This interacts with the higher layer protocols usually user-initiated functions such as SIM card attachment or removal, explicit request for user registration/deregistration, explicit request for group attachment/detachment, MS power on/off, user request for energy economy mode change, and on demand authentication.
- **MM with Network-MM peer.** This interacts with the network through the exchange of protocol data units and performs the MM status query and update of relevant status information, which is stored in the network database. The status information includes location update, terminal enable/disable status, attachment status, and energy mode.

MM protocol functions under the above three categories are summarised below.

8.5.4 MM at the LMM-SAP

The whole action of mobile cellular communication is initiated upon the activation of the MS-MLE by the MM protocol entity. The MLE, which is responsible for cell scanning, preferred cell ranking, and handover decisions, will actually stay inactive until the MM issues an MLE-ACTIVATE request primitive to the MLE entity. In other words, the LMM-SAP acts as the master SAP, enabling and disabling service provision at the other access points including circuit switched connections and packet data services.

The initial activation of the MLE is usually triggered when the mobile is switched on or when a SIM card is inserted. This status is routed from the user application down to the MM protocol entity through the TNMM-SAP. The MLE-DEACTIVATE primitive, on the other hand, does the opposite of suspending the radio link. Table 8.8 lists some of the service primitives available for the LMM-SAP with brief descriptions of their functions.

The detailed functions handled through MM-MLE interaction are as follows:

- *Intervention.* This takes place whenever the MM is required to take over control of the roaming or migration processes either by performing registration/authentication or by performing activation. Interventions by the MM protocol entity should be evident from the primitives listed in Table 8.9.
- *Roaming.* This is to perform registration with the change of a location area (LA) within the TETRA network.
- *Forwarding.* This is to perform forward registration with the change of an LA within the TETRA network while still being attached to the previous LA.
- *Migrating.* This is to perform registration of an LA from one TETRA network to another network.

Table 8.8 Selected service primitives for LMM-SAP

Service primitives	Primitive function
MLE-ACTIVATE request	a request to initiate the selection of a cell for communications; always invoked on after power on and may be made at any time thereafter.
MLE-ACTIVATE confirm	a confirmation to the MM entity that a cell has been selected with the required characteristics.
MLE-ACTIVATE indication	an indication to the MM to react when no suitable cell is available.
MLE-DEACTIVATE request	a request by the MM entity to deactivate all MLE procedures and to return to the NULL state. No communication resources are available for use after this primitive has been issued.
MLE-OPEN request	a request by the MM entity to instruct the MLE to provide access to communication resources for other layer 3 entities after successful registration.
MLE-CLOSE request	a request by the MM entity to instruct the MLE to remove access to communication resources for the other layer 3 entities, but keeping access to the communication resources for the MM entity.
MLE-LINK indication	An indication by the MLE to the MM entity that the MS has selected or is about to select a cell outside the Registered Area.
MLE-UPDATE request	a request by the MM entity to inform the MLE about new criteria concerned with the monitoring of other possible cells.

8.5.5 MM at the TNMM-SAP

The MM primitives at the TNMM-SAP originate from the user application and they are reported to the network through peer interaction of the mobile and network MM protocols. The detailed primitive functions at the TNMM-SAP are listed below:

- **ITSI attach.** An attachment of ITSI, after previously being detached, should be reported to the MM so that a registration can be initiated with the network. This event may be triggered upon SIM card insertion, mobile reset, or power switch on.
- **ITSI detach.** A detachment of ITSI, after previously being attached, should be reported to the MM. Examples of events that should cause an ITSI detachment are removal of the SIM card and power down of MS.
- **Registration on demand.** Registration may be requested by the user to force the registration and/or the authentication procedure. This request is reported to the MM via the TNMM-SAP which will then trigger registration on demand.
- **Network authentication on demand.** Authentication can be requested by the user to force an authentication of the TETRA infrastructure. This may be invoked by an "authentication button" which can be provided on the user terminal or easily simulated by the user application.
- **Energy economy mode change.** The change of an energy saving scheme can be requested by the user. This is reported to the mobility management peer entity in the infrastructure.

8.5.6 MM with Network-MM Peer

This aspect of the MM protocol is concerned with the detailed functions outlined below. A selection of illustrative examples is presented in the following section.

- **Location update command**: used to initiate registration
- **Location acceptance**: used to accept or reject a location registration.
- **Network initiated user authentication**: to initiate authentication independent of registration.
- **Temporary disable**: used to temporary disable a MS.
- **Enable**: used to enable a MS temporary disabled.
- **Permanent disable**: used to permanently disable a MS.
- **Downloading of group identities**: performs attachment/detachment to group identities so that group calls can be routed to mobiles belonging to that group.
- **Periodic registration**: The need for periodic registration has been addressed in Section 5.9. Periodic registration is based on a timer within a MS.

8.5.7 Overview of MM Procedures

The implementation of the MM functions summarised in Section 8.5.1 and subsequently described with respect to the MM SAPs up to Section 8.5.6, should be straightforward based on the service primitives and PDUs which are specified by the TETRA standard. Some of these are summarised in Table 8.8 and in the following section in Table 8.9.

In the remaining part of the MM topic, we will describe illustrative MM procedures based on the peer-to-peer PDU exchanges between the MS and the infrastructure which have been outlined in Section 8.5.6. This approach is intended to serve two purposes: first, to illustrate the protocol mechanisms that conceptually apply to the other MM functions, and second, to highlight important MM protocol functions and associated system parameters.

8.5.8 Description of Selected MM PDUs

Figure 8.16 shows some of the important peer-to-peer MM procedures between the MS and the infrastructure. Table 8.9 summarises the MM PDUs used for exchange of commands and parameters between the MS and the infrastructure. A selection of these PDUs, shaded in Table 8.9, will be described.

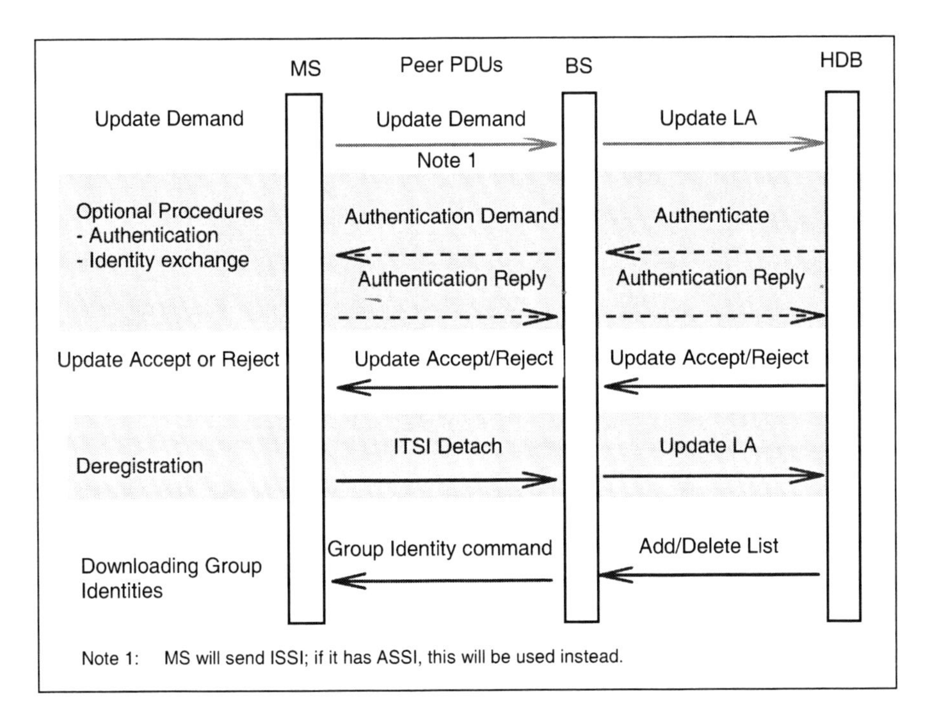

Figure 8.16 MM registration procedures

Since the aim here is to provide an insight into MM parameters and protocol mechanisms rather than an exhaustive description, the reader should be aware that in some of the cases below only a partial description of a PDU is given. Furthermore, a PDU may contain an optional element, which may or may not be transmitted with a mandatory part of the information.

Table 8.9 MM to network PDU types

Binary ID	MM PDU type (downlink)	MM PDU type (uplink)
0000	D-OTAR (note 1)	U-AUTHENTICATION RESPONSE
0001	D-AUTHENTICATION DEMAND	U-ITSI DETACH
0010	D-AUTHENTICATION REJECT	U-LOCATION UPDATE DEMAND
0011	D-DISABLE	U-STATUS
0100	D-ENABLE	U-AUTHENTICATION DEMAND
0101	D-LOCATION UPDATE ACCEPT	U-OTAR (note 1)
0110	D-LOCATION UPDATE COMMAND	U-AUTHENTICATION RESULT
0111	D-LOCATION UPDATE REJECT	U-ATTACH/DETACH GROUP IDENTITY
1000	D-AUTHENTICATION RESPONSE	U-ATTACH/DETACH GROUP IDENTITY ACK
1001	D-LOCATION UPDATE PROCEEDING	U-TEI PROVIDE
1010	D-ATTACH/DETACH GROUP IDENTITY	U-AUTHENTICATION REJECT
1011	D-ATTACH/DETACH GROUP IDENTITY ACK	U-DISABLE STATUS
1100	D-ENERGY SAVING	Reserved
1101	D-STATUS	Reserved
1110	D-AUTHENTICATION RESULT	Reserved
1111	Reserved	Reserved

Note 1: Over The Air Rekeying (OTAR) is a technique for encryption key management (Section 9.2.10)

8.5.8.1 U-LOCATION UPDATE DEMAND

This is the most common PDU message sent by the MS to the infrastructure to request an update of its location registration. This PDU, depicted in Figure 8.17, reveals a number MM information, including location type update, cipher parameters for voice encryption, details of MS capability, notified by the *Class of MS field*, and energy saving mode. Some of the PDU elements are commented upon in Figure 8.17, with bitwise encoding also shown for some of the elements.

The PDU element for *location updating type* indicates that there are up to eight types of location updates as summarised with bit encoding in the figure, most of which are too obvious from the discussions on MLE in Section 8.4 and on MM in the preceding part of this section. The infrastructure can also demand the MS for location updating (see D-LOCATION UPDATE DEMAND below). When the MS responds to this demand, it will set its location update type to *demand location updating* as one of the possible update types. This PDU element also makes it clear that the MS, with the use of a location update procedure, is required to update its status on being disabled.

Section 5.10 has introduced the concepts of enlarging a registration area and the use of periodic registration for maintaining a more accurate status of the mobile. Both of these features are controlled with this PDU as remarked in Figure 8.17.

PDU type	Location update type	Request to append LA	Cipher control	Ciphering parameters	Class of MS	Energy saving mode	LA info.	and others...
(4)	(3)	(1)	(1)	(10)	(24)	(3)	38	[...]

PDU type: U-LOCATION UPDATE DEMAND = 0010_2

Location update type: as defined by the binary code shown below.

000 Roaming location updating	100 Call restoration roaming location updating
001 Migrating location updating	101 Call restoration migrating location updating
010 Periodic location updating	110 Demand location updating
011 ITSI attach	111 Disabled MS updating

Request to append LA: 0 : No request to append LA; 1: Request to append LA

Cipher control: 0: Cipher off; 1: Cipher on

Ciphering parameters: this PDU element is present only if the **Cipher control** is set to on. (For explanation, refer to main text)

Class of MS: (refer to main text)

Energy saving mode: as defined by the binary code shown below.

000 Stay Alive	100 Economy mode 4
001 Economy mode 1	101 Economy mode 5
010 Economy mode 2	110 Economy mode 6
011 Economy mode 3	111 Economy mode 7

LA information: LA, LA country code (LACC) and LA network code (LANC)

Figure 8.17 MS PDU for location update demand

The PDU elements for *cipher*[3] *control* and *ciphering parameters* refer to the control information for TETRA encryption for network security. Encryption principles and TETRA network security management are presented in Chapter 9; here only the significance of these parameters will be briefly explained. The *cipher control* as a binary flag indicates whether the air interface encryption is on (enabled) or off (disabled). If enabled, then the *ciphering parameters* indicate the necessary detail of what type of encryption should be in operation.

Section 9.2.8 makes it clear that there are two types of encryption keys used for individual TETRA calls: the derived cipher key (DCK), or the static cipher key (SCK) which will be one of 32 keys stored in a MS. There are up to 16 key stream generators to choose from (8 for TETRA and 8 proprietary). The 10-bit ciphering parameters therefore encode whether DCK or SCK should be used (1-bit), a choice of 1 out of 16 possible key stream generators (4-bit), and if appropriate a choice of 1 out of 32 SCKs stored in a mobile (5-bit).

[3] A *cipher* is a method or algorithm for encrypting or decrypting information.

The *Class of MS* PDU element informs the infrastructure with the characteristics of the mobile terminal. Some of these parameters have been presented in Section 5.3.5 in the context of the different classes and capabilities of the TETRA MS. This PDU element forms a more comprehensive list with additional information pertaining to various protocols as summarised in Table 8.10.

The PDU element for *energy saving mode* designates one of eight energy economy levels, with *mode 0* indicating disabled energy saving, and modes 1 to 7 indicating varying levels of energy saving that a mobile may request. The need for energy saving as part of MM arises from the fact that low battery life is expected with the (approximately) linear power amplifier characteristics necessitated by *Π*/4-DQPSK modulation (Section 1.7.8). The infrastructure responds to the energy saving request either with the D-ENERGY SAVING PDU specific to energy management or upon location update through the U-LOCATION UPDATE ACCEPT PDU which is described below.

Table 8.10 Contents of the *Class of MS* PDU element

Note: *Unless indicated otherwise, the following information elements are encoded with a single bit flag, thus (0) or (1) as shown in the Remarks column.*

PDU information sub-element	Remarks
Frequency simplex/duplex operation	(0) simplex; (1) duplex
Single/multi-slot support	(0) Single slot; (1) Multi-slot
Concurrent multi-carrier operation	(0) Single; (1) Multi-carrier
Voice support	(0) No; (1) Yes
End-to-end encryption	(0) No; (1) Yes
Circuit mode data	(0) No; (1) Yes
SCLNP (connectionless data protocol)	(0) No; (1) Yes
CONP (connection-oriented protocol)	(0) No; (1) Yes
Air interface encryption	(0) No; (1) Yes
CLCH needed on carrier change (Note 1 below)	(0) No; (1) Yes
Concurrent channels (for concurrent services)	(0) No; (1) Yes
Advanced link	(0) No; (1) Yes
Minimum mode	(0) No; (1) Yes
Carrier specific signalling channel	(0) No; (1) Yes
Reserved	2-bit for future expansion
TETRA air interface standard version number	3-bit version number
Reserved	5-bit for future expansion

Note1: CLCH (Common Linearisation Channel) is explained in Section 4.5.5.

8.5.8.2 D-LOCATION UPDATE PROCEEDING

This PDU message is sent to the MS by the infrastructure in response to the U-LOCATION UPDATE DEMAND. Of significance here, also remarked in Figure 8.18, is the allocation of an alias short subscriber identity (ASSI), or V-ASSI in a visited network, which is specified to support subscriber confidentiality as part of the TETRA network security. Background on the TETRA address definition, including the address extension element shown in Figure 8.18, appears in Section 5.8.4.

PDU type	SSI	Address extension	Proprietary optional information
(4)	(24)	(24)	[...]

PDU type: U-LOCATION UPDATE PROCEEDING = 1101_2

SSI: ASSI or V-ASSI that the MS shall use in subsequent communication with the network.

Address Extension: this indicates the extended part of the subscriber identity which is formed from a 10-bit mobile country code (MCC) and 14-bit mobile network code (MNC).

Figure 8.18 Infrastructure to MS PDU for allocating an alias subscriber address

8.5.8.3 U-LOCATION UPDATE ACCEPT

This PDU is sent by the infrastructure to indicate that updating in the network has been completed. Most of the information elements for this PDU are evident from the preceding discussion as remarked in Figure 8.19. The *energy saving information* element represents confirmation of the energy saving request by the MS. The 14-bit encoding for this element is made up of a 3-bit selection for energy economy mode (see also Figure 8.17) and energy saving starting point specified in terms of a 5-bit TDMA frame number and 6-bit multiframe number.

PDU type	Location update type	SSI	Address extension	Subscriber class	Energy saving info	and others...
(4)	(3)	(24)	(24)	(10)	(14)	[...]

PDU type: U-LOCATION UPDATE ACCEPT = 0101_2

Location update type: as defined for U-LOCATION UPDATE DEMAND above.

SSI: as assigned with the U-LOCATION UPDATE PROCEEDING PDU above.

Address Extension: as assigned with the U-LOCATION UPDATE PROCEEDING PDU above.

Subscriber class: the subscriber class is intended to offer a population subdivision of some sort in up to 16 classes; The operator defines the values and meaning of each class. The MS can obtain this information through the broadcast message transmitted in a cell.

Figure 8.19 Infrastructure PDU for location update accept

8.5.8.4 D-LOCATION UPDATE REJECT

This PDU is sent by the infrastructure to indicate that updating in the network is not accepted, with the *reject cause* information element indicating the reason. Currently defined reject causes are shown in Table 8.11 and these are all self-explanatory.

PDU type	Location update type	Reject cause	Cipher control	Ciphering parameters	Address extension	Proprietary
(4)	(3)	(5)	(1)	(10)	(24)	[...]

PDU type: D-LOCATION UPDATE REJECT = 0111_2

Location update type: as defined for U-LOCATION UPDATE DEMAND above.

Reject cause: as assigned with the U-LOCATION UPDATE PROCEEDING PDU above.

Cipher control: as defined for U-LOCATION UPDATE DEMAND above.

Ciphering parameters: as defined for U-LOCATION UPDATE DEMAND above.

Ciphering parameters: as defined for U-LOCATION UPDATE DEMAND above.

Address Extension: as defined for U-LOCATION UPDATE PROCEEDING above.

Figure 8.20 Infrastructure PDU for location update reject

Table 8.11 Contents of *Reject cause* PDU element

Reject code	Reject cause	Reject code	Reject cause
00000	Reserved	01011	Roaming not supported
00001	ITSI unknown	01100	Migration not supported
00010	Illegal MS	01101	No cipher KSG
00011	LA not allowed	01110	Identified cipher KSG not supported
00100	LA unknown	01111	Requested cipher key type not available
00101	Network failure	10000	Identified cipher key not available
00110	Network congestion	10001	Incompatible service
00111	Service not supported	10010	Ciphering required
01000	Service not subscribed	10011	Reserved
01001	Mandatory element error (note 1)	...etc.	...etc.
01010	Message consistency error (note 1)	11111	Reserved

Note 1: This refers to an error associated with a PDU message being replied to.

8.5.8.5 D-LOCATION UPDATE COMMAND

This PDU is sent by the infrastructure to initiate a location update by the MS. Following this, a U-LOCATION UPDATE DEMAND by the MS is therefore expected. The only new PDU element that needs explanation here is the *group identity report* which is a flag indicating whether or not the MS should report with all its active group identities which will be used for establishing group calls. As an MS roams into a different network, for instance, it may be necessary for the network to determine how current group calls are configured. One way of doing this is by interrogating the MS for group identity report.

PDU type	Location update type	Group identity report	Cipher control	Ciphering parameters	Address extension	Proprietary
(4)	(3)	(1)	(1)	(10)	(24)	[...]

PDU type: D-LOCATION UPDATE COMMAND = 0110_2

Location update type: as defined for U-LOCATION UPDATE DEMAND above.

Group identity report: a flag for no report request (0) or report request (1).

Cipher control: as defined for U-LOCATION UPDATE DEMAND above.

Ciphering parameters: as defined for U-LOCATION UPDATE DEMAND above.

Ciphering parameters: as defined for U-LOCATION UPDATE DEMAND above.

Address Extension: as defined for U-LOCATION UPDATE PROCEEDING above.

Figure 8.21 Infrastructure PDU for location update reject

8.5.8.6 U-GROUP ATTACH/DETACH IDENTITY

This message is sent by the MS to indicate attachment/detachment of group identities in the MS. The infrastructure responds with the D-GROUP ATTACH/DETACH IDENTITY ACK PDU, with *Group identity accept/reject* element substituted for *Group identity report* element to flag if the request is accepted or rejected.

PDU type	Group identity report	Attach detach mode	Proprietary	Group identity uplink
(4)	(1)	(1)	[...]	[...]

PDU type: U-GROUP ATTACH/DETACH IDENTITY = 0111_2

Group identity report: a flag for no report request (0) or report request (1).

Attach detach mode: used to indicate group identity attachment (0) or detachment (1). The group identities to be attached or detached are specified within the **Group identity uplink** element.

Figure 8.22 Infrastructure PDU for location update reject

8.5.9 Concluding Remarking on MM Protocols

Like the radio link management functions handled by the MLE, the MM functions are too extensive to cover. The above protocol functionalities are selected to demonstrate some of the parameters and control mechanisms within the mobility management. The above examples have highlighted how location updating is performed and the information exchanged during the update process. Types of location update, MS characteristics, ciphering parameters, energy saving mode, on-registration alias subscriber identity download, and group identity attachment/detachment are some of the important MM functions identified with the above selected PDUs. There are MM functions not mentioned here, including MS enable/disable, subscriber detach, and status report by MS and infrastructure. As the names suggest, these functions are relatively simple with a straightforward protocol implementation.

8.6 CIRCUIT MODE CONNECTION ENTITY

8.6.1 Overview

The CMCE protocol entity is a network layer protocol that provides services to an end user application in call control (CC) service, supplementary service (SS), and short data service (SDS). The TETRA standard covers the CMCE protocol only on the terminal side (i.e. for the MS and the LS) and not the network side. However, a peer-to-peer relationship between the layers on the terminal side and the network side can be assumed.

As depicted by Figure 8.23, the CMCE function relies on the underlying MLE which implements the radio link management as detailed in Section 8.4. Services from voice and data MLE are obtained through the LCMC-SAP and these will be described in the subsections to follow. On the user side, the CMCE provides bearer services to higher layer applications through service primitives defined for three service access points (SAPs), namely:

- TNCC-SAP for CC services;
- TNSS-SAP for SS services;
- TNSDS-SAP for SDS services.

Understanding of the LCMC-SAP provided by the MLE and the SAPs on the application side should provide sufficient ground of how the CMCE function facilitates the circuit mode control of bearer services.

8.6.2 MLE Functions at CMCE SAP

The CMCE protocol entity obtains services from the underlying voice and data MLE through the LCMC-SAP (Figure 8.23). The CMCE is internally subdivided into four different sub-entities: CC, SS, SDS, and the protocol control (PC). The PC essentially coordinates the interaction of CMCE sub-entities with the underlying MLE protocol entity and becomes relevant only for operational description. Table 8.12 and Table 8.13 summarise the peer-to-peer PDUs between the MS and the infrastructure with a brief description of each of the PDUs. The functions of the CMCE sub-entities are summarised below.

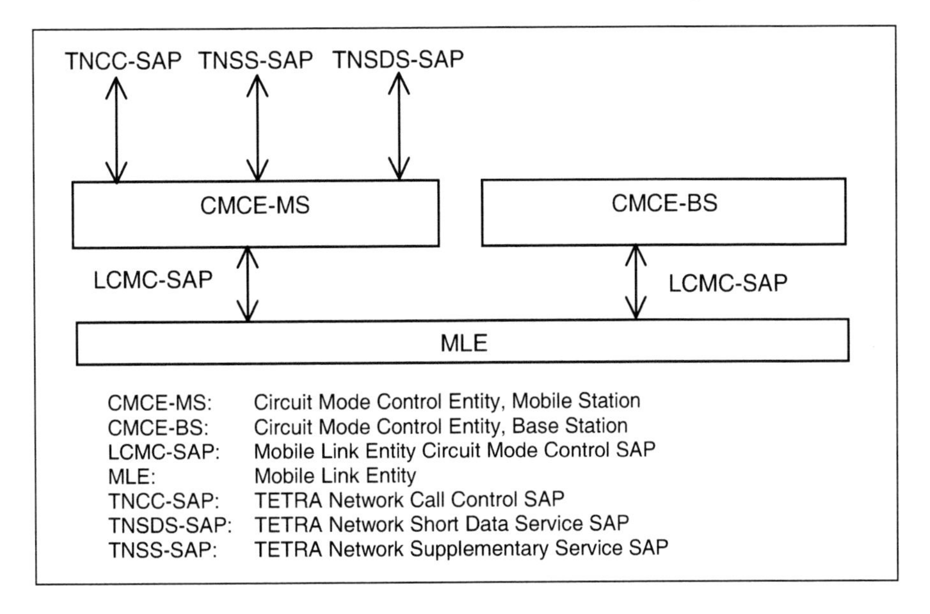

Figure 8.23 CMCE protocol organisation

8.6.2.1 Call control service

This is concerned with:

- basic call set-up (with attributes);
- call maintenance;
- dual tone multi-frequency (DTMF) encoding and sending;
- PTT (press-to-talk) requests, grants, and information;
- call clearance;
- change of teleservice/bearer service within a call.

8.6.2.2 SS service

These are services at the TNSS-SAP and consist of the following:

- invocation of a SS;
- activation/deactivation of a SS;
- definition of a SS;
- cancellation of a SS;
- interrogation of a SS;
- registration of a user to a supplementary service;
- reception of supplementary service messages.

8.6.2.3 Short data service

These are services at the TNSDS-SAP transferring free text messages containing up to 256 characters and up to 65536 status messages. The short data functional entity provides the following mobile originated and mobile terminated services:

- user defined short message reception and transmission both for individual message and group message;
- predefined short message reception and transmission both for individual message and group message.

Table 8.12 Summary of CMCE PDUs from the SwMI to terminal[1]

PDU	Brief description and comments
D-ALERT	Indication to the originating terminal that the call is proceeding and the connecting party has been alerted.
D-CALL-PROCEEDING	Acknowledgement from the infrastructure to call set-up request indicating that the call is proceeding.
D-CALL-RESTORE	Indication to a terminal that a call has been restored after a temporary break of the call. Refer to cell selection in Section 8.4.3 for the significance of this.
D-CONNECT	Indication to the calling terminal to through-connect[2] in response to a U-SETUP PDU.
D-CONNECT ACKNOWLEDGE	Indication to the called terminal to through-connect in response to a U-CONNECT PDU.
D-DISCONNECT	Disconnect request message sent from the SwMI to the terminal.
D-INFO	General information message to the terminal.
D-RELEASE	Message from the infrastructure to the terminal to inform that the connection has been released.
D-SDS-DATA	Used for receiving user defined SDS data.
D-STATUS	Used for receiving a pre-coded status message.
D-SETUP	Used for sending a call setup message to the called terminal.
D-TX-CEASED	Used to send a message from the SwMI to all terminals within a call that a transmission has ceased.
D-TX-CONTINUE	Used to send a message from the SwMI to the terminal that the interruption of the call has ceased.
D-TX-GRANTED	Used to inform the terminal concerned with a call that permission to transmit has been granted by the SwMI to one terminal, and to inform that one terminal that it has been granted permission to transmit.
D-TX-INTERRUPT	A message from the SwMI indicating that a permission to transmit has been withdrawn.
D-TX-WAIT	A message from the SwMI that the call is being interrupted.
D-FACILITY	Used to send call unrelated SS information to the terminal.

1. *terminal* refers to both MS and LS.

2. *through-connect* refers to a direct call-connection regardless of whether the terminal is on-hook or off-hook.

Table 8.13: Summary of CMCE PDUs from terminal (MS/LS) to the SwMI

PDU	Brief description and comments
U-ALERT	Acknowledgement from the called MS/LS that the called user has been alerted.
U-CALL-RESTORE	Indication from the terminal for restoration of a specific call after a temporary break of the call.
U-CONNECT	Acknowledgement to the SwMI that the called MS/LS is ready for through-connection.
U-DISCONNECT	Request from the MS/LS to the SwMI to disconnect a call.
U-FACILITY	Used to send call unrelated SS information to the SwMI
U-INFO	General information message from the terminal.
U-STATUS	Used for sending a pre-coded status message to the SwMI.
U-SDS-DATA	Used for sending user defined SDS data.
U-RELEASE	Acknowledgement to a disconnection request D-DISCONNECT.
U-SETUP	Request for a call set-up from a terminal.
U-TX-CEASED	Message to the SwMI that a transmission has ceased.
U-TX-DEMAND	Message to the SwMI that a transmission is requested.

8.7 PACKET MODE DATA SERVICES

8.7.1 Introduction

Packet mode data services are supported both by V+D and PDO TETRA systems. Two types of packet mode data services are envisaged depending on user application.

- *Connection oriented packet data service.* This is a data service which transfers X.25 packets from one source node to another by using a multiphase protocol that establishes and releases logical connections or virtual circuits between end users.
- *Connectionless packet data service.* This is a data service which transfers a single packet of data from one source node to one or more destination nodes in a single phase without establishing a virtual circuit. Standard ISO and TETRA-specific protocols are supported for this mode of data service.

At the user side, the appropriate protocol stacks for the above services can be identified with regard to the TETRA reference points described in Section 5.6. The implementation of the above data services takes one of two forms:

- Packet-mode X.25 data communication protocol (connection oriented or connectionless) running on R1 reference point.
- Character-mode X.28 data communication protocol at R4 reference point with integrated PAD (packet assembler and dissembler).

8.7.2 Overview of X.25 Data Services

X.25 represents the most widely implemented OSI network layer standard, with its original approval as far back as 1976. This standard defines procedures for connection of data terminal equipment (DTE) such as computers and data terminals, to packet switched networks via data communication equipment (DCE) as shown in Figure 8.24. With its revised specifications, X.25 is also used for packet switching over ISDN.

8.7.2.1 X.25 protocol layers

X.25 specifies three layers (Figure 8.25) for the DTE/DCE interface based on:

- the physical layer based on X.21 and V.24;
- the data link layer standard known as LAP-B (link access protocol – balanced); and
- the network layer (or packet layer in X.25 parlance) called packet level protocol (PLP).

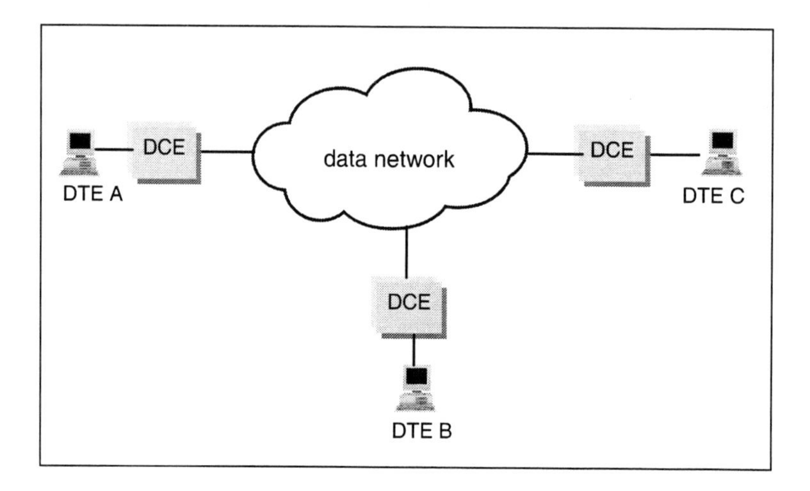

Figure 8.24 X.25 Network interface

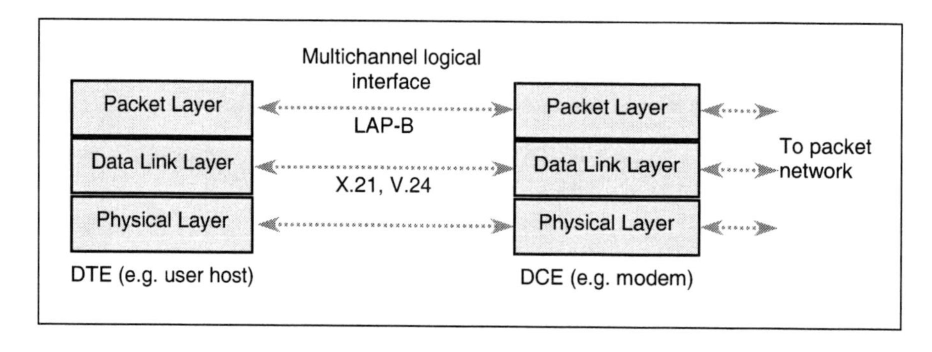

Figure 8.25 X.25 protocol layers

8.7.2.2 X.25 Virtual circuit services

X.25 network layer supports two types of virtual circuits: a permanent virtual circuit (PVC) and a virtual call (VC) also known as switched virtual call (SVC). Permanent virtual circuit (PVC) data service is like a leased or private telephone line which is permanently set up. On the other hand, virtual calls are like dial telephone calls and require connect and disconnect phases.

8.7.2.3 X.25 Inter-networking: the X.75 standard

The X.75 standard specifies the interface and protocol required for interconnecting X.25 networks. X.75 interface is essentially like the DTE/DCE X.25 link except that a special type of DCE called *signalling terminal exchange* (STE) is used on each side of the link between two X.25 networks (Figure 8.26).

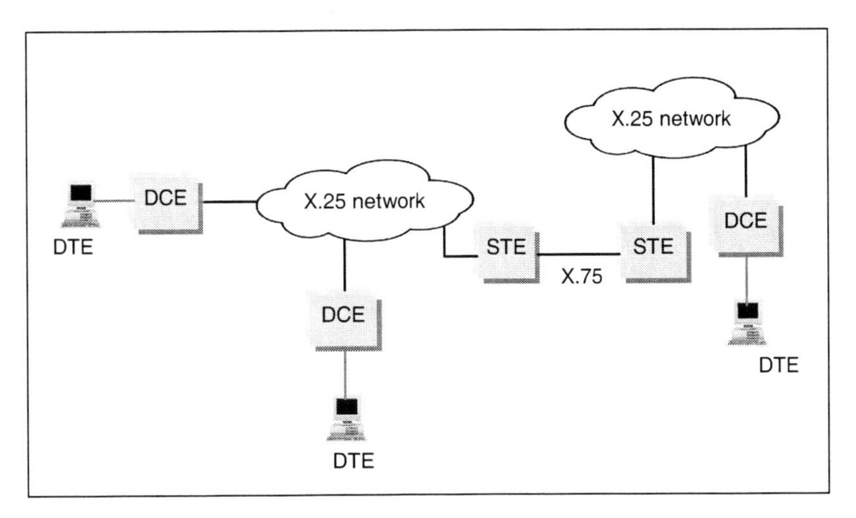

Figure 8.26 X.25/X.75 inter-networking

8.7.2.4 X.28 Simple character terminal connection

Simple character terminals with minimal functions are unable to handle the full synchronous X.25 protocols. Consequently, additional standards have been defined to allow simple asynchronous character terminals to be interfaced with X.25 connection. The interface function between a simple (character mode) terminal and X.25 is provided by a packet assembler/disassembler (PAD) under the X.3 standard which is usually implemented in software within the DCE.

Standard X.28 defines the interface between X.3 PAD and character mode DTE including data and control exchanges. A related standard, X.29, defines the protocol interface that allows parameters in X.3 PAD to be set by a remote intelligent packet mode DTE. The three standards X.3, X.28 and X.29 are collectively known as *Triple X standard*.

8.7.3 Overview of TETRA Data Services

For the TETRA LS interface, a peer-to-peer X.25 protocol is assumed from the network side as shown in Figure 8.27. The interface is typically provided by X.21 physical layer over the ISDN line. If a character mode interface is instead to be supported (over V.24 physical interface), then a packet assembler/disassembler (PAD) can be used between the character mode terminal and X.25 connection as shown in Figure 8.28. The X.21 physical layer interface is an ITU-T specification for synchronous operation over public data networks.

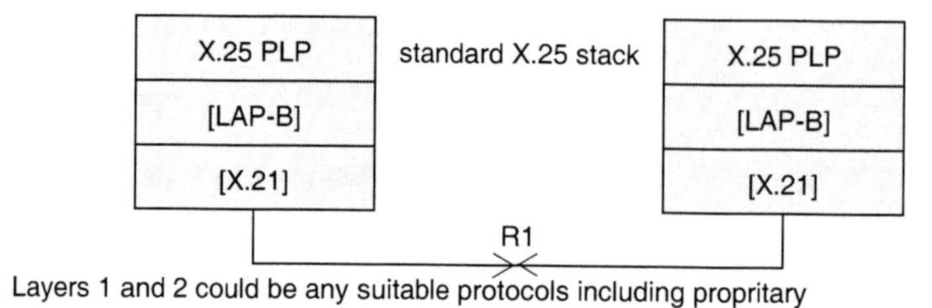

Figure 8.27 A connection-oriented protocol stack at reference point R1

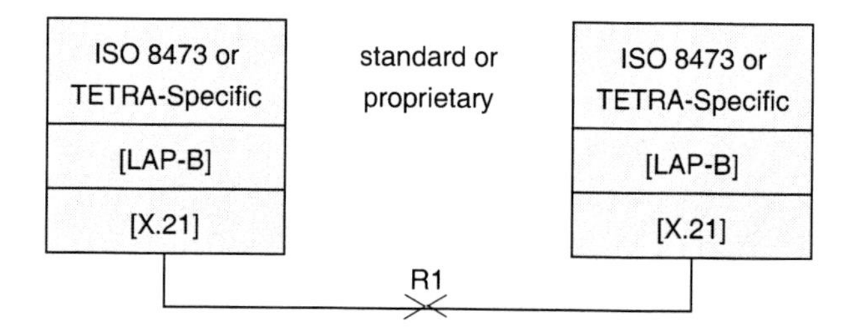

Figure 8.28 A connectionless protocol stack at reference point R1

The preceding descriptions for TETRA LS also apply to TETRA MS except that additional air interface protocols will be used at the reference point R2 between the mobile and the network before data is formatted for X.25 packet mode bearer service at reference point R1. This may be available as character mode interface through a PAD (possibly integrated with a mobile termination unit) as described under TETRA reference configurations in Chapter 5.

The air interface side of packet mode data services is handled by the CONP and CNLS sub-entities as introduced in Section 8.3.3. To convert the air interface protocols to X.25 packet services for connection to generic data terminal equipment (as for LS), an inter-working unit would be required. This function is actually served by the interface designated as MTU2 in the mobile functional grouping.

Based on the above overview, the packet mode bearer services can be grouped as follows:

- Connection mode data service on the air interface – the CONP protocol entity.
- Connectionless mode data service on the air interface – the CNLS protocol entity.
- Connection mode data service at the user interface – X.25 protocol.
- Connectionless data service at the user interface – ISO 8473 or TETRA specific protocols.

It should be noted that the first two are applicable only to the MS at R2 reference point whereas the last two apply both to MS and LS at the R1 and R4 reference points.

8.7.4 Specific Connectionless Network Protocol (SCLNP)

This data service protocol is so named due to its functionally similar services to the standard ISO 8473 connectionless network protocol but different in its implementation for optimal packet size. There are various TETRA network services that can be built around connectionless data protocols but as telecommunications application, most of these services require minimum delay and packet overhead. The SCLNP is therefore intended to meet these requirements, which otherwise may be constrained with the standard connectionless protocol. Some of the services using SCLNP include call control (call priority, call queuing, etc), numerous supplementary services and interfaces.

8.7.5 TETRA Inter-network Protocols

The above sections have described the TETRA protocol stacks that exits on the user side of the network. These are contained within the R0 to R4 reference configuration (Section 5.5). Recalling the six defined interfaces of TETRA, there are protocol stacks that apply to the inter-networking interfaces. These are identified with the reference points are summarised below:

- R5 reference point between a TETRA network and a Network Management Unit (NMU);
- R6 reference point between one TETRA network and another TETRA network;
- R7 reference point between one TETRA network and a non-TETRA packet data network.

The protocol implementation aspects of these inter-networking interfaces are similar to those already described in the Section 8.7.3, with the exception of the protocol type applicable at the interface. As an example, Figure 8.29 shows the protocol stack at the R6 reference point (TETRA to TETRA interface) based on X.25 or X.75 interface standard.

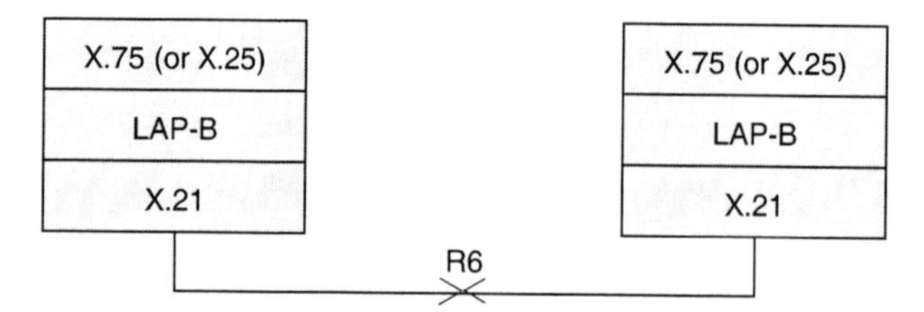

Figure 8.29 Protocol stacks at the R6 reference point

8.8 LAYERS 4 TO 7

The realm of TETRA standardisation is layer 1 to 3 of the OSI protocol architecture as the materials covered so far have attempted to illuminate. In fact, protocols in layers 4 to 7 have virtually nothing to do with the infrastructure directly, since these concern only the communicating entities such as end terminals or user applications. They are essentially shielded from the infrastructure by layer 3 protocols that provide network bearer services.

The higher layer protocols have some extra job to do, for instance providing a reliable communication link through end to end acknowledgement, and/or for manipulating user data for some communication objective (e.g. data compression, data encryption, etc). All these activities are totally transparent to the layer 3 protocol. For instance, as far as layer 3 protocol is concerned acknowledgement data could appear just like any other data with distinction made only by higher layer protocols. Similarly, it is not for a layer 3 protocol to know if a data is compressed or encrypted; its job is merely to route the data to its destination according to some communication policy defined within its protocol function.

In this section, we present a brief overview of higher layer protocols simply to highlight their significance in relation to the bearer service protocols described in this chapter. For a thorough treatment of higher layer protocols the reader is referred to [3] and [4].

8.8.1 Transport Layer Protocol

User applications built over bearer services generally require end-to-end transmissions of data packets with a specific quality of service and possibly with some translations for local requirement for an application. In particular, the transport and session layers work together for supervising the end-to-end delivery of data.
The transport layer is charged with the management of end-to-end information transfer, with the following being some of main functions:

- data packet fragmentation, re-sequencing, and flow control;
- exchanging connection parameters;
- multiplexing of tandem connections or virtual ports (for instance, to emulate several user ports on a single data pipeline when connected to a data network such as like X.25);

- discarding delayed copies of packets when the delay reaches some maximum threshold.

Transport control protocol (TCP) and user datagram protocol (UDP) are the most commonly used transport layer protocols. The TCP implements a reliable, connection-oriented communication service on the top of the IP (Internet Protocol), a datagram network protocol for Internet. UDP is the connectionless version of the TCP.

8.8.2 Session Layer Protocol

This layer is not as intensive as the network or transport layer and it could even be null for some communication services. Its task is mainly to do with the co-ordination of sessions between terminals. It essentially establishes connection and supervises the transfer of data. Session set-up may require some synchronisation, say with start-stop-resume type procedures. It is possible that synchronisation functions for multimedia frames (e.g. video conferencing) or end-to-end encryption of TETRA voice could be implemented with session layer functionalities. Contrary to this approach, it is also possible that the desired synchronisation functionalities could be implemented within the application layer. The decision at which layer to implement becomes important usually when inter-working with other hardware/software systems is an issue.

8.8.3 Presentation Layer Protocol

The presentation layer performs the syntax conversion, data encryption, and data compression. Examples include data encryption and decryption protocols. For instance, the decryption function of a TETRA end-to-end encrypted voice could be handled at this layer or possibly at the application layer.

8.8.4 Application Layer Protocol

Application layer, as the name suggests, defines the task or application to be performed and represents the culmination of all the layers to provide a communication service immediately close to the user (or user application). For example, a TETRA voice service or data service represents an application layer protocol. Other examples include, SDS messaging service, email, FTP, and the simple network management protocol (SNMP) widely used for network monitoring and control.

8.9 CONCLUSION

This chapter has detailed the main functions of the TETRA network layer protocols for the air interface. Protocol stacks for the TETRA air interface of the V+D, PDO and DMO systems have been outlined and important aspects of the protocols identified for further discussion. The network layer provides a convergence layer for the TETRA air interface protocols, the V+D, PDO and DMO.

The Mobile/base Link Entity (MLE) protocol entity, dealing with radio link management, represents the most comprehensive TETRA protocol functions. It has various mechanisms at its disposal for signal quality assessment and new cell reselection procedures and these have been described with some detail.

The mobility management procedures on the other hand are based on the interaction of three protocol entities – the user application (for user initiated authentication and registration) the MLE (for location update procedures automatically triggered by MLE measurements) and the network side (which can demand authentication and registration). MS-network peer PDUs have been used to illustrate MM procedures. Internetworking protocols use standard data communication protocols and these have been briefly described.

REFERENCES

[1] ETSI Technical Standard ETS 300 392-2: "Radio Equipment and Systems (RES); Trans-European Trunked Radio (TETRA); Voice plus Data (V+D); Part 2: Air Interface (AI)", 1996.

[2] ETSI Technical Report ETR 300-1: "Terrestrial Trunked Radio (TETRA); Voice plus Data (V+D); Designers' Guide; Part 1: Overview, Technical Description and Radio Aspects", 1997.

[3] Stallings, W., *Data and Computer Communications*, (5th Edn.), Prentice-Hall, Inc., Upper Saddle River, N. J., 1997.

[4] Tanenbaum, A. S., *Computer Networks*, (3rd Edn.), Prentice Hall PTR, Upper Saddle River, N. J., 1996.

9

Operational Aspects of the TETRA Network

9.1 INTRODUCTION

This chapter will introduce some of the topics which, in addition to the networking layer functions presented in the preceding three chapters, will provide the underlying concepts of an operational TETRA system. The following section introduces network security management based on authentication and encryption schemes, followed by TETRA specific encryption key management and protocol mechanisms for implementing secure communication. This is followed by the TETRA inter-system interface (ISI) signalling protocol architecture which provides an overview of how network services such as basic call control, supplementary services, and security management functions are provided over an interconnection of networks. The TETRA ISI signalling is based on a standard widely used in corporate networks and is built around the private integrated services network (PISN) architecture, which is the fixed network counterpart of the PMR. Finally, an introduction of network management concepts based on the ITU-T management framework is presented with an outline of recommended system structure and network management functionalities for TETRA.

9.2 NETWORK SECURITY MANAGEMENT

9.2.1 Overview of TETRA Network Security

The need for secure communication was one of the factors that instigated the standardisation of TETRA, and security management is therefore the characteristic feature of the system. Several mechanisms for secure communication are supported by the standard [1], [2] and, in fact, a significant portion of the standards document is devoted to describing these issues. TETRA security encompasses the following features for implementing secure communication at various levels [1]-[3].

- authentication of the user with an identity stored in a SIM;
- authentication of the a MS with its unique equipment number;
- authentication of the network and network management system;
- individual and group user identity confidentiality;
- data integrity and data origin authentication for signalling data;
- signalling information confidentiality;
- secure functions for air interface key management.

The standard specifies a number of protection mechanisms at various levels of the radio communication protocol layers, from the low-level air interface to high-level end-to-end user applications. These are summarised below.

- **Multi-level authentication.** This ensures that users intending to connect to the network have valid access permission. The authentication mechanisms may take place at various levels including user to terminal, terminal to network, network to terminal, network to network, and user to user. Most of these authentication procedures are carried out through the intra-system MM schemes described in Section 5.10 and inter-system MM schemes mentioned later in this section.
- **Air interface encryption.** This protects the radio path between the terminal and the base station. This is essentially to combat "eavesdropping of information" over the airwave which if not adequately protected may be possible with the use of decoders based on radio scanners.
- **User anonymity with alias addresses.** TETRA has a provision for disguising the identities of subscribers involved in a call in order to prevent location disclosure. This is achieved with the use of an alias address, the ATSI identity (Section 5.8.4).
- **Terminal enable/disable.** TETRA terminals can be enabled or disabled by the infrastructure over the air, and this can provide an added protection mechanism, for instance, for disabling terminals that present some breach of access permission.
- **End-to-end encryption.** This is intended for most critical applications where protection is required through the entire network. End-to-end encryption protects against eavesdropping and fraudulent access.
- **Frequency hopping.** The TDMA slot structure of TETRA permits a form of slow frequency hopping which can be used to overcome frequency jamming. In certain situations, this, too, can be employed as a protection mechanism against intentional or unintentional disturbance in the radio environment.

Although the TETRA specifications are defined to support the above protection mechanisms, it is ultimately an operator option for implementing these mechanisms for secure communications. The standard is concerned only with the specification of how the *security mechanisms* are integrated into the TETRA protocols; it is not concerned with the specification of requirements for cryptographic algorithms or with the implementation details.

The standard, however, stipulates general requirements which are essential for implementing secure communication if demanded by the TETRA infrastructure. The notable ones are listed below:

- The use of a different and totally independent algorithm for each security service, e.g. air interface versus end-to-end interface;
- The capability of TETRA terminals (MS and LS) to support the network security services if demanded by the network;
- No significant reduction in the quality of service as a result of network security services being in operation. This, for example, means that enabling the encryption algorithms should not significantly degrade performance as a result of slowing down the normal processing speed of the mobile terminal.

9.2.2 Principles of Encryption and Authentication

In this section, an overview of encryption and authentication techniques is presented. This is intended to provide some appreciation of the basic principles that will be helpful in understanding the protocol mechanisms specified for TETRA authentication and encryption. The emphasis is therefore not to discuss *how* encryption is performed or implemented, but rather to outline *what* mechanisms are involved in order to appreciate the network security protocol and management issues. It is also worth noting that encryption algorithms are not part of the TETRA standard, but protocol mechanisms are. For detailed treatment of the subject, the reader is referred to good sources on the subject [4], [5].

Basic Concepts

Encryption is a technique of transforming plain digital information referred to as *plaintext (P)* into an unreadable form called *ciphertext* (*C*) with the use of an encryption key. The reverse process, decryption, operates on the ciphertext (*C*) to produce the original plain information. The same encryption key is usually used to decrypt the ciphertext, which means that provided the decryption algorithm is known, the secret key is all the receiver needs to know. Figure 9.1 depicts a conceptual representation of the encryption scheme, as an encoding function $E_k(P)$ of the plaintext P, and the decryption scheme as the reverse process of a decoding function $D_k(C)$ of ciphertext C.

This arrangement presupposes that there must be many possible encoding functions $E_k(P)$ with inverse functions $D_k(C)$, otherwise a determined code breaker could easily learn how to decode C. The security of the encryption scheme relies on two factors. On the one hand there is the secrecy of the key itself. For instance, how the key is transported to the genuine users and stored for subsequent use, or how it is changed from time to time for better security will have an impact on the overall security provided by the encryption scheme. These issues are part of the cryptographic key management. On the other hand there is the robustness of the encryption technique itself, in terms of its resilience against attacks by the determined to crack the code. These issues relate to the cryptographic algorithms. The main interest here is the key management aspect rather than the cryptographic algorithms, as it is this aspect which is addressed in the TETRA network security protocols.

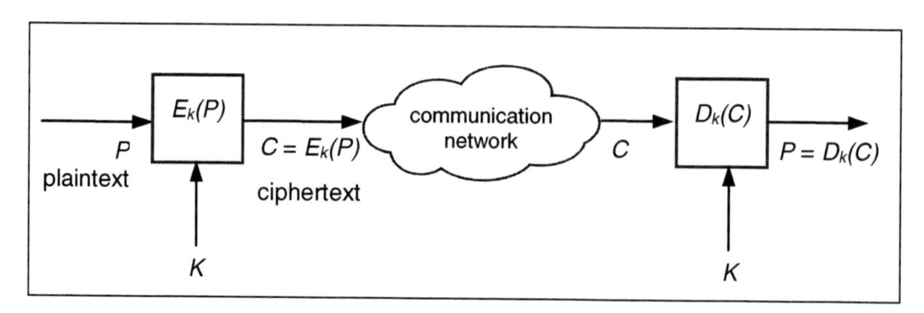

Figure 9.1 A generalised encryption system

Depending on the algorithm used, encryption techniques can be open to various threats, the classic example being attacks based on symbol frequency in the ciphertext (e.g. such as the letter e in the English text). Fortunately for mobile communications, such security attacks can be made very difficult by making use of timing signals inherent in the system. For instance, based on the TDMA time slots and higher hierarchies thereof (frames and super frames), encryption parameters can be dynamically derived and also periodically changed. The repetition period for a unique TDMA frame number could be hours (in fact, several days for TETRA!) and this mechanism will play a crucial role in the design of an encryption system in mobile communications. Such time dependent encryption techniques can prevent code breaking through "recording and replay" of an encrypted bit stream.

To achieve a desired level of security regardless of the key management technique (such as changing a key periodically), the encryption algorithm should be what cryptography experts refer to as a one-way or "trap-door" function. That is, the computation of C from the encryption key K and P should be easy, whereas the computation of K, knowing P and C, should be as complex as possible such that it will be prohibitively lengthy to do the reverse process. This is to say, if K is a shared key between a subscriber and the network, and (P,C) is an *(input, output)* pair of the authentication procedure, then, even with several pairs of (P,C) from the same subscriber, the computation of K should remain very difficult.

The trap-door encryption concept can be easily illustrated with a rudimentary example of an algebraic function, say, $y = 69x^{75} + 9x^9 + 11x^3 + 5$. Given x, it is straightforward to compute y. Given y however it is much more difficult to compute x in reverse – hence the "trap-door" function property. In encryption algorithms there are various techniques for making the trap-door even tighter, for instance by using properties from number theory such as one based on the difficulty of finding the prime factors of a large number [4].

If different encryption and decryption keys are used, such computational complexities can even allow the encryption key and the algorithm to be made public with only the decryption key kept secret. This makes a public key cryptography possible as exemplified by the RSA public key algorithm [6]. It has been claimed that factoring a 200-digit number would theoretically require several billion years, not an encouraging prospect even if the speed of a computer were to increase by a factor of a billion.

The security of an encryption algorithm is usually a relative one, depending on the "value" and/or the lifetime of information being protected. An encryption algorithm is said to be *computationally secure* if it meets one or both of the following criteria [4]:

- The cost of breaking the cipher exceeds the value of the encrypted information;
- The time required to break the cipher exceeds the useful lifetime of the information.

9.2.3 Types of Encryption

Encryption schemes fall into two categories: block cipher and stream cipher. These are summarised below.

Block Cipher

With the block cipher the plaintext is segmented into blocks of fixed sizes and each block is encrypted independently from the others. One of the block cipher standards adopted for encryption, the DES (data encryption standard), uses a combination of substitution and transposition on data blocks of 64 bits with a 56-bit key word. The International Data Encryption Algorithm (IDEA), on the other hand, uses a 128-bit key word for the same block size of 64 bits as for DES. The RSA is also a block cipher.

Stream Cipher

With stream encoding there is no fixed block size; each bit of the plaintext is encrypted with a sequence of "key stream" symbols generated with the encoding key. The encryption is periodic if the key stream repeats itself after a fixed number of symbols or characters, otherwise it is non-periodic. Stream encryption can be further divided into synchronous and self-synchronous. In the synchronous stream encryption, the key stream is generated independently of the message, so that a lost character during transmission necessitates a re-synchronisation of the transmission and receiver key generators. The starting state of the key generator is initialised with a known input as depicted in Figure 9.2.

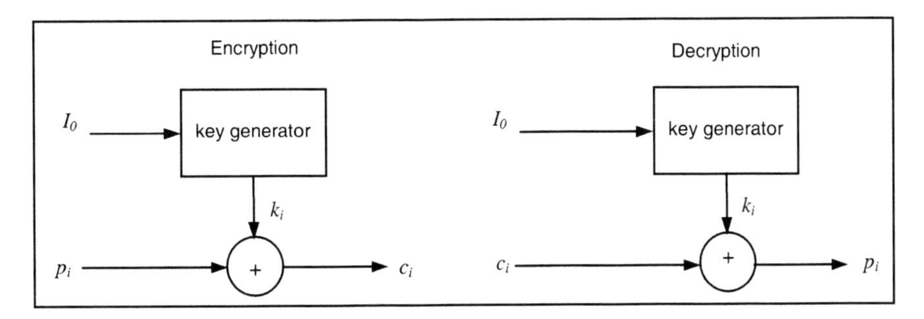

Figure 9.2 Synchronous stream encryption and decryption

In a self-synchronous stream cipher, shown in Figure 9.3, each key character is derived from a fixed number N of preceding cipher text characters. If a ciphertext is lost during transmission, the error propagates forward for N characters but the system re-synchronises itself after N correct cipher text characters are received. The main advantage of this scheme is a non-repeating key that can be easily generated due to the feedback arrangement.

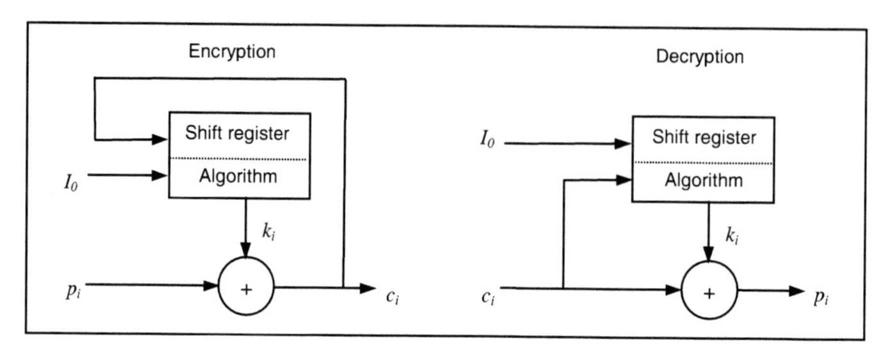

Figure 9.3 Self-synchronous stream encryption and decryption

9.2.4 Principles of Authentication

Authentication is the process of verifying the identity or legitimacy of an entity (a person, object or system). Although authentication based on a password or a personal identity number (PIN) code is widely in use, the level of protection achieved by such method in the radio environment is very low. Listening once to the personal code is enough to break the protection. The encryption principle described above can be used to provide a more robust authentication procedure. With this method the authenticating parties determine an outcome (or a response) that involves a secret key and a random number (or a challenge). This scheme is known as a *Three-Pass Challenge-Response-Result* procedure as illustrated in Figure 9.4.

The *Challenge-Response* authentication method essentially decrypts a random number (received as a challenge) with a secret key (or a derived session key) as a response. Authentication is achieved by both parties (indirectly) proving to each other knowledge of the shared secret key. In mobile communications, the secret key is usually stored in a protected way within a SIM card so that even the user has no direct access to this secret key. With this method, one *secret* key is shared by each of the authenticating parties and for this reason the scheme is known as *symmetric key cryptography*. There are asymmetric schemes where a single secret key is not used, for example public-key cryptography based on *public* and *private* keys.

The procedure in Figure 9.4 shows that different algorithms are used for generating the *session key* (*KS*) from a random seed (*RS*), and for encrypting a *random challenge* (*RAND*). *RS* and *RAND* make up a challenge pair (*RS*, *RAND*) sent to the party being authenticated. For the authentication to succeed, both parties should arrive at the same computed response value.

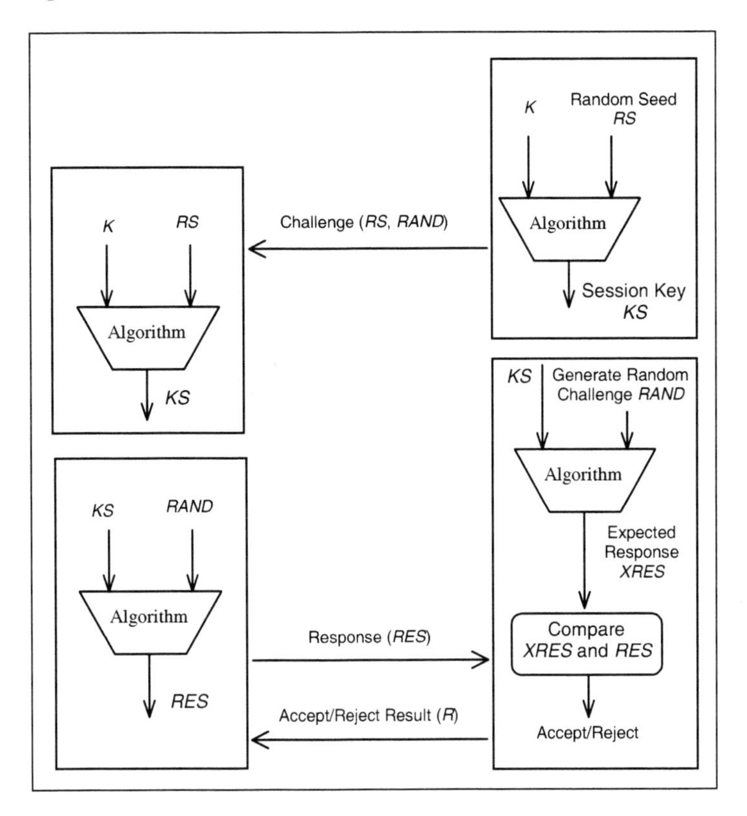

Figure 9.4 A Three-Pass Challenge-Response-Result authentication procedure

9.2.5 TETRA Encryption Algorithms

TETRA uses standard encryption algorithms, details of which are not part of the specification. The standardisation of the encryption algorithms is carried out under a special ETSI expert committee, the Security Algorithms Group of Experts (SAGE). SAGE has developed two standard air interface encryption algorithms, one for the public safety organisations and one for other (commercial) TETRA users. In addition, a set of algorithms for authentication and key management has been developed for general use within TETRA systems. These encryption and authentication algorithms are available only on the need to know basis, which means that they are restricted to responsible organisations that need to evaluate the suitability of the algorithms.

9.2.6 Protection Differences in TETRA Systems

Before authentication and encryption keys are described, the following points summarise the basic differences that exist with V+D, PDO and DMO systems.

- **V+D system.** As should be expected, the TETRA V+D system uses several keys for authentication and air interface encryption. Most of the descriptions below therefore apply to this system.
- **PDO system.** Only authentication is specified for TETRA PDO and the use of encryption is not specified although this can be easily provided by the application using the PDO as a transport and network service. Furthermore, PDO has no voice encryption cipher keys since no voice service is supported. The authentication procedure described below for V+D system will also apply to a PDO system with the generation of the voice cipher keys being redundant.
- **DMO system.** DMO uses only static cipher key (SCK) for the air encryption part. No explicit authentication is employed as for PDO or V+D system and therefore authentication between DMO MSs is only implicit in that a successful communication implies that the DMO MSs are using a valid common SCK for the air interface. Each DMO MS is provided with up to 32 SCKs in a SCK set, and the keys are distributed either manually or by a standardised mechanism over the air as described in Section 9.2.10. At least one common SCK is provided for a group of MSs to allow a group call setup.

Having identified the commonalties and differences of the protection mechanisms for the three types of TETRA, the following discussion will be mainly confined to the V+D system.

9.2.7 Authentication in TETRA

Authentication in TETRA is understood to be between the infrastructure and a user since this is the most important and the one explicitly defined in the standard. Authentication of TETRA Equipment Identity (TEI), on the other hand, is not (explicitly) defined by the standard although protocol mechanisms do exist if the infrastructure demands an MS to provide its TEI in an encrypted form (for instance, during location registration procedure). TETRA authentication is based on a symmetric key algorithm with the challenge-response procedure already described in Section 9.2.4 above.

Generation of Authentication Key

When a user is registered for the first time, he or she will be assigned with a user authentication key (UAK), which at the network side will be stored in the Authentication Centre database, and at the user side stored in a SIM card. Given the UAK code, the authentication key, K, is generated in one of three ways as shown in Figure 9.5. It can be derived from the UAK, either directly from the UAK alone by using an algorithm shown as TB2, or from a combination of UAK and user PIN by using an algorithm shown as TB3. There is also a provision for generating K with a manual entry of an authentication code (AC) which is then passed through an algorithm shown as TB1.

Authentication Procedures

Three authentication procedures are possible:

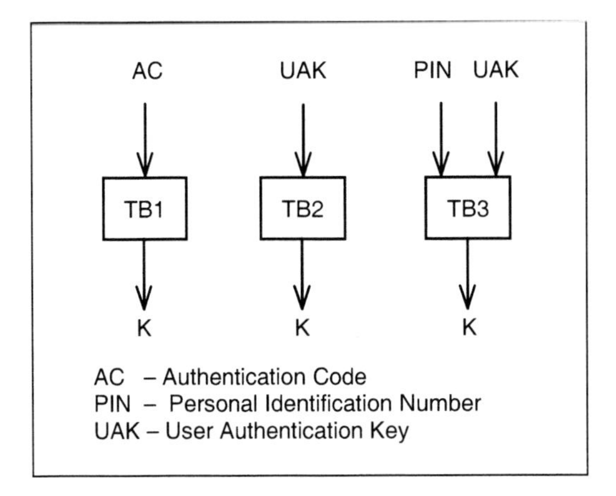

Figure 9.5 Generation of the authentication key

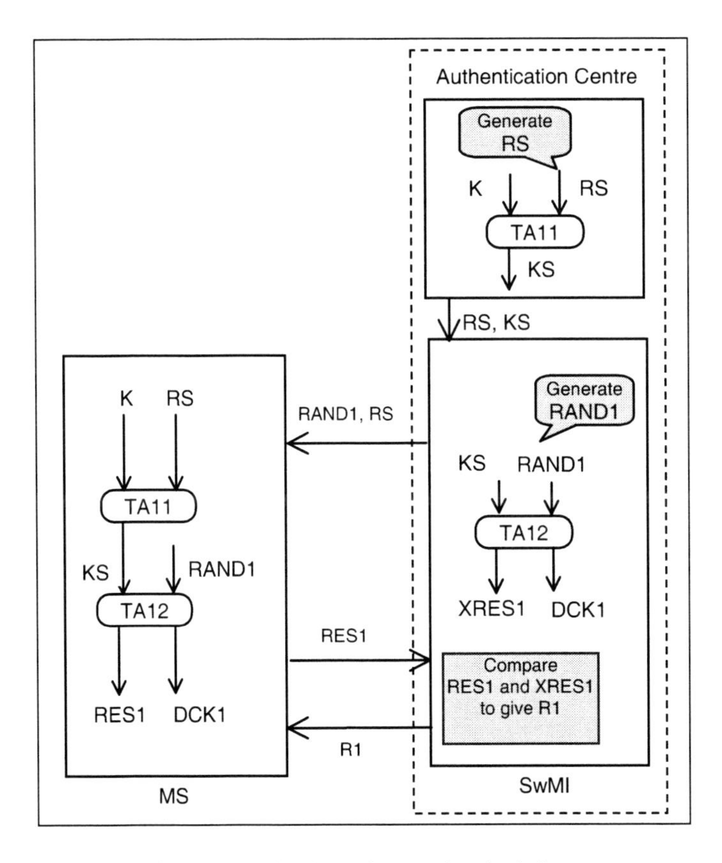

Figure 9.6 Authentication of a user by the infrastructure

- *Authentication of a user by the infrastructure.* Algorithm TA11 is used for generating a session authentication key, and algorithm TA12 used for the computation of the response. This is illustrated in Figure 9.6. For the V+D TETRA system, this procedure will also produce intermediate keys (DCK1 and DCK2) for generating a derived cipher key (DCK) which may subsequently be used for air interface encryption. V+D air interface cipher keys, including DCK, are summarised in the next Section.

- *Authentication of a user by the infrastructure.* Effectively the above role is reversed, but for this case algorithm TA21 is used for generating a session authentication key and algorithm TA22 used for computing the response.
- *Mutual authentication of user and infrastructure.* This procedure is achieved by combining the unilateral three-pass authentication mechanisms just described. The decision to make the authentication mutual is made by the first party to be challenged, not the initial challenging party. If the first authentication of the mutual pair fails, the second authentication will be abandoned. This authentication scheme will result in a *four-pass procedure* since some of the authentication results are exchanged in tandem as shown in Figure 9.7.

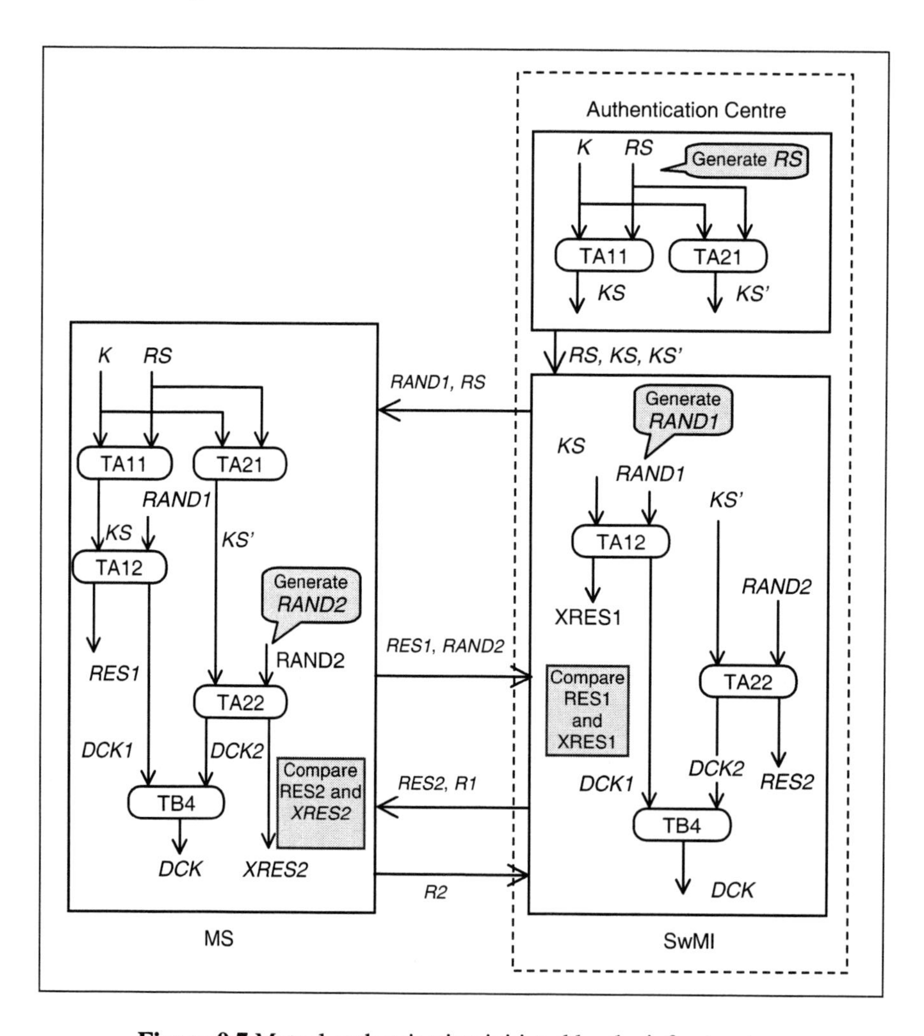

Figure 9.7 Mutual authentication initiated by the infrastructure

9.2.8 Authentication PDU Examples

Authentication in TETRA is carried out as part of the mobility management procedures as detailed in Section 8.5. PDUs for implementing the authentication schemes introduced above also appear in Table 8.9. Three PDUs are described below for the purpose of demonstrating the protocol mechanisms.

9.2.8.1 D-AUTHENTICATION DEMAND

This PDU is used by the infrastructure to initiate an authentication of an MS and could be sent by the infrastructure at any time or most commonly in response to the location update demand by the MS (details in Section 8.5.8). In response, the U-AUTHENTICATION RESPONSE message from the MS is expected.

PDU type	Random Challenge (RAND1)	Random Seed (RS)	Proprietary element
(4)	(80)	(80)	[...]

PDU type: D-AUTHENTICATION DEMAND = 0001_2

Random Challenge (RAND1): an 80-bit number used as the input to the authentication algorithm, from which a response is calculated (see Figures 9.6 and 9.7).
Random Seed (RS): an 80-bit number used as the input to the session key generation algorithm (see Figures 9.6 and 9.7).

Proprietary element: an optional, variable length element for proprietary defined information.

Figure 9.8 Infrastructure PDU for authentication demand

9.2.8.2 U-AUTHENTICATION RESPONSE

This PDU is by the MS to respond to an authentication demand from the infrastructure.

PDU type	Response Value (RES1)	MAF	Random Challenge (RAND2)	Proprietary element
(4)	(32)	(1)	(80)	[...]

PDU type: U-AUTHENTICATION RESPONSE = 0000_2

Response Value (RES1): a 32-bit value returned by the challenged party, calculated from the random challenge. (see Figures 9.6 and 9.7).
MAF (Mutual Authentication Flag): a flag used to indicate whether or not the PDU is part of a mutual authentication exchange between the MS and infrastructure (see Figure 9.7).
Random Challenge (RAND2): an 80-bit number used as the input to mutual authentication algorithm, but its presence is conditional on the MAF element being set.

Figure 9.9 MS PDU for authentication response

9.2.8.3 D-AUTHENTICATION RESULT

This PDU, by the infrastructure, is used to report the result of an authentication to the MS. If authentication fails, the MS will be informed of the cause with the D-AUTHENTICATION REJECT procedure.

The above illustrative procedures are for a network initiated authentication. The necessary PDUs are defined for implementing a mobile initiated authentication in a similar way. In either case, mutual authentication can be implemented by enabling this feature through the Mutual Authentication Flag as remarked in Figure 9.9 and Figure 9.10.

PDU type	Authentication Result (R1)	MAF	Response Value (RES2)	Proprietary element
(4)	(1)	(1)	(32)	[...]

PDU type: D-AUTHENTICATION RESULT = 1110_2

Authentication Result (R1): a flag indicating the success (1) or failure (0) of an authentication.

MAF (Mutual Authentication Flag): as for U-AUTHENTICATION RESPONSE to indicate mutual authentication procedure.

Response Value (RES2): a 32-bit value returned by the challenged party, calculated from the random challenge, but its presence is conditional on the MAF element being set.

Figure 9.10 MS PDU for authentication response

9.2.9 Encryption Keys

Figure 9.11 summarises the various encryption keys and their associated algorithm names, while Table 9.1 summarises the bit length of the encryption parameters referred to directly or indirectly throughout Section 9.2. The following outlines the encryption keys used in the V+D TETRA system.

- The derived cipher key (DCK) is generated during the authentication procedure and its successful use subsequently therefore will provide an extended implicit authentication. DCK is used to protect voice, data, and signalling during the call, and it is derived from DCK1 and DCK2 with algorithm TB4 (Figure 9.7). For a unilateral authentication, only DCK1 or DCK2 exits (Figure 9.6) and in this case the missing parameter is set to 0.

- The common cipher key (CCK) is generated by the infrastructure and distributed to MSs by "sealing" it with the DCK. When a key is transferred over the air interface, known as Over The Air Re-keying (OTAR), it is further encrypted (or sealed) for extra security. The OTAR mechanism, shown with a block diagram in Figure 9.12, is used to convey 3 keys – CCK, GCK, and SCK. Every location area will be provided by a CCK and a MS may obtain this when registering in the LA. It is therefore efficient to use this key for encryption of messages for group addressed calls.

- The group cipher key (GCK) is used by a specific user group and it is generated by the infrastructure and distributed to the MSs of a group. It is not used on its own but modified by the CCK to produce MGCK which is used for encryption of group calls. If GCK is not defined then CCK will be used instead of MGCK.

- The static cipher key (SCK) allows encrypted operation prior to authentication. These keys are not changed by an authentication procedure and hence the name "static". TETRA supports the use of up to 32 SCKs and these can be distributed in a similar way to the GCKs. The SCK is also used for encryption (and implicit authentication) in DMO as remarked in Section 9.2.6.

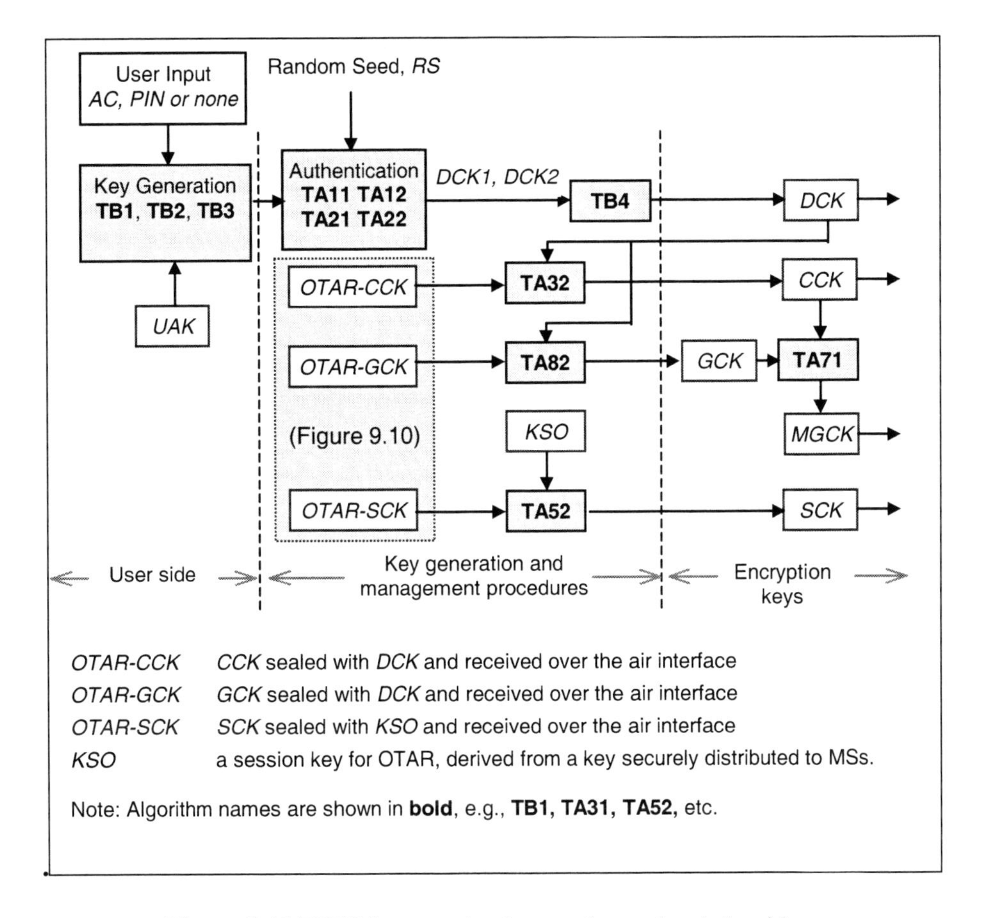

Figure 9.11 TETRA encryption keys and associated algorithms

9.2.10 Over The Air Re-keying

Over The Air Re-keying (OTAR) is a mechanism for transferring an encryption parameter by sealing it with another key. This mechanism is used for transferring encryption keys CCK, GCK, and SCK over the air interface.

CCK and GCK are both sealed with the DCK which is derived on successful authentication (Figure 9.7). The static cipher key, SCK, on the other hand is sealed with a session key known as KSO which is derived from a key securely distributed to mobiles for this purpose.

Table 9.1 Sizes of keys and related parameters mentioned in Section 9.2

Parameter	Bits	Parameter	Bits
AC	16-32	RAND1	80
CK	80	RAND2	80
CCK	80	RES1	32
DCK1	80	RES2	32
DCK2	80	RS	80
DCK	80	SCCK	120
GCK	80	SCK	80
K	128	SGCK	120
KS	128	SSCK	120
KSO	128	UAK	128
MGCK	80	XRES1	32
PIN	16-32	XRES2	32

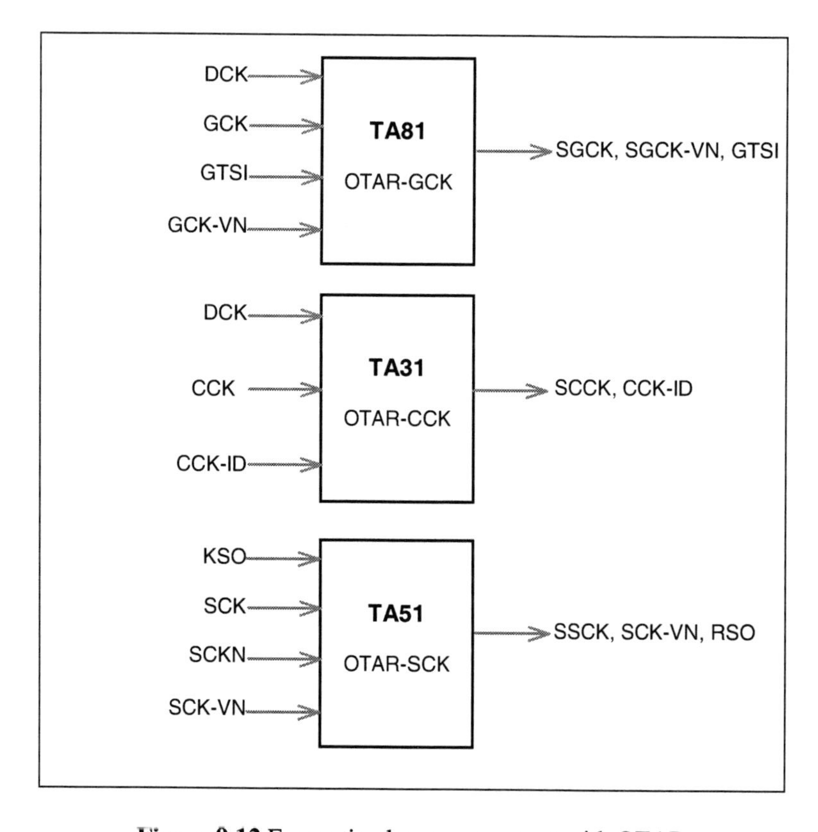

Figure 9.12 Encryption key management with OTAR

9.2.11 OTAR PDU Examples

The following PDUs are chosen to illustrate the procedure of how encryption keys CCK, GCK and SCK are exchanged between a MS and the network. The example below is for CCK key transfer; SCK and GCK are transferred in a similar manner.

9.2.11.1 U-OTAR CCK Demand

This OTAR PDU is used by an MS to request CCK from the infrastructure for a location area (LA). In response, D-OTAR CCK Provide, shown below is expected from the infrastructure.

PDU type	OTAR subtype	Location Area	Proprietary element
(4)	(3)	(14)	[...]

PDU type: U-OTAR = 1101_2

OTAR subtype: a 3-bit code indicating whether the PDU is a demand for one of the three encryption keys (CCK, GCK, SCK) or the result of key transfer. This is shown in Table 9.2.

Figure 9.13 MS PDU for authentication response

Table 9.2 OTAR encoding for key transfer indication

OTAR subtype	Remark
000	CCK Demand from Uplink or CCK Provide to Downlink
001	CCK Result
010	SCK Demand from Uplink or SCK Provide to Downlink Reject
011	SCK Result
100	GCK Demand from Uplink or GCK Provide to Downlink Reject
101	GCK Result
110	Reserved
111	Reserved

9.2.11.2 D-OTAR CCK Provide

This OTAR PDU is sent by the infrastructure to provide CCK to a MS.

PDU type	OTAR subtype	CCK provision indicator	Optional and proprietary element
(4)	(3)	(2)	[...]

PDU type: D-OTAR = 0000_2

OTAR subtype: Figure 9.13 for OTAR subtype definition and Table 9.2 for its encoding.

CCK provision indicator: a 2-bit code indicating the usage scope of a CCK as shown below.

00 No CCK in use or CCK not known;
01 CCK provided for this LA;
10 CCK provided for other LAs (in which case optional elements provide LAs)
11 System wide CCK provided.

Figure 9.14 MS PDU for authentication response

9.2.12 Transfer of Authentication Information

When a user roams another TETRA network, the visited TETRA network will need to obtain authentication information from the user's home network, to be able to perform authentication and generate and/or provide encryption keys. The most straightforward method is to simply transfer the authentication key K to the visited network but this approach is not advisable if a very tight security has to be maintained (as security may somehow be compromised by the intervening networks, for instance). An improvement over this is the transfer of information which can be used for a single authentication procedure. This method is used in GSM and can provide good level of security but at the expense of some overhead resulting from a transfer of information for every authentication procedure.

TETRA supports an approach that attempts to combine both security and efficiency. With this approach, a home network is required to transfer a session authentication key for a MS only once, which then can be used for repeated authentication in a visited network. This does not reveal the original authentication key, hence secure, and avoids a transfer of information per authentication procedure, hence efficient.

9.2.13 Overview of Air Interface Encryption

Air interface encryption is handled within the data link layer by using a stream encryption. The following presents an overview of this method.

Air interface encryption is realised by using an encryption algorithm based on a Key Stream Generator (KSG), which has two inputs: an Initial Value (IV)[1] and a cipher key as shown in Figure 9.15. The encrypted ciphertext bits are obtained by modulo-2 addition (exclusive OR logical operation) of the KSG stream bits with plaintext bits in data, speech and control channels. There is one important exception in that the MAC header bits are

[1] The initial value is a 29-bit data derived from numbers representing a slot (2 bits), frame (5 bits), multiframe (6 bits), hyperframe (15 bits), and uplink/downlink direction (1 bit).

not encrypted since these are used for channel allocation purposes. The cipher key input to the KSG is one of the encryption keys described in Section 9.2.9 and the choice of an appropriate one can be made as part of signalling for key management, for instance during registration or call set-up. For the DMO this procedure is already predetermined due to the use of static cipher keys.

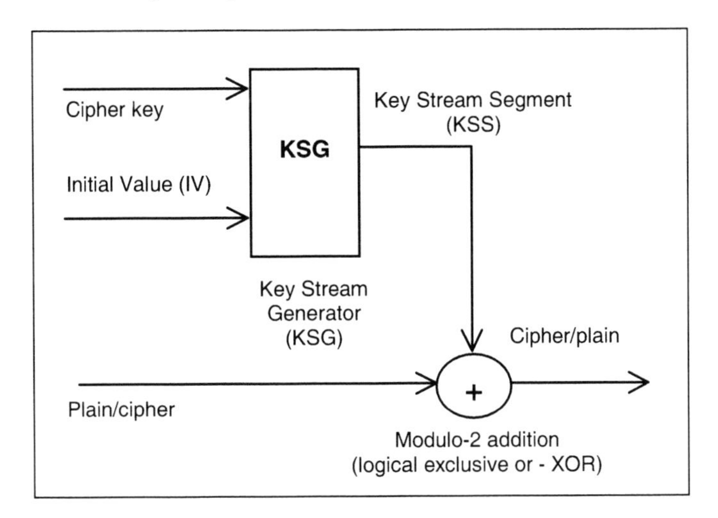

Figure 9.15 Speech and control information encryption

9.2.14 Overview of End-to-End Encryption

End-to-end encryption algorithms and key management are not specified in the TETRA standard. The standard describes only the mechanism for synchronisation of the encryption system, which is based on a synchronous stream cipher and applies only to the U-plane traffic [1]. The functional diagram of voice encryption and decryption mechanisms is shown in the Figure 9.16.

- An End-to-end Key Stream Generator (EKSG), similar to that of the air interface encryption, is used to generate a key stream from a cipher key (*CK*) and an initialisation value (*IV*). The *IV* is required to be a time variant parameter (e.g. a sequence number or a timestamp) that is used to initialise synchronisation of the encryption units. This is to provide protection against "recording and replay". For the same requirement, time variance of the cipher key may be achieved by deriving a key for each encrypted call.

- Function F_1 combines the plaintext (*P*) bit stream and EKSS (End-to-end Key Stream Segment) resulting in an encrypted ciphertext (*C*) bit stream. Function F_1^{-1} is the inverse of F_1 and combines the bit streams *C* and EKSS resulting in the decrypted bit stream *P*. Function F_2 replaces a half slot of *C* with a synchronisation frame provided by the "sync control" functional unit. Function F_3 recognises a synchronisation frame in the received *C*, and supplies them to "sync detect" functional unit.

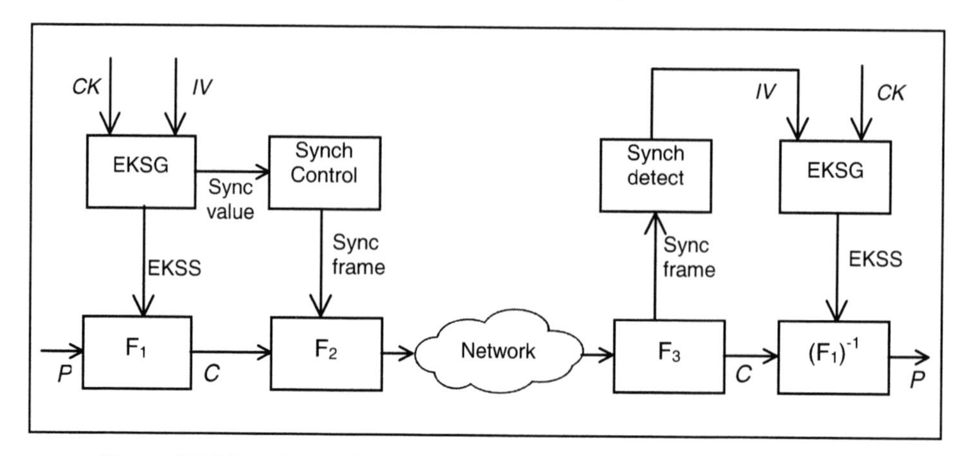

Figure 9.16 Functional diagram of end-to-end voice encryption/decryption

9.2.15 Lawful Interception

The TETRA network security management provides a means for lawful access to network information and communication by Law Enforcement Agencies. This will be achieved by the TETRA interface specification referred to as *Lawful Interception Interface* (LII), which at the time of writing was being developed [7]. The provision of the TETRA LII is a national option and where implemented specific regulatory requirements will be achieved with a "mediation function" that lies between a TETRA network and a law enforcement agency.

9.3 TETRA INTER-SYSTEM SIGNALLING

TETRA defines a standard interface, identified as I3 (see Sections 4.3 and 5.4), for inter-system interface (ISI) between two independent TETRA networks. This interface is required to provide the necessary bandwidth and efficient inter-system signalling so that extended functions such as inter-system mobility management and user authentication, and therefore user roaming, can be supported over an interconnected network. For instance, a number of additional network functions will be required to extend services such as group calls over an interconnected network. These additional network functions rely on the signalling architecture on which the ISI communication services are built. This section introduces operational aspects of the TETRA ISI from a signalling protocol point of view.

9.3.1 The TETRA ISI Reference Configuration

The reference configuration for the TETRA ISI follows directly that of the PISN for the reason that should be apparent. Such a reference configuration has already been specified by ECMA and adopted as an ISO standard [8]. The conceptual representation of the TETRA ISI reference configuration is shown in Figure 9.17.

The significance of this representation is that, in addition to the user-to-ISDN reference configuration specified in ITU-T standard [9] (Section 5.6.1), it specifies a reference configuration which specifies the functions that apply to the control of connections and calls between any two exchanges of a given PISN. For this, notice the two interfaces shown to the right as the public ISDN and ISI in Figure 9.17. The concept of the ISI is based on the assumption that inter-PISN exchange connections are routed through an intervening network (IVN), which can be of any nature, provided it offers the required capabilities for carrying user and control information between the PISN exchanges.

9.3.2 Q and C Reference Points

Within the PISN, reference point Q defines the boundary between the Switching (SW) functional grouping and the Mapping (MP) functional grouping which is required to provide the necessary adaptation of the ISI side interface to the SW. It corresponds to a conceptual point where the inter-exchange call control functions and signalling information flows are specified.

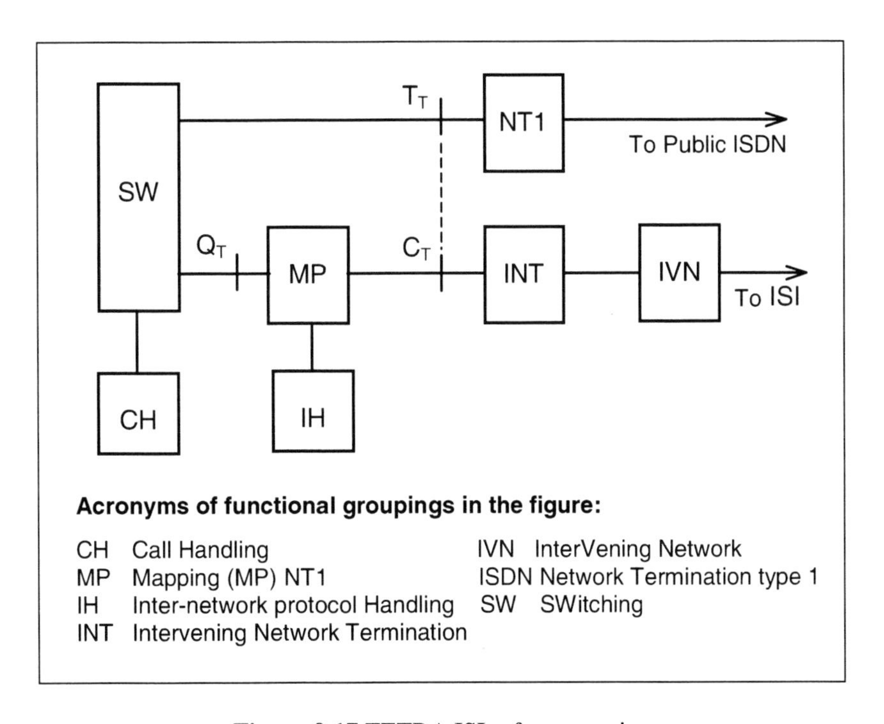

Figure 9.17 TETRA ISI reference points

Reference point C defines the boundary between MP and IVN, and it represents the physical interface point for inter-PINX connections, which is in contrast to the (logical) call control functions in the case of Q reference point. As briefly remarked earlier, various IVN can be envisaged (e.g. leased line, dedicated or switched connection, analogue or digital link, LAN, ISDN, etc.) and consequently the nature of the interface at the C reference point (hence protocol) may vary from one system to another. In the reference configuration, some symmetry can be observed between the public ISDN side and the ISI side; in this case the INV is a public ISDN, the C reference point coincides with the T reference point of ISDN as shown with a dotted vertical line.

The TETRA ISI application is built on top of the Private Signalling System 1 (PSS1) protocol specifically developed for interconnecting PINXs (Private Integrated Network EXchanges) in a PISN. The PSS1 is the ISO term, which in ECMA and ETSI documents is also known as QSIG protocol (for Q reference point signalling). The PSS1 series of standards uses ISDN concepts as developed by ITU-T and conforms to the framework of the OSI model as defined by ISO.

9.3.3 Private Signalling System 1 for TETRA

The PSS1 is a powerful and intelligent inter-PINX signalling system designed specifically to meet requirements of sophisticated business communications. The standard functionalities defined for PSS1 fall under two major categories:

- signalling protocol for the support of circuit-mode basic services – defined as an ISO standard [10]; and
- generic functional protocol, defined for the support of supplementary services – also defined as an ISO standard [11].

Among numerous features, for instance, PSS1 provides enhanced supplementary services not supported by the public ISDN, such as:

- call intrusion;
- call completion on no reply;
- mobility services.

Since TETRA networks are in the realm of corporate or private organisations, this means that they too can benefit directly from the features provided by PSS1. Some of the special operational requirements are evident from the supplementary services specified for TETRA, and the PSS1 signalling architecture provides full capability to support these and additional functionalities required to convey ISI services. Based on the PSS1 signalling architecture, the TETRA ISI specifications comprise two major components: the *generic functional protocol* component and the *additional network features* component.

Generic Functional Protocol

Generic functional protocol (GFP) is an intrinsic feature of PSS1 and represents a general protocol mechanism that is concerned with the concepts of interworking between private networks including TETRA. It specifies how supplementary services and additional network features, described below, are provided over an interconnected network.

Additional Network Feature

Additional network features (ANFs) are network capabilities defined on the top of the GFP. They represent services over and above basic services provided by the network, but unlike supplementary services, not directly provided to the network user. Noting this basic difference, the ANFs can otherwise be regarded just like supplementary services, particularly for the purpose of understanding the PSS1 protocol described here.

The TETRA ANFs are required to facilitate essential services between the interconnected networks. These comprise of the following ISI services:

- **Mobility Management.** This ANF defines additional MM services to the SwMIs, and complements the intra-SwMI-MM (Section 8.5), authentication and key management services (Section 9.2).
- **Individual Call.** This enables calls to be set-up by a user registered in one TETRA network to another user registered in another TETRA network, operating at the ISI of both SwMIs. It also supports call restoration when a user has migrated to another TETRA network during an established call. Layer 3 protocol of PSS1 is utilised for basic call control.
- **Group Call.** This is concerned with ISI group calls and handled by the same call control protocol entity for the individual call.
- **Short Data Service.** In addition to the normal user messaging services, the SDS also internally provides information transport for the various ANF and SS procedures.
- **Supplementary Services Control (SS-C).** This is in fact already part of the GFP since the GFP was, in the first place, specified to provide SS capabilities within the PSS1 signalling architecture.

It should be noted that the above protocols are inter-system, as opposed to intra-system protocols which have been described in Chapter 8.

9.3.4 PSS1 Protocol Stack

Figure 9.18 shows the conceptual model for the PSS1 signalling protocol. It depicts the interrelationship of the bottom layer 3 basic call defined in [10], the GFP defined in [11], and the various SS and ANF services accessible at the application layer.

Many supplementary services and ANFs involve remote interactions between a pair of application entities (e.g. user application and network application). Such remote interactions are handled by the remote operations service entity (ROSE) which provides a set of services such as the transfer of protocol data units and responses between applications. The ROSE protocol is defined in ITU-T Recommendation X.229 [12].

The function of each of the ISI ANFs is therefore based on the PSS1 signalling and ROSE remote procedure call, to facilitate inter-system operation with functionality similar to that of intra-system capability. For instance, the ANF-ISIMM will essentially control inter-system mobility management with user authentication and OTAR key management services. Similarly, the ANF-ISIGC will be in charge of group call reconfiguration as group members migrate from one network to another.

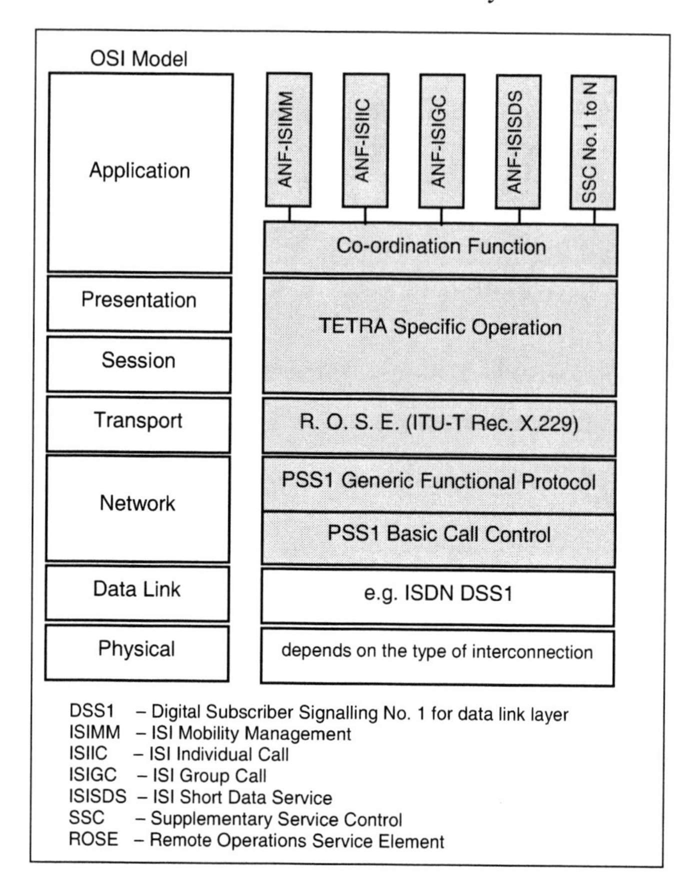

Figure 9.18 PSS1 protocol stack for TETRA

The essential difference from the intra-system operation is therefore the remote invocation and signalling dictated by a remote TETRA network. For remote procedure calls required by each of the defined ANFs, protocol data units (conveyed by the ROSE) will be used in much the same way as the various protocol mechanisms described so far, e.g. mobility management. Although the TETRA inter-system operation strives to provide performance comparable to that of intra-system operation, there are some limitations that should be noted, with the most important ones mentioned below.

- Forward registration is not supported by ANF-ISIMM over TETRA ISI. Forward registration is intended to provide seamless handover and necessitates handover channel allocation on a new cell by currently serving cell. This is the most demanding task performed by TETRA's radio link management as detailed in Section 8.4.3.

- Similarly, ANF-ISIMM does not support the assignment of group cipher key (GCK) over ISI. Intra-system group key transfer over the air, for instance, is only possible if the user is explicitly authenticated so that the resulting derived cipher key (DCK) is then used for sealing GCK (see for instance Figure 9.12).

- *Discreet listening* supplementary service is not supported at ISI, for the reason is all too obvious if the ISI is to form a logical demarcation for organisational networks.

9.4 TETRA NETWORK MANAGEMENT

9.4.1 Overview

Implementing a network is a major task that requires the design of complex hardware and software systems. To ensure that a network delivers the best possible service at all times, the use of automated network management tools becomes vital. For instance, it will be necessary to monitor network behaviour regularly and identify problems and clear them where possible before system degradation or service disruption.

Some aspects of the network behaviour can be monitored by adding software to specific nodes of the network. Other aspects require specialised hardware such as network analysers for gathering network statistics. In addition to performance monitoring, system malfunction indications or alarms are also valuable and these would involve different tools and techniques. After analysing network traffic statistics, network reconfiguration may be necessary to improve network performance. In general a sophisticated set of network management tools is required to deal with the various network events coupled with the performance monitoring. It is also essential that the way network information is gathered, analysed, and acted upon is systematic with clearly defined procedures that are understood by all network entities.

The range of management activities also extend beyond the technical aspects of performance monitoring. Network monitoring of subscribers' activities, for instance, will be as important to a network provider for accounting and security management of the network system. It will be highly desirable to carry out all of the management tasks from a central point and with management components as an integral part of the network being managed.

With rapid developments in network technology and increasing need to manage installed networks in a systematic fashion, network management principles now represent a well-established discipline to be part of any network system design. The TETRA *Designers' Guide* to TETRA Network Management [13] provides the following definition of Network management:

> *Network management provides a distributed application enabling monitoring and control of network resources, in order to control the overall environment in an orderly fashion. It has to interface across all of the physical elements in a network, and to this end it is increasingly important that open standards are adopted as the norm in all network elements, enabling the control of the network to be undertaken by an integrated network management system.*

To enable the development of network management expressed by the above sentiment, the TETRA standard caters for the following support.

- It provides a standardised network management interface, I5, which is intended to facilitate inter-working between systems from different manufacturers. The I5 interface standard has been described in Section 4.3.
- It specifies the TETRA Management Identity (TMI) scheme for secure addressing of all managed network entities within the infrastructure. The TMI is described in Section 5.8.3.
- Based on the ITU-T Recommendation M.3400 [14], it provides functional requirements and guidance for implementing a standard TETRA network management scheme [15], [16].

The following section introduces some basic concepts of standardised network management framework, namely the ITU-T telecommunications management network (TMN) which provides the basis for TETRA network functional requirements. This is followed by an introduction to implementation aspects of network management based on the widely adopted simple network management protocol (SNMP). The remainder of this section is devoted to TETRA-specific network management aspects.

9.4.2 Principles of Network Management

Network management involves systematic activities in network monitoring and control for various objectives, ranging from the day to day efficient use of the available resources to a long term planning for network expansion. The typical features of a network management system may include:

- service provision;
- traffic management;
- network monitoring;
- configuration management;
- fault management;
- subscriber management; and
- planning.

To accomplish the above tasks, network management applications view the entire network as a collection of "managed objects" with addresses and names assigned to the managed objects. For instance, to a network management application, it is possible that a remotely controlled base station and a gateway are merely managed objects (identified with their TMI addresses), and distinguished only with their management attributes. Such a view of a network by the management application makes the following four key elements apparent:

- The managed objects or nodes being managed, i.e. the physical network resources being monitored and controlled;

- The management station or manager carrying out the network management tasks which usually is a general-purpose computers running special management software;
- Management information base (MIB) which is a structured, object-oriented representation of a network as a collection of *network elements*[2], i.e. the object-oriented information model of network elements as a virtual network system; and
- The protocol for implementing the management function by providing a communication link between the management station and the network elements or *managed nodes* i.e. the mechanism for the manager and managed nodes to communicate and interact.

While the first two items above are very generic, the last two items represent software models whose implementation could vary depending on some target objective. The methodology by which these software models are implemented determines the type of network management, often referred to as the *management framework*. At present, there are two main management frameworks in existence:

- the ITU-T TMN framework standardised as M.3400 [14]; and
- the simple network management protocol or SNMP [17].

These management frameworks essentially provide an inter-operable interface to achieve interconnection between various types of equipment, communicating via a defined management protocol (described below). Only the ITU-T TMN will be discussed here as this provides the basis for TETRA network management functional requirements.

The ITU-T TMN describes network management in five broad management functional areas (MFAs) as summarised in Figure 9.19.

Configuration Management	**Performance Management**	**Fault Management**	**Accounting Management**	**Security Management**
- physical configuration	- monitoring	- Testing & fault localisation	- Data usage collection	- Audit trail
- subscriber management	- control	- Alarm management	- Billing & tariffing	- Intrusion
- database management	- analysis	- Trouble management	- Report generation	- Access rights
- ...etc	- ...etc	- ...etc	- ...etc	- ...etc

Figure 9.19 Network management functional areas

The broad and hefty categorisation of the TMN methodology means that, while still maintaining the management methodology, a simpler and clearer way of MFA definition may be employed for small to medium scale networks. In fact, this is the case for TETRA with its defined MFAs [15] as follows:

[2] The terms *managed objects, managed nodes*, or *network elements* are synonymously used.

- subscriber management
- configuration management
- traffic measurement
- performance measurement
- security management
- accounting administration
- fault management

For ITU-T TMN, for instance, it can be observed that subscriber management is part of configuration management; this is not the case for the modified TETRA MFA listed above.

9.4.3 Network Management Protocols

Network management protocols exist in two major categories:

- the common management information protocol (CMIP) and its variants identified with their connection-oriented data services – characteristic of telecommunication networks); and
- the simple network management protocol or SNMP[3] identified with its connectionless data services – characteristic of computer networks.

The TETRA network standard guide [15] recommends the use of SNMP owing to the fact that this protocol places minimal requirements on the network elements within small to medium size networks (relative to GSM or ISDN network) which is typical of a TETRA network. The SNMP will be described next to illustrate the basic principles of implementing network management.

9.4.4 The SNMP Model

The SNMP model of a "managed network" is depicted in Figure 9.20. The managed nodes can be network elements such as switches, base stations, gateways, network interface cards, or any other devices capable of communicating status information with the management station. To be managed directly by SNMP, a node must be capable of running an SNMP management process referred to as an *SNMP management agent*. The management agent within a network element can be configured to carry out a number of tasks such as monitoring events and report these to the network manager. It is also possible for an agent to be configured in order to carry out some limited tasks on behalf of the network manager.

[3] The name SNMP is used both for management framework and management protocol.

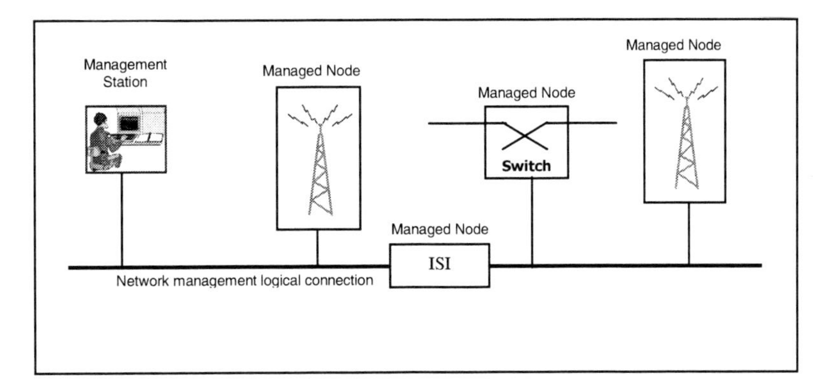

Figure 9.20 Components of the SNMP management model

Most networks are multi-vendor-sourced, for instance with network switches from one manufacturer and base stations from another. In order to allow the management station to talk to multi-vendor nodes, the information maintained by all the nodes must be rigidly specified. Each node maintains one or more variables that describe its state, called *objects*. As remarked in Section 9.4.2 earlier, the collections of these objects constitute the management information base (MIB) which is essentially the data structure for the management application. The SNMP protocol allows the management station to interrogate the state of these objects and change them if necessary. To express the MIB and the format of SNMP packets independent of specific executable machine languages and equipment vendors, a formal description standardised by ISO, the Abstract Syntax Notation 1 (ANS.1), is almost invariably used.

SNMP Protocol Stack

SNMP is an application layer protocol and relies on the user datagram protocol (UDP) at the transport layer. The SNMP protocol is shown in relation to the OSI reference model as depicted in Figure 9.21. This can be easily implemented with TETRA's connectionless data service (Section 8.7.4).

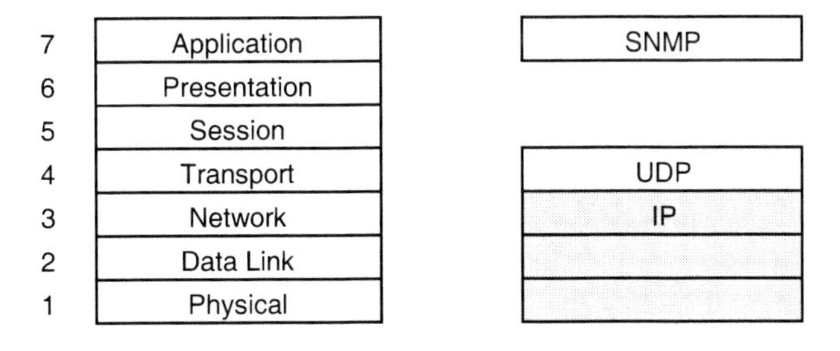

Figure 9.21 SNMP protocol stack in relation to the OSI model

SNMP PDUs

SNMP, or more specifically version 2, uses a set of seven protocol data units (PDUs) which are summarised in Table 9.3.

Table 9.3 SNMP protocol data units

Protocol Data Unit	Function
GetRequest	Retrieves the value of one or more of agent's variable
GetNextRequest	Sequences through the agent's MIB table.
SetRequest	Provides the manager with the ability to alter and update variables
GetResponse	Agent's response to a request
Trap	Used by an agent to alert a manager
GetBulkRequest	Retrieves multiple rows of data from an agent's MIB.
InformRequest	Enables one manager to transmit unsolicited information other managers, to permit support for a distributed network management.

9.4.5 TETRA Network Management Methodology

To preserve the open standard, multi-vendor philosophy of TETRA at various levels (networking as well as management), and ensure inter-working of systems from different manufacturers, the TETRA network management methodology recommends the use a hierarchical interconnection built around the following entities.

- A local network management (LNM) for managing each individual TETRA network system;
- A central network management (CNM) for managing a group of LNM, i.e. each LNM will be regarded as a client to the CNM.

This is depicted in Figure 9.22 below. Another notable feature of this structure is that the network management does not compete with the network bandwidth, in that a separate, possibly dedicated, standard interface I5 is used for interconnecting the LNMs. The individual TETRA networks are connected with the ISI interface, which is TETRA interface standard I3.

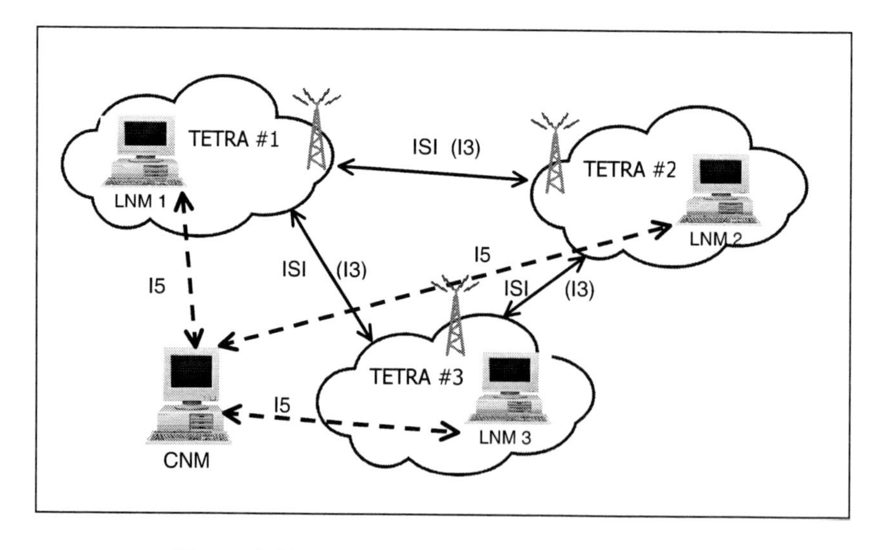

Figure 9.22 TETRA network management structure

9.5 CONCLUSIONS

This chapter has introduced three important topics that are essential to the operation of a TETRA system, viz., network security management, inter-system interface and signalling protocols, and network management. Basic principles of TETRA's authentication and encryption have been introduced with various types of authentication and encryption keys identified. Encryption algorithms are not part of the TETRA standardisation and the various algorithms are simply referred to by their designated names such as TA11 and TA12. The network security network management standard is therefore only concerned with protocol definitions for authentication procedures, encryption key (and related parameter) management, and mechanisms for end-to-end encryption. One important aspect of the encryption key management is the OTAR mechanism which allows the transfer of encryption keys over the air interface. Examples based on protocol data units have been used for illustrating authentication and OTAR key management schemes.

The inter-system interface (ISI) specification extends network coverage and mobility beyond a single TETRA network and this is achieved with the PSS1 signalling protocol architecture which is widely used in private integrated networks also commonly referred to as corporate networks. TETRA network management is defined on the ITU-T TMN principles with some simplifications made to suit the requirements of a small to medium network operation, which is a feature of TETRA network. The simple network management protocol has been used to highlight some of the operational aspects of a network management protocol.

REFERENCES

[1] ETSI Technical Standard ETS 300 392-7: "Radio Equipment and Systems (RES); Terrestrial Trunked Radio (TETRA); Voice plus Data (V+D); Part 7: Security", December 1996.

[2] ETSI Technical Report ETR 086-3: "Terrestrial Trunked Radio (TETRA); Technical requirements specification Part 3: Security aspects", January 1994.

[3] Roelofsen, G., "TETRA Security – the fundament of a high performance system", TETRA Conference 1997.

[4] Stallings, W., *Cryptography and Network Security* (2nd Edn.), Prentice-Hall, Inc., Upper Saddle River, N. J., 1999.

[5] Schneier, B., *Applied Cryptography: Protocols, Algorithms And Source Code,* John Wiley and Sons Ltd., 1995.

[6] Rivest, R., Shamir, A., and Adleman, L. "A Method for Obtaining Digital Signatures and Public Key Cryptosystems", *Communications of the ACM*, Feb. 1978.

[7] ETSI Draft EN 301 040 V2.0.0: "Terrestrial Trunked Radio (TETRA); Security; Lawful Interception Interface", July 1998.

[8] ISO/IEC 11579-1 (1994), "Information technology – Telecommunications and information exchange between systems – Private Integrated Services Network – Part 1: Reference configurations for PISN Exchanges (PINX)".

[9] CCITT Recommendation I.411: "ISDN User-Network Interfaces – Reference Configurations".

[10] ISO/IEC 11572: "Information technology – Telecommunications and information exchange between systems – Private Integrated Services Network – Circuit mode bearer services – Inter-exchange signalling procedures and protocol", 1997.

[11] ISO/IEC 11582: "Information technology – Telecommunications and information exchange between systems - Private Integrated Services Network – Generic functional protocol for the support of supplementary services - Inter-exchange signalling procedures and protocol", 1995.

[12] ITU-T Recommendation X.229: "Remote operations: Protocol specification", 1988.

[13] ETSI Technical Report ETR 300-4: "Terrestrial Trunked Radio (TETRA); Voice plus Data (V + D); Designers' guide; Part 4: Network management", Dec. 1997.

[14] ITU-T Recommendation M.3400: "TMN management functions".

[15] ETSI Technical Report ETR 292: "TETRA voice plus data, Technical requirements Specifications for Network Management", July 1997.

[16] ETSI Technical Standard ETS 300 392-1: "Radio Equipment and Systems (RES); Terrestrial Trunked radio (TETRA); Voice Plus Data (V+D); Part 1: General Network Design, Clause 7", Feb. 1996.

[17] Rose, M. T. and McCloghrie, K., *How to Manage your Network Using SNMP*, Prentice-Hall, Englewood Cliffs, N. J., 1995.

[18] Stallings, W., *Data and Computer Communications*, (5th Edn.), Prentice-Hall, Inc., Upper Saddle River, N. J., 1997.

[19] Tanenbaum, A. S., *Computer Networks*, (3rd Edn.), Prentice-Hall PTR, Upper Saddle River, N. J., 1996.

Appendices

A.1 THE TETRA TECHNICAL SPECIFICATIONS

(a) TETRA Technical Specification series, ETS 300 39x

ETS 300-392-x: TETRA Voice plus Data (V+D)
ETS 300-393-x: TETRA Packet Data Optimised (PDO)
ETS 300-394-x: TETRA Conformance Testing Specification
ETS 300-395-x: TETRA Speech Codec
ETS 300-396-x: TETRA Direct Mode Operation (DMO)

(b) Main Technical Specifications, TETRA V+D, ETR 300 392-xx Series

Part 1: "General network design".
Part 2: "Air Interface (AI)".
Part 3: "Inter-working – Basic Operation", (DE/RES-06001-3).
Part 4: "Gateways for Basic Services", (DE/RES-06001-4).
Part 5: "Terminal equipment interface", (DE/RES-06001-5).
Part 6: "Line connected stations", (DE/RES-06001-6).
Part 7: "Security".
Part 8: "Management services", (DE/RES-06001-8).
Part 9: "Performance objectives", (DE/RES-06001-9).
Part 10: "Supplementary Services (SS) Stage 1".
Part 11: "Supplementary Services (SS) Stage 2", (DE/RES-06001-11).
Part 12: "Supplementary Services (SS) Stage 3", (DE/RES-06001-12).
Part 13: "SDL Model of the Air Interface", (DE/RES-06001-13).
Part 14: "PICS Proforma", (DE/RES-06001-14).
Part 15: "Inter-working – Extended Operations", (DE/RES-06001-15).
Part 16: "Gateways for Supplementary Services", (DE/RES-06001-16).

(c) Other Related Technical Specifications

ETS 300 812 – Terrestrial Trunked Radio (TETRA); Security aspects; Subscriber Identity Module to Mobile Equipment (SIM-ME) Interface

(d) Designers' Guide (ETR 300-x Series)

[1] ETR 300-1: Terrestrial Trunked Radio (TETRA); Voice plus Data (V+D); Designers' guide; Part 1: Overview, technical description and radio aspects', May 1997.
[2] ETR 300-2: Terrestrial Trunked Radio (TETRA); Voice plus Data (V+D); Designers' guide; Part 2: Radio channels, network protocols and service performance', May 1997.
[3] ETR 300-4: Terrestrial Trunked Radio (TETRA); Voice plus Data (V + D); Designers' guide; Part 4: Network management, July 1997
[4] ETR 300-5: Terrestrial Trunked Radio (TETRA); Voice plus Data (V+D); Designers' guide; Part 5: Dialling and addressing, July 1998.

A.2 TETRA MoU MEMBERS

The TETRA Memorandum of Understanding (MoU) was established in December 1994 to create a forum which could act on behalf of all interested parties, representing users, manufacturers, operators, test houses and telecom agencies. As of May 1999 the TETRA MoU represented 58 organisations from 18 countries.

Information courtesy of TETRA MoU, web site: http://www.tetramou.com

A.2.1 Manufacturers

ATMEL ES2
Zone Industrielle, F-13106 Rousset
France
Tel: +33 4 4253 6194
Fax: +33 4 4253 6323

BESCom Elektronik GmbH
Hammer Deich 63, D-20537 Hamburg
Germany
Tel: +49 40 2111 1911
Fax: +49 40 2111 9123

CLEARTONE Telecoms Plc.
Pontyfelin Industrial Estate, New Inn,
Pontypool, South Wales NP4 ODQ
UK
Tel: +44 1495 752 255
Fax: +44 1495 752 323

CONDAT
Alt-Moabit 91d, D-10559 Berlin
Germany
Tel: +49 30 390 94153
Fax: +49 30 390 94300

Consumer Microcircuits Ltd.
1 Wheaton Road, Witham, Essex CM8 3TD
UK
Tel: +44 137 651 3833
Fax: +44 137 651 8247

Damm Cellular Systems A/S
Moellegade 64, DK-6400 Soenderborg,
Denmark
Tel: +45 74 42 35 00
Fax: +45 74 42 32 30

DeTeWe Funkwerk Köpenick
Wendenschloßstraße 142, D-12557 Berlin
Germany
Tel: +49 30 6104 2702
Fax: +49 30 6104 2701

Ericsson Inc.
Private Radio Systems, Region Europe,
1 Mountain View Road, Lynchburg, Virginia 245002
USA
Tel: +1 804 592 7058
Fax: +1 804 592 3606

ETELM 9 Avenue Des Deux Lacs P.A. Villejust, 91971 Courtaboeuf Cedex France	Tel: Fax:	+33 1 6931 2284 +33 1 6931 2261
FERCOM Ltd. Besci ut. 85, Budapest 1037 Hungary	Tel: Fax:	+36 1 250 49 10 +36 1 250 49 09
FREQUENTIS Nachrichtentechnik GmbH Spittelbreitengasse 34, A-1120 Wien Austria	Tel: Fax:	+43 1 811 500 +43 1 811 509
GEC-Marconi Communications New Street, Chelmsford, CM1 1PL, Essex UK	Tel: Fax:	+44 1245 353 221 +44 1245 287 125
HIDALGO Ltd. Cambridge House, 91 High Street, Longstation, Cambridgeshire CB4 5BS UK	Tel: Fax:	+44 195 420 6225 +44 195 420 6223
HITACHI DENSHI Ltd. 14 Garrick Industrial Centre, Irwing Way, Hendon, London NW9 6AQ UK	Tel: Fax:	+44 20 8202 4311 +44 20 8202 2451
Icom Inc. French Liaison Office Zac De La Plaine, Rue Brindejonc Des Moulinais BP5804 France	Tel: Fax:	+33 5 6136 0316 +33 5 6136 0317
IFR Ltd. P.O.Box 10, Six Hills Way, Stevenage, SG1 2AN UK	Tel: Fax:	+44 143 874 2200 +44 143 872 7601
Infomatrix Software Engineering Consultants The Old School, High Street, Fen Drayton Cambridge CB4 5SJ UK	Tel: Fax:	+44 195 423 2010 +44 195 423 0031
ITALTEL Palazzo Quadrifoglio CLRC, 20019 Settimo Milanese (MI), Italy	Tel: Fax:	+39 0 2 4388 8345 +39 0 2 4388 8012

Kenwood Electronics UK Ltd Kenwood House, Dwight Road, Watford, WDI 8EB UK	Tel: Fax:	+44 192 321 2044 +44 192 365 5297
Marconi Communications – Mobile Networks OTE S.p.a, Via E. Barsanti 8, 50127 Firenze Italy	Tel: Fax:	+39 0 55 4381 424 +39 0 55 4381 387
MOTOROLA Radio Network Solutions Group, Jays Close, Viables Industrial Estate, Basingstoke, RG22 4PD UK	Tel: Fax:	+44 125 648 4599 +44 125 648 4474
Nokia Telecommunications P.O. BOX 350, FIN-00045 Nokia Group Finland	Tel: Fax:	+358 9 5112 8477 +358 9 5112 8310
Panasonic Deutschland GmbH Winsbergring 15, 22525 Hamburg Germany	Tel: Fax:	+49 408 549 0 +49 408 549 2555
Radio Holland Electronics B.V. Microfoonstraat 5, 1322 BN Almere, The Netherlands	Tel: Fax:	+31 36 546 26 00 +31 36 546 26 01
Rohde & Schwarz BICK Mobilfunk GmbH Postfach 2062, D-31844 Bad Münder Germany	Tel: Fax:	+49 5042 998 201 +49 5042 998 105
Sigma Wireless Technologies Ltd. McKee Avenue, Finglas, Dublin 11 Ireland	Tel: Fax:	+353 1 814 2050 +353 1 814 2051
Simoco P.O. Box 24, St Andrews Rd. Cambridge CB4 1DP UK	Tel: Fax:	+44 122 335 8985 +44 122 331 3834
Sinclair Technologies Ltd. William James House, Cowley Rd. Cambridge CB4 4WX UK	Tel: Fax:	+44 122 342 0303 +44 122 342 0606
Tait Electronics Ltd. 558 Wairakei Road, Christchurch, P.O. Box 1645 New Zealand	Tel: Fax:	+64 3 358 3399 +64 3 358 0432

Teltronic, s.a., Radiocomunicaciones Leopoldo Romeo, 18, E-50002 Zaragoza Spain	Tel: Fax:	+34 976 41 80 66 +34 976 59 26 85
TERRAFIX Ltd 23c Newfield Industrial Estate, High Street Tunstall, Stoke-on-Trent, ST6 5PD UK	Tel: Fax:	+44 178 257 7015 +44 178 283 5667
TetraNed vof Oranje Buitensingel 6, 2511 VE The Hague, The Netherlands	Tel: Fax:	+31 70 343 6834 +31 70 343 7431
Uniden America Corporation 4700 Amon Carter Blvd., Fort Worth, TX 76155 USA	Tel: Fax:	+1 800 445 5017 +1 817 858 3306

A.2.2 Users

BAPCO British Association of Public Communications Officers, P.O.Box 374, Lincoln, LN1 1FY UK	Tel: Fax:	+44 152 257 5542 +44 152 257 5542
BT Radio Engineering Services PP 113, 207 Old Street, London EC1V 9NR UK	Tel: Fax:	+44 20 7250 7496 +44 20 7250 7449
Comlog N.V. c/o ITP Group, 25 Lechi Str. 51200 Bnei Brak Israel	Tel: Fax:	+972 3 578 0904 +972 3 578 0911
Dolphin Telecommunications Ltd. The Crescent, Jays Close, Basingstoke Hampshire RG22 4BS UK	Tel: Fax:	+44 125 636 7190 +44 125 636 8223
Dutch Home Office, ITO Informatieen Communicatie Technologie Organisatie P.O.Box 238, 3970 AE Driebergen, The Netherlands	Tel: Fax:	+31 343 534 701 +31 343 534 799
ElTele Øst AS Sandakerveien 114b, 0483 Oslo Norway	Tel: Fax:	+47 23 18 13 83 +47 23 18 10 03

Gobierno Vasco, Departemento de Interior
Gran Via 81, 4 - Dep 6,
E-48011 Bilbao
Spain
Tel: +34 94 427 3979
Fax: +34 94 441 5249

Home Office, Police Department
Horseferry House, Dean Ryle Street
London SW1P 2AW
UK
Tel: +44 20 7217 8040
Fax: +44 20 7976 5427

ICCC
International Computer & Communication Consultants,
Maaløv Hovedgade 88
DK-2760 Maaløv
Denmark
Tel: +45 4465 0400
Fax: +45 4468 1525

London Underground Limited
Box 2/8, 30 The South Collonade,
Canary Wharf, London E14 5EU
UK
Tel: +44 20 7308 2327
Fax: +44 20 7308 2135

Nexus Media Limited
Nexus House, Swanley, Kent BR8 8HY
UK
Tel: +44 132 266 0070
Fax: +44 132 266 1257

NTL Radio Communications
Crawley Court, Winchester,
Hampshire SO21 2QA
UK
Tel: +44 196 282 2527
Fax: +44 196 282 2474

Radio Red
Manuel Tovar 35, 28034 Madrid
Spain
Tel: +34 91 334 9199
Fax: +34 91 334 9171

Securicor Information Systems Limited
Marshfield, Chippenham, Wiltshire SN14 8NN
UK
Tel: +44 122 589 4118
Fax: +44 122 589 1440

SingTel Paging,
Singapore Telecommunications Academy
1 Hillcrest Road #09-00, Tower 2,
Singapore 288893
Singapore
Tel: +65 462 7136
Fax: +65 469 1312

SONOFON Dansk Mobil Telefon I/S
Skelagervej 1, 9100 Aalborg
Denmark
Tel: +45 9936 7000
Fax: +45 9936 7070

Speedway Telecom Co., Ltd.
No. 168, Jun-Min Road, San Min Dist., Kaohsiung,
Taiwan, 807
Republic of China
Tel: +886 7 3867879
Fax: +886 7 385 0573

Telia MobiTel AB, Radio Services
Box 2004,
S-42102 V. Frölunda
Sweden
Tel: +46 3189 7600
Fax: +46 3189 7540

TETRA DANMARK A.M.B.A.
c/o Telelaboratoriet, 2 Telegade,
DK-2630 Taastrup
Denmark
Tel: +45 4334 5601
Fax: +45 4352 8076

TETRA Forum Norge
NTNU, 7034 Trondheim,
Norway
Tel: +47 73 59 27 41
Fax: +47 73 59 69 73

T-Mobil,
DeTeMobil Deutsche Telekom MobilNet GmbH
Landrabenweg 151, D-53227 Bonn
Germany
Tel: +49 228 936 7424
Fax: +49 228 936 7409

WESTEL RádióTelefon Kft.
P.O.Box 295, 1300 Budapest
Hungary
Tel: +36 1 166 5620
Fax: +36 1 265 8417

A.2.3 Regulators

Home Office, RFCPU
Horseferry House, Dean Ryle Street,
London SW1P 2AW
UK
Tel: +44 20 7217 8106
Fax: +44 20 7630 9633

A.2.4 Accredited Test Houses

KTL Arnhem bv
P.O.Box 60004, 6800 JA Arnhem,
The Netherlands
Tel: +31 263 780 780
Fax: +31 263 780 789

Telelaboratoriet, Tele Danmark A/S
Telegade 2, DK 2630 Taastrup
Denmark
Tel: +45 4334 5535
Fax: +45 4371 0848

A.3 ERLANG B TABLE (FOR BLOCKED-CALLS-CLEARED)

A.3.1 Overview of Basic Concepts

The Erlang B table is based on Erlang's Loss Formula, also called Erlang B Formula (B for *blocked*), given by equation A.1. The formula provides an estimate of a call blocking probability, P_B, used as a measure of grade of service (GoS) in telephone systems when a mean traffic load of A Erlangs is applied to a trunked system of C channels.

$$P_B = \frac{\left(\dfrac{A^C}{C!}\right)}{\left(\displaystyle\sum_{k=0}^{C} \frac{A^k}{k!}\right)} \tag{A.1}$$

The total applied traffic A in Erlangs is λT where λ is the arrival rate in *calls/hour* and T is the *average* call duration or call holding time. For instance, an emergency centre receiving150 calls/hour (λ), with an average call holding time T of 30 sec is handling a load traffic 150 x (30 s/3600 s/hour) = 1.25 Erlangs. While calculating the GoS is straightforward from equation A.1, often, the exercise is to determine the other parameter(s), i.e. offered traffic A and/or trunk size C, for a given value of GoS. A typical problem could therefore be stated as: What trunk size would support a 1% GoS? The example used in Section 1.21 illustrates this point. In this situation calculating for A or C becomes cumbersome and the table of values as shown on the following pages will be useful.

The reader should however be aware of the underlying assumptions for this formula, as caution should be exercised in its application for capacity estimation:

- The call arrivals λ are assumed to be random and independent. For instance, the Erlang B formula could overestimate the capacity for systems with correlated calls, e.g. TETRA emergency calls that ensue an incident, as opposed to random calls.
- The number of traffic sources are assumed sufficiently large compared to the number of channels C. This condition is the basis for assuming a constant arrival rate λ regardless of number of call requests. A finite number of call requests therefore tends to provide a conservative estimate of the blocking probability.

The table on the following pages gives the offered traffic load A corresponding to call blocking probability P_B and number of channels C in the trunk. The actual carried load is $A.(1\text{-} P_B)$ which should always be less than C even if the offered load $A > C$. The trunking efficiency, η_t, is therefore given by the ratio of the actual traffic carried to the trunk size.

$$\eta_t = \frac{A.(1\text{-}P_B)}{C} \tag{A.2}$$

The ratio $\rho = A/C$ represents the offered traffic per channel and is known as the *channel occupancy*.

A.3.2 Erlang B Table

The table gives the offered traffic load A in *Erlangs* corresponding to the number of traffic channels C in a trunk (column 1) and call blocking probability P_B in percentage (top row).

Channels	Blocking Probability, P_B										
C	0.5%	1%	1.5%	2%	2.5%	3%	5%	7%	10%	15%	20%
1	0.01	0.01	0.02	0.02	0.03	0.03	0.05	0.08	0.11	0.18	0.25
2	0.11	0.15	0.19	0.22	0.25	0.28	0.38	0.47	0.60	0.80	1.00
3	0.35	0.46	0.54	0.60	0.66	0.72	0.90	1.06	1.27	1.60	1.93
4	0.70	0.87	0.99	1.09	1.18	1.26	1.52	1.75	2.05	2.50	2.95
5	1.13	1.36	1.52	1.66	1.77	1.88	2.22	2.50	2.88	3.45	4.01
6	1.62	1.91	2.11	2.28	2.42	2.54	2.96	3.30	3.76	4.44	5.11
7	2.16	2.50	2.74	2.94	3.10	3.25	3.74	4.14	4.67	5.46	6.23
8	2.73	3.13	3.40	3.63	3.82	3.99	4.54	5.00	5.60	6.50	7.37
9	3.33	3.78	4.09	4.34	4.56	4.75	5.37	5.88	6.55	7.55	8.52
10	3.96	4.46	4.81	5.08	5.32	5.53	6.22	6.78	7.51	8.62	9.68
11	4.61	5.16	5.54	5.84	6.10	6.33	7.08	7.69	8.49	9.69	10.9
12	5.28	5.88	6.29	6.61	6.89	7.14	7.95	8.61	9.47	10.8	12.0
13	5.96	6.61	7.05	7.40	7.70	7.97	8.83	9.54	10.5	11.9	13.2
14	6.66	7.35	7.82	8.20	8.52	8.80	9.73	10.5	11.5	13.0	14.4
15	7.38	8.11	8.61	9.01	9.35	9.65	10.6	11.4	12.5	14.1	15.6
16	8.10	8.88	9.41	9.83	10.2	10.5	11.5	12.4	13.5	15.2	16.8
17	8.83	9.65	10.2	10.7	11.0	11.4	12.5	13.4	14.5	16.3	18.0
18	9.58	10.4	11.0	11.5	11.9	12.2	13.4	14.3	15.5	17.4	19.2
19	10.3	11.2	11.8	12.3	12.7	13.1	14.3	15.3	16.6	18.5	20.4
20	11.1	12.0	12.7	13.2	13.6	14.0	15.2	16.3	17.6	19.6	21.6
21	11.9	12.8	13.5	14.0	14.5	14.9	16.2	17.3	18.7	20.8	22.8
22	12.6	13.7	14.3	14.9	15.4	15.8	17.1	18.2	19.7	21.9	24.1
23	13.4	14.5	15.2	15.8	16.2	16.7	18.1	19.2	20.7	23.0	25.3
24	14.2	15.3	16.0	16.6	17.1	17.6	19.0	20.2	21.8	24.2	26.5
25	15.0	16.1	16.9	17.5	18.0	18.5	20.0	21.2	22.8	25.3	27.7
26	15.8	17.0	17.8	18.4	18.9	19.4	20.9	22.2	23.9	26.4	28.9
27	16.6	17.8	18.6	19.3	19.8	20.3	21.9	23.2	24.9	27.6	30.2
28	17.4	18.6	19.5	20.2	20.7	21.2	22.9	24.2	26.0	28.7	31.4
29	18.2	19.5	20.4	21.0	21.6	22.1	23.8	25.2	27.1	29.9	32.6
30	19.0	20.3	21.2	21.9	22.5	23.1	24.8	26.2	28.1	31.0	33.8
31	19.9	21.2	22.1	22.8	23.4	24.0	25.8	27.2	29.2	32.1	35.1
32	20.7	22.0	23.0	23.7	24.4	24.9	26.7	28.2	30.2	33.3	36.3
33	21.5	22.9	23.9	24.6	25.3	25.8	27.7	29.3	31.3	34.4	37.5
34	22.3	23.8	24.8	25.5	26.2	26.8	28.7	30.3	32.4	35.6	38.8
35	23.2	24.6	25.6	26.4	27.1	27.7	29.7	31.3	33.4	36.7	40.0
36	24.0	25.5	26.5	27.3	28.0	28.6	30.7	32.3	34.5	37.9	41.2
37	24.8	26.4	27.4	28.3	29.0	29.6	31.6	33.3	35.6	39.0	42.4
38	25.7	27.3	28.3	29.2	29.9	30.5	32.6	34.4	36.6	40.2	43.7
39	26.5	28.1	29.2	30.1	30.8	31.5	33.6	35.4	37.7	41.3	44.9
40	27.4	29.0	30.1	31.0	31.7	32.4	34.6	36.4	38.8	42.5	46.1
41	28.2	29.9	31.0	31.9	32.7	33.4	35.6	37.4	39.9	43.6	47.4
42	29.1	30.8	31.9	32.8	33.6	34.3	36.6	38.4	40.9	44.8	48.6
43	29.9	31.7	32.8	33.8	34.6	35.3	37.6	39.5	42.0	45.9	49.9
44	30.8	32.5	33.7	34.7	35.5	36.2	38.6	40.5	43.1	47.1	51.1
45	31.7	33.4	34.6	35.6	36.4	37.2	39.6	41.5	44.2	48.2	52.3
46	32.5	34.3	35.6	36.5	37.4	38.1	40.5	42.6	45.2	49.4	53.6
47	33.4	35.2	36.5	37.5	38.3	39.1	41.5	43.6	46.3	50.6	54.8
48	34.2	36.1	37.4	38.4	39.3	40.0	42.5	44.6	47.4	51.7	56.0
49	35.1	37.0	38.3	39.3	40.2	41.0	43.5	45.7	48.5	52.9	57.3
50	36.0	37.9	39.2	40.3	41.1	41.9	44.5	46.7	49.6	54.0	58.5

Erlang B Table (continued from page 422)

Channels	Blocking Probability, P_B									
C	0.5%	1%	1.5%	2%	2.5%	3%	5%	7%	10%	15%	20%
51	36.9	38.8	40.1	41.2	42.1	42.9	45.5	47.7	50.6	55.2	59.7
52	37.7	39.7	41.0	42.1	43.0	43.9	46.5	48.8	51.7	56.3	61.0
53	38.6	40.6	42.0	43.1	44.0	44.8	47.5	49.8	52.8	57.5	62.2
54	39.5	41.5	42.9	44.0	44.9	45.8	48.5	50.8	53.9	58.7	63.5
55	40.4	42.4	43.8	44.9	45.9	46.7	49.5	51.9	55.0	59.8	64.7
56	41.2	43.3	44.7	45.9	46.8	47.7	50.5	52.9	56.1	61.0	65.9
57	42.1	44.2	45.7	46.8	47.8	48.7	51.5	53.9	57.1	62.1	67.2
58	43.0	45.1	46.6	47.8	48.8	49.6	52.6	55.0	58.2	63.3	68.4
59	43.9	46.0	47.5	48.7	49.7	50.6	53.6	56.0	59.3	64.5	69.7
60	44.8	46.9	48.4	49.6	50.7	51.6	54.6	57.1	60.4	65.6	70.9
61	45.6	47.9	49.4	50.6	51.6	52.5	55.6	58.1	61.5	66.8	72.1
62	46.5	48.8	50.3	51.5	52.6	53.5	56.6	59.1	62.6	68.0	73.4
63	47.4	49.7	51.2	52.5	53.5	54.5	57.6	60.2	63.7	69.1	74.6
64	48.3	50.6	52.2	53.4	54.5	55.4	58.6	61.2	64.8	70.3	75.9
65	49.2	51.5	53.1	54.4	55.5	56.4	59.6	62.3	65.8	71.4	77.1
66	50.1	52.4	54.0	55.3	56.4	57.4	60.6	63.3	66.9	72.6	78.3
67	51.0	53.4	55.0	56.3	57.4	58.4	61.6	64.4	68.0	73.8	79.6
68	51.9	54.3	55.9	57.2	58.3	59.3	62.6	65.4	69.1	74.9	80.8
69	52.8	55.2	56.9	58.2	59.3	60.3	63.7	66.4	70.2	76.1	82.1
70	53.7	56.1	57.8	59.1	60.3	61.3	64.7	67.5	71.3	77.3	83.3
71	54.6	57.0	58.7	60.1	61.2	62.3	65.7	68.5	72.4	78.4	84.6
72	55.5	58.0	59.7	61.0	62.2	63.2	66.7	69.6	73.5	79.6	85.8
73	56.4	58.9	60.6	62.0	63.2	64.2	67.7	70.6	74.6	80.8	87.0
74	57.3	59.8	61.6	62.9	64.1	65.2	68.7	71.7	75.6	81.9	88.3
75	58.2	60.7	62.5	63.9	65.1	66.2	69.7	72.7	76.7	83.1	89.5
76	59.1	61.7	63.4	64.9	66.1	67.2	70.8	73.8	77.8	84.2	90.8
77	60.0	62.6	64.4	65.8	67.0	68.1	71.8	74.8	78.9	85.4	92.0
78	60.9	63.5	65.3	66.8	68.0	69.1	72.8	75.9	80.0	86.6	93.3
79	61.8	64.4	66.3	67.7	69.0	70.1	73.8	76.9	81.1	87.7	94.5
80	62.7	65.4	67.2	68.7	70.0	71.1	74.8	78.0	82.2	88.9	95.7
81	63.6	66.3	68.2	69.6	70.9	72.1	75.8	79.0	83.3	90.1	97.0
82	64.5	67.2	69.1	70.6	71.9	73.0	76.9	80.1	84.4	91.2	98.2
83	65.4	68.2	70.1	71.6	72.9	74.0	77.9	81.1	85.5	92.4	99.5
84	66.3	69.1	71.0	72.5	73.8	75.0	78.9	82.2	86.6	93.6	100.7
85	67.2	70.0	71.9	73.5	74.8	76.0	79.9	83.2	87.7	94.7	102.0
86	68.1	70.9	72.9	74.5	75.8	77.0	80.9	84.3	88.8	95.9	103.2
87	69.0	71.9	73.8	75.4	76.8	78.0	82.0	85.3	89.9	97.1	104.5
88	69.9	72.8	74.8	76.4	77.7	78.9	83.0	86.4	91.0	98.2	105.7
89	70.8	73.7	75.7	77.3	78.7	79.9	84.0	87.4	92.1	99.4	106.9
90	71.8	74.7	76.7	78.3	79.7	80.9	85.0	88.5	93.1	100.6	108.2
91	72.7	75.6	77.6	79.3	80.7	81.9	86.0	89.5	94.2	101.7	109.4
92	73.6	76.6	78.6	80.2	81.6	82.9	87.1	90.6	95.3	102.9	110.7
93	74.5	77.5	79.6	81.2	82.6	83.9	88.1	91.6	96.4	104.1	111.9
94	75.4	78.4	80.5	82.2	83.6	84.9	89.1	92.7	97.5	105.3	113.2
95	76.3	79.4	81.5	83.1	84.6	85.8	90.1	93.7	98.6	106.4	114.4
96	77.2	80.3	82.4	84.1	85.5	86.8	91.1	94.8	99.7	107.6	115.7
97	78.2	81.2	83.4	85.1	86.5	87.8	92.2	95.8	100.8	108.8	116.9
98	79.1	82.2	84.3	86.0	87.5	88.8	93.2	96.9	101.9	109.9	118.2
99	80.0	83.1	85.3	87.0	88.5	89.8	94.2	97.9	103.0	111.1	119.4
100	80.9	84.1	86.2	88.0	89.5	90.8	95.2	99.0	104.1	112.3	120.6

A.4 ERLANG C TABLE (FOR BLOCKED-CALLS- QUEUED)

A.4.1 Overview of Basic Concepts

Call request queuing is supported within TETRA and Erlang C table is therefore useful for trunking capacity estimation under call queuing strategy. With the Erlang C system, a call arriving at a busy system is queued rather than blocked immediately as in the Erlang B system. A queued call will only be lost if it cannot be connected within a certain queuing time T_Q which is usually imposed by the switching system. The queued call requests do not, unlike Erlang B, immediately contribute to a new arrival process and therefore a modified equation applies to Erlang C formula as shown below.

$$P_Q = \frac{A^C}{A^C + C!\left(1 - \frac{A}{C}\right)\left(\sum_{k=0}^{C-1} \frac{A^k}{k!}\right)} \tag{A.3}$$

Note that the Erlang C equation gives the probability of queue, P_Q, that an arriving call finds a busy system and has to wait. (1- P_Q is the probability that call requests are being serviced). The equation does not say how long the wait time is and for this another equation is employed. The probability that a *queued call* waits for more than t is given by

$$P_W = \text{Prob}(T_Q > t) = e^{-\left(\frac{C-A}{T}\right)t} \tag{A.4}$$

where A, C and T are as defined for Erlang B formula. On the other hand, the probability that *all calls*, including those serviced immediately, are delayed for more than t is given by $P_Q.P_W$. The underlying assumptions of traffic source behaviour for Erlang B also apply to Erlang C, with some additional points as noted below.

- The queue size is assumed to be sufficiently large (ideally infinite) and queued calls are served on first-come first-served basis.
- Erlang C function makes sense only if $\rho < 1$ (or $A < C$) otherwise the queue would overflow as the system cannot cope with the call requests.

Example: An 8-channel trunked TETRA system operates under a call queuing strategy with up to 5% of call connections delayed. Determine the additional number of channels required for lowering the delayed calls to 1% or less. What percent of the delayed calls waits beyond the mean call holding time T, before and after increasing the channel?

Solution: Given $C = 8$ and $P_Q = 0.05$ (5%), from the Erlang C table on the next page, the offered traffic can be found to be $A = 3.87$. If P_Q has to be reduced to 0.01 (1%), for approximately the same offered traffic the number of channels can be found to be 10. This corresponds to the nearest safe value of $A = 4.08$ from the table. Hence two additional channels would be required. The probability of queued calls waiting for more than T is given by equation A.4. Since the waiting time is equal to the mean call holding time T, equation A.4 reduces to $e^{-(C-A)}$ giving, $P_W = 0.016$ for $C = 8$, and $P_W = 0.0022$ for $C = 10$. Hence, with 8 channels 1.6% of the queued calls wait for more than the mean call holding time. As expected, this figure is reduced to 0.22% after the channel is increased to 10.

A.4.2 Erlang C Table

The table gives the offered traffic load A in *Erlangs* corresponding to the number of traffic channels C in a trunk (column 1) and call queuing probability P_Q in percentage (top row).

Channels	Delay Probability, P_Q									
C	0.5%	1%	1.5%	2%	2.5%	3%	5%	7%	10%	15%	20%
1	0.00	0.01	0.02	0.02	0.02	0.03	0.05	0.07	0.10	0.15	0.20
2	0.10	0.15	0.18	0.21	0.24	0.26	0.34	0.41	0.50	0.63	0.74
3	0.33	0.43	0.50	0.55	0.60	0.65	0.79	0.90	1.04	1.23	1.39
4	0.66	0.81	0.91	0.99	1.06	1.12	1.32	1.47	1.65	1.90	2.10
5	1.07	1.26	1.39	1.50	1.59	1.66	1.91	2.09	2.31	2.61	2.85
6	1.52	1.76	1.92	2.05	2.15	2.24	2.53	2.75	3.01	3.34	3.62
7	2.01	2.30	2.49	2.63	2.75	2.86	3.19	3.43	3.73	4.10	4.41
8	2.54	2.87	3.08	3.25	3.38	3.50	3.87	4.14	4.46	4.88	5.21
9	3.10	3.46	3.70	3.88	4.04	4.17	4.57	4.87	5.22	5.67	6.03
10	3.68	4.08	4.34	4.54	4.71	4.85	5.29	5.61	5.99	6.47	6.85
11	4.28	4.71	5.00	5.21	5.39	5.55	6.02	6.36	6.76	7.28	7.69
12	4.90	5.36	5.67	5.90	6.09	6.26	6.76	7.13	7.55	8.10	8.53
13	5.53	6.03	6.35	6.60	6.81	6.98	7.51	7.90	8.35	8.93	9.38
14	6.17	6.71	7.05	7.31	7.53	7.71	8.27	8.68	9.16	9.76	10.2
15	6.83	7.39	7.76	8.04	8.26	8.46	9.04	9.47	9.97	10.6	11.1
16	7.50	8.09	8.48	8.77	9.00	9.21	9.82	10.3	10.8	11.4	12.0
17	8.18	8.80	9.20	9.51	9.75	9.97	10.6	11.1	11.6	12.3	12.8
18	8.87	9.52	9.94	10.3	10.5	10.7	11.4	11.9	12.4	13.1	13.7
19	9.57	10.2	10.7	11.0	11.3	11.5	12.2	12.7	13.3	14.0	14.6
20	10.3	11.0	11.4	11.8	12.0	12.3	13.0	13.5	14.1	14.9	15.5
21	11.0	11.7	12.2	12.5	12.8	13.1	13.8	14.3	15.0	15.7	16.3
22	11.7	12.5	12.9	13.3	13.6	13.9	14.6	15.2	15.8	16.6	17.2
23	12.4	13.2	13.7	14.1	14.4	14.6	15.4	16.0	16.7	17.5	18.1
24	13.2	14.0	14.5	14.9	15.2	15.4	16.3	16.8	17.5	18.3	19.0
25	13.9	14.7	15.2	15.6	16.0	16.2	17.1	17.7	18.4	19.2	19.9
26	14.6	15.5	16.0	16.4	16.8	17.1	17.9	18.5	19.2	20.1	20.8
27	15.4	16.3	16.8	17.2	17.6	17.9	18.7	19.4	20.1	21.0	21.7
28	16.1	17.0	17.6	18.0	18.4	18.7	19.6	20.2	20.9	21.9	22.6
29	16.9	17.8	18.4	18.8	19.2	19.5	20.4	21.1	21.8	22.8	23.5
30	17.7	18.6	19.2	19.6	20.0	20.3	21.2	21.9	22.7	23.6	24.4
31	18.4	19.4	20.0	20.4	20.8	21.1	22.1	22.8	23.6	24.5	25.3
32	19.2	20.2	20.8	21.3	21.6	22.0	22.9	23.6	24.4	25.4	26.2
33	20.0	21.0	21.6	22.1	22.5	22.8	23.8	24.5	25.3	26.3	27.1
34	20.7	21.7	22.4	22.9	23.3	23.6	24.6	25.4	26.2	27.2	28.0
35	21.5	22.5	23.2	23.7	24.1	24.5	25.5	26.2	27.1	28.1	28.9
36	22.3	23.3	24.0	24.5	24.9	25.3	26.3	27.1	27.9	29.0	29.8
37	23.1	24.2	24.8	25.4	25.8	26.1	27.2	28.0	28.8	29.9	30.7
38	23.9	25.0	25.7	26.2	26.6	27.0	28.0	28.8	29.7	30.8	31.7
39	24.6	25.8	26.5	27.0	27.4	27.8	28.9	29.7	30.6	31.7	32.6
40	25.4	26.6	27.3	27.8	28.3	28.7	29.8	30.6	31.5	32.6	33.5
41	26.2	27.4	28.1	28.7	29.1	29.5	30.6	31.4	32.4	33.5	34.4
42	27.0	28.2	29.0	29.5	30.0	30.3	31.5	32.3	33.3	34.4	35.3
43	27.8	29.0	29.8	30.3	30.8	31.2	32.4	33.2	34.1	35.3	36.2
44	28.6	29.8	30.6	31.2	31.7	32.1	33.2	34.1	35.0	36.2	37.2
45	29.4	30.7	31.4	32.0	32.5	32.9	34.1	35.0	35.9	37.1	38.1
46	30.2	31.5	32.3	32.9	33.4	33.8	35.0	35.8	36.8	38.1	39.0
47	31.1	32.3	33.1	33.7	34.2	34.6	35.8	36.7	37.7	39.0	39.9
48	31.9	33.1	34.0	34.6	35.1	35.5	36.7	37.6	38.6	39.9	40.8
49	32.7	34.0	34.8	35.4	35.9	36.3	37.6	38.5	39.5	40.8	41.8
50	33.5	34.8	35.6	36.3	36.8	37.2	38.5	39.4	40.4	41.7	42.7

Erlang C Table (continued from page 425)

Channels	Delay Probability, P_Q										
C	0.5%	1%	1.5%	2%	2.5%	3%	5%	7%	10%	15%	20%
51	34.3	35.6	36.5	37.1	37.6	38.1	39.3	40.3	41.3	42.6	43.6
52	35.1	36.5	37.3	38.0	38.5	38.9	40.2	41.2	42.2	43.5	44.5
53	35.9	37.3	38.2	38.8	39.3	39.8	41.1	42.0	43.1	44.4	45.5
54	36.8	38.1	39.0	39.7	40.2	40.7	42.0	42.9	44.0	45.4	46.4
55	37.6	39.0	39.9	40.5	41.1	41.5	42.9	43.8	44.9	46.3	47.3
56	38.4	39.8	40.7	41.4	41.9	42.4	43.8	44.7	45.8	47.2	48.2
57	39.2	40.7	41.6	42.2	42.8	43.3	44.6	45.6	46.7	48.1	49.2
58	40.1	41.5	42.4	43.1	43.7	44.1	45.5	46.5	47.6	49.0	50.1
59	40.9	42.4	43.3	44.0	44.5	45.0	46.4	47.4	48.5	50.0	51.0
60	41.7	43.2	44.1	44.8	45.4	45.9	47.3	48.3	49.5	50.9	52.0
61	42.6	44.0	45.0	45.7	46.3	46.7	48.2	49.2	50.4	51.8	52.9
62	43.4	44.9	45.8	46.6	47.1	47.6	49.1	50.1	51.3	52.7	53.8
63	44.2	45.7	46.7	47.4	48.0	48.5	50.0	51.0	52.2	53.6	54.8
64	45.1	46.6	47.6	48.3	48.9	49.4	50.9	51.9	53.1	54.6	55.7
65	45.9	47.5	48.4	49.2	49.8	50.3	51.7	52.8	54.0	55.5	56.6
66	46.7	48.3	49.3	50.0	50.6	51.1	52.6	53.7	54.9	56.4	57.6
67	47.6	49.2	50.2	50.9	51.5	52.0	53.5	54.6	55.8	57.3	58.5
68	48.4	50.0	51.0	51.8	52.4	52.9	54.4	55.5	56.7	58.3	59.4
69	49.3	50.9	51.9	52.6	53.3	53.8	55.3	56.4	57.7	59.2	60.4
70	50.1	51.7	52.8	53.5	54.1	54.7	56.2	57.3	58.6	60.1	61.3
71	50.9	52.6	53.6	54.4	55.0	55.5	57.1	58.2	59.5	61.0	62.2
72	51.8	53.4	54.5	55.3	55.9	56.4	58.0	59.1	60.4	62.0	63.2
73	52.6	54.3	55.4	56.1	56.8	57.3	58.9	60.0	61.3	62.9	64.1
74	53.5	55.2	56.2	57.0	57.7	58.2	59.8	60.9	62.2	63.8	65.1
75	54.3	56.0	57.1	57.9	58.5	59.1	60.7	61.9	63.2	64.8	66.0
76	55.2	56.9	58.0	58.8	59.4	60.0	61.6	62.8	64.1	65.7	66.9
77	56.0	57.8	58.8	59.7	60.3	60.9	62.5	63.7	65.0	66.6	67.9
78	56.9	58.6	59.7	60.5	61.2	61.7	63.4	64.6	65.9	67.6	68.8
79	57.7	59.5	60.6	61.4	62.1	62.6	64.3	65.5	66.8	68.5	69.8
80	58.6	60.4	61.5	62.3	63.0	63.5	65.2	66.4	67.8	69.4	70.7
81	59.4	61.2	62.3	63.2	63.9	64.4	66.1	67.3	68.7	70.4	71.6
82	60.3	62.1	63.2	64.1	64.7	65.3	67.0	68.2	69.6	71.3	72.6
83	61.2	63.0	64.1	64.9	65.6	66.2	67.9	69.1	70.5	72.2	73.5
84	62.0	63.8	65.0	65.8	66.5	67.1	68.8	70.0	71.4	73.1	74.5
85	62.9	64.7	65.9	66.7	67.4	68.0	69.7	71.0	72.4	74.1	75.4
86	63.7	65.6	66.7	67.6	68.3	68.9	70.6	71.9	73.3	75.0	76.3
87	64.6	66.4	67.6	68.5	69.2	69.8	71.5	72.8	74.2	76.0	77.3
88	65.5	67.3	68.5	69.4	70.1	70.7	72.4	73.7	75.1	76.9	78.2
89	66.3	68.2	69.4	70.3	71.0	71.6	73.4	74.6	76.1	77.8	79.2
90	67.2	69.1	70.3	71.1	71.9	72.5	74.3	75.5	77.0	78.8	80.1
91	68.0	69.9	71.1	72.0	72.8	73.4	75.2	76.5	77.9	79.7	81.1
92	68.9	70.8	72.0	72.9	73.6	74.3	76.1	77.4	78.8	80.6	82.0
93	69.8	71.7	72.9	73.8	74.5	75.2	77.0	78.3	79.8	81.6	83.0
94	70.6	72.6	73.8	74.7	75.4	76.1	77.9	79.2	80.7	82.5	83.9
95	71.5	73.5	74.7	75.6	76.3	77.0	78.8	80.1	81.6	83.4	84.8
96	72.4	74.3	75.6	76.5	77.2	77.9	79.7	81.0	82.5	84.4	85.8
97	73.2	75.2	76.5	77.4	78.1	78.8	80.6	82.0	83.5	85.3	86.7
98	74.1	76.1	77.3	78.3	79.0	79.7	81.5	82.9	84.4	86.3	87.7
99	75.0	77.0	78.2	79.2	79.9	80.6	82.5	83.8	85.3	87.2	88.6
100	75.8	77.8	79.1	80.1	80.8	81.5	83.4	84.7	86.3	88.1	89.6

Index